Tungsten

Properties, Chemistry, Technology of the Element, Alloys, and Chemical Compounds

Tungsten

Properties, Chemistry, Technology of the Element, Alloys, and Chemical Compounds

Erik Lassner and
Wolf-Dieter Schubert
Vienna University of Technology
Vienna, Austria

Kluwer Academic / Plenum Publishers
New York, Boston, Dordrecht, London, Moscow

Library of Congress Cataloging-in-Publication Data

Lassner, Erik.
Tungsten : properties, chemistry, technology of the element, alloys, and chemical compounds / Erik Lassner and Wolf-Dieter Schubert.
p. cm.
Includes bibliographical references and index.
ISBN 0-306-45053-4
1. Tungsten. I. Schubert, Wolf-Dieter. II. Title.
QD181.W1L37 1998
620.1'8934--dc21 98-45787
CIP

ISBN 0-306-45053-4

233 Spring Street, New York, N.Y. 10013

10 9 8 7 6 5 4 3 2 1

A C.I.P. record for this book is available from the Library of Congress.

Printed in the United States of America

To a group of pioneers in the field of Power Metallurgy and Tungsten Science: Friedrich Benesovsky, Gustav F. Hüttig, Richard Kieffer, Tivadar Millner, Hans Novotny, Paul Schwarzkopf, and Karl Sedlatschek. They educated a generation of scientists and technicians who spread their knowledge throughout the world.

And to our friend, the late Bernhard F. Kieffer.

Preface

Why does someone write a book about Tungsten? There are several reasons and precedents for this, the most important of which is that the last book on tungsten was written more than 20 years ago, in 1977, by St. W. H. Yih and Ch T. Wang. During the intervening period there have been many new scientific and technological developments and innovations, so it was not only our opinion but the view of many other members of the "tungsten family" that it was time to start writing a new book about tungsten. Preparations of the new book began in 1994.

Further impetus to the project was provided by the realization that in spite of this new knowledge having been presented at seminars or published in the technical press, a general acknowledgement of it by the majority of technicians and scientists is still far from being realized. It is our hope that this book will significantly contribute to a broader acceptance of recent scientific and technological innovations.

An important prerequisite for such a project is the availability of a recently retired, experienced person willing to devote his time and talents to the tedious part of the exercise. Erik, who retired in 1993, was both highly motivated and eager to start; and it was a relatively easy task for Erik to persuade Wolf-Dieter to participate in the project. The fact that both of us have for many years enjoyed a close cooperation in tungsten-related research projects leading to the publication of several joint papers only facilitated the decision. Moreover, as authors of "Tungsten, Tungsten Alloys, and Tungsten Compounds" in the latest edition of *Ullmann's Encyclopedia of Industrial Chemistry*, we have also had some common experience in writing.

Another important prerequisite was that we both had extensive experience in tungsten-related research and technology during our professional work. Wolf-Dieter's scientific career and extensive research activities combined with Erik's long-term industrial and development knowledge contributed to a fruitful cooperation in preparing the manuscript for this book.

Nevertheless, in the writing of this book we also learned a lot more about tungsten. Richard Kieffer (the father of Bernhard Kieffer and for many years one of the leading experts on tungsten) always said: "If you want to inform yourself only roughly about something, present a paper about the subject; but, if you want to gain an in-depth knowledge about it — you have to write a book." So it seems that during the writing of this book (it is now February 1998) we grew into experts.

Encouragement, a very important aid in a project of this kind, was given to us first of all by Bernhard F. Kieffer, who unfortunately passed away prematurely. Bernhard not only encouraged us to write this book, but also encouraged Plenum Publishing Corporation to publish it.

Sometimes it was not easy to decide what should be included and what should be omitted; or the extent of detail to which each item should be discussed. We decided in general to emphasize the technologically related matters and to refrain from discussing purely scientific principles. Naturally, these decisions are always subjective and chapters dealing with those subjects with which we are more familiar are undoubtedly treated more precisely than others. This is natural, because otherwise the individual chapters should each have been written by an expert on that particular topic. As our original manuscript already exceeded the maximum allowed by the publisher, such an expert treatise would have exceeded the publisher's maximum several times over. Nevertheless, we hope that our book gets a good reception from the reader.

Returning once again to the original question: "Why write a book?" somebody once said it is to establish a "monument" to the author(s). If this should be the truth, we want many people to come and glance at it. However, besides constructing a monument, it is our intention to summarize the current status of the science and technology of tungsten for those interested in extending their knowledge of this subject. A book always provides for easier and more efficient learning than searching for a large number of original publications, which must first of all be identified.

The preparation of this manuscript has sometimes exacted its toll on our families, we want to especially thank our wives for their patience and understanding.

Vienna and Graz, February 1998

Acknowledgments

Without the support and help of numerous colleagues and friends as well as companies, it would not have been possible to write the book in its present form.

For valuable information, literature, photographs, data sheets, etc., we are kindly indebted to many people in many countries:

Austria
K. Voigt and R. Cantz, Böhlerit, Kapfenberg
M. Schwarzkopf, G. Leichtfried, A. Singer, and R. Pürcher, Plansee, Reutte
H. Wöhrle, Plansee Tizit, Reutte
P. Putz, Sandvik Austria, Vienna
F. Sattler, Tamrock Voest-Alpine Bergtechnik, Zeltweg
M. Spross, B. Zeiler, O. Grau, and E. Moos,
Wolfram Bergbau-und Hüttenges., St. Peter i.S
I. Begsteiger, Porzellanfabrik Frauenthal GmbH
Th. Nagl, Austrian Energy and Environment
F. Koch, Böhler Edelstahl GmbH, Kapfenberg
G. Groboth, Austrian Research Center Seibersdorf
E. Hengge, Technical University Graz
B. Lux, P. Ettmayer, H. Danninger, and R. Haubner, Vienna University of Technology

Belgium
D. Lison, University of Louvain, Brussels

China
Zhao Quinsheng and Zou Zhiqiang, Central South University of Technology, Changsha

Czech Republic
V. Dufek, Prague
D. Rafaja, Charles University Prague

Germany
K. Dreyer, Widia, Essen
D. Ermel, ALD Vacuum Technologies, Erlensee
H. Frick, Kemmer Präzision, Schwäbisch Gmünd

J. Holz, Industrievertretungen GmbH., Wächtersbach
H. Kolaska, Fachverband Pulvermetallurgie, Hagen
E. Kübel, WI Hartmetall, Münsingen
G. Marsen, Osram, Schwabmünchen
H. Palme, University of Cologne
D. Pratschke, Gebr. Leitz, Oberkochen
G. Gille, Starck, Goslar
H. Westermann, United Hardmetal, Horb;
Elino Industrieofenbau, Düren

Hungary
L. Bartha, O. Horacsek, and J. Neugebauer, Hungarian Academy of Science, Budapest

Israel
R. Gero, Ashot Ashkelon, Ashkelon
R. Porat and M. Leiderman, ISCAR; Tefen

Japan
T. Tanase, Mitsubishi Materials Corporation, Omiya
T. Nomura, Sumitomo Electric Industries, Itami
Y. Yamamoto, Tokyo Tungsten, Toyamaken
K. Kitamura, Toshiba Tungaloy, Iwaki City

Luxemburg
J. P. Lanners, Cerametal, Mamer

Principality of Liechtenstein
Th. Kraus, Triesen

Sweden
L. Rohlin, AB Sandvik Coromant, Stockholm
U. Fischer, AB Sandvik Rock Tools, Sandviken
B. Uhrenius, Sandvik Hard Materials, Stockholm
C. G. Granquist, University of Uppsala

The Netherlands
F. Mertens and G.vd. Kerkhof, Philips Lighting, Maarheeze

UK
B. Williams, EPMA, Shrewsbury
S. Jones, Royal Ordnance Speciality Metals, Wolverhampton
M. Maby, Secretary General, International Tungsten Industry Association, London

USA
B. North, M. Greenfield, and G. J. Wolfe, Kennametal, Latrobe
B. F. Kieffer, M. Ostermann, J. Oakes, and K. Horten,
Teledyne Advanced Materials, Huntsville
E. Rudy, SINTEX, Oregon
C. L. Conner, The Dow Chemical Company, Midland

E. A. Amey, USBM
J. Stone, Denver
Ch. W. Miller, Jr., Harper International, Lancaster

We gratefully acknowledge the critical reading of various parts of the manuscript by the following people:

M. Spross and B. Zeiler, Wolfram Bergbau- und Hüttenges., St. Peter i.S., Austria

G. Leichtfried, Plansee, Reutte, Austria

K. Mereiter, E. Zobetz, H. Mayer, P. Mohn, and *J. Redinger*, Vienna University of Technology, Austria

W. Kiesl, R. Stickler, and R. Podloucky, University of Vienna, Austria

M. Shale, William Rowland, Sheffield, UK

E. Bennett, NPL, Teddington, UK

H. C. Starck, Goslar, Germany

O. Foglar, Schladming, Austria

A particular vote of thanks is due to *A. Bartl* and *W. Prohaska* (Institute for Chemical Technology of Inorganic Materials, Vienna, Austria) for the preparation of numerous drawings, tables, copies, and photographs.

We owe special thanks to *Mrs. Ann Stanly* of Teledyne Metalworking Products for correcting the manuscript in regard to the English, which is not our mother tongue.

We gratefully acknowledge the financial support of *Teledyne Advanced Materials*, Huntsville, AL, USA

Contents

CHAPTER 3. Important Aspects of Tungsten Chemistry

CHAPTER 4. Tungsten Compounds and Their Application

CHAPTER 5. Industrial Production

CHAPTER 11. Tungsten Scrap Recycling

CHAPTER 12. Ecology

CHAPTER 13. Economy

CHAPTER 14. Tungsten and Living Organisms

Tungsten

Properties, Chemistry, Technology of the Element, Alloys, and Chemical Compounds

1

The Element Tungsten

Its Properties

Tungsten is a metallic transition element. Its position in the Periodic Table is characterized by:

Period	6
Group	6* (6 A)†
Atomic Number	74
Average Relative Atomic Mass	183.85 ± 0.03 (not regarding geological exceptional samples) ($^{12}C = 12.0000$)

1.1. ANALOGOUS TO ATOM RELATED PHYSICAL PROPERTIES [1.1, 1.2]

Important figures related to the atom are:

Atomic Radius	Metallic	137.0 pm (coordination number 8)
	Covalent	125.0 pm (single bond; valence 6)
		121.0 pm (double bond; valence 6)
Ionization Potential		7.98 eV
Electron Affinity ($M \rightarrow M^-$)		0.816 ± 0.008 eV
Ionic Radii	W^-	226.5 pm
	W^+	136 pm
	W^{2+}	117 pm
	W^{3+}	101 pm
	W^{4+}	90 pm, 66 pm (CN 6) [1.3]
	W^{5+}	80 pm, 62 pm (CN 6) [1.3]
	W^{6+}	74 pm (crystal, CN 6) [1.3]
		69 pm (ionic, CN 6) [1.3]
Atomic Volume (W/D)		9.2

* New IUPAC proposal 1985, which is used henceforth exclusively.
† Old IUPAC recommendation.

Electronegativity	1.7 (Pauling)
	1.40 (Allred)
	4.40 eV (Pearson; absolute)
Effective Nuclear Charge	4.35 (Slater)
	9.85 (Clementi)
	14.22 (Froese–Fischer)

1.1.1. Nucleus

By definition, tungsten contains 74 protons in its nucleus besides 84 to 116 neutrons. Thirty-five isotopes (including isomers) are known. Five of them are naturally occurring; the rest can be formed artificially and are unstable. Their half-life varies between milliseconds and more than 200 days. The characteristics of the natural and two of the more important artificial isotopes are listed in Table 1.1. More detailed information about the tungsten radioisotopes can be found elsewhere [1.4].

The cross-section for absorption of thermal neutrons is 18.4 barn ($v = 2200\ \mathrm{m \cdot s^{-1}}$).

The Nuclear Magnetic Resonance characteristics for ^{183}W are:

(Relative Sensitivity ($^1H = 1.00$)	7.20×10^{-4}
Recepticity ($^{13}C = 1.00$)	0.0589
Gyromagnetic Ratio/rad $T^{-1} \cdot s^{-1}$	1.1145×10^7
Frequency ($^1H = 100$ MHz; 2.3488T)/MHz	4.161
Reference: WF_6	

1.1.2. Electron Configuration

The electron configuration of the unexcited tungsten atom is defined by [Xe] $4F^{14}5d^46s^2$. The term symbol is $^5D_0$, which means that K, L, M, and N shells are completed while, O and P shells are incomplete. The neutral atom contains 74 electrons. Their energy states and quantum numbers are given in Table 1.2.

Ionization energy in eV [1.5].

$$W^0 \xrightarrow{7.89} W^+ \xrightarrow{17.7} W^{2+} \xrightarrow{(24)} W^{3+} \xrightarrow{(35)} W^{4+} \xrightarrow{(48)} W^{5+} \xrightarrow{(61)} W^{6+}$$

TABLE 1.1. Important Tungsten Isotopes [1.1, 1.2]

Symbol	Number of protons	Number of neutrons	Atomic mass	Natural abundance	$T_{1/2}$	Decay type energy (keV)	Nucl. spin	Magnetic nuclear momentum	Application
^{180}W	74	106	179.946701	0.13	Stable	—	0+	—	—
^{182}W	74	108	181.948202	26.3	Stable	—	0+	—	—
^{183}W	74	109	182.950220	14.3	Stable	—	1/2−	0.11778	NMR
^{184}W	74	110	183.950928	30.67	Stable	—	0+	—	—
^{185}W	74	111	184.953416	0	75.1 d	$\beta^-(0.433); \gamma$	3/2−	—	Tracer
^{186}W	74	112	185.954357	28.6	Stable	—	0+	—	—
^{187}W	74	113	186.957153	0	23.9 h	$\beta^-(1.312); \gamma$	3/2−	0.688	Tracer

TABLE 1.2. Energy States of Electrons in the Neutral Tungsten Atom

Electron shell	Energy states			
	Spectral name	Quantum number n	Quantum number k_j	Number electrons
K	1s	1	1_1	2
L	2s	2	1_1	2
	2p	2	2_1, 2_2	2, 4
M	3s	3	1_1	2
	3p	3	$2_1 2_2$	2, 4
	3d	3	3_2, 3_3	4, 6
N	4s	4	1_1	2
	4p	4	2_1, 2_2	2, 4
	4d	4	3_2, 3_3	4, 6
	4f	4	4_3, 4_4	6, 8
O	5s	5	1_1	2
	5p	5	2_1, 2_2	2, 4
	5d	5	3_2, 3_3	4
P	6s	6	1_1	2

The electron configuration, especially the $5d^4$ niveaus, is responsible for the typical tungsten-related physical and chemical properties.

1.1.3. Spectra

1.1.3.1. X-Ray Emission [1.6–1.8]. A typical X-ray emission spectrum from a W target X-ray tube at 50 kV is shown in Fig. 1.1. It consists of a continuous X-ray spectrum, or white radiation ("Bremsstrahlung"), on which are superposed a few characteristic lines (L lines), resulting from the direct ionization by the impinging electrons. Table 1.3 shows the strongest characteristic lines of the extended X-ray spectrum. For more details and the O spectra, the reader is referred elsewhere [1.1].

The X-ray spectrum is used for the analytical determination of the element, preferably in X-ray fluorescence analysis and electron-beam microanalysis (L and M lines). These methods have gained much importance, primarily because there is no necessity for any prior chemical separations. Moreover, the matrix influences are much less compared to other methods.

Due to its high atomic number, tungsten gives an excellent X-ray output, making it a preferred material for stationary and rotating anodes of X-ray tubes, which are used for medical diagnosis (see also Section 7.6).

1.1.3.2. X-Ray Absorption [1.1, 1.8]. The attenuation of X-radiation, which passes through a material of thickness t (cm), is given by $I = I_0 \exp(-\mu t)$, with I_0 being the intensity of the incident beam, I the intensity of the transmitted beam, and μ the linear absorption (attenuation) coefficient. Most tabulations refer to the mass absorption coefficient (μ/ρ), with ρ being the density of the absorber and having units cm^2/g.

The mass absorption coefficient for X-rays in tungsten is given in Fig. 1.2 as a function of X-ray wavelengths. The coefficient varies with wavelength, exhibiting sharp

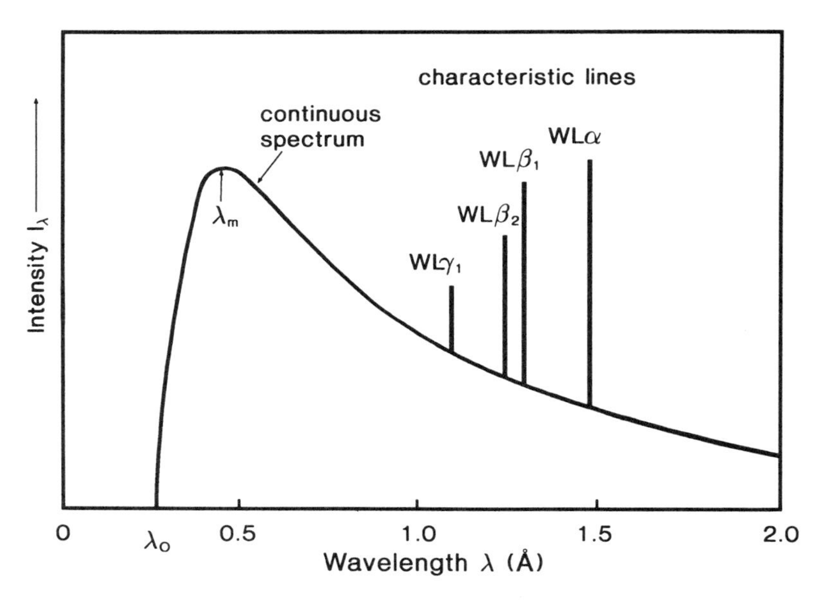

FIGURE 1.1. Typical emission spectrum from a W target X-ray tube at 50 kV [1.8].

discontinuities in the otherwise smooth curve (*absorption edges*). These correspond to photon energies, which are determined by the energies of the respective K, L, and M shells. The absorption edges are at 0.17837 Å (69.509 keV) for the K series, at 1.2155 Å (10.200 keV) for the L series, and at 6.83Å (1.814 keV) for the M series. The complete set of absorption wavelengths (energies) for tungsten is shown in Table 1.4.

Due to its high density, tungsten is a very effective X-ray absorber. Therefore, tungsten and some of its alloys are important materials for X-ray radiation shielding.

TABLE 1.3. Wavelengths of Important K-, L-[a], and M-Emission[b] Series

Spectral line name	Transition	λ (Å)	Energy (keV)
$K\alpha_2$	K–L_2	0.213813	57.99
$K\alpha_1$	K–L_3	0.208992	59.32
$K\beta_1$	K–M_3	0.184363	67.25
$K\beta_2$	K–N_3, K–N_2	0.17950	69.07
$L\alpha_2$	L_3–M_4	1.48742	8.3356
$L\alpha_1$	L_3–M_5	1.47635	8.3981
$L\beta_1$	L_2–M_4	1.28176	9.6730
$L\beta_2$	L_3–N_5	1.24458	9.9620
$M\alpha_2$	M_5–N_6	6.978	1.7768
$M\alpha_1$	M_5–N_7	6.969	1.7791
$M\beta$	M_4–N_6	6.743	1.8387

[a] *International Tables for Crystallography* [1.6].
[b] R. Jenkins (1973) [1.7].

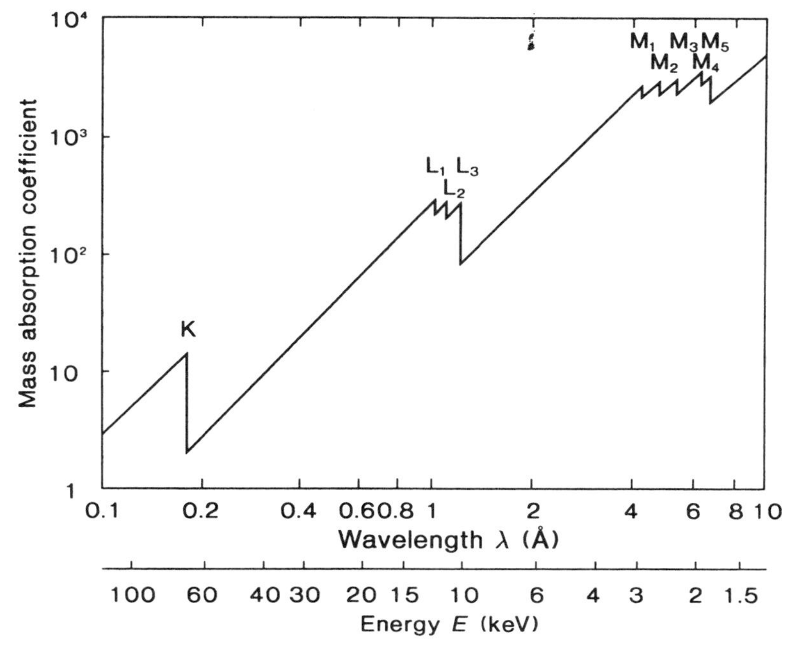

FIGURE 1.2. Mass absorption coefficient of tungsten as a function of wavelength and radiation energy [1.8].

Values of (μ/ρ) for tungsten are listed below for several important X-radiations [1.6].

$$\begin{aligned}
&\textit{Mo}\ K\alpha\ (\lambda = 0.7107\ \text{Å}): && 94\ \text{cm}^2 \cdot \text{g}^{-1}\\
&\textit{Cu}\ K\alpha\ (\lambda = 1.5418\ \text{Å}): && 168\ \text{cm}^2 \cdot \text{g}^{-1}\\
&\textit{Co}\ K\alpha\ (\lambda = 1.7905\ \text{Å}): && 246\ \text{cm}^2 \cdot \text{g}^{-1}\\
&\textit{Fe}\ K\alpha\ (\lambda = 1.9373\ \text{Å}): && 301\ \text{cm}^2 \cdot \text{g}^{-1}
\end{aligned}$$

TABLE 1.4. X-Ray Absorption Wavelengths for Tungsten [1.1]

Series	Name	Wavelength (Å)	Energy (keV)
K		0.17837	69.508
L	L_1	1.02467	12.100
	L_1	1.07450	11.538
	L_3	1.21550	10.200
M	M_1	4.407	2.813
	M_2	4.815	2.575
	M_3	5.435	2.281
	M_4	6.590	1.880
	M_5	6.830	1.814

1.1.3.3. Auger Transitions [1.9]. The electron energies (eV) of the principal Auger peaks for tungsten are shown in Fig. 1.3 for an operating voltage of 5 keV.

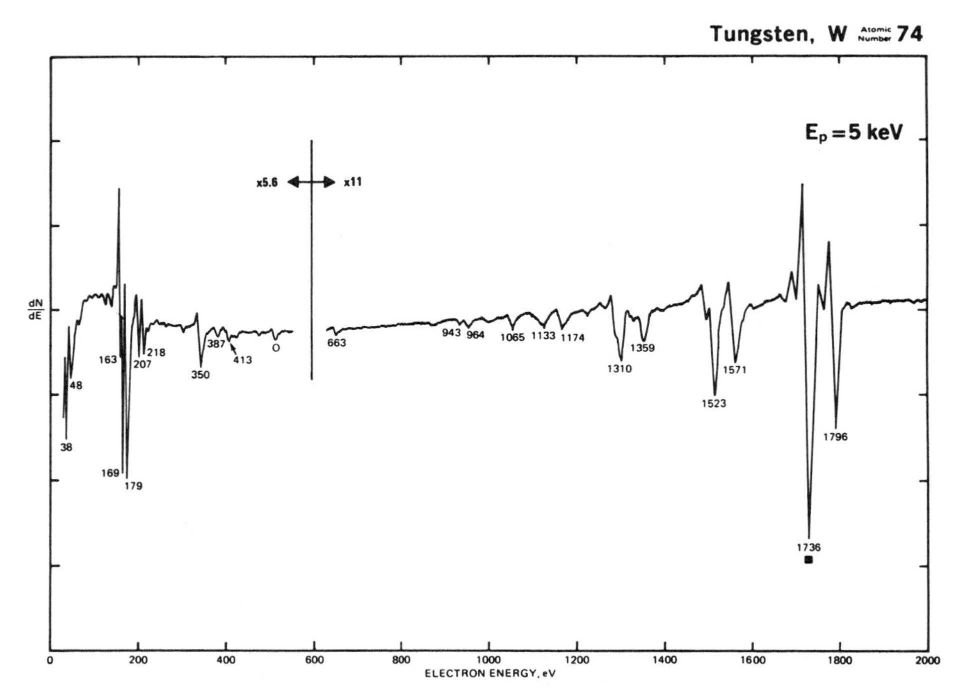

FIGURE 1.3. Electron energies of the principal Auger peaks of tungsten [1.9].

1.1.3.4. Atom Spectrum [1.1]. The atom spectrum of tungsten is very complex. It consists of approximately 6800 lines between 2000 and 10,000 Å. The strongest lines are tabulated below.

Wavelength (nm)	Species
429.461	I
407.436	I
400.875	I
255.135	I (used in AAS)
207.911	II
202.998	II

These lines are used for the spectrographic determination of the element (AAS, DC arc).

Due to the large number of lines, spectral analysis of trace impurities in a tungsten matrix is only possible in a reducing atmosphere after conversion into tungsten carbide and when using a carrier distillation method.

1.1.4. Thermodynamic Functions [1.10]

Values for heat capacity, Gibbs energy function, entropy, and enthalpy increment have been calculated for an ideal gas of monoatomic tungsten. They are listed in Table 1.5.

TABLE 1.5. Thermodynamic Functions (Tungsten Gas) [1.1, 1.10]

Temperature T (K)	Heat capacity C_p ($J \cdot K^{-1} \cdot mol^{-1}$)	Enthalpy increment $H - H_{298}$ ($kJ \cdot mol^{-1}$)	Gibbs energy function $-(G - H_{298})/T$ ($J \cdot K^{-1} \cdot mol^{-1}$)	Entropy S ($J \cdot K^{-1} \cdot mol^{-1}$)
298.15	21.304	0.000	173.951	173.951
400.00	23.169	2.262	174.809	180.464
600.00	30.366	7.556	178.504	191.097
800.00	37.738	14.409	182.892	200.903
1000.00	41.226	22.387	187.403	209.790
1500.00	37.684	42.480	197.817	226.137
2000.00	32.614	59.955	206.249	236.227
2500.00	30.299	75.569	212.977	243.205
3000.00	30.150	90.618	218.487	248.693
3500.00	31.392	105.966	223.146	253.422
4000.00	33.326	122.127	227.203	257.735
4500.00	35.474	139.327	230.823	261.784
5000.00	37.511	157.582	234.113	265.629
5500.00	39.285	176.793	237.146	269.290
6000.00	40.775	196.818	239.971	272.774

Values for 298.15 K are named *standard Enthalpy* and *standard entropy*, the latter valid at a pressure of 1 bar; to convert J to cal, divide by 4.184.

1.2. BULK TUNGSTEN METAL RELATED PHYSICAL PROPERTIES

1.2.1. Electronic Structure and Bonding [1.1, 1.11–1.13]

Tungsten has been of keen theoretical interest for electron band-structure calculations [1.14–1.25], not only because of its important technical use but also because it exhibits many interesting properties. Density functional theory [1.11], based on the *ab initio* (nonempirical) principle, was used to determine the electronic part of the total energy of the metal and its cohesive energy on a strict quantitative level. It provides information on structural and elastic properties of the metal, such as the lattice parameter, the equilibrium volume, the bulk modulus, and the elastic constants. Investigations have been performed for both the stable (bcc) as well as hypothetical lattice configurations (fcc, hcp, tetragonal distortion).

1.2.1.1. Total Energy. The bulk equilibrium total energy (including all 74 electrons) was calculated to be −439,457 eV [1.17, 1.20], and the total energy of the isolated atom −439,447 eV [1.17, 1.20].

1.2.1.2. Cohesive Energy. Cohesive energies can be obtained from *ab initio* calculations as the difference between crystalline bulk and the total energy of the free atom (d^4s^2 configuration). Tungsten has the largest cohesive energy of all elements (Fig. 1.4) [1.12], including diamond (carbon). The cohesive energy can be interpreted as the *bonding energy* of the metal in the bcc crystal lattice at absolute zero ($T = 0$ K). Values obtained from different calculations vary between 7.9 eV/atom [1.14] and 10.09 eV/atom

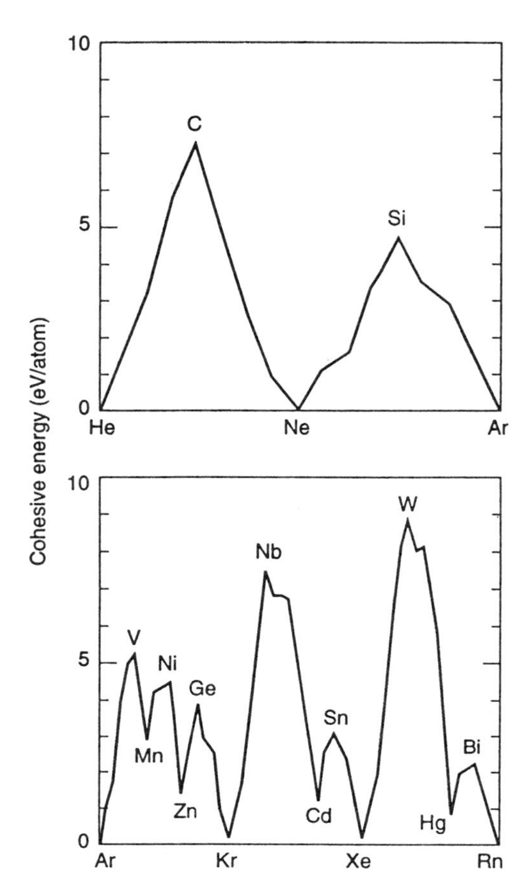

FIGURE 1.4. Cohesive energy across the short periods (upper panel) and long periods (lower panel) [1.12].

[1.20], depending on the theoretical approach [1.14–1.16, 1.19–1.21]. The "experimental" value of the cohesive energy, which is derived from extrapolations of the sublimation enthalpy (ΔH_s) to absolute zero ($T = 0$ K), is quoted as 8.9 eV/atom ($\approx$859 kJ/mol) [1.1].

1.2.1.3. Structural Energy (Lattice Stability). The total energy of W was calculated for the bcc, fcc and hcp structures at various atomic volumes close to the experimental volume. The result of such a calculation is shown in Fig. 1.5 [1.22]. From this, it follows that the bcc modification has the lowest energy and thus is the stable crystal structure of W. The cohesive energy decreases in the order of bcc $\rightarrow$ fcc $\approx$ hcp. The difference $\Delta E = E_{tot}(\text{fcc}) - E_{tot}(\text{bcc})$, or $\Delta E = E_{tot}(\text{hcp}) - E_{tot}(\text{bcc})$ is named *structural energy* or *lattice stability*. Values for ΔE are in the range of 0.45–0.57 eV/atom [1.17, 1.19, 1.22] for fcc–bcc, which corresponds to an enthalpy difference $H^{\text{fcc}} - H^{\text{bcc}}$ between 44 and 55 kJ/mol. The value for hcp–bcc (0.6 eV/atom; 57.9 kJ/mol) is only slightly higher [1.22]. The similarity between the fcc and hcp results is due to the close relation between these two lattices.

Lattice stabilities can also be obtained from computer calculations of phase diagrams (Pt–W, Ir–W; so-called CALPHAD assessments) [1.25–1.27]. This approach is semi-

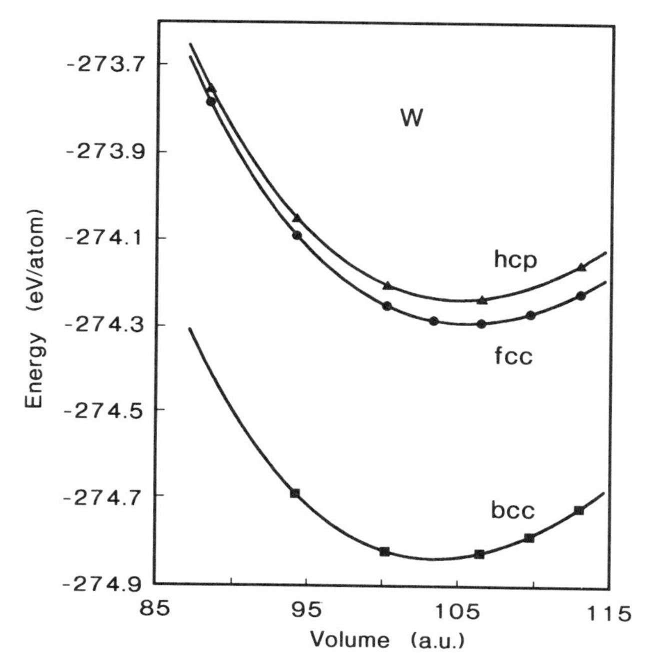

FIGURE 1.5. Total energy of W as a function of atomic volumes for bcc, fcc, and hcp (with ideal c/a ratio) structures [1.22]. The energy is given in the pseudopotential formalism by restricting the calculations to the valence electrons only. Note that bcc W has the smallest volume per atom (i.e., the strongest bonding). 1 au = 0.529 Å.

empirical and has its roots in chemical thermodynamics. Early calculations have given misleading low results for the enthalpy difference ($H^{\mathrm{fcc/bcc}} = 10\,\mathrm{kJ/mol}$), but recent work resulted in more realistic estimates, and a generally better agreement with *ab initio* results ($H^{\mathrm{fcc/bcc}} = 30\,\mathrm{kJ/mol}$) [1.26]. Differing from the *ab initio* calculations, which refer to absolute zero, CALPHAD calculations are based on high-temperature data.

1.2.1.4. The Density of States. When the free atoms bind together to form a crystal, their energy levels are broadened and form the density of states, which can be seen as the solid-state analogy to the free-atom energy spectrum. The densities of states (DOS), as obtained from electronic structure calculations, are shown in Fig. 1.6 [1.19] for both the stable bcc and hypothetical fcc W. The DOS is not uniform throughout the band but displays considerable structure that is characteristic of the given crystal lattice [1.12]. This can be analyzed into low-lying bonding and antibonding states at higher energy. In bcc tungsten, all the bonding states are filled with electrons and all the antibonding states are empty. The Fermi energy E_{f}, which is the energy of the highest occupied state, lies exactly in between the bonding and antibonding states. This bimodal behavior is characteristic for the strong unsaturated covalent bonds between the valence 5d orbitals [1.12]. It also reflects the strong stability of the bcc lattice with respect to the fcc structure, where this clear decomposition between bonding and antibonding states is not present.

1.2.1.5. The Nature of Metallic Bonding. The bonding in tungsten arises from the strong unsaturated, covalent bonds between the valence 5d orbitals (localized "t_{2g}"-like

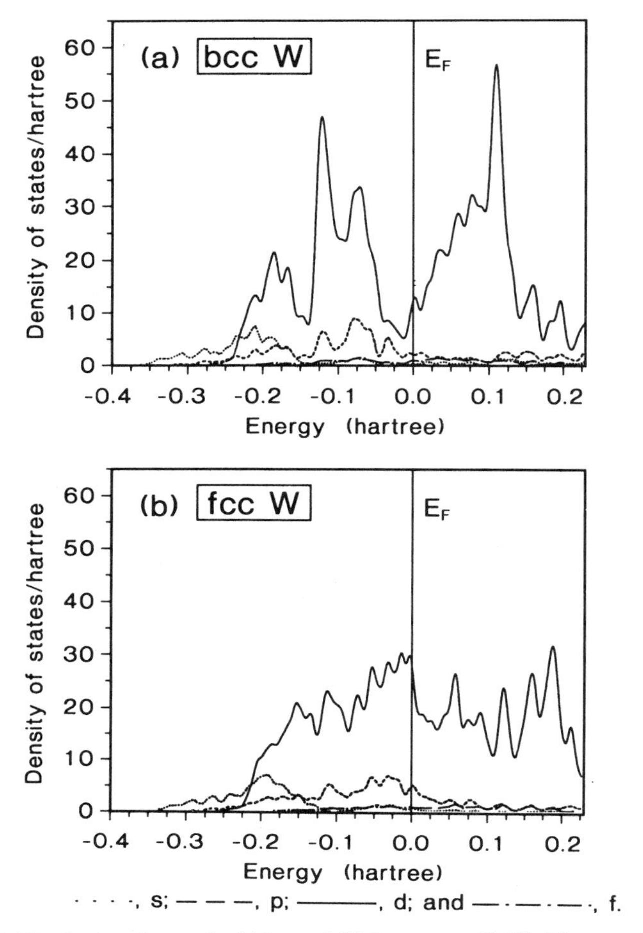

FIGURE 1.6. The density of states for (a) bcc and (b) fcc tungsten [1.19]; 1 hartree = 27.211 eV.

states); while, s and p bonds have lower contributions, forming a diffuse (delocalized, nearly free electron-like) background. According to the local spin density theory the ground-state configuration in the bcc tungsten crystal is d^5s^1 rather than d^4s^2 in the free atom [1.1, 1.14], demonstrating the half-filled d band in the solid.

1.2.1.6. Ab Initio Calculations of Structural and Elastic Properties. Equilibrium lattice constants, equilibrium volumes, as well as bulk and shear moduli can be assessed based on *ab initio* electron-structure calculations. They are obtained from the calculated total energies as a function of volume in the bcc or fcc crystal structure and from respective volume-conserving distortions of the lattice. In most cases, they agree well with experiments (Table 1.6).

Hypothetical properties can be calculated for the fcc modification (Table 1.6) [1.17, 1.21, 1.27]. It should be noted that both C' and C_{44} for the fcc structure are negative, which means that the fcc lattice is unstable with respect to the respective distortions.

TABLE 1.6. Comparison of Equilibrium Lattice Constants and Bulk/Shear Moduli of bcc and fcc Tungsten

Reference	Lattice parameter (Å)	Bulk modulus (GPa)	Shear modulus (GPa)
	bcc tungsten		
Experimental [1.6, 1.30, 1.31]	3.16524	305–310	$C' = 163.8$; $C_{44} = 163.1$
Bylander *et al.* [1.16]	3.163	297	
Jansen *et al.* [1.17]	3.15	345	
Wei *et al.* [1.20]	3.164	318	
Mattheiss *et al.* [1.21]	3.16	340	
Chan *et al.* [1.22]	3.13	333	
Guillermet *et al.* [1.25]			$C' = 178$; $C_{44} = 155$
	fcc tungsten		
Jansen *et al.* [1.17]	3.99	274	
Mattheiss *et al.* [1.21]	4.00	309	
Guillermet *et al.* [1.25]			$C' = -232$; $C_{44} = -256$

Theoretical predictions can be made of structural phase transitions under very high pressure (compressions down to 30% of the ambient volume), and bcc tungsten was demonstrated to be the only stable modification over the whole pressure range [1.28]. Recently, electronic structure calculations were also applied to assess the linear coefficient of thermal expansion [1.29].

1.2.1.7. Experimental Determination of Binding Energies. High-accuracy determinations of core level binding energies were carried out on clean tungsten surfaces by ESCA measurements [1.32]. The obtained binding energies (taken as the peak maxima) are summarized in Table 1.7. Experimental valence-band spectra were obtained by high-resolution angle-resolved X-ray photoemission spectroscopy [1.33]. At room temperature, the following peaks were observed below the Fermi level: −4.8 eV (s-like character), −3.2 and −2.3 eV (mainly d-like character), and 0.6 eV. Good agreement was obtained between experimental values and theoretical predictions [1.33].

TABLE 1.7. Binding Energies for Tungsten in eV[a]

4s	$4p_{1/2}$	$4p_{3/2}$	$4d_{3/2}$	$4d_{5/2}$
594.3(8)	490.8(5)	423.7(3)	256.0(3)	243.5(3)
5s	$5p_{1/2}$	$5p_{3/2}$	$5f_{5/2}$	$5f_{7/2}$
75.5(8)	47(1)	37.2(4)	33.48(10)	31.32(10)

[a] Calibrated against the Au $4f_{7/2}$ level (84.00 eV) [1.32].

1.2.2. Structural Properties

1.2.2.1. Crystallographic Properties [1.1, 1.6]. Besides amorphous tungsten, three modifications (α-, β-, and γ-tungsten) are known. Table 1.8 provides crystallographic parameters and related data, such as density, molar volume, and specific volume.

TABLE 1.8. Polymorphism of Tungsten: Crystallographic and Related Data [1.1]

Modification	α-W	β-W	γ-W
Model type	A2 $Z = 2$ atoms in 000 and $\frac{1}{2}, \frac{1}{2}, \frac{1}{2}$ position	A15 (Cr_3Si) $Z = 8$ atoms, 2 in 2a sites: 000 and $\frac{1}{2}, \frac{1}{2}, \frac{1}{2}$ position; 6 atoms in 6c sites: $0, \frac{1}{4}, \frac{1}{2}$; $\frac{1}{2}$; $0, \frac{3}{4}, \frac{1}{2}$; $\frac{1}{2}, 0, \frac{1}{4}$; $\frac{1}{2}, 0, \frac{3}{4}$; $\frac{1}{4}, \frac{1}{2}, 0$; $\frac{3}{4}, \frac{1}{2}, 0$	A1
Space group	O_n^9 – Im3m (No. 229)	O_h^3 – Pm3n	
Shortest interatomic distance	2.741	25.2_5 and 2.82_6 Å	
Lattice parameter[a] (25 °C)	3.16524 Å ± 0.00004	5.05 Å (5.037–5.09) depending on preparation method	4.13 Å
Density calc.	19.246 g/cm^3	18.9 g/cm^3	15.8 g/cm^3
pycnometric	19.250 g/cm^3	19.1 g/cm^3	
Specific volume at melting point	0.057–0.062 cm^3		
Molar volume	9.53 cm^3		

[a] *International Tables for Crystallography* (1995) [1.6].

α-Tungsten is the only stable modification. It has a body-centered cubic lattice of space group O_h^9 – Im3m (No. 229). A diffraction pattern is shown in Fig. 1.7, together with a crystal structure model.

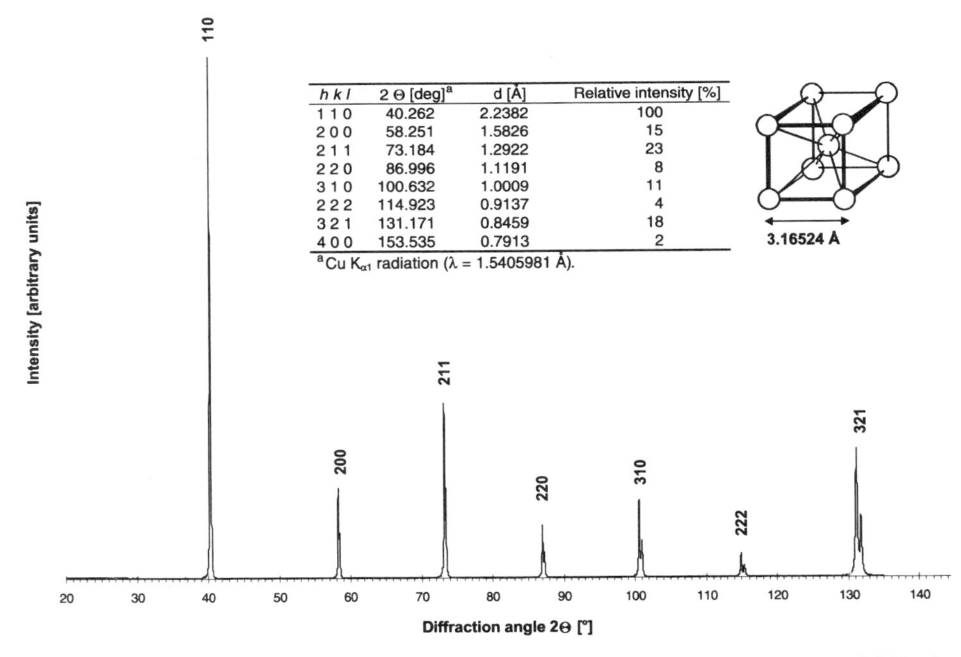

h k l	2 Θ [deg][a]	d [Å]	Relative intensity [%]
1 1 0	40.262	2.2382	100
2 0 0	58.251	1.5826	15
2 1 1	73.184	1.2922	23
2 2 0	86.996	1.1191	8
3 1 0	100.632	1.0009	11
2 2 2	114.923	0.9137	4
3 2 1	131.171	0.8459	18
4 0 0	153.535	0.7913	2

[a] Cu $K_{\alpha 1}$ radiation (λ = 1.5405981 Å).

FIGURE 1.7. X-ray diffraction diagram of α-tungsten; 2θ values (Cu $K\alpha_1$); according to *International Tables for Crystallography* [1.6].

β-Tungsten is a metastable phase and converts to α-tungsten when heated above 600 to 700 °C. It has a cubic A 15 lattice (space group O_h^3 – Pm3n; structure type Cr_3Si) and was first characterized structurally as a suboxide (W_3O) [1.34]. As such, it is still listed in the International Powder Diffraction File (41-1230). A more recent diffraction diagram is shown in Fig. 1.8 together with the proposed crystal structure [1.35, 1.36]. β-Tungsten forms at least partially during low-temperature hydrogen reduction of tungsten oxides, tungsten bronzes, or ammonium paratungstate besides α-tungsten. Several foreign elements, such as P, As, Al, K, etc., promote its formation and also have a stabilizing effect at higher temperatures. Moreover, β-tungsten forms during electrolytic reduction of WO_3 in phosphate melts and in thin films, produced by evaporation, sputtering, or chemical vapor deposition.

γ-Tungsen (face-centered cubic structure of A1 type) was only found in thin sputtered layers and amorphous tungsten at the very beginning of sputtering. γ-tungsten too will form the α-modification when heated ≥700 °C.

1.2.2.2. Density [1.35]. The density of tungsten, equal to that of gold, is among the highest of all metals. Measured densities of sintered materials may vary significantly, since full density is only achieved after heavy working. Exact measurements were therefore carried out on foils, thin wires, single crystals, etc. A summary of the respective values, obtained by different methods, is given below.

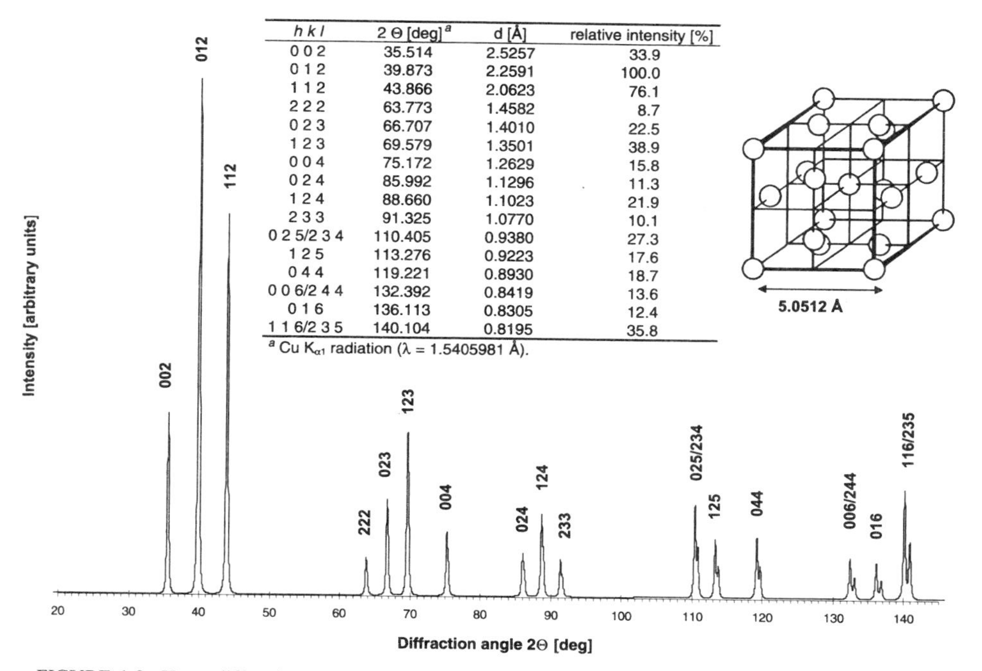

h k l	2 Θ [deg][a]	d [Å]	relative intensity [%]
0 0 2	35.514	2.5257	33.9
0 1 2	39.873	2.2591	100.0
1 1 2	43.866	2.0623	76.1
2 2 2	63.773	1.4582	8.7
0 2 3	66.707	1.4010	22.5
1 2 3	69.579	1.3501	38.9
0 0 4	75.172	1.2629	15.8
0 2 4	85.992	1.1296	11.3
1 2 4	88.660	1.1023	21.9
2 3 3	91.325	1.0770	10.1
0 2 5/2 3 4	110.405	0.9380	27.3
1 2 5	113.276	0.9223	17.6
0 4 4	119.221	0.8930	18.7
0 0 6/2 4 4	132.392	0.8419	13.6
0 1 6	136.113	0.8305	12.4
1 1 6/2 3 5	140.104	0.8195	35.8

[a] Cu $K_{\alpha 1}$ radiation (λ = 1.5405981 Å).

FIGURE 1.8. X-ray diffraction diagram of β-tungsten; 2θ values (Cu $K\alpha_1$); according to A. Bartl [1.36].

Density of α-W *in* g/cm^3

Calculated from X-ray lattice parameter measurements:

19.246 ± 0.003 (25 °C),
19.254 ± 0.0046 (25 °C)
19.256 ± 0.02 (25 °C)
19.316 (77 K)

Extrapolated from 40–180 K to 298 K:

19.250

Measured by hydrostatic weighting:

19.250 ± 0.004 (20 °C) on zone refined single crystals,
19.253 ± 0.11 (25 °C) on the same material,
19.3615 (25 °C) on less pure samples.

Measured by X-ray reflection and X-ray refraction:

19.262 (room temperature)

At the melting temperature:

16.74 ± 0.6.

Density of β-*W*

19.1 (Pycnometrically),
18.9 (calculated from X-ray parameter).

Density of γ-*W film*

15.8 (grazing X-ray reflection).

The apparent density is used: (a) in case of deagglomerated powders for a rough and simple grain-size estimation; (b) in case of compacts (green density) or sintered specimens to determine the residual porosity.

1.2.2.3. Lattice Defects [1.1, 1.37]. Generally, lattice imperfections can be divided into four groups:

point defects (vacancies, self interstitial atoms, impurity atoms),
line defects (dislocations),
plane defects (stacking faults, grain boundaries, twins), and
volume defects (clusters, voids, and bubbles; segregations, microcracks, second phases and domains, etc.).

Even very pure and long-time annealed tungsten always contains lattice defects, such as vacancies and dislocations. The concentration of defects depends on temperature and deformation of the material, but also on the purity.

Vacancies. The formation energy of a monovacancy in tungsten E_f is between 2.801 and 4.47 eV/atom (270–430 $kJ \cdot mol^{-1}$). The vacancy concentration is given by $c = 760 \cdot \exp(-72,500/RT)$ and reaches 3.4 % at the melting point. Vacancy migration starts at 350–400 °C and measured as well as calculated values for the activation energy are between 1.3 and 2.3 eV/atom (125–222 kJ · mol^{-1}).

Interstitial impurities. In a bcc lattice two interstitial sites can be occupied: the octahedral and the tetrahedral. Oxygen, nitrogen, and carbon occupy the octahedral position and hydrogen the tetrahedral position in the tungsten host lattice. Interstitial impurities induce lattice distortions by that altering the mechanical properties.

Dislocations. Dislocation densities (dislocation lines/cm^2) in electron-beam-zone refined tungsten have been measured to be between 5×10^3 and 10^6. A diminution from the center of a rod to its surface was determined. In cold-worked foils, filings, rods, and thin wires (as worked or annealed) the densities vary between 10^8 and 10^{12}.

A very complete compilation of the respective literature on all kinds of lattice imperfections in tungsten is given elsewhere [1.1].

1.2.2.4. Other Structural Properties: Slip, twinning, and cleavage of single crystals. [1.35]. Single crystal deformation mechanisms are slip (gliding, shear), twinning, and cleavage with fracture. The dislocations contained are the fundamental carriers of the plastic deformation in general.

Slip or gliding. Slip is the basic mechanism in the plastic deformation of tungsten single crystals. Slip occurs in the most densely packed [111] direction and either in {110} or in {122} planes. At elevated temperature also {111} occurs in addition.

Twinning. Twinning is a less dominant deformation mechanism and occurs in high-purity tungsten only. Twins are mainly formed on {112} planes in [111] direction. Twins have been observed in tension below 0 °C and, after rolling, compressing, swaging, and explosion deformation, up to 1500 °C.

Cleavage. The usual cleavage plane for bcc metals is {110}, but cleavage may also happen along {110} planes and twin interfaces.

Surface free energy/surface tension [1.37, 1.38]. Between the surface tension (γ) and the surface free energy (σ) the following relation exists: $\gamma = \sigma + A(d\sigma/dA)$, for an area A.

Cracking, zero creep, and field emission microscopy yield equal values for γ and σ. By inert gas bubble precipitation, grain boundary grooving, and scratch relaxation rate studies σ can be determined, and γ by contact angle and multiphase equilibrium studies.

The calculated as well as experimental values for tungsten scatter largely depending on method and material. Values of σ measured at single crystal samples are around 4.3 J · m^{-2} for (110) planes and scatter between 1.7 and 6.3 J · m^{-2} for (100) planes. Polycrystalline samples showed σ values between 1.8 and 5.38 J · m^{-2}. The temperature dependence of σ was determined to be 5.38 J · m^{-2} at 1200 K, decreasing linearly to 3.88 J · m^{-2} at 1800 K. Theoretical calculations resulted in 4.435 J · m^{-2} for 0 K and 3.938 J · m^{-2} for 2100 K.

Values of γ at the melting temperature were measured between 2.316 and 2.545 J · m^{-2}.

Surface diffusion [1.37]. The data for the surface diffusion of tungsten atoms on their own lattice in a general direction are $D_0 = 4$ cm^2/s and $Q = 3.14$ eV/atom (~300 kJ · mol^{-1}). The activation energy for migration on {100}, {110}, and {211} planes is 2.79 eV/atom (~270 kJ · mol^{-1}).

1.2.3. Mechanical Properties [1.35, 1.39–1.43]

The mechanical properties of tungsten are only intrinsic as long as highly pure single crystals are considered. Polycrystalline tungsten, from a technical standpoint, is the most important form and, by far, the biggest amount of the metal used. These properties are also strongly influenced by two main factors:

- microstructure,
- presence, concentration, and combination of impurity elements.

Microstructure. The microstructure of a tungsten sample, which influences the mechanical properties within a wide range, depends on the type of preparation (powder metallurgy, arc cast, electron beam melted, zone refined, chemical vapor deposition) and the subsequent working (deformation, annealing, recrystallization).

In case of powder metallurgical processing (P/M tungsten), the crystal size and shape can be regulated by the sintering conditions, the type and degree of deformation, and the intermediate and final annealing processes. In this way it becomes possible to produce material of desired mechanical properties. Examples for the variety of different microstructures of polycrystaline tungsten are shown in Fig. 1.9.

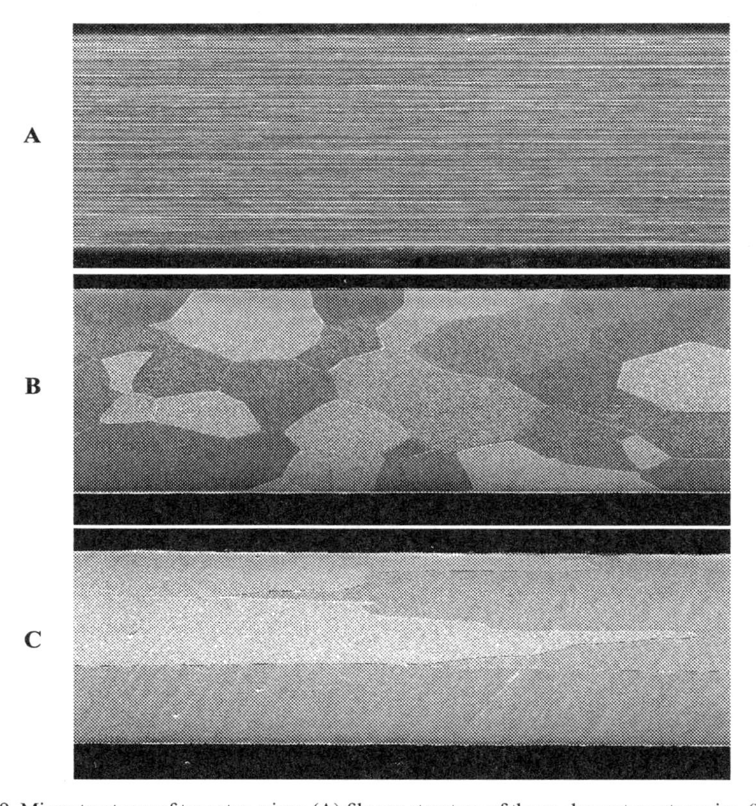

FIGURE 1.9 Microstructures of tungsten wires: (A) fibrous structure of the as-drawn tungsten wire, (B) equiaxed grain structure of recrystallized pure wire, (C) interlocking grain structure of recrystallized NS-doped wire. By courtesy of Philips Lighting B.V., Eindhoven, The Netherlands.

Generally, the following rules exist:

- a wrought (as-worked) structure, consisting of deformation direction elongated crystals, is stronger than an annealed or recrystallized one. In particular, this is valid for the strength in deformation direction ("texture strengthening"),
- the lower the working temperature and the more energy is stored during deformation, the higher is the strength ("work hardening"),
- the smaller the grain size, the higher is the strength ("fine-grain strengthening"),
- different fabrication processes (extrusion, swaging, rolling, drawing) yield different properties, and even after final working, there remain origin-related differences in the properties of arc-melted and P/M materials.

Impurities. Foreign elements can be contained in tungsten in different modes. Homogeneously or quasi-homogeneously distributed foreign elements can occur either in solid solution or segregated. In the first case, one has to distinguish between substitution of tungsten atoms by atoms of the foreign element in the crystal lattice and atoms at interstitial positions. The solubility for interstitial elements in tungsten, such as hydrogen, nitrogen, oxygen, silicon, and carbon at room temperature, is very low ($< 0.1\ \mu g/g$). Their solubility at eutectic temperature ranges between 100 and $1000\ \mu g/g$. During cooling, the difference between the two concentrations will be segregated mainly at the grain boundaries, thereby significantly weakening the grain boundary strength. Generally, in case of segregations, the impurities can be located either at grain boundaries or within the crystals (at dislocations or subgrain boundaries).

During heavy working of tungsten, additional low-energy sites (dislocation tangles) are formed. Segregation of interstitials to these sites increases the grain boundary strength, thus increasing both ductility and strength [1.41].

The analytical determination of the foreign element concentration and distribution in tungsten is linked with highly sophisticated equipment (Auger spectroscopy, secondary ion mass spectrometry) and with complicated pretreatment of the samples. This may be one reason why our knowledge about foreign element influences is still quite incomplete.

Heterogeneous impurities (particles of foreign matter in the raw materials) cause locally foreign phases or voids, often combined with diffusion zones.

In any case, the consequences of the impurities present are various types of structural defects. Their size can be quite different and can range from atomic distances upward to some micrometers. The concentration of the impurity plays an important role. In some cases, traces of an impurity element show a considerable influence on the mechanical properties. Finally, the combination of impurity elements could change completely the effect they would cause if present alone. Taking this into account, one may understand that a comparison of older values of mechanical properties with those obtained today may sometimes show bigger differences. The reason for that is that, generally, the impurity content of today's tungsten is quite lower than 40 or 50 years ago.

Generally speaking, the indication of absolute values for the mechanical properties of polycrystalline tungsten is not appropriate as long as the related structure, structural history, type of impurity elements, their concentration, and kind of distribution cannot be precisely defined. However, for technical samples, this is especially impossible due to the high cost and the time required for analyses of such type. Therefore, any values given for the mechanical properties should be regarded as lines instead of points and as planes instead of curves.

The outstanding mechanical properties of polycrystalline tungsten are:

- the high strength and yield point at elevated temperature and the high creep resistance.

1.2.3.1. Elastic Properties [1.30, 1.31, 1.35]. In regard to elasticity, at least below room temperature, tungsten behaves nearly isotropically; the anisotropy coefficient at 24 °C is $A = 1.010$ [1.35]. The elastic constants for polycrystalline tungsten at 20 °C are given below. Their temperature dependence as well as the respective values for single-crystal elastic constants are shown in Fig. 1.10 [1.40], based on ultrasonic measurements [1.30, 1.31].

Young's modulus, $E = 390\text{–}410$ GPa
shear modulus, $G = 156\text{–}177$ GPa
bulk modulus, $K = 305\text{–}310$ GPa
Poisson's ratio, $\nu = 0.280\text{–}0.30$

The *compressional modulus* L of polycrystalline tungsten can be derived from [1.30]

$$L = (5.2415 \times 10^{12}) - (3.7399 \times 10^{8})T - (4.598 \times 10^{4})T^2$$

(dyn · cm^{-2}; for conversion to GPa, multiply by 100; T in °C).

1.2.3.2. Hardness. Hardness measurement is easy to perform; this is why the determination is used worldwide in production control as a measure of strength. As long as the measures are taken in one plant, any comparison might be suitable. However, comparison between different testing facilities and on different samples, respectively, measured at different load, might be sometimes doubtful and misleading, especially if methods are not standardized. Furthermore, different methods are in use to determine the hardness (Vickers, Brinell, Rockwell, Knoop, and microhardness), and the results cannot easily be compared with each other.

Today, Vickers hardness is the most common method of determination. Table 1.9 compares the results of Vickers microhardness measurements obtained from two differently produced types of single crystals and several polycrystalline tungsten samples.

The following conclusions can be drawn:

- there exists a load dependence (the higher the load, the lower the hardness),
- measures on single crystals exhibit a slight dependence on crystal orientation; the average of the three crystal orientation measures compares very well with those of polycrystalline samples (annealed or recrystallized),
- the hardness of swaged rods "as-worked" is significantly higher than that of the subsequently annealed material (463 kg · mm^{-2} ⇎ 357 kg · mm^{-2})

Macro Vickers hardness for polycrystalline tungsten is given as (HV_{30}) [1.39]:

0 °C	450
recrystallized	300
worked/deformed	up to 650
400 °C	240
800 °C	190

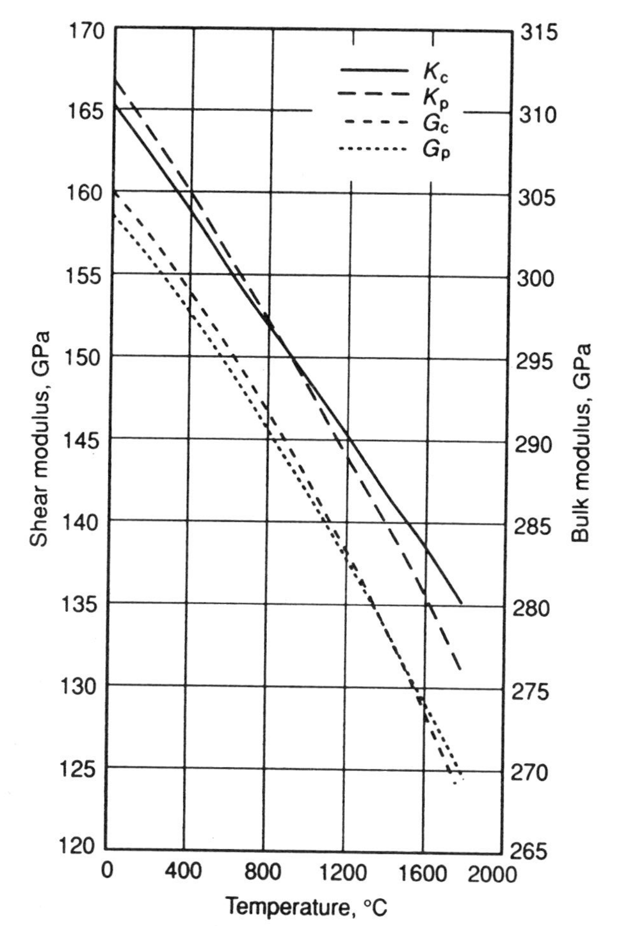

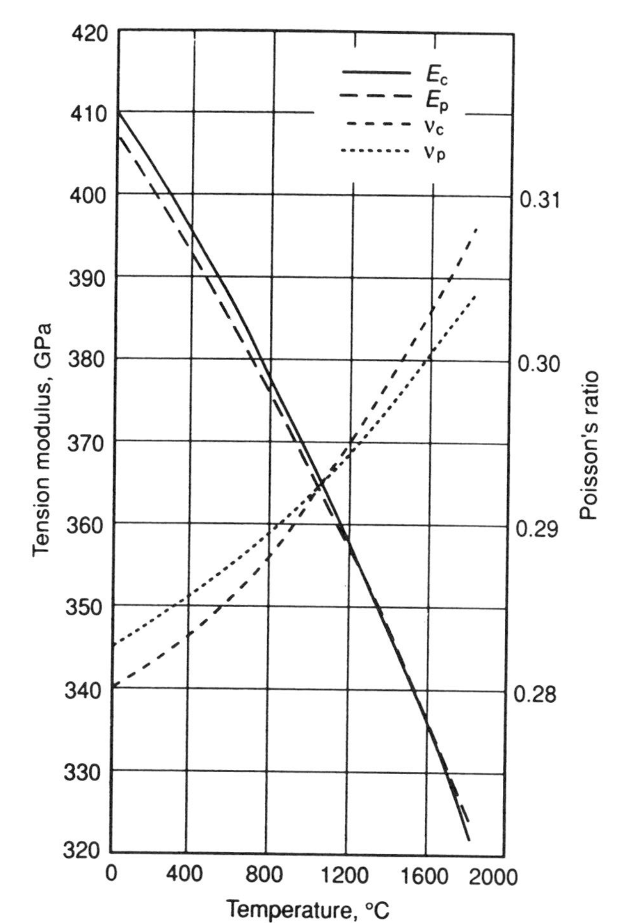

FIGURE 1.10. Shear modulus G, bulk modulus K, Young's modulus E, and Poisson's ratio ν of tungsten vs. temperature, as calculated from single-crystal elastic constants (G_c, K_c, E_c, and ν_c) [1.31], and from measurements on polycrystalline tungsten (G_c, K_c, E_c, and ν_c). [1.30] Taken from *Metals Handbook* [1.40].

TABLE 1.9. Microhardness ($kg \cdot mm^{-2}$) of Tungsten of Different Manufacturing History [1.39]

Material	Plane of indentation	Load			
		100 g	200 g	300 g	400 g
	Single-crystal tungsten				
Grown on a Pintch wire	(100)	360	362	352	350
	(110)	395	389	385	380
	(111)	408	410	394	380
	Average	392	379	363	357
Zone melted	(100)	361	349	350	340
	(110)	405	390	383	373
	(111)	396	382	393	387
	Average	387	377	379	367
	Polycrystalline tungsten				
Electron beam melted		397	386	374	363
Swaged as-worked		498	474	475	463
Swaged and annealed		392	379	363	357
Rolled and recrystallized		401	393	384	372

The hardness of polycrystalline tungsten increases appreciably as the grain size is reduced. It was shown that hardness and average grain size follow a Hall–Petch relationship:

$$H = H_0 + K_H l^{-1/2}$$

with $H_0 = 350\,kg \cdot mm^{-2}$, $K_H = 10\,kg \cdot mm^{-3/2}$, and l being the average grain diameter. Hardness values of up to $1200\,kg \cdot mm^{-2}$ (2000 g load) were obtained on submicron-grained materials [1.44].

The temperature dependence for work-hardened and annealed tungsten is shown in Fig. 1.11 [1.40, 1.45]. At about 1600 °C, both curves coincide, i.e., at this temperature, the tungsten completely recrystallizes as it is heated to the testing temperature.

Doped tungsten (AKS-, Thoria-doped) retains deformation hardness up to higher temperatures. Thoria-doped tungsten may even have $450\,kg \cdot mm^{-2}$ at 1900 °C [1.37].

1.2.3.3. Low-Temperature Brittleness. At low temperatures, tungsten behaves in a brittle manner during loading, if subjected to tensile stresses. This behavior changes above a certain transition temperature, which is called the ductile-to-brittle transition temperature (DBTT). This temperature is not a sharp, reproducible value, but a certain range (for example 200–300 °C), which depends on mechanical, structural, and chemical conditions. The most important influencing parameters are summarized in Table 1.10. The DBTT is an important material characteristic for the "low temperature" use and fabricability (shearing, punching, bending, etc. operations) of the metal.

The brittleness of polycrystalline tungsten, at low temperatures, is attributed to the weakness of the grain boundaries, which leads to initiation of cracking in both wrought and recrystallized tungsten. Pure, single crystalline tungsten (free of grain boundaries)

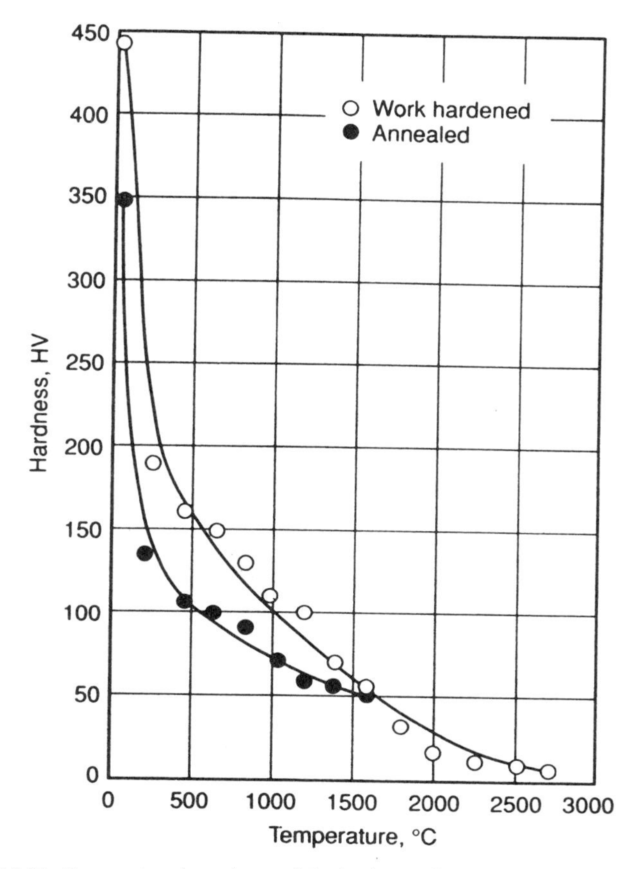

FIGURE 1.11. Temperature dependence of the hardness of tungsten [1.40]; source [1.45].

remains ductile down to at least 20 K. In this form, its fracture behavior is very sensitive to notches or other surface defects.

Originally, it was assumed that cracking only occurs along grain boundaries (i.e., intergranular), but later, a high percentage of cleavage (transgranular fracture) was also observed. Today, we know that, depending upon deformation history, purity, and stress state, the fracture will be more or less on one side or the other.

TABLE 1.10. Parameters which Modify the Ductile-to-Brittle Transition Temperature [1.43]

Mechanical	Microstructural	Chemical
Stress state	Grain size/grain area	Distribution of interstitial/substititial
(tensile/compressive/shear stresses)	Cold work	impurities (C, N, O, P, S, Re)
	Texture; grain shape	
Strain rate	Twins; subboundaries	
	Dislocations	
	Surface roughness	
	(noches, surface cracks)	

The main structural influences can be described as follows. Any plastic deformation decreases the transition temperature by fining the structure. The higher the degree of deformation, the lower is the DBTT. An "as- rolled" tungsten sheet shows a DBTT of about 100–200 °C. Fine wires or foils are even ductile at room temperature. Any annealing increases the transition temperature, and recrystallized samples show a maximum, mainly due to the coarsening of the structure (360 °C).

Grain coarsening has a very important consequence. Simultaneously with grain growth, the internal grain boundary area is reduced, and hence the concentration of the grain boundary impurities is raised. Since the grain boundaries are stronger when the segregations are distributed on a larger area, the ductility decreases.

Today it is assumed that the brittle to ductile transition occurs in two phases:

grain interior:	30–40 °C
grain boundaries:	280–330 °C

In general, the transformation temperature is increased by the presence of small amounts of interstitially soluble elements, such as O, C, and N, which lead to intergranular precipitations, thereby further weakening the grain boundary strength. However, in regard to the respective impurities and their influence on the DBTT, a lot of contradicting statements exist. In the beginning, oxygen segregations at grain boundaries were thought to be most detrimental. Hydrogen and nitrogen came in addition, and later, in particular phosphorus and carbon were made responsible for embrittlement. Some metallic elements were said to favor the ductility. Unfortunately, most of these investigations disregarded that, in principle, not just one but several elements are present at grain boundaries. Only their concentrations vary within a wide range. Now it seems well established that the element combination and their mutual concentrations are of importance. This means, for example, that if phosphorus is present with no additional metallic element, the grain boundary strength is low. However, in presence of metallic elements and of phosphorus, samples show a high percentage of cleavage (a consequence of high boundary strength).

Additions of rhenium, iridium, and ruthenium can cause the transformation temperature to fall below room temperature, even for slightly deformed products. The ductilizing effect of Re is explained by alloy softening for low Re alloys and through scavenging effects (oxygen removal from grain boundaries and dislocations) for high Re alloys [1.31, 1.35].

Drawn wires and rolled sheet exhibit elongated fibrous grains, which can alleviate the embrittlement caused by interstitials. However, "delamination" can occur in heavily worked materials during bending along the fiber boundaries. The splitting mode can be related to contaminated {100} cleavage planes [1.41, 1.43].

Noncoherent dispersions, such as thoria or zirconia, were shown to have a slightly ductilizing effect at low temperature, due to their grain-refining effect. Dispersion strengthened NS–W wires (see Section 6.2.1) behave in a ductile manner during bending, even in the recrystallized condition as a result of their interlocking, elongated grain structure [1.42].

1.2.3.4. Plasticity, Strength, and Creep [1.35, 1.46, 1.47]. Strength, plastic, and creep properties of tungsten are of paramount interest for its widespread use as structural material. They have therefore been the subject of a large number of investigations. Most of the fundamental work on the deformation of pure tungsten single crystals and polycrystalline tungsten was carried out in the late sixties and early seventies, but recent work based on dislocation theory of plasticity and work hardening (considering its structural defects) has significantly contributed to a better understanding of the deformation behavior [1.35, 1.46–1.49]. It is well beyond the scope of this book to discuss the manifold, complex metal physical principles which govern the deformation of bcc tungsten; therefore, only the general principles are discussed below. For more information, the reader is referred to other excellent compilations [1.35, 1.46, 1.47].

Since the early days of tungsten metal manufacture, the metal was considered to be brittle and difficult to work. Inherent brittleness was at first attributed to the low degree of purity which could be obtained by P/M techniques and the strong influence of interstitials on the fracture strength of the metal. In contrast, high-purity single crystals, obtained by electron beam zone refining, behaved in a ductile manner during deformation, even at the temperature of liquid helium.

Although these findings are correct in principle, they are only part of the "truth" in terms of the characteristic deformation behavior of bcc tungsten. Strictly speaking, tungsten cannot be regarded as "ideally" ductile, in the same way as the fcc metals [1.48]. This is well reflected through Fig. 1.12, which shows the temperature dependence of the flow stress of a pure tungsten single crystal (orientation ⟨110⟩) [1.48]. The strong increase in the flow stress with temperature above a certain critical temperature T_c, as indicated in the figure, is characteristic for all bcc metals [1.47–1.50]. Even for pure tungsten, T_c is well above room temperature (about 600–800 K ≅ 330–530 °C) [1.35, 1.46–1.49], which explains the difficulty in metal forming at ambient temperatures. It must be regarded as intrinsic to the bcc crystal lattice and its structural defects.

Above the critical temperature, tungsten behaves during deformation similarly to fcc metals, showing a characteristic three-stage or four-stage work hardening ("high-temperature work hardening") [1.49]. Below this temperature, parabolic stress–strain curves

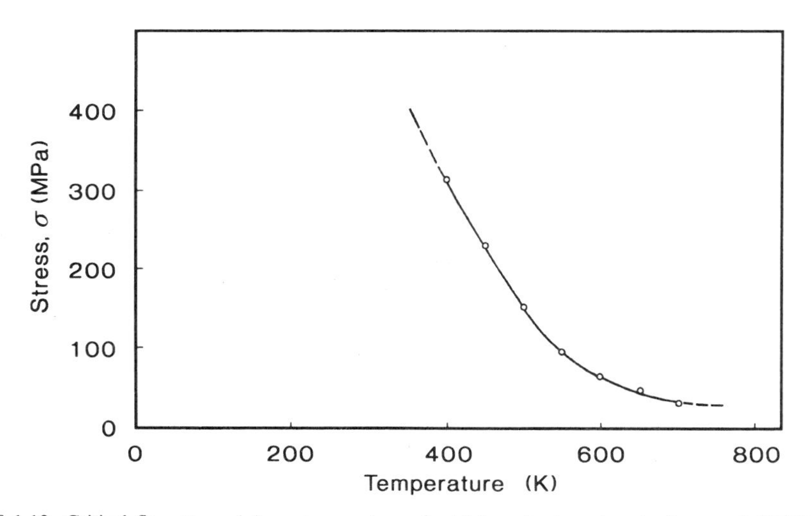

FIGURE 1.12. Critical flow stress (σ) vs. temperature of a high-purity tungsten single crystal (⟨110⟩) [1.48]; tensile deformation rate, $1.6 \times 10^{-5}\ s^{-1}$.

TABLE 1.11. Effect of Orientation and Temperature on the Critical Resolved Shear Stresses for Slip of Tungsten Single Crystals in Tension [1.50]

Temperature (K)	Orientation of tensile axis	Initial yield stress ($kg \cdot mm^{-2}$)	Critical resolved shear stress on observed slip system ($kg \cdot mm^{-2}$)	Observed slip systems	Systems with the highest resolved shear stress
448	[001]	14.3	6.1	$(101)[\bar{1}11]$	$(112)[\bar{1}\bar{1}1]$ and
293	[001]	23.2	9.5	and equivalent	three other
199	[001]	38.0	15.5	systems. Also	equivalent systems
77	[001]	60.0	24.6	$(112)[\bar{1}\bar{1}1]$	
423	[011]	25.4	6.6	$(\bar{1}12)[\bar{1}\bar{1}1]$	$(112)[\bar{1}\bar{1}1]$ and
293	[011]	46.5	11.0	and three other	one other equivalent
199	[011]	84.5	20.0	equivalent systems	system
77	[011]	124.0	29.3		
293	[111]	34.5	9.4	$(110)[\bar{1}11]$	$(211)[\bar{1}\bar{1}1]$ and
199	[111]	50.7	16.0	and equivalent	two other
77	[111]	81.0	25.6	systems	equivalent systems
293	[112]	23.1	10.8	$(\bar{1}\bar{1}2)[\bar{1}11]$	$(011)[1\bar{1}1]$ and one other equivalent system
293	"C"	22.9	10.8	$(110)[\bar{1}11]$	$(\bar{1}01)[111]$

("low-temperature work hardening") are obtained, as shown in Fig. 1.13 for W single crystals with various orientations [1.50]. It is striking, that the yield stress and critical shear stress vary for different orientations, which is a further characteristic difference between bcc and fcc metals [1.49]. Values for both parameters are summarized in Table 1.11 for different temperatures, together with the operating slip systems [1.50]..

Microplasticity and dislocation concepts. Considerable dislocation motion can occur at stresses far below the conventional macroscopic yield stress, which can be followed by high-sensitivity strain transducers. Based on the hysteresis loops in load–unload cycling experiments, the true elastic limit for pure, polycrystalline (arc-cast) tungsten was measured as $\sigma_E 3.43 \pm 0.8$ MPa [1.51]. It is widely agreed today that the movement of edge and non-screw dislocations is responsible for this microdeformation (microplasticity). Above the yield point, $\langle 111 \rangle$ screw dislocations contribute to the macroscopic deformation (work hardening; double-kink mechanism). The increased activation energy for the gliding of $\langle 111 \rangle$ screw dislocations, as compared to edge dislocations, is an important characteristics of bcc tungsten. Its activation threshold value is responsible for the strong increase in flow strength with decreasing temperature [1.48]. Thus, at low deformation temperatures, the screw dislocations control the macroscopic flow stress. The rate-determining step is believed to be the formation of kink pairs in screw dislocations [1.47–1.49].

Deformation of polycrystalline tungsten. Stress/strain curves of high-purity single and polycrystalline tungsten (swaged and subsequently annealed for 1 h at 2000 °C), obtained at room temperature and 400 °C are shown in Fig. 1.14 [1.52]. Whereas the polycrystalline material exhibited pronounced ductility at 400 °C (elongation 53 %), it

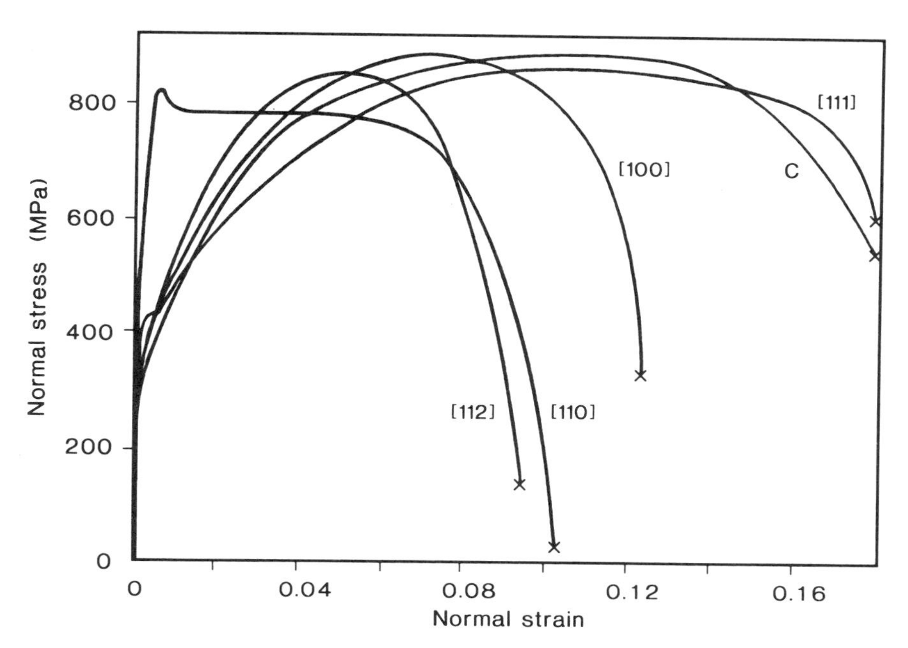

FIGURE 1.13. Stress–strain curves of tungsten single crystals with various orientations, tested at room temperature [1.50].

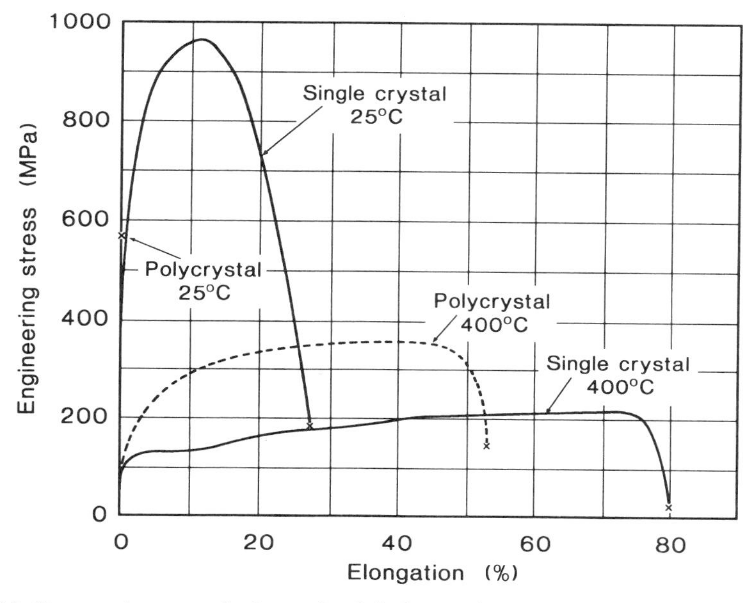

FIGURE 1.14. Stress–strain curves of polycrystal and single-crystal tungsten at 25 °C and 400 °C [1.52]; strain rate 2 % · min^{-1}

behaved in a completely brittle manner (elongation 0.0–0.3 %) during RT testing. In contrast, the single crystals behaved in a ductile manner under both test conditions (elongation 27 %, resp. 80 %). These results indicate that the presence of grain boundaries significantly lowers the DBTT and remarkably deteriorates the low-temperature plastic properties of the metal.

Deformation curves of heavily worked tungsten ribbons are presented in Fig. 1.15 [1.53] and show a quite different picture. In this case, plastic flow occurred even at 200 K, indicating a pronounced decrease in the DBTT as a result of cold working ("ductile tungsten"). This behavior is due to changes in microstructure (texture formation), but also to the formation of a large amount of edge dislocations, which move rather easily during deformation and thus provide better ductility [1.48]. Again, an increase in yield stress with decreasing temperature is obvious, as is characteristic for bcc metals.

The grain size and grain structure have a great influence on both strength and ductility (elongation). These can be controlled by sintering conditions, type of deformation, degree of deformation, and annealing processes both during and after machining (intermediate annealing, stress-relieve annealing, soft annealing, recrystallization annealing).

For maximum deformability, tungsten has to be partly recrystallized (primary recrystallization), forming a columnar structure with good retention of strength and ductility. Higher temperatures must be avoided since secondary recrystallization occurs with significant grain coarsening, leading to a pronounced embrittlement [1.54]. The temperature for onset of this coarsening is very sensitive to the nature of the starting material and its deformation history [1.55].

The hot strength of tungsten, like its elongation, depends on its deformation and annealed condition, and can vary widely at a given temperature (Fig. 1.16) [1.56]. The

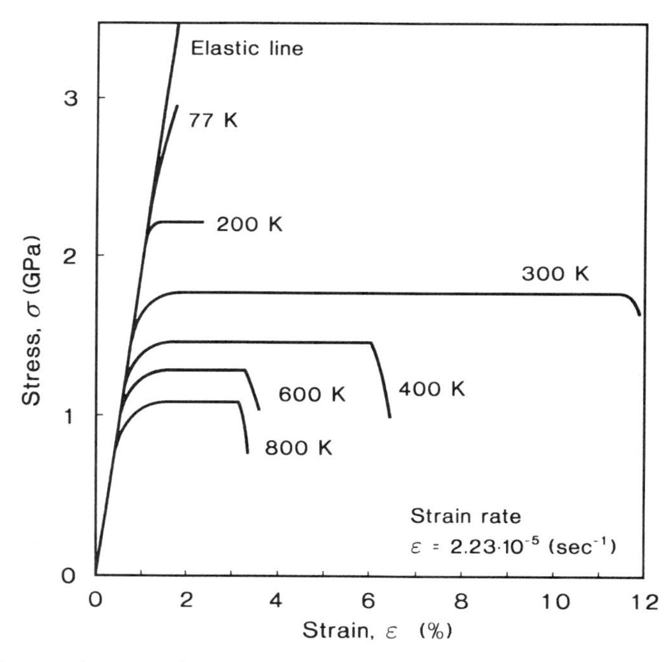

FIGURE 1.15. Stress–strain curves of as-received tungsten ribbons at various deformation temperatures [1.53]; strain rate $2.23 \times 10^{-5}\ s^{-1}$.

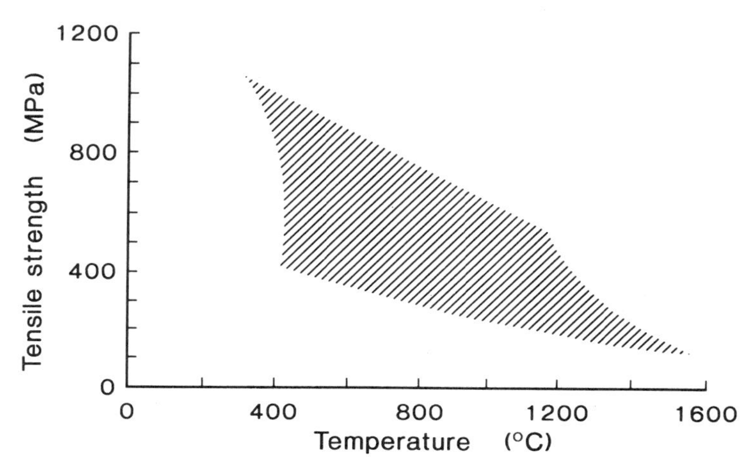

FIGURE 1.16. Hot tensile strength of tungsten sheet [1.56]; thickness 1 mm, strain rate $15\,\% \cdot \text{min}^{-1}$.

upper limit corresponds to a stress-relieved sheet, and the lower limit to a recrystallized sheet. At ~1500 °C, both limiting curves coincide, i.e., at this temperature the tungsten completely recrystallizes as it is heated to the tensile testing temperature.

The hot strength of pure tungsten is exceeded only by that of rhenium. It can be significantly further increased by dispersion or precipitation hardening. Alloys based on W–Re–HfC are the strongest human-made metallic materials at temperatures up to 2700 K.

Strength of tungsten wires, whiskers, and theoretical strength. The room-temperature ultimate tensile strength (UTS) of polycrystalline rods is commonly in the range of 580–1470 MPa [1.37]. Cold deformation may increase it significantly. The strength of wires (exhibiting a characteristic fiber structure in the as-drawn condition) varies with diameter. The following strengths values are quoted [1.37]: 1765 MPa (600 μm); 2450 MPa (300 μm); 2940 MPa (100 μm); 3920 MPa (20 μm).

The tensile strength of tungsten whiskers, grown by vapor-phase transport, was reported as up to 27.5 GPa [1.57].

For the *theoretical strength*, the stress required to cause decohesion in a perfect single crystal can be assessed by

$$\sigma_c = \sqrt{\frac{E\gamma_s}{a_0}}$$

where E is Youngs's modulus, γ_s the specific surface energy, and a_0 the interatomic spacing [1.46]. Calculations of the ideal strength of tungsten (tension along ⟨100⟩) revealed 90.8 GPa at 0 K and 61 GPa at room temperature. The ideal shear strength of tungsten (shear plane {110} and the direction ⟨111⟩) was calculated as 16.5–19 GPa [1.58, 1.59]. Both values well reflect the high bonding energy of W in the bcc lattice.

Influence of impurities/alloying. Impurities, in particular interstitials (C, N, O), are partly responsible for the high DBTT of tungsten. On the other hand, certain amounts of interstitials or solutes (Re, Ir) can improve the low-temperature ductiliy. Both these effects are explained by the interaction of the foreign atoms with the dislocation network, which can shift the critical temperature T_c to lower or higher values. It is assumed that the propagation of dislocations by kink nucleation and lateral motion along the screw dislocation line is either retarded by the distortion field of the foreign atoms (solid-solution hardening), or enhanced, by generating new kink pairs for their movement (solid-solution softening) [1.47, 1.49].

The most effective way to improve the low-temperature plastic properties of tungsten is by alloying with rhenium [1.60]. For more information see section 6.1.1.

Creep. Creep properties of polycrystalline tungsten, in particular of thin wires, are of utmost importance for their high-temperature application.

For polycrystalline tungsten, the creep rate can be expressed by [1.61–1.64]

$$\varepsilon = B\sigma^n \exp\left(-\frac{\Delta H}{kT}\right)$$

where ε is the steady-state creep rate, σ the applied stress, ΔH the activation enthalpy, T the temperature (K), and B, n, and k are constants.

Values for B, n, and ΔH vary significantly in the literature, depending on the respective material and testing condition [1.62–1.64]. They are characteristic for the respective dominating creep mechanism.

For large samples of pure tungsten at temperatures >2200 °C (0.65 T_m), the stress exponent $n = 5$ and $\Delta H = 585\,\text{kJ}\cdot\text{mol}^{-1}$. This activation enthalpy is similar to that of self-diffusion in tungsten. It was therefore assumed that under these conditions the deformation of the grains by dislocation climb or glide processes is the rate-controlling reaction [1.63].

For thin wires, having a grain size comparable to the creep specimen diameter ("bamboo"-grain structure), a stress exponent of $n = 2$, and an activation enthalpy of $276\,\text{kJ}\cdot\text{mol}^{-1}$ were obtained, as characteristic for grain boundary diffusion. In this case, the creep deformation occurred almost entirely by grain boundary sliding (offsetting) without extensive grain deformation [1.62, 1.63].

In general, different independent deformation mechanisms will simultaneously contribute to the macroscopic deformation, such as dislocation glide (at lower temperatures), dislocation creep (at $> 0.05\ T_m$), and diffusional creep (Coble creep, or Nabarro–Herring creep), depending on temperature, magnitude of applied stress, grain size, grain morphology, purity, and specimen diameter.

Competition between the various mechanisms can be described by so-called deformation-mechanism maps, as shown in Fig. 1.17 for pure W with a grain size of 10 μm [1.35, 1.66]. In industrial practice these maps, however, are of limited use, because the predominance areas for the respective mechanisms alter significantly with changes in microstructure (grain size, subgrain size, grain aspect ratio), which may even occur during deformation [1.64].

High temperature creep resistant tungsten alloys. The microstructure is a crucial parameter in determining the magnitude of creep. It can be remarkably influenced by the presence of second phases (ThO_2, Y_2O_3, La_2O_3, HfC, potassium bubbles), as demonstrated in Fig. 1.18 [1.62] for both non-sag doped and thoriated tungsten. These additives

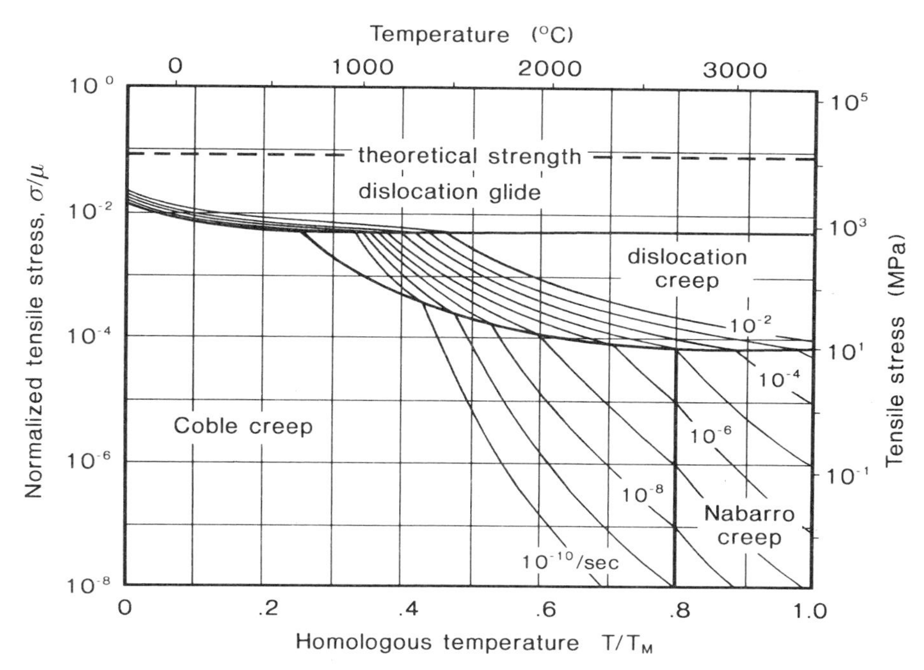

FIGURE 1.17. Deformation-mechanism map of pure tungsten with grain size of 10 μm ($\rho = 4 \times 10^{10}$ cm^{-2}); normalized tensile stress refers to the shear modulus (μ); T_M is the melting point [1.65].

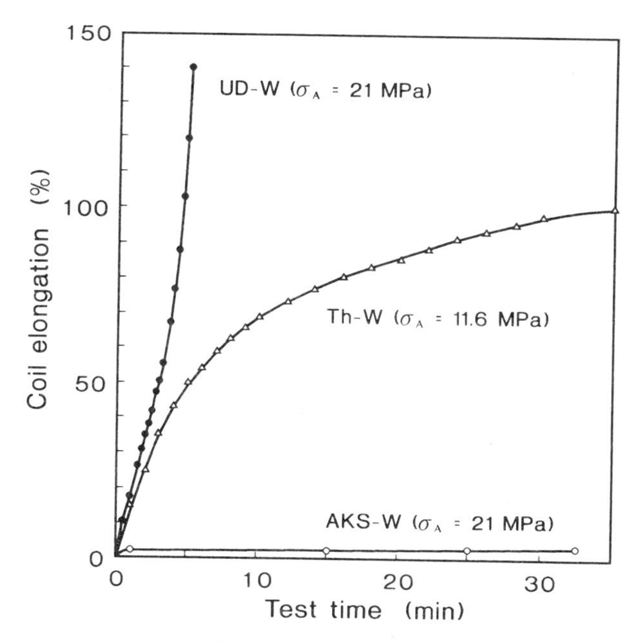

FIGURE 1.18. Creep curves for pure tungsten, thoriated tungsten, and nonsag tungsten wires at 2500 °C, showing the outstanding creep resistance of nonsag tungsten [1.62].

affect the recrystallization grain morphology and the retained dislocation substructure. In particular, non-sag tungsten is significantly more creep-resistant than pure tungsten, mainly as a result of the interlocking grain structure, which forms on recrystallization and which prevents both grain boundary sliding and diffusional creep. In addition, the fine dispersion of potassium bubbles contributes to the outstanding creep resistance through pinning of dislocations [1.62, 1.66].

Non-sag tungsten wires are the most creep-resistant wires, with the exception of monocrystalline tungsten. They are therefore used for sag-free lamp filaments at service temperatures of up to 3000 °C (0.88 T_m), and shear stresses in the range of 0.5 to 10 MPa [1.66].

1.2.4. Thermal Properties

Melting point: $3422 \pm 15\,^\circ$C (3695 K) [1.67],
$3390 \pm 40\,^\circ$C (3663 K) [1.68],
$3423 \pm 30\,^\circ$C (3696) [1.69].

The high melting point (highest of all metals) is the most prominent and important property in regard to all applications as refractory metal. It is a consequence of the electron density of states. Small amounts of impurities, such as carbon, lower the melting point.

The molar volume increases by 8 % on melting. This is the largest expansion observed for bcc metals [1.70].

The melting curve of W has been determined to 5 GPa [1.71].

Enthalpy of fusion: $46 \pm 4\,\mathrm{kJ \cdot mol^{-1}}$ [1.68].
Entropy of fusion: $14\,\mathrm{J \cdot mol^{-1} \cdot K^{-1}}$ [1.70].
Enthalpy of sublimination: $858.9 \pm 4.6\,\mathrm{kJ \cdot mol^{-1}}$ [1.35].

Vapor Pressure. Tungsten has the lowest vapor pressure of all metals. Within the temperature range from 2600 to 3100 K, it obeys the following equation [1.76]:

$$\log p[\mathrm{Pa}] = -45395T^{-1} + 12.8767$$

At 2000 °C, the vapor pressure is 8.15×10^{-8} Pa; at 3000 °C it is 10^{-1} Pa. Experimental data for p over liquid W are not available.

The *rate of evaporation* in vacuum is about $6.2 \times 10^{-11}\,\mathrm{g \cdot cm^{-2} \cdot s^{-1}}$ at 2000 °C, about $7 \times 10^{-8}\,\mathrm{g \cdot cm^{-2} \cdot s^{-1}}$ at 2500 °C, and about $2.5 \times 10^{-5}\,\mathrm{g \cdot cm^{-2} \cdot s^{-1}}$ at 3000 °C [1.72]. It is markedly reduced by an inert gas atmosphere (Ar, Kr). Therefore, modern incandescent lamps contain inert gas fillings to avoid enhanced wall-blackening (the rate of evaporation in vacuum is about 500 times larger as compared to an Ar atmosphere of 1.2 bar) [1.73].

Boiling Point: calculated from rates of evaporation of solid tungsten,

5663 °C (5936 K) [1.74],
$5700 \pm 200\,^\circ$C [1.64].

Critical temperature: 13400 ± 1400 K [1.75].
Critical pressure: $(3.37 \pm 0.85) \times 10^8$ Pa [1.75].
Critical density: 4.31 $g \cdot cm^{-3}$ [1.35].

Thermal expansion. At room temperature, values between 4.32 and 4.68 $\times 10^{-6} K^{-1}$ were obtained for the linear coefficient of expansion α, depending on the material (P/M sheet, arc-cast sheet, etc.) and the type of measurement. Values for low and high temperatures are listed in Table 1.12. The linear coefficient of expansion can also be calculated according to the following equations [1.76]:

temperature range: 293–1395 K

$$\alpha = 4.266 \times 10^{-6}(T - 293) + 8.479 \times 10^{-10}(T - 293)^2 - 1.974 \times 10^{-13}(T - 293)^3$$

temperature range 1395–2495 K

$$\alpha = 0.00548 + 5.416 \times 10^{-6}(T - 1395) + 1.952 \times 10^{-10}(T - 1395)^2 + 4.422 \times 10^{-13}(T - 1395)^3$$

temperature range 2495–3600 K

$$\alpha = 0.01226 + 7.451 \times 10^{-6}(T - 2495) + 1.654 \times 10^{-9}(T - 2495)^2 + 7.568 \times 10^{-14}(T - 2495)^3$$

The very low thermal expansion of tungsten makes it compatible with glass and ceramics in high temperature applications.

Thermodynamic functions [1.35]. Thermodynamic functions for solid tungsten are listed in Table 1.13. For more details and data for liquid tungsten, see elsewhere [1.10].

TABLE 1.12. Thermal Expansion Coefficient for Low and High Temperatures [1.35]

T (K)	$10^6 \cdot \alpha(K^{-1})$	T (K)	$10^6 \cdot \alpha(K^{-1})$
10	0.006	160	3.82
15	0.019	190	4.06
20	0.048	220	4.20
25	0.102	260	4.32
30	0.20	300	4.49
40	0.53	600	4.75
50	0.90	1000	5.02
60	1.43	1400	5.46
70	1.88	1800	6.11
80	2.30	2200	6.89
100	2.82	2600	7.76
130	3.42	3000	9.05
		3400	11.60

TABLE 1.13. Thermodynamic Functions [1.35]

T (K)	$H_T - H_{298.5}$ ($kJ \cdot mol^{-1}$)	$H_T - H_0$ ($kJ \cdot mol^{-1}$)	S ($kJ \cdot mol^{-1} \cdot K^{-1}$)	$S - S_0$ ($kJ \cdot mol^{-1} \cdot K^{-1}$)	C_p ($kJ \cdot mol^{-1} \cdot K^{-1}$)	C_v ($kJ \cdot mol^{-1} \cdot K^{-1}$)
298.15	0	4.970	32.640	32.618	24.10–24.42	23.96
500	5.022–5.024	10.007	45.516	45.457	24.33–25.44	25.16
1000	18.23–18.25	23.142	63.658	63.686	27.19–27.60	26.70
1500	32.33–32.62	37.249	75.068	75.099	29.23–29.86	28.12
2000	47.59–48.11	52.599	83.881	83.585	31.37–32.13	29.80
2500	64.52–64.78	69.476	91.399	91.397	34.67–36.00	—
3000	83.10–83.88	88.310	98.254	98.442	39.25–41.80	35.91
3500	105.19	109.845	104.880	—	46.49–50.85	—
3600	110.5–111.8	114.574	106.212	106.89	48.11–54.68	—

Analyses of thermodynamic properties of tungsten at high temperatures are available [1.77, 1.78].

The heat capacity at low temperatures is:

T in K	25	30	35	40	50	60	70	80	100
C_p ($J \cdot mol^{-1} \cdot K^{-1}$)	0.73	1.35	2.22	3.30	5.82	8.39	10.74	12.81	16.04
T in K	120	140	160	180	200	220	260	300	
C_p ($J \cdot mol^{-1} \cdot K^{-1}$)	18.28	19.87	21.01	21.86	22.54	23.04	23.81	24.35	

The heat capacity of liquid tungsten is $35.564\,J\,mol^{-1} \cdot K^{-1}$ [1.10]

Self-diffusion [1.35]. At a certain temperature, the diffusion process is characterized by the diffusion coefficient D. Its temperature dependency is given by the Arrhenius equation $D = D_0 \exp(-Q/R \cdot T)$ with D_0 a constant in $cm^2 \cdot g^{-1}$, T the absolute temperature, R the molar gas constant in $J \cdot K^{-1} \cdot mol^{-1}$, and Q the activation enthalpy in $kJ \cdot mol^{-1}$.

Activation enthalpies for the lattice (volume) diffusion were derived between 586 and $628\,kJ \cdot mol^{-1}$ for single crystals and between 502 and $586\,kJ \cdot mol^{-1}$ for polycrystalline tungsten. Accordingly, the following equations were set up [1.79]:

Self-diffusion in tungsten single crystals: $D = 42.8 \exp(-640/RT)$.
Self diffusion in polycrystalline tungsten: $D = 54 \exp(-504/RT)$.

Over the range of 1900 to 2800 °C, a linear relationship between $\log D$ and $1/T$ was obtained for polycrystalline tungsten.

The self-diffusion parameters D_0 and Q are influenced by the impurity content of the diffusion zone. Higher values for both were obtained for impure tungsten. This effect is more pronounced at low temperatures and vanishes above 2043 K. (It is assumed that impurities attract vacancies and the higher vacancy concentration disturbs the diffusion process).

The fact that lower Q values were obtained for polycrystalline tungsten than for single crystals, especially for $T < 0.7T_m$, is due to a more or less significant contribution of grain boundary diffusion to the total bulk diffusion. Observed rates of volume diffusion, obtained on polycrystalline tungsten, will therefore always depend on the microstructure of the material under investigation.

Grain boundary diffusion is the dominant mechanism in polycrystalline tungsten below 2100 °C. For grain boundary diffusion, activation enthalpies between 377 and 460 kJ · mol^{-1} were measured.

Thermal conductivity and diffusivity: [1.35, 1.80, 1.81]. At room temperature, the thermal conductivity coefficient λ is about 1.75 W · cm^{-1} · K^{-1}. Its temperature dependence is shown in Fig. 1.19. Some typical data are given in Table 1.14. In the low-temperature range, it is strongly dependent on the RRR (residual resistivity ratio ρ_0/ρ_T), which is also an indication of the purity. Table 1.15 shows the differences between samples of varying purity.

Between 1200 and 2800 K, λ obeys the following equation [1.80]:

$$\lambda(\mathrm{W \cdot cm^{-1} \cdot K^{-1}}) = 1.0834 - 1.052 \times 10^{-4} T + \frac{234.199}{T}$$

At the melting point, λ drops from 0.895 (solid) to 0.705 (liquid) [1.35].

Sintered and arc-cast tungsten specimens differ in λ values at elevated temperatures. The thermal conductivity of single crystals and polycrystalline tungsten, both having the same degree of purity, coincide at high temperature.

There exist certified NBS standards of arc-cast as well as sintered tungsten as reference material for the temperature range of 4 to 3000 K [1.83].

The high thermal conductivity of tungsten (very suitable for use as heat sinks) in combination with the low specific heat results in high cooling rates during hot-working of the metal, which makes the handling during working more difficult.

The *thermal diffusivity* of tungsten at 300 K is 0.662 $cm^2 \cdot s^{-1}$, and decreases to 0.246 $cm^2 \cdot s^{-1}$ at the melting point. At low temperatures (<200 K) the values are strongly

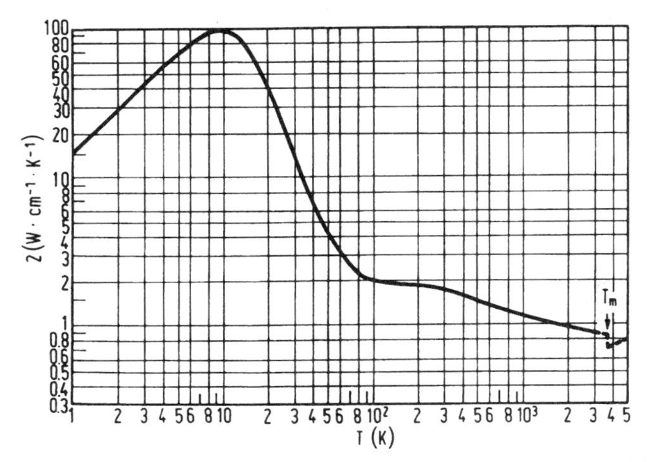

FIGURE 1.19. Temperature dependence of the thermal conductivity λ of well-annealed high-purity tungsten [1.35]; source [1.81].

TABLE 1.14. Thermal Conductivity Coefficient at Different Temperatures [1.35, 1.81]

Temperature (K)	λ ($W \cdot cm^{-1} \cdot K^{-1}$)
0	0
10	97.1
50	4.28
100	2.08
300	1.74
500	1.46
1000	1.18
2000	1.00
3400	0.90

dependent on the impurity level and the density of lattice defects. For more information see elsewhere [1.82].

TABLE 1.15. Thermal Conductivity Coefficient in the Low-Temperature Range and Dependence on the Residual Resistivity Ratio [1.35, 1.80]

	λ($W \cdot cm^{-1} \cdot K^{-1}$)	
Temperature (K)	RRR 100	RRR 300
1	0.50	1.51
5	2.49	7.49
10	4.88	14.04
20	7.99	17.34
50	3.57	3.98
100	2.17	2.24
150	1.97	2.01
200	1.89	1.91
300	1.74	1.76

1.2.5. Electromagnetic Properties [1.35]

Electrical Resistivity ($\mu\Omega \cdot cm$). In the low-temperature range <0.07 K the electrical resistivity ρ_0 is independent of the temperature and will be only determined by the impurity level.

In order to calculate ρ within different temperature ranges, equations have been set up:

up to 40 K

$$\rho = 1.5 \times 10^{-5} + 7 \times 10^{-7} T^2 + 5.2 \times 10^{-10} T^5$$

40–90 K

$$\rho = 0.14407 - 1.16651 \times 10^{-2}T + 2.41437 \times 10^{-4}T^2 - 3.66335 \times 10^{-9}T^4$$

90–750 K

$$\rho = 1.06871 + 2.06884 \times 10^{-2}T + 1.27971 \times 10^{-6}T^2 + 8.53101 \times 10^{-9}T^3 - 5.14195 \times 10^{-12}T^4$$

300–1240 K

$$\rho = 4.33471 \times 10^{-12}T^2 + 2.19691 \times 10^{-8}T - 1.64011 \times 10^{-6}$$

1240–2570 K

$$\rho = 4.06012 \times 10^{-12}T^2 + 4.67093 \times 10^{-8}T - 1.94101 \times 10^{-5}$$

Moreover, the resistivity is influenced by all kinds of lattice defects (vacancies, dislocations, grain boundaries, etc.) and impurities.

Some selected values near and above room temperature are listed in Table 1.16. The resistivity of liquid tungsten close to the melting point and at 5000 K was given as 131 μΩ · cm and 160 μΩ · cm, respectively [1.84].

The resistivity of thin films depends on microstructure, impurity content, and surface roughness. These properties are the consequence of deposition conditions, substrate temperature, and annealing. Therefore, a wide scatter of values was observed, for example, between 6 and 20 μΩ · cm at 30 K [1.35]. Under certain deposition conditions, metastable β-W forms, which has a significantly higher bulk resistivity (approximately 10 times that of the thermodynamically stable α-W phase).

Superconductivity [1.35]. Tungsten is a Type I superconductor with a transition temperature of 0.0154 ± 0.0005 K. The critical magnetic field strength $H_c(T \rightarrow 0)$ is 1.15 ± 0.03 Oe. ($91.5\ \mathrm{A \cdot m^{-1}}$). Impurities only show a minor influence on the transition

TABLE 1.16. Electrical Resistivity of Tungsten at Near-Room and at Elevated Temperature [1.35]

T (K)	ρ(μΩ · cm)	T (K)	ρ(μΩ · cm)
273	4.82	400	8.05
293	5.28	500	10.70
298	5.40	600	13.35
300	5.54	800	18.85
		1000	24.75
		1200	30.90
		1400	37.20

temperature:

RRR 57000	0.0160 K
RRR 7500	0.0154 K

In thin films, due to structural influences or the presence of β- or γ-tungsten, much higher critical temperatures (up to 6 K) were found.

Thermoelectric effects. A temperature gradient generates in tungsten a small potential difference (the Thomson effect). This difference, called the thermoelectric power S (μV/K), drops with increasing temperature. Its temperature dependence is tabulated in Table 1.17.

Thermocouples. For temperature measurements up to 2300 °C, thermocouples consisting of tungsten and tungsten–rhenium alloys are employed widely. The most common combinations are W versus W26Re, W5Re versus W26Re, and W3Re versus W25Re. Electromotive forces built up by each combination for different temperatures are shown in Table 1.18.

W–Re thermocouples can only be used in neutral or reducing atmospheres. Other combinations, such as W/Pt, W/Ta, or W/Mo, have gained no commercial importance.

Magnetoelectric effects [1.35]. The *Hall coefficient* at room temperature varies between 10 and 12×10^{-11} $m^3 \cdot A^{-1} \cdot s^{-1}$.

The *Magnetic susceptibility* is given either as χ_{mol} (10^{-6} $cm^3 \cdot mol^{-1}$) or as χ_g (10^{-6} $cm^3 \cdot g^{-1}$). It increases with temperature. Values for different temperatures are given in Table 1.19.

1.2.6. Optical Properties

The optical properties of tungsten have been studied in more detail than those of any other metal or material, because tungsten is not only used as an incandescent lamp filament, but also as a comparison temperature standard in specially constructed strip lamps.

TABLE 1.17. Temperature Dependence of Thermoelectric Power of Tungsten [1.35]

T (K)	S (μV · K)	T (K)	S (μV · K)
10	+ 0.05	400	4.62
20	− 0.28	600	10.75
50	− 2.78	800	15.51
80	− 3.70	1000	18.46
100	− 4.04	1200	20.06
150	− 2.45	1400	20.63
200	− 1.41	1500	20.70
250	− 0.10	1600	20.61
273	+ 0.56	1800	19.15
300	+ 1.44		

TABLE 1.18. Tungsten and Tungsten–Rhenium Thermocouple Electromotive Forces at Various Temperatures for a Reference Junction at 0 °C [1.39]

	Electromotive force (mV)		
Temperature (°C)	W vs. W26Re	W5Re vs. W26Re	W3Re vs. W25Re
0	0.000	0.000	0.000
100	0.344	1.381	1.146
200	1.006	2.988	2.604
300	1.985	4.768	4.289
400	3.282	6.655	6.129
500	4.793	8.573	8.098
600	6.488	10.508	10.092
700	8.331	12.450	12.129
800	10.300	14.374	14.184
900	12.319	16.266	16.226
1000	14.393	18.120	18.242
1100	16.497	19.994	20.229
1200	18.648	21.724	22.192
1300	20.767	23.424	24.082
1400	22.814	25.033	25.896
1500	24.841	26.583	27.686
1600	26.849	28.078	29.450
1700	28.842	29.528	31.182
1800	30.814	30.922	32.874
1900	32.589	32.298	34.359
2000	34.246	33.632	35.723
2100	35.851	34.915	37.037
2200	37.435	36.089	38.307
2300	38.896	36.928	39.350

Spectral emissivity ϵ_λ [1.37, 1.85–1.88]. The spectral emissivity of tungsten as a function of temperature is shown in Fig. 1.20. In the wavelength range of 0.3–0.5 μm, the emissivity has a maximum. The X point of tungsten, where all the emissivity wavelength isotherms cross, corresponds to $\lambda_X = 1.27\,\mu\text{m}\ \epsilon = 0.33$).

Total emissivity e_t. The experimental results of e_t measurements can be well

TABLE 1.19. Temperature Dependence of the Magnetic Susceptibility of Tungsten [1.35]

Temperature (°C)	$\chi_{mol}(10^{-6}\ \text{cm}^3 \cdot \text{mol}^{-1})$
−193	51.1
−78	53.7
Room temperature	
Single Crystal	54.0
Polycrystalline	53.3
500	58.0
1600	63
1850	68

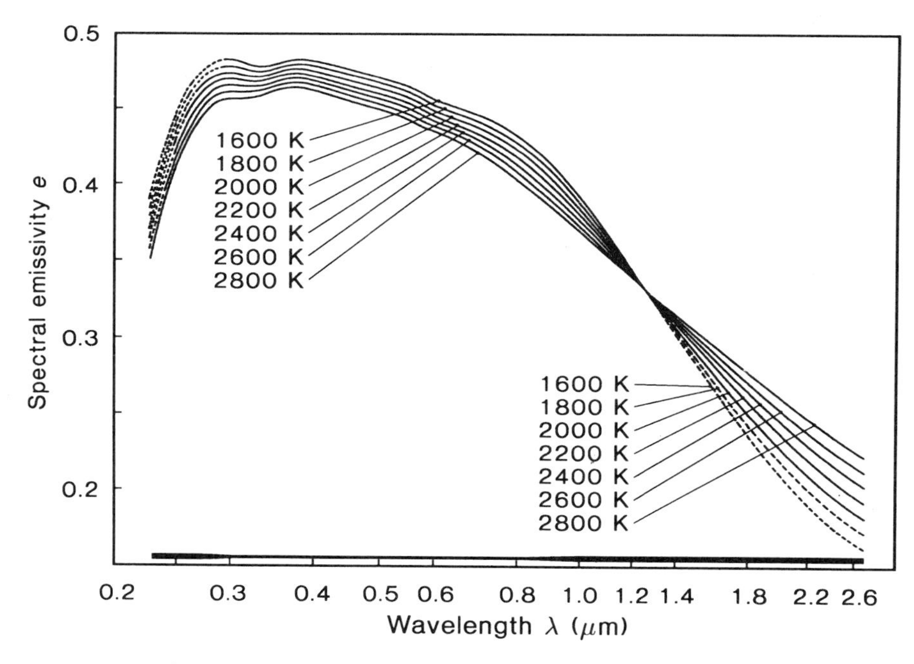

FIGURE 1.20. Spectral emissivity of tungsten according to de Vos [1.37].

represented by the following empirical equation (valid from 400 to 3600 K) [1.88]:

$$e_t = (-2.6875 \times 10^{-2}) + (1.819696 \times 10^{-4})T - (2.1946163 \times 10^{-8})T^2$$

A comparison between measured and calculated results is given in Table 1.20. Values in the temperature range from 1200 to 3200 K are compared for single crystals and polycrystalline tungsten in Table 1.21.

Pyrometry. It was shown that the emissivity of tungsten depends on wavelength. It is

TABLE 1.20. Total Emissivity and Radiated Heat of Tungsten at Various Temperatures [1.39]

True temperature		Radiated heat	Emissivity	
(°C)	(K)	$P(\mathrm{W \cdot cm^{-2}})$	Calculated	Measured
3327	3600	327.4	0.344	0.354
2927	3200	197.0	0.331	0.341
2527	2800	108.2	0.311	0.323
2127	2400	53.3	0.283	0.296
1727	2000	22.6	0.249	0.260
1327	1600	7.72	0.208	0.207
927	1200	1.87	0.160	0.143
527	800	0.238	0.105	0.088
127	400	0.0042	0.0424	0.042

TABLE 1.21. Temperature Dependence of Total Tungsten Emissivity for Single Crystals and Polycrystalline Material [1.35]

Temperature (K)	Single crystal (e_t)	Polycrystal (e_t)
1200	0.126	0.139
1400	0.163	0.173
1600	1.196	0.204
1800	0.226	0.233
2000	0.251	0.257
2200	0.273	0.278
2400	0.292	0.298
2600	0.307	0.312
2800	0.319	0.324
3000	0.329	0.333
3200	0.337	

higher in the blue and less in the red, in comparison to a uniform gray body. Therefore, the brightness temperature of tungsten at 0.65 μm (where an optical pyrometer reads temperature) is higher than the true temperature of a gray surface in equilibrium with tungsten. Related corrections must be applied. Table 1.22 shows a comparison between brightness temperature and true temperature as well as color temperature. The last represents the true temperature of a black body radiating with the same intensity distribution of wavelength as tungsten (it would exhibit the same color). The more intense radiation (emissivity) in the blue is the reason why the color temperature is higher than the true temperature.

TABLE 1.22. Temperature Scale for Tungsten [1.39]

Brightness temperature (K)	True temperature (K)	Color temperature (K)
1300	1577	—
1400	1674	—
1500	1771	—
1600	1866	—
1700	1961	—
1800	2055	—
1900	2149	2080
2000	2242	2200
2100	2335	2320
2200	2427	2450
2320	2520	2580
2400	2612	2700
2500	2703	2840
2600	2793	2960
2700	2886	3100
2800	2977	3240

TABLE 1.23. Work Function for Different Metals and Tungsten/Adsorbate Combinations [1.90] and Electronegativity of the Metals According to Pauling

Metal	Electronegativity	Work function (eV)	Metal/adsorbate	Work function (eV)
Cs	0.79	1.81	Cs on W	1.36
Ca	1.00	3.2	Ba on W	1.56
Th[a]	1.33	3.31	Th on W	2.6
Ta	1.50	4.19	La on W	2.70
W	1.70	4.52	Ce on W	2.70
Pt	2.28	5.32	Zr on W[b]	3.13

[a] From Ref. 1.93.
[b] From Ref. 1.89.

1.2.7. Electron Emission

The energy required by an electron to escape from the surface of a crystalline solid is called the *work function* (φ) of the material. It is a characteristic parameter for its electron emission behavior. The work function of metals is in the range of ~2 to 6 eV and correlates mainly with the electronegativity of the element (Table 1.23).

For polycrystalline tungsten, it is commonly quoted as 4.50–4.56 eV [1.38, 1.89, 1.90]. In case of single crystals, it varies for different crystallographic orientations and ranges between 4.21 eV for the (310) plane and 5.79 eV for the (110) plane [1.90]. The higher the atom density of the emitting plane, the higher is its work function ($\varphi_{110} > \varphi_{100} > \varphi_{111}$). Calculated single-plane work function data are presented elsewhere [1.38].

The temperature coefficient of the work function of polycrystalline tungsten ($d\varphi/dT$) is in the range of 60–110 $\mu eV \cdot K^{-1}$ [1.38].

Thermionic emission. The number of electrons which escape from the metal surface increases rapidly with temperature (thermionic emission). In general, the higher the temperature and the lower the work function, the higher is the electron emissivity. The current density can be calculated by the Richardson–Dushman equation (in the absence of an external electrical field), according to $i = AT^2 \exp(-\varphi/kT)$, where A is the Richardson constant ($A \cdot cm^{-2} \cdot K^{-2}$), T is the temperature (K), and φ is the work function (eV). For pure tungsten $A = 60.2$ ($A \cdot cm^{-2} \cdot K^{-2}$) [1.91]. The thermionic current ($A \cdot cm^{-2}$) can then be calculated as: $i = 60.2T^2 \exp(-52230/T)$ [1.37].

The work function of tungsten is considerably changed by adsorption of foreign elements on the crystal surface. It can be both increased, as in case of oxygen contamination, or considerably decreased, as in case of Th, Ba, and Cs additions (Table 1.23). In general, the work function is increased if an electron transfer occurs from tungsten to the adsorbate (i.e., if the adsorbate is more electronegative than tungsten), or, vice versa, it is decreased if the adsorbate is less electronegative and electrons are thus transferred from the adsorbate to tungsten. In the latter case, a minimum in the work function is obtained for a monolayer of the adsorbate, covering the tungsten surface [1.92]. For example, the work function of thoriated tungsten in the activated form (where a monolayer of thorium metal forms on the tungsten surface as a result of a high-temperature treatment) is significantly lower (2.6 eV) than that of bulk thorium metal (3.38 eV) [1.93].

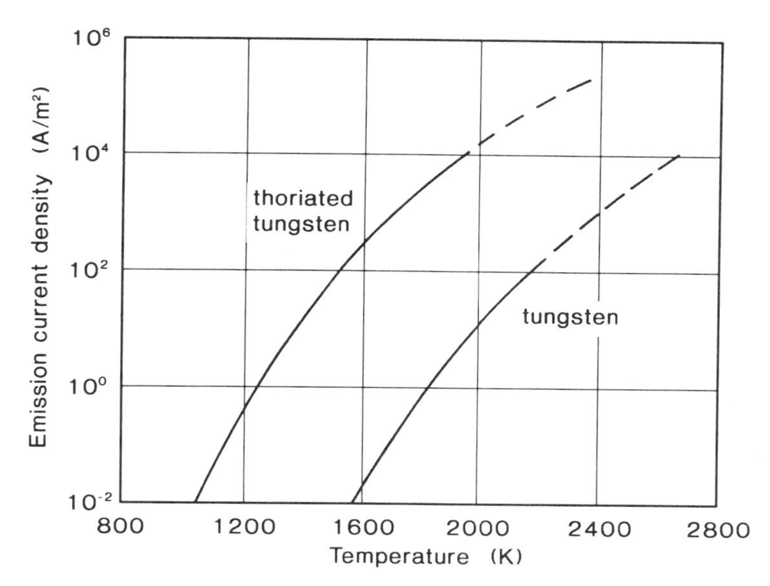

FIGURE 1.21. Variation of emission current density with tungsten for pure and thoriated tungsten electrodes [1.93]; according to Ref. 1.92.

Due to the lower work function of W–Th, W–Cs, or W–Ba, much higher thermionic current densities can be obtained as compared to pure tungsten (Fig. 1.21). This is of high practical importance for the widespread use of tungsten as an electron emitter material. On the one hand, significantly higher current densities can be achieved at a certain service temperature. On the other hand, significantly lower service temperatures can be applied for an equal current density, which results in considerably longer lifetimes of the respective emitter parts.

Enhancement of the thermionic emission of tungsten by addition of thoria was first observed by J. Langmuir in 1914 [1.94]. Already, it was assumed that metallic thorium is responsible for this improvement. How this metallic layer forms on the W surface has never been fully clarified, although several explanations have been given in the past (see also Section 6.2.2). In common with all these theories is the fact that the material must be activated prior to use. This is done by heating it to above 2000 °C and then holding it at a somewhat lower temperature. During this treatment, metallic thorium is formed and thorium atoms diffuse from the bulk to the surface, forming a monolayer. The emission efficiency can be enhanced by heating the material in a hydrocarbon vapor. In this case, it is assumed that W_2C is acting as a reducing agent for the formation of metallic thorium.

Tungsten is the most important metal for thermoemission applications, not only because of its high emissivity but because of its high thermal and chemical stability (extremely low vapor pressure at service temperature; high hot strength and rigidity; excellent corrosion resistance against metal and oxide vapors).

Tungsten-base materials are used for cathodes in power grit tubes, radio valves, X-ray tubes, as electrode material for gas-discharge lamps, electric arc welding, electron guns, electron microscopes, and plasma generators. The electron emission characteristic is also an important property for their use as electrical contact materials. Finally, tungsten is a potential candidate for crucial parts in thermionic energy conversion systems and cesium-ion engines for extraterrestrial applications [1.95].

In most of these applications, either thoriated tungsten is used (service temperature 1700–1800 °C), or impregnated cathodes, based on $BaO/Al_2O_3/CaO$ additions (so-called dispenser cathodes; service temperature 900–1250 °C). However, because of its radioactive nature, thoriated tungsten is currently being more and more replaced by tungsten with rare earth (RE) oxide additions or W–ZrO_2 (see also Section 6.2.2).

Field emission. For only a few specific applications (for example, in field emission electron microscopy) is tungsten used as a cold emitter. In this case, the electrons are extracted from the tungsten surface by applying an external electrical field in the order of 5×10^7 V/cm. This field is obtained by using tungsten single crystal wires with tip radii of about 100 nm (produced by electrolytic etching) and applying extraction tensions of up to 6 kV. Field emission cathodes can only be used in ultrahigh vacuum (10^{-8} Pa).

Photoelectric emission. Electron emission can also be caused by incident photons (photoelectric emission). The energy required to remove an electron by light energy in terms of wavelength is called the threshold wavelength. For thoriated tungsten, it has been quoted as 319 and 438 nm, respectively, depending on the measurement device [1.96].

For more information on electron emission, including interactions of tungsten with charged particles and atoms (secondary emission), the reader is referred to the compilation in a recently published handbook [1.38].

1.2.8. Acoustic Properties [1.97]

The sonic velocities in tungsten are $5180\,m \cdot s^{-1}$ for longitudinal (compression) waves, $2870\,m \cdot s^{-1}$ for transverse (shear) waves, and $2650\,m \cdot s^{-1}$ for surface waves. The acoustic impedance for longitudinal waves is $9.98\,g \cdot cm^{-2} \cdot s^{-1}$.

1.3. CHEMICAL PROPERTIES OF TUNGSTEN METAL

1.3.1. General Remarks

Although the main directing properties in tungsten's applications are of a pure physical nature (high melting point, high density, low vapor pressure, etc.), the chemical properties too are of special importance because they determine and limit the application fields of the metal in regard to diverse environments. It must be stated in advance that these properties are quite opposite.

On the one hand, tungsten can be considered as a rather inert metal, which is resistant to many elements and compounds. It is compatible with most ceramics and glasses up to high temperatures and shows good resistance to many molten metals. Tungsten is stable to mineral acids in the cold and is only slightly attacked at higher temperatures.

On the other hand, it reacts with numerous agents as well as elements and chemical compounds. For example, at room temperature, it is strongly attacked by fluorine. Below 100 °C, it dissolves in hydrofluoric–nitric acid mixtures, aqua regia, and alkali solutions containing oxidizing agents. Hydrogen peroxide is a good solvent for tungsten powder. At elevated temperatures, the number of reacting agents increases. At 250 °C, it reacts with chlorine, phosphoric acid, potassium hydroxide, and sodium nitrate or nitrite. At 500 °C, the attack by oxygen or hydrogen chloride becomes vigorous. At 800 °C it reacts with ammonia and at 900 °C with carbon monoxide, bromine, iodine, and carbon disulfide. With carbon or carbon-containing compounds, reactions occur above 1000 °C. Reactions of this type are important in the preparation of tungsten carbide (in regard to quantity, the most important tungsten compound) and are treated in more detail in Section 3.6.

Tungsten metal is stable in dry and humid air only at moderate temperature. It starts to oxidize at about 400 °C. The oxide layer is not dense and does not offer any protection against further oxidation. Above 700 °C the oxidation rate increases rapidly, and above 900 °C, sublimation of the oxide takes place, resulting in catastrophic oxidation of the metal. Any moisture content of the air enhances the volatility of the oxide.

Although tungsten is the metal with the highest melting point, its sensitivity toward oxidation is a big disadvantage. Therefore, all high-temperature applications are limited to a protective atmosphere or vacuum.

Bulk tungsten does not react with water but will be oxidized by water vapor at elevated temperature, e.g., at 600 °C.

Due to the importance of the reactions with oxygen, air, and water, Sections 3.1 and 3.2 have been specially devoted to this subject.

This chapter deals primarily with the reactions of bulk tungsten and tungsten powder. A vast literature exists about their reactions. However, it is not so much the reactions of the bulk metal which attracted the interest of scientists and technicians as its reactions at the surface. Scientific as well as practical reasons boosted that research owing to the widespread application of tungsten as a catalyst and as an electron or ion emitting source. It would be far beyond the scope of this book to deal with surface reactions in detail. In this regard, reference is made to the very complete compilation elsewhere [1.99–1.101].

Figure 1.22 presents a Periodic Table of the elements with information about the reactivity of tungsten toward elements, formation of compounds, and solubility of tungsten in metal melts. As can be seen, tungsten reacts with the majority of nonmetallic elements, with the exception of hydrogen, inert gases, and nitrogen. Molecular nitrogen, due to its high bonding strength, does not react with tungsten directly, but tungsten nitrides can be prepared via reaction with dry ammonia. Reactions with nonmetals are described in the following subsection.

The oxidation state of tungsten in compounds may vary between -2 and $+6$, the latter being the most common. Lower oxidation state compounds exhibit basic properties while higher ones are acidic. The maximum coordination number is 8, but it may attain 13 in coordination compounds with cyclic organic ligands. Chapter 4 treats tungsten compounds in detail.

However, in regard to the reactions of tungsten with metallic elements, the situation is quite different. A large number of metals exist which fail to react with tungsten, like the alkali metals, the alkaline earth metals with the exception of beryllium, the rare earth metals with the exception of cerium, and especially the elements scandium, yttrium, lanthanum, copper, silver, gold, zinc, cadmium, mercury, indium, thallium, tin, lead, antimony, bismuth, and polonium.

"Not reacting with tungsten" does not imply surface reactions, such as surface bonding of metals, which occurs with all metals. Some of these layers may change considerably specific surface-related properties of tungsten, like the work function, as in the case of thorium, lanthanum, barium, etc. This is of great technical interest for high-energy electron-emitting cathodes or welding electrodes.

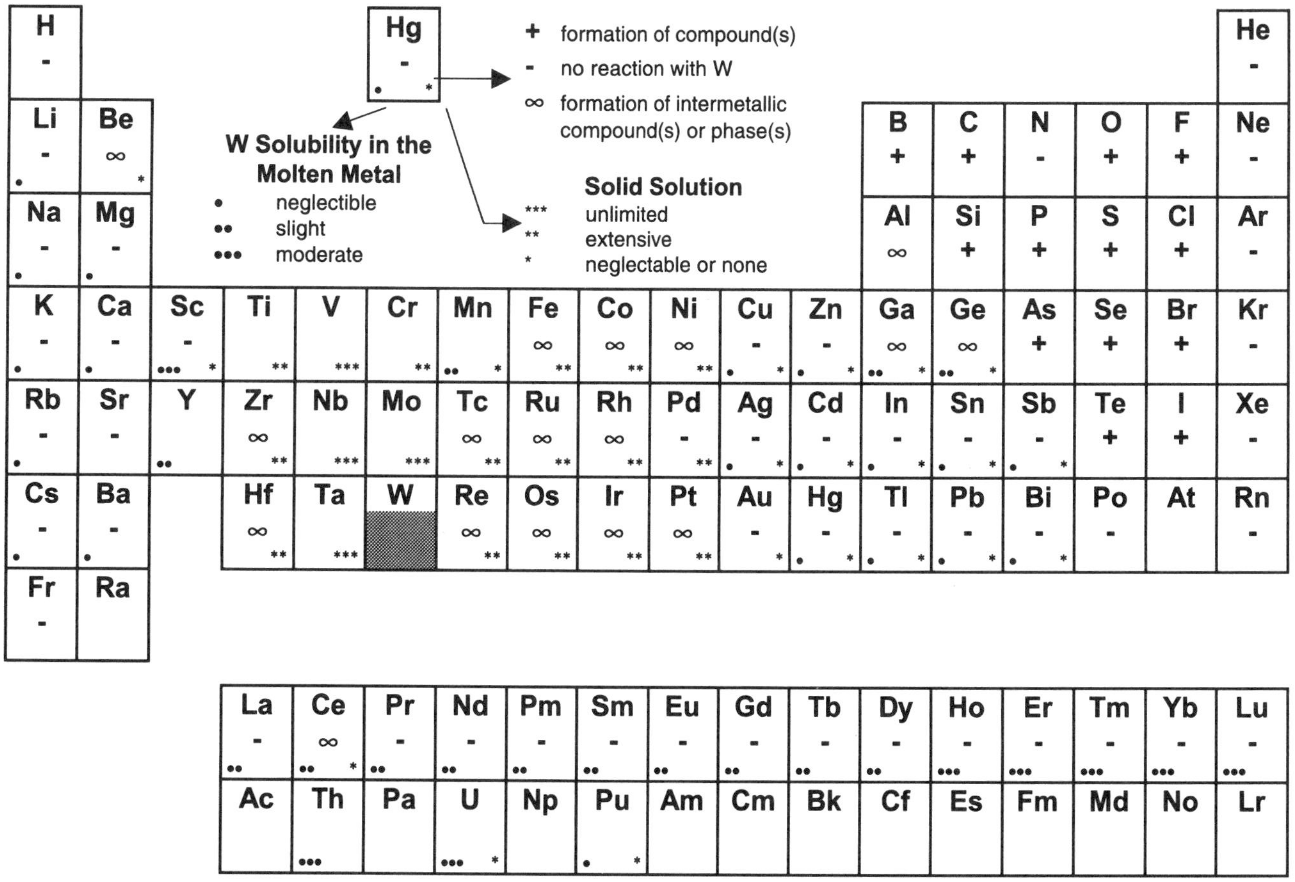

Hg
- • *

+ formation of compound(s)
- no reaction with W
∞ formation of intermetallic compound(s) or phase(s)

W Solubility in the Molten Metal
• neglectible
•• slight
••• moderate

Solid Solution
*** unlimited
** extensive
* neglectable or none

H - •																	He -
Li - •	Be ∞ *											B +	C +	N -	O +	F +	Ne -
Na - •	Mg - •											Al ∞	Si +	P +	S +	Cl +	Ar -
K - •	Ca - •	Sc - ••• *	Ti **	V ***	Cr **	Mn •• *	Fe ∞ **	Co ∞ **	Ni ∞ **	Cu - • *	Zn - • *	Ga ∞ •• *	Ge ∞ •• *	As +	Se +	Br +	Kr -
Rb - •	Sr - •	Y ••	Zr ∞ **	Nb ***	Mo ***	Tc ∞ **	Ru ∞ **	Rh ∞ **	Pd - **	Ag - • *	Cd - • *	In - • *	Sn - • *	Sb - • *	Te +	I +	Xe -
Cs - •	Ba - •		Hf ∞ **	Ta ***	W	Re ∞ **	Os ∞ **	Ir ∞ **	Pt ∞ **	Au - *	Hg - • *	Tl - • *	Pb - • *	Bi - • *	Po -	At	Rn -
Fr -	Ra																

La - ••	Ce ∞ •• *	Pr - ••	Nd - ••	Pm - ••	Sm - ••	Eu - ••	Gd - ••	Tb - ••	Dy - ••	Ho - •••	Er - •••	Tm - •••	Yb - •••	Lu - •••
Ac	Th •••	Pa	U ••• *	Np	Pu • *	Am	Cm	Bk	Cf	Es	Fm	Md	No	Lr

FIGURE 1.22. Periodic Table; reactions of W with the elements.

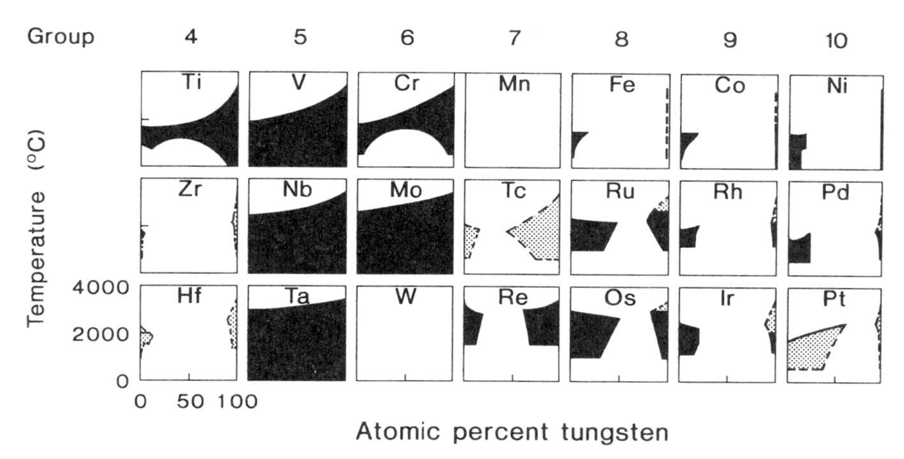

FIGURE 1.23. Solid solubilities in the transition metal–tungsten allosy [1.101].

Metallic elements reacting with tungsten besides beryllium and cerium are concentrated in groups 4 to 7 of the Periodic Table. Here, we find binary intermetallic compounds with close homogeneity ranges and phases with very broad ranges, besides solid solubility ranges toward the element-rich sides of the respective binary system. In some cases, especially within groups 4 to 6, only a continuous series of solid solubility exists. A survey on solid solubilities is presented in Fig. 1.23.

Table 1.24 presents data on compound types in the various binary systems. Reactions with metals are dealt with in more detail in Section 1.3.3.

TABLE 1.24. Types of Compounds and Intermetallic Phases in Binary Tungsten Systems[a]

Formula	at % W	Number of systems	Element
$M_{22}W$	4.3	1	Be
$M_{12}W$	7.7	2	Al, Be
M_6C	14	3	Br, Cl, F
M_5W	17	4	Al, Cl, Br, F
M_4W	18–20	8	Al, B, Cl, Br, F, I, Ni, Rh
M_3W	25	8	Br, I, Co, Os, Pd, Re, Rh
M_5W_2	28–31	1	B
M_2W	33.3	13	As, Be, Br, Cl, Fe, Ge, I, P, Pt, S. Se, Si, Te
M_3W_2	40	1	As
M_7W_6 (μ)	42–46	2	Co, Fe
MW	50	10	B, C, Fe, Ir, Ni, P, Pt, Re, Ru, Tc
$M_{1-x}W$	62	1	C
M_3W_5	62.5	2	Ge, Si
MW_2	66.7	5	B, C, Hf, Ni, Zr
MW_3	75	2	Cr, P
σ		5	Ir, Os, Re, Ru, Tc
ε		3	Ir, Pt, Rh

[a] Excluding nitrogen and oxygen.

TABLE 1.25. Reaction of Tungsten Metal with Acids and Alkalis

	Temperature	
Reagent	20 °C	100–110 °C
HF	None	None
HNO_3	Slight attack	Oxidation
H_2SO_4	None	Slight attack
HCl	None	Slight attack
H_3PO_4	None	Slight attack
H_2O_2	None	Dissolution
NH_4OH	None	None
KOH	None	None
NaOH	None	None
HCl + HNO_3	Oxidation	Dissolution
HF + HNO_3	Dissolution	Dissolution
KOH + H_2O_2	Slight attack	Dissolution

Finally, Sections 1.3.4–1.3.6 are concerned with reactions of tungsten with compounds, aqueous solutions, and miscellaneous substances, which are of importance for diverse applications. The reactivity of tungsten in aqueous solutions is interesting in relation to analytical procedures and chemical engineering. It will be shown in Table 1.25 that aqueous solutions of nonoxidizing acids and alkalis, as well as ammonia, do not attack tungsten. On the other hand, the presence of oxidizing agents in combination with acids or alkalis (nitric acid, sodium nitrate, sodium nitrite, hydrogen peroxide, potassium peroxide) results in rapid dissolution.

Tungsten resists the attack of molten alkalis. Here, too, the addition of oxidizing compounds leads to vigorous dissolution.

Due to its complexity, the chemistry of aqueous solutions is treated in Section 3.7.

1.3.2. Reactions with Nonmetals [1.39, 1.99, 1.102]

Table 1.26 presents a review of the reacting elements, the related reaction conditions, and the products formed.

Reference is made to Chapter 4 for preparation possibilities concerning the compounds described here and further possible tungsten compounds, their properties, and their technical importance.

1.3.3. Reactions with Metals [1.39, 1.99–1.101]

This section not only includes reactions where chemical compounds of close homogeneity ranges are formed (intermetallic binary compounds), but also cases in which phases with broad homogeneity ranges exist or only mutual solid solubility was observed.

Group 1: Alkali Metals (Li, Na, K, Rb, Cs). These metals do not react with tungsten and do not form any binary compounds. Due to the very low solubility of tungsten in molten alkali metals, only a slight attack of tungsten metal specimens was observed in

TABLE 1.26. Reactions of Tungsten Metal with Nonmetals

Element	Reaction conditions	Remarks	Products
Boron	W + B (amorphous) powder mixture compacts 500 °C in H_2/1 hour 800–1200 °C in Ar/2 hours	Individual boride formation depends on W/B ratio	W_2B, WB, W_2B_5, WB_4
Carbon	Reaction between 800–1900 °C 800 °C 1050 °C 1550 °C 1900 °C	Depending on reactant form and atmosphere • CVD carbon layer on W foil • Fine powder mixture • Polycryst. Wire in C powder • Monocryst. W wire in C powder	W_2C, WC
Silicon	W/Si powder mixtures at elevated temperature and protective atmosphere	Properties of starting materials and experimental conditions determine reaction products	WSi_2, W_3Si_2, W_2Si_3, W_5Si_3
Phorphorus	W/P (red) mixtures at 700–950 °C	Under vacuum	WP_2
Arsenic	W/As mixture at 620 °C	Protective atmosphere	WAs_2
Oxygen	Reaction starts at ~400 °C	Above 500 °C oxide layer forms cracks, at 800 °C sublimation of WO_3	WO_3
Sulfur	Molten S or vapor attacks slowly, powder mixtures start reacting at 400 °C		WS_2
Selenium	W/Se mixtures start reacting at 480–520 °C	Exothermic	WSe_2
Tellurium	Mixture react slowly at 600–700 °C	In vacuum	WTe_2
Fluorine	Reacts already at room temperature	Reaction products volatile, no formation of protective layer	Main product WF_6
		In presence of oxygen	WOF_4
Chlorine	Attack starts at 250–300 °C		Main product at low and medium temp.: WCl_6
	above 2000 K		WCl_2 is dominant
		In presence of O or H_2O	$WOCl_4$, WO_2Cl_2
Bromine	Reaction starts at 450–500 °C	Reaction product depends on reaction conditions	WBr_6, WBr_5, WBr_4
		In presence of O or H_2O	$WOBr_4$, WO_2Br_2
Iodine	Iodine vapor starts reacting between 550–700 °C	O and H_2O enhance the attack	W_6I_{12}

long-time corrosion tests. The corrosion is temperature-dependent and is further influenced by the presence of oxygen traces, which lead to an enhanced dissolution.

Tungsten exhibits an excellent chemical stability against cesium vapor at elevated temperature up to 1900 °C. This is of importance for the application of tungsten in ionic propulsion systems in spacecrafts.

Group 11 (Cu, Ag, Au). These metals do not react with tungsten, and no binary compounds are formed. The solubility of tungsten in the molten metals is negligible.

Pseudo-alloys of Cu–W and Ag–W, combining the beneficial properties of both metals, play an important role as electrical contact materials and electrode materials for spark erosion machines.

Group 2: Alkaline Earth Metals (Be, Mg, Ca, Sr, Ba). Between 750 and 1000 °C, tungsten reacts with Beryllium forming WBe_{12} and WBe_{22}. Above 1000 °C, only WBe_{22} is stable. Reaction of solid tungsten with Be vapor leads to diffusion zones, with WBe_2 formed in the inner zone and both WBe_{12} and WBe_{22} in the outer zone.

Magnesium, calcium, strontium, and barium do not react with tungsten, and no binary compounds are known. The solubility of tungsten in the respective molten metal is very low.

BaO- resp. $BaCO_3$-doped porous tungsten play an important role as cathodes in high-energy electron tubes. The emitting element is metallic Ba, which forms upon heating above 1000 °C.

Group 12 (Zn, Cd, Hg). These elements do not react with tungsten, and no binary compounds are known.

Group 3 (Sc, Y, La) and Rare Earth Metals. These metals do not form intermediates with tungsten. Only cerium is said to form an intermetallic compound of the composition W_2Ce with tungsten, but this is still doubtful.

At higher temperature, tungsten exhibits a certain solubility in the respective metal melt. The solubility in RE metals at 2200 K decreases with increasing atomic radius from Sc 4.5 at% to La<0.1 at%.

Group 13 (Al, Ga, In, Tl). Tungsten dissolves in molten aluminum. Three binary compounds WAL_{12}, WAL_5, and WAL_4 are known and form by peritectic reactions.

No intermediates are formed with gallium under atmospheric pressure. At 400–800 °C and increased pressure (77 kbar), W_2Ga_5 and W_2Ga_3 form. Tungsten exhibits a slight solubility in molten Ga at 900–1100 °C.

Indium and thallium do not react with tungsten, and no compounds were found. Solubility in the respective metal melt is negligible.

Group 4 (Ti, Zr, Hf) and Th. No intermetallic phases form with titanium. Tungsten is only slightly soluble in α-Ti, but ~25 at% Ti dissolve in W at 740 °C. β-Ti forms a continuous solid solution with W at temperatures above 1250 °C. Below 1250 °C, a miscibility gap occurs.

Tungsten reacts with zirconium to form ZrW_2. The solid solubility of Zr in W is 3.5–10.0 at% (≥2000 °C), depending on the respective investigation. The maximum solubility of W in α-Zr is 0.5 at% and in β-Zr 4 at%.

Also, hafnium forms an intermetallic compound HfW_2. Within the system, the following solubilities were determined: 0.9 at% W in α-Hf (1560 °C), 13.5 at% W in β-Hf (1950 °C), 4 at% Hf in W (800 °C), and 9 at% Hf in W (2512 °C)

No compounds are known with thorium. The solid solubilities of W in Th and of Th in W are negligible. The solubility of W in liquid Th corresponds to 1.2 at% at 1695 °C and to 8.0 at% at 2509 °C.

Group 14 (Ge, Sn, Pb). Germanium reacts with tungsten at elevated temperature only under increased pressure (77 kbar, 1500–2500 °C). Two binary compounds are known: W_5Ge_3 and WGe_2. No solid solubility exists between the two metals and the solubility of W in molten Ge is negligible.

Tin and lead do not react with tungsten and do not form compounds, while the solubility of tungsten in the respective metal is negligible.

Group 5 (V, Nb, Ta). There exists unrestricted solid solubility in all three cases, and no binary compounds are formed.

Group 15 (Sb, Bi). Antimony reacts sluggishly, and no compounds are formed. With bismuth there is no corrosion of tungsten at 1000 °C, and no intermetallic compound is formed.

Group 6 (Cr, Mo) and U. In the case of chromium and molybdenum unrestricted solid solubility exists, while it is very restricted ($\leq$1 at%) for uranium. Below 1677 °C, a miscibility gap occurs in the system Cr–W. No compounds are known with one of the three metals.

Group 7 (Mn, Tc, Re). No binary compounds are formed with manganese, and no homogeneous product could be prepared. Technetium reacts with W, forming a solid solution and a σ-phase (WTc_3). Tungsten–technetium alloys are of practical interest as superconductors.

The solid solubility of W in rhenium is 11 at% at 1600 °C and 20 at% at 2825 °C. The solid solubility of Re in W is 27.5 at% at 1600 °C and 37 at% at 3000°C. Besides the solid solutions, two intermetallic phases exist: a tetragonal σ-phase with a broad homogeneity range and a cubic χ-phase with a narrow homogeneity range.

Solid solution alloys of W and Re in the range of 2–25 wt% Re play an important role as rotating anodes in X-ray tubes for diagnostic purposes and as wires for thermocouples. These materials are produced by powder metallurgical techniques. Inappropriate mixing of the two powder components may result in a partial σ-phase formation, which is undesirable due to embrittlement.

Groups 8–10 (Fe, Co, Ni, Ru, Rh, Pd, Os, Ir, Pt)

Iron group metals. Iron and tungsten form two stable intermediate phases: Fe_7W_6 (μ) and FeW (δ), as well as a metastable transitional phase, Fe_2W (λ). The solubility of W in fcc Fe is <1.5 at% but up to 14.3 at% in bcc Fe at the peritectic temperature of 1548 °C. The maximum solubility of Fe in bcc W is ~2.6 at%. Above 2 wt% W, no γ-Fe (fcc) exists.

The element-rich solid solubility alloys are of importance in the so-called “heavy metal” alloys. The tungsten phase formed during liquid-phase sintering is saturated with iron (and nickel) and, vice versa, the nickel–iron binder phase with tungsten.

Cobalt and W form two intermetallic compounds: Co_3W and Co_7W_6. The maximum solubility of W in fcc Co is 17.5 at% (at 1471 °C) and <0.5 at% in hcp Co. The maximum solubility of Co in W is about 0.9 at%.

Three intermediate phases are known in the nickel–tungsten system: NiW_2, NiW, and Ni_4W. Up to 17.5 at% W dissolves in fcc Ni (at 1495 °C). The solubility of W in Ni is about 0.3 at% W at 1495 °C.

Platinum metals. Tungsten reacts with ruthenium, rhodium, palladium, osmium, iridium, and platinum, forming extended areas of solid solutions. The maximum solubility of tungsten in the respective metal decreases from 55 at% (Os) to 48 at% (Ru), 25 at% (Pt), 22 at% (Pd) to 19 at% (Rh, Ir). The solubility of the metals in tungsten decreases from Ru (23 at%) to Os (18.5 at%), Ir (10 at%), Rh (6 at%), Pd (5 at%) to Pt (<3.5 at%).

The following intermediate phases exist: W_3Ru_2 (σ), Rh_3W (ε), OsW_3 (δ). Furthermore, two phases are known to form with platinum [a tetragonal γ-phase (33–37 at% W) and a hexagonal ε-phase (45–52 at% W)] as well as with iridium [an ε-phase (homogeneity range between 22 and 47 at% W) besides a δ-phase (IrW_3)].

Tungsten/palladium alloys between 26 and 44 at% W are superconducting.

1.3.4. Reactions with Compounds [1.4]

1.3.4.1. Reactions with Nonmetal Compounds

Water (H_2O). Water reacts with tungsten from room temperature up to 2000 °C. At higher temperatures it dissociates, and oxygen is the reactant.

In principle, tungsten is oxidized by water. The reaction rate is determined by the temperature and the pH_2O/pH_2 ratio. As a reaction product, the presence of H_2 always has to be considered. The reaction rate as well as the O:W ratio of the oxide formed increases with temperature and pH_2O/pH_2 ratio. Water increases the volatility of the tungsten oxides by the formation of the volatile oxide hydrate $WO_2(OH)_2$. For more details, see Sections 3.2 and 3.3.

Ammonia (NH_3). At room temperature, ammonia does not react with tungsten. Ammonia containing moist air shows enhanced corrosion in comparison to ammonia free. A quantity of 1–2 μm tungsten powder in contact with a fast stream of ammonia shows an uptake of up to 3.5 wt% N at 600 to 700 °C, increasing to 3.7 wt% N at 800 °C. The reaction products are W_2N and γ-W_3N_4, both fcc.

At lower temperatures the reaction is restricted to the surface, and at higher temperatures the rather unstable nitrides decompose. There is no attack by ammonia from there up to the melting point. Due to the very low solubility of nitrogen in the tungsten crystal lattice, volume reactions proceed mainly via grain boundary diffusion.

Tungsten is an efficient catalyst for both synthesis and decomposition of ammonia. At elevated temperature, partial or complete dissociation of NH_3 occurs, and hydrogen, and at high temperature also nitrogen, are released into the gas phase.

Nitrogen oxides (N_2O, NO, NO_2). They behave as oxidants and at elevated temperature form tungsten oxides, whereby nitrogen is released.

Hydrogen fluoride (HF). Tungsten powder reacts only slightly with HF at 300 °C, and the reaction product is undefined. Up to 600 °C, there is only little corrosion. The oxidic passivation layer on tungsten converts into low-valent, nonvolatile fluoride which protects the surface. At higher temperature, they disproportionate forming volatile fluorides, and the corrosion rate is markedly increased.

Hydrogen chloride (HCl). Tungsten metal is stable against dry HCl up to 700 °C and even higher.

Hydrogen bromide (HBr). The reaction of HBr with tungsten plays an important role in modern incandescent lamps (see Chapter 7).

Hydrogen sulfide (H_2S). Tungsten powder reacts at 350–500 °C according to $W + 2H_2S \rightarrow WS_2 + 2H_2$. At low temperature, the hexagonal modification forms.

Sulfur dioxide (SO_2). SO_2 behaves as an oxidant. At red heat tungsten is oxidized to lower oxides, whereby elemental sulfur is formed: $W + 2SO_2 \rightarrow 1/n\,(WO_3) + \frac{1}{2}O_2 + S_2$.

Sulfur hexafluoride (SF_6). At 500–700 °C the ternary compound WSF_4 is formed, besides tungsten tetrafluoride: $3W + 2SF_6 \rightarrow 2WSF_4 + WF_4$.

Disulphur dichloride (S_2Cl_2). In presence of air, $WOCl_4$ and, under exclusion, WCl_6 are formed.

Hydrogen selenide (H_2Se). At 300–400 °C, WSe is formed and, between 500 and 800 °C, the diselenide. In both cases hydrogen is released. Above 850 °C, decomposition of the diselenide starts.

Boron nitride (BN). BN reacts with tungsten powder at 1500 °C and with dense tungsten at 1600 °C to form W_2B and WB.

Boron trichloride (BCl_3). It boronizes tungsten at 900 to 1000 °C.

Carbon monoxide (CO). Between 80 and 200 °C, tungsten hexacarbonyl forms. From 1000 °C upward, bulk tungsten is carburized: $W + 2CO \rightarrow WC + CO_2$ resp. $2W + 2CO \rightarrow W_2C + CO_2$. Nanocrystalline powders start to carburize already at about 550 °C.

Carbon dioxide (CO_2). At elevated temperature, CO_2 acts as an oxidizing agent and, depending on temperature and partial pressure, lower or higher tungsten oxides will be formed: $xW + yCO_2 \rightarrow W_xO_y + yCO$.

Hydrogen cyanide (HCN). From 730 to 2200 °C tungsten reacts as follows: $W + HCN \rightarrow WC + \frac{1}{2}H_2 + \frac{1}{2}N_2$. Above 2200 °C, W_2C and, at 2600 °C, in addition graphite are formed.

Carbon tetrafluoride (CF_4). At elevated temperature, WF_6 is formed.

Carbon tetrachloride (CCl_4). At 600 °C, tungsten powder reacts to WCl_4 and WCl_6.

Carbon disulfide (DS_2). Tungsten reacts at elevated temperature to WS_2.

Boron carbide (B_4C). In the contact zone between the two components, tungsten boride layers can be detected after heat treatment between 1100 and 1600 °C. Depending on the reaction conditions, α-WB or W_2B is formed. Additions to NaF as activator.

Silicon dioxide (SiO_2). If the protecting atmosphere is free of oxidizing species, tungsten is very stable against liquid SiO_2 up to 2500 °C. Above 2000 °C, slight oxidation starts: $2W + 3SiO_2 \rightarrow 2WO_3 + 3Si$.

Trisilicon tetranitride (Si_3N_4). In vacuum at 1300 °C, tungsten silicides are formed.

Silicon tetrachloride ($SiCl_4$). Tungsten reacts at 1000–1200 °C in hydrogen atmosphere according to: $W + 2SiCl_4 + 4H_2 \rightarrow WSi_2 + 8HCl$. The reaction product forms a protective layer on tungsten.

Silicon tetrabromide ($SiBr_4$). Reacts analogously.

Silicon carbide (SiC). Between 1100 and 1900 °C, WC and tungsten silicides are formed.

Hydrogen phosphide (PH_3). Tungsten reacts at 850 °C to WP.

Phosphoric acid concentrated (H_3PO_4). Tungsten shows a slight attack depending on temperature and time.

1.3.4.2. Reactions with Metal Compounds.

Alkali hydroxide melts. These compounds do not attack as long as any oxidizing material (including air) is excluded. Any oxidic scale on the tungsten metal surface will be dissolved. The presence of oxidizing agents like nitrates, nitrites, chlorates, etc. leads to a vigorous attack.

For example, for technical purposes, the dissolution of tungsten scrap material mixtures of NaOH with sodium nitrate or sodium nitrite have been widely used. The dissolution occurs due to the following reactions:

$$W + 6NaNO_2 \rightarrow Na_2WO_4 + 2Na_2O + 6NO$$
$$W + 6NaNO_3 \rightarrow Na_2WO_4 + 2Na_2O + 6NO_2$$

Because NO as well as NO_2 react with the excessive NaOH to form nitrite, less nitrogen oxides are evolved than expected.

Sodium oxide (Na_2O). Above 500 °C, it reacts with tungsten according to: $W + 4Na_2O \rightarrow Na_2WO_4 + 6Na$.

Sodium peroxide (Na_2O_2). Gives a violent raction.

Sodium nitrate ($NaNO_3$). Gives a violent raction.

Sodium nitrite ($NaNo_2$). Gives a violent reaction.

Sodium chloride (NaCl). Gives a slight attack.

Sodium sulfide (Na_2S). There is no reaction up to 1500 °C.

Sodium sulfate (Na_2SO_4). Reacts from 320 °C upwards: $3W + 2Na_2SO_4 \rightarrow 2Na_2WO_4 + WS_2$.

Sodium borides. Form borides.

Sodium carbonate (Na_2CO_3). From 950 °C onward tungstate is slowly formed: $W + 3Na_2CO_3 \rightarrow Na_2WO_4 + 2Na_2O + 3\ CO$.

Potassium nitrate (KNO_3). Reacts like $NaNO_3$.

Potassium perchlorate ($KClO_3$). Gives a violent reaction.

Beryllium oxide (BeO). Oxidizes tungsten above 1800 °C.

Magnesium oxide (MgO). Reacts with tungsten from 2000 °C onward. The reaction products are Mg, $MgWO_3$, $MgWO_4$, WO_2, and WO_3.

Calcium oxide (CaO). Reacts with tungsten at $\geq$ 1900 °C under formation of $CaWO_3$ and $CaWO_4$.

Stronium oxide (SrO). Reacts with tungsten at $\geq$ 850 °C under formation of Sr_3WO_6 and $SrWO_4$. Metallic Sr forms according to the equation: $2\ Sr_3WO_6 + W \rightarrow 3SrWO_4 + 3Sr$.

Barium oxide (BaO). At $\geq$ 800 °C Ba_3WO_6 is formed: $6BaO + W \rightarrow Ba_3WO_6 + 3Ba$. Above 1000 °C, the reaction ceases, because the bronze compound dissociates thermally into WO_3 and Ba, which evaporates. Above 1400 °C, the tungsten surface is clean.

Barium carbonate ($BaCO_3$). Reacts with tungsten in the temperature range between 600 and 800 °C according to: $3BaCO_3 + W \rightarrow Ba_3WO_6 + 3CO$.

Aluminum oxide (Al_2O_3). Compatibility is given up to 1900 °C. No or very slow reaction occurs with molten alumina according to: $3W + Al_2O_3 \rightarrow 3WO_2 + 4Al$ resp.$W + Al_2O_3 \rightarrow WO_3 + 2Al$. The reaction products are in both cases volatile. Seamless tungsten crucibles are used to grow large sapphire single crystals.

Zirconium oxide (ZrO_2). Is compatible with tungsten up to 1600 °C in vacuum and up to 1400–1500 °C in protective atmospheres. Starting temperature indications in the literature are quite different due to differences in zirconia stabilization (2000–2390 °C).

Thorium oxide (ThO_2). Is compatible up to 2200 °C in vacuum and up to 2000 °C in protective atmosphere. Indications about the stability at higher temperatures are quite different. The reaction: $2W + \alpha\text{-}ThO_2 \rightarrow 2WO_x + \alpha\text{-}Th$ takes place at a temperature of >2000 K [1.37].

Alkali chromates. Binary powder mixtures of tungsten and potassium dichromate are easy to ignite (650–660 °C) and burn over a wide range of compositioin (20–90 % W). There is only little change in weight (gasless pyrotechnical system): $W + K_2Cr_2O_7 \rightarrow K_2WO_4 + Cr_2O_3$.

Uranium dioxide (UO_2). No reaction with tungsten occurs up to the melting point of uranium dioxide (approx. 2760 °C). The molten oxide penetrates tungsten along grain boundaries.

1.3.5. Reactions with Aqueous Solutions

Many aqueous solutions of chemical compounds only attack the surface of tungsten metal slightly. This may be compared with the attack of pure water, especially if diluted solutions are considered. Typical examples of this group of compounds are nitric acid, hydrochloric acid, sulfuric acid, sodium and potassium hydroxide, sodium chloride or ammonium chloride (see Table 1.27), and many others. A completely different behavior is shown with compounds containing a complexing anion (forming very stable coordination compounds with tungsten). A prominent member of this group is hydrofluoric acid (see Table 1.27). The attack is about one magnitude higher than in the case of hydrochloric acid. Solutions of compounds which combine complexing and oxidizing ability are very useful dissolution reagents, as, for example, hydrogen peroxide or mixtures of hydrofluoric and nitric acid.

TABLE 1.27. Corrosion of Tungsten by Diverse Aqueous Solutions [a,b]

	HNO_3 10 %	HF 3 %	HCl 10 %	H_2SO_4 10 %	KOH 10 %	NaCl 3 %	NH_4Cl
Boiling reflux nonaerated highly agitated	0.20	2.80	0.13	0.15	0.41	0.38	0.03
75 °C reflux slightly aerated slightly agitated	0.25	2.00	0.08	0.65	5.20	0.80	0.72
20 °C static immersed normally aerated	0.002	0.032	0.005	0.0055	0.55	0.0041	0.0066
20 °C dipping highly aerated highly agitated	0.02	0.34	0.008	0	2.14	0.020	0.022

[a] *Gmelin Handbook of Inorganic Chemistry*, 8th ed., Syst. No. 54, Tungsten, Suppl. Vol. A7, Springer-Verlag, Heidelberg (1987).
[b] Corrosion rate in $g \cdot m^{-2} \cdot d^{-1}$.

Hydrogen peroxide (H_2O_2). Tungsten dissolves without inhibition. The dissolution rate is proportional to the peroxide concentration. The dissolution rate for bulk tungsten corresponds to 1.6 mg W/l · min at 20 °C and 14 g H_2O_2/l. Under the same conditions, tungsten powder dissolves at a rate of 300 mg/l · min. Increased temperature enhances the dissolution rate up to 60 °C. Higher temperatures slow down the reaction because of peroxide decomposition. Addition of nitric acid, hydrofluoric acid, sulfuric acid, phosphoric acid, acetic acid, or sodium hydroxide does not influence the reaction. The solution shows a bright yellow color. The reaction is said to proceed as follows:

$$W + 2H_2O_2 \rightarrow WO_2 + 2H_2O$$
$$WO_2 + H_2O_2 \rightarrow H_2WO_4$$
$$3H_2WO_4 + 2H_2O_2 \rightarrow H_2W_3O_{12} + 4H_2O$$

or

$$2WO_2 + 6H_2O_2 \rightarrow H_2W_2O_{11} + 5H_2O$$
$$3H_2W_2O_{11} \rightarrow 2H_2W_3O_{12} + H_2O$$

NaOH solutions containing 5 vol% of hydrogen peroxide are proposed as etchants.

1.3.5.1. Acids

Nitric acid (HNO_3). At room temperature, tungsten is rather resistant against all concentrations and also against hot concentrated acid. Hot and diluted nitric acid attacks slowly (see Table 1.27).

Hydrofluoric acid (HF). The dissolution rate is about one magnitude higher than that for nitric acid and the attack is faster in hot solution. Any addition of oxidizing agents increases the dissolution rate considerably (see Table 1.27).

Hydrochloric acid (HCl). Tungsten shows rather good resistivity against HCl at all concentrations and temperatures. The corrosion rate is somewhat smaller than for nitric acid (see Table 1.27).

Hydrobromic acid (HBr). Shows no significant effect.

Hydroiodic acid (HJ). Tungsten is dissolved slowly in concentrated solutions.

Periodic acid (HJO_4) Does not react with tungsten.

Sulfuric acid (H_2SO_4). The resistivity of tungsten is comparable to that for hydrochloric acid over a wide range of concentration and temperature. Increased attack occurs by fuming sulfuric acid (see Table 1.27).

Phosphoric acid (H_3PO_4). Tungsten is highly resistant to diluted phosphoric acid as long as air is excluded. It dissolves easily in concentrated acid.

1.3.5.2. Important Acid Mixtures to Dissolve Tungsten

40 vol% conc. HNO_3 and 60 vol% conc. HF. Already at room temperature, a vigorous attack of tungsten takes place. The metal surface is oxidized by NO_2 and the tungsten oxide dissolves in HF, whereby tungsten fluoride or oxofluoride ions are formed which show a high solubility.

HNO_3–HCl Mixture (aqua regia). Tungsten is attacked only slowly in cold, but rapidly in hot, solutions. Due to the fact that there is no complexing agent like fluoride as in the former acid mixture, an oxide layer is formed which slows down the reaction.

HF–H_2SO_4–HNO_3 mixtures. Tungsten is dissolved quickly at all concentrations.

1.3.5.3. Aklaline Solutions. In general, tungsten shows less resistivity to alkaline than to acidic solutions, especially in the presence of air. Corrosion rate increases with increasing OH^- concentration. Oxidizing agents including oxygen or air raise the dissolution rate.

Sodium hydroxide (NaOH). In the absence of oxygen, dilute solutions do not attack tungsten at temperatures up to 10 °C. Concentrated solutions show a slight attack.

Potassium hydroxide (KOH). Reacts analogously to NaOH (see Table 1.27).

Ammonium hydroxide (NH_4OH). Is more corrosive than KOH and NaOH.

1.3.5.4. Salts. Sodium chloride (NaCl). Slightly corrodes (see Table 1.27).

Sodium hypochlorite (NaClO). Dissolves tungsten slowly according to: $W + H_2O + 3ClO^- \rightarrow WO_4^{2-} + 3Cl^- + 2H^+$.

Ammonium chloride (NH_4Cl). Slightly corrodes (see Table 1.27).

Iron trichloride ($FeCl_3$). Is strongly corrosive for tungsten.

Potassium hexacyanoferrate(III) ($K_3[Fe(CN)_6]$). Alkaline solutions are in use as etchants in metallography and for meal finishing (Murakami etch):

$$W + 6K_3[Fe(CN)_6] + 8KOH \rightarrow K_2WO_4 + 6K_4[Fe(CN)_6] + 4H_2O$$

The $K_3[Fe(CN)_6]$:KOH molar ratio for polishing action is ≥ 2.5, and for etching (revealing details in microstructure) <2.5.

Cu^{2+} salts. Show an etching effect in ammoniacal solution preferentially on W(100) faces (Millner–Sass solution).

1.3.6. Miscellaneous

Ceramics [1.56]. Compatibility is given for the following ceramic materials up to the indicated temperature (valid in vacuum, in protective atmosphere 100–200 °C lower):

Alumina (Al_2O_3)	~1900 °C
Beryllia (BeO)	~2000 °C
Magnesia (MgO)	~1500 °C
Thoria (ThO_2)	~2200 °C
Zirconia (ZrO_2)	~1600 °C
Sillimanit	~1700 °C
Magnesit (brick)	~1600 °C
Fire Brick	~1200 °C
$3BeO \cdot 2CaO$	1400–1500 °C
$3BeO \cdot 2MgO$	1800–1900 °C
$Al_2O_3 \cdot 4BeO \cdot 4MgO$	1600–1700 °C
$Al_2O_3 \cdot 4BeO \cdot MgO$	~1900 °C

Glass melts [1.56, 1.101]. Tungsten is very stable against glass melts up to 1400 °C. Above 1400 °C, a slight attack occurs. The wetting behavior of glass melts besides glass quality and temperature depends on the atmosphere. Oxidizing conditions cause yellow (WO_3) or yellow-green colored glasses or opal glasses due to crystallization.

Due to its excellent corrosion resistance, tungsten is used as electrode material for melting kaolin.

Rock and magmatic melts [1.101]. Tungsten is severely oxidized and dissolved by those melts: The solubility at 1500 K in Bandelier tuff melts is 0.4–1.6 wt% and is similar for Janez basalt at 1900 K and 3.5 wt% at 1900 K for other basalts. The solubility in basalts increases with increasing Fe^{3+}/Fe^{2+} ratio. Tholeitic basalt lava from 1971 Kilauea eruption oxidizes tungsten strongly in an H_2O, CO_2, CO, SO_3 atmosphere.

1.3.7. Reactions with Organic Compounds

The numerous reactions which yield in the formation of organometallic compounds are not considered here (see Chapter 4).

In general, organic compounds like hydrocarbons react with tungsten at elevated temperature via dissociation of the organic molecule and subsequent carbide formation.

Methane is the most prominent compound in this field and should act as a representative for all the others here. It also plays a very important role in the technical carburization process of tungsten as an intermediate which enables the gaseous transport of carbon.

Methane (CH_4). Bulk tungsten reacts with methane at temperatures above 1000 K forming W_2C and WC according to: $2W + CH_4 \rightarrow W_2C + 2H_2$ or $W + CH_4 \rightarrow WC + 2H_2$.

The thermochemical data for the reactions in the W–CH_4–H_2 system are:

$$2W + CH_4 \rightarrow W_2C + 2H_2$$
$$\Delta G^\circ = 13470 - 25.7T$$
$$\log p(CH_4 = 2 \cdot \log p(H_2) - 89.5 + 2490/T$$
$$CH_4 \rightarrow C\ (\text{in W}) + 2H_2$$
$$\Delta G^\circ = 34270 - 25.8T$$
$$\log p(CH_4) = 2 \cdot \log p(H_2) + \log c(C) - 10.52 + 9460/T$$

the units of ΔG° are $\text{cal} \cdots \text{mol}^{-1} \cdot \text{K}^{-1}$, p is in Torr, and c in at%.

The equilibrium pressure of the methane dissociation is only slightly higher than that of the carbide formation. In order to suppress excessive free-carbon formation, the CH_4 pressure has to be maintained above the minimum pressure for the carburization, but below that resulting in free-carbon formation.

REFERENCES FOR CHAPTER 1

1.1. *Gmelin Handbook of Inorganic Chemistry*, 8 ed., Syst. No. 54, *Tungsten*, Suppl. Vol. A2, Springer-Verlag, Heidelberg (1987).
1.2. J. Emsley, *Die Elemente*, Walter de Gruyter, Berlin, New York (1994).
1.3. R. D. Shannon, *Acta Crystallogr.* **A32** (1976), 751–767.
1.4. J. C. Bailar Jr, H. J. Emeleus, R. Nyholm, and A. F. Trotman-Dickenson, Eds. *Comprehensive Inorganic Chemistry*, Vol. 3, p. 746, Pergamon Press, Oxford (1973).
1.5. G. V. Samsonov, *Handbook of the Physiochemical Properties of the Elements*, Plenum Press, New York (1968).
1.6. A. J. C. Wilson, ed., *International Tables for Crystallography*, Vol. C, Kluwer Academic Publishers, Dordrecht/Boston/London (1995).
1.7. R. Jenkins, *X-Ray Spectrom.* **2** (1973), 207.
1.8. R. Tertian and F. Claisse, *Principles of Quantitative X-ray Fluorescence Anlaysis*, Heyden, London (1982).

1.9. L. E. Davis, N. C. MacDonald, P. W. Palmberg, G. E. Riach, and R. E. Weber, *Handbook of Auger Electron Spectroscopy*, Physical Electronics Industries, Inc., Eden Prairie, Minnesota 55343 (1976).
1.10. I. Barin, *Thermochemical Data of Pure Substances*, Vol. II, VCH Verlagges.m.b.H., Weinheim (1995).
1.11. P. Hohenberg and W. Kohn, *Phys. Rev. B* **136** (1964), 864.
1.12. D. Pettifor, *Bonding and Structure of Molecules and Solids*, Clarendon Press, Oxford (1995).
1.13. W. A. Harrison, *Electronic Structure and the Properties of Solids*, W.H. Freeman and Company, San Francisco (1980).
1.14. A. Zunger and M. Cohen, *Phys. Rev. B* **19** (1979), 568–583.
1.15. D. M. Bylander and L. Kleinman, *Phys. Rev. B* **27** (1982), 3152–3159.
1.16. D. M. Bylander and L. Kleinman, *Phys. Rev. B* **29** (1984), 1534–1539.
1.17. H. J. F. Jansen and A. J. Freeman, *Phys. Rev. B* **30** (1984), 561–569.
1.18. H. L. Skriver, *Phys. Rev. B* **31** (1985), 1909–1923.
1.19. J. W. Davenport, M. Weinert, and R. E. Watson, *Phys. Rev. B* **32** (1985), 4876–4882.
1.20. S. H. Wei, H. Krakauer, and M. Weinert, *Phys. Rev. B* **32** (1985), 7792–7797.
1.21. L. F. Mattheiss and D. R. Hamann, *Phys. Rev. B* **33** (1986), 823–840.
1.22. C.T. Chan, D. Vanderbilt, and S. G. Louie, *Phys. Rev. B* **33** (1986), 7941–7946.
1.23. M. Körling and J. Häglund, *Phys. Rev. B* **45** (1992), 13293–13297.
1.24. V. Ozolins and M. Körling, *Phys. Rev. B* **48** (1993), 18304–18307.
1.25. A. F. Guillermet, V. Ozolins, G. Grimvall, and M. Körling, *Phys. Rev. B* **51** (1995), 10364–10373.
1.26. G. Grimvall, M. Tiessen, and A. F. Guillermet, *Phys. Rev. B* **36** (1987), 7817–7826.
1.27. A. F. Guillermet and G. Grimvall, *Phys. Rev. B* **44** (1991), 4332–4340.
1.28. P. Söderlind, R. Ahuja, O. Eriksson, B., Johansson, and J. M. Wills, *Phys. Rev. B* **49** (1994), 9365–9371.
1.29. P. Söderlind and B. Johansson, *Thermochim. Acta* **218** (1993), 145–153.
1.30. R. Lowrie and A. M. Gonas, *J. Appl. Phys.* **36** (1965), 4505–4509.
1.31. R. Lowrie and A. M. Gonas, *J. Appl. Phys.* **38** (1967), 2189–2192.
1.32. R. Nyholm, A. Berndtsson, and N. Mårtensson, *J. Phys. C.* **13** (1980), L1091–L1094.
1.33. Z. Hussain, C. S. Fadley, S. Kono, and L. F. Wagner, *Phys. Rev. B* **22** (1980), 3750–3765.
1.34. G. Hägg and N. Schönberg, *Acta Crystallogr.* **7** (1954), 351.
1.35. *Gmelin Handbook of Inorganic Chemistry*, 8 ed., Syst. No. 54, *Tungsten*, Suppl. Vol. A3, Springer-Verlag, Heidelberg (1989).
1.36. A. Bartl, *Doctoral Thesis*, TU Vienna, Austria (1997).
1.37. G. D. Rieck, *Tungsten and Its Compounds*, Pergamon Press, Oxford (1967).
1.38. *Gmelin Handbook of Inorganic Chemistry*, 8th ed., No. 54, *Tungsten*, Suppl. Vol. A4, Springer-Verlag, Heidelberg (1993).
1.39. S. W. H. Yih and C. T. Wang, *Tungsten*, Plenum Press, New York, London (1979).
1.40. *Metals Handbook*, 10th ed., Vol. 2, pp. 1169–1172, ASM International, Metals Park, Ohio (1995).
1.41. J. P. Wittenauer, T. G. Nieh, and J. Wadsworth, *Advanced Materials & Processes* **9** (1992), 28–37.
1.42. E. Pink and R. Eck, "Refractory Metals and their Alloys," in: *Materials Science and Technology*, Vol. 9 (R. W. Cahn, P. Haasen, and E. J. Kramer, eds.), VCH, Weinheim (1992).
1.43. J. P. Morniroli, "Low Temperature Embrittlement of Undoped and Doped Tungsten," in: *The Metallurgy of Doped/Non-sag Tungsten* (E. Pink and L. Barha, eds.), Elsevier, London (1989).
1.44. U. K. Vashi, R. W. Armstrong, and G. E. Zima, *Metall. Trans.* **1** (1970), 1769–1771.
1.45. G. S. Pisarenko, V. A. Borisenko, and Y. A. Kashtalyan, *Sov. Powder Metall. Met. Ceram.* **5** (1962), 371–374.
1.46. H. Mughrabi, "Microstructure and Mechanical Properties," in: *Materials Science and Technology* (R. W. Cahn, P. Hassen, and E. J. Kramer, eds.), pp. 1–17, VCH, Weinheim (1993).
1.47. H. Neuhäuser and C. Schwink, "Solid Solution Strengthening," in: *Materials Science and Technology* (R. W. Cahn, P. Haasen, and E. J. Kramer, eds.), pp. 191–250, VCH, Weinheim (1993).
1.48. H. Schultz, *Z. Metallkd.* **78** (1987), 469–475.
1.49. B. Sestak and A. Seeger, *Z. Metallkd.* **69** (1978), 195–202, 355–363, 425–432.
1.50. A. S. Argon and S. R. Maloof, *Acta Metall.* **14** (1966), 1449–1462.
1.51. J. D. Meakin, *Can. J. Phys.* **45** (1967), 1121–1134.
1.52. B. C. Allen, D. J. Maykuth, and R. I. Jaffee, *J. Inst. Met.* **90** (1961), 120–128.
1.53. O. Boser, *J. Less-Common Met.* **23** (1971), 427–435.
1.54. E. Lugscheider, R. Eck, and P. Ettmayer, *Radex Rundsch.* **1/2** (1983), 52–84.
1.55. G. L. Davis, *Metallurgia* **58** (1958), 177–184.

1.56. *Wolfram*, company brochure, Metallwerk Plansee, Reutte, Austria (1980).
1.57. A. G. Starliper and H. Kenworthy, *Chem. Abstr.* **69**, 21516 (1968).
1.58. A. Kelly, W. R. Tyson, and A. H. Cottrell, *Philos. Mag.* [8] **15** (1967), 567–586.
1.59. N. H. McMillan, *J. Mater. Sci.* **7** (1972), 239–254.
1.60. P. L. Raffo, *J. Less-Common Met.* **17** (1969), 133–149.
1.61. D. M. Moon and R. Stickler, *Philos. Mag.* **24** (1971), 1087–1094.
1.62. D. M. Moon and R. Stickler, *High Temp. High Pressures* **3** (1971), 503–518.
1.63. S. L. Robinson and O. D. Sherby, *Acta Metall.* **17** (1969), 109–125.
1.64. E. Pink, in: *Proceedings of 10th Plansee Seminar* (H. M. Ortner, ed.), Vol. 1, pp. 221–235, Metallwerk Plansee, Reutte, Austria (1981).
1.65. M. F. Ashby, *Acta Metall.* **20** (1972), 887–893.
1.66. E. Pink and I. Gaal, "Mechanical Properties and Deformation of Non-Sag Tungsten Wires," in: *The Metallurgy of Doped/Non-Sag Tungsten* (E. Pink and L. Bartha, eds.), pp. 209–233, Elsevier, London, (1989).
1.67. A. Cezairliyan, *High Temp. Sci.* **4** (1972), 248–252.
1.68. G. Wouch, E. L. Gray, R. T. Frost, and A. E. Lord, *High Temp. Sci.* **10** (1978), 241–259.
1.69. E. Rudy, St. Windisch, and J. R. Hoffman, *AFML-TR-65-2*, Part I, Vol. VI, Air Force Materials Laboratory, Res. And Techn. Div., Dayton, Ohio (1966).
1.70. P. Gustafson, *Int. J. Thermophys.* **6** (1985), 395.
1.71. D. A. Young, *Phase Diagrams of the Elements*, University of California Press, Berkeley, p. 174 (1991).
1.72. H. A. Pastor, "Microstructure-Independent Physical Properties of Tungsten," in: *The Metallurgy of Doped/Non-Sag Tungsten*, (E. Pink and L. Bartha, Eds.), Elsevier, London (1989).
1.73. B. Kühl, "Lichterzeugung," in *Ullmanns Encyclopaedie der technischen Chemie*, 4th ed., **16** (1978), 227.
1.74. D. R. Stull and H. Prophet, *JANAF Thermochemical Tables*, 2nd ed., NSRDS-NBS37, NBS, Washington D.C. (June 1971).
1.75. W. Fucke and U. Seydel, *High Temp. High Pressures* **12** (1980), 419–432.
1.76. J. A. Mullendore, "Tungsten and Tungsten Alloys," in Kirk-Otmer: *Encyclopedia of Chemical Technology*, Vol. 23, 3rd ed., Wiley, New York (1983).
1.77. G. Grimvall, M. Thiessen, and A. F. Guillermet, *Phys. Rev. B* **36** (1987), 7816–7826.
1.78. A. F. Guillermet and G. Grimvall, *Phys. Rev. B* **44** (1991), 4332–4340.
1.79. R. L. Andelin, J. D., Knight, and M. Kahn, *Trans. AIME* **233** (1965), 19–24.
1.80. C. Y. Ho, R. W. Powell, and P. E. Liley, *J. Phys. Chem. Ref. Data* **3**, Suppl. 1 (1974).
1.81. C. Y. Ho, R. W. Powell, and P. E. Liley, *J. Phys. Chem. Ref. Data* **1** (1972), 279–421.
1.82. Y. S. Touloukian, R. W. Powell, C. Y. Ho, and M. C. Nicoladu, in: *Thermophysical Properties of Matter*, Vol. 10, Plenum Press, New York (1973).
1.83. J. G. Hust, *High Temp. High Pressures* **8** (1976), 377–390.
1.84. P. D. Desai, T. K. Chu, H. M. James, and C. Y. Ho, *J. Phys. Chem. Ref. Data* **13** (1984), 1069–1096.
1.85. L. N. Latyev, V. Ya. Chekhovskoi, and E. N. Shestakov, *High Temp. High Pressures* **4** (1972), 679–686.
1.86. N. Aksyutov and A. K. Pavlyukov, *Zhr. Prikl, Spektrosk*, **32** (1980), 813–820.
1.87. N. Aksyutov, *Zhr. Prikl. Spektrosk.* **26** (1977), 914–918.
1.88. P. Hiernaut, R. Beukers, M. Hoch, and M. Matsui, *High Temp. High Pressures* **18** (1986), 627–633.
1.89. L. W. Swanson and L. C. Crouser, *Physl Rev.* **163** (1967), 622–641.
1.90. G. E. R. Schulze, *Metallphysik*, Akademieverlag Berlin GmbH (1967).
1.91. C. Zwikker, *Physica* **5** (1925), 249.
1.92. R. L. Sproull, *Modern Physics: The Quantum Physics of the Atom, Solid and Nucleus*, 2nd ed., pp. 435–444, Wiley, London (1963).
1.93. *Gmelin Handbook of Inorganic Chemistry*, 8th ed., Syst. No. 44, *Thorium*, Vol. A3, pp. 77–84, Springer-Verlag, Berlin (1988).
1.94. *Gmelinls Handbuch der anorganischen Chemie*, Thorium and Isotope, Systemnummer 44, Verlag Chemie GmbH, Weinheim (1955).
1.95. R. N. Wall, D. L. Jacobson, and D. R. Bosch, *Metall. Trans.* **24A** (1993), 951–958.
1.96. J. M. Heras, J. Borrajo, and F. E. Mola, *An. Soc. Cient. Argent.* **191** (1971), 179–190.
1.97. *Metals Handbook*, 9th ed., Vol. 17, p. 235, ASM International, Metals Park, Ohio.
1.98. *Gmelin Handbook of Inorganic Chemistry*, 8th ed. Syst. No. 54, *Tungsten*, Suppl. Vol. A5b, Springer-Verlag, Heidelberg (1993).

1.99. *Gmelin Handbook of Inorganic Chemistry*, 8th ed. Syst. No. 54, *Tungsten*, Suppl. Vol. A6a, Springer-Verlag, Heidelberg (1991).
1.100. *Gmelin Handbook of Inorganic Chemistry*, 8th ed. Syst. No. 54, *Tungsten*, Suppl. Vol. A6b, Springer-Verlag, Heidelberg (1998).
1.101. S. V. Nagender and P. Rama Rao, eds., *Phase Diagrams of Binary Tungsten Alloys*, Indian Institute of Metals, Calcutta (1991).
1.102. see folio 51

2

Tungsten History

From Genesis to the 20th Century Products

2.1. THE FORMATION OF TUNGSTEN ATOMS [2.1–2.5]

Articles concerned with tungsten history usually start by describing the discovery of tungsten ores, compounds, and the element. In reality, however, the history of tungsten began with the formation of the tungsten atoms, a very long time ago.

The explanation—how elements have been formed and are still formed today—will be among the outstanding achievements in the history of natural science of the 20th century [2.1]. Therefore, it seems worthwhile to discuss the formation of heavier elements, like tungsten.

The rough schedule in Table 2.1 should give an idea as to the period of the universal evolution in which the element tungsten present in the solar system was formed. It was within the long period of 10 billion years, between the occurrence of the first stars in our galaxy and the formation of the solar system.

The formation of elements is a long and complicated path of diverse nuclear reactions, which took place during the evolution of massive stars (>8 times the solar mass). The "raw material" is always hydrogen, which was formed in addition to the smaller amounts of deuterium and helium shortly after the Big Bang. During the lifetime

TABLE 2.1. Important Events in the Universe's Evolution (Time Table)

Years after the Big Bang ($\times 10^9$)	Years ago ($\times 10^9$)	Event	
0	15	Big Bang	
0.0003	14.997	Formation of H and He atoms	
0.5	14.5	Formation of galaxies	
1	14	Early stars in Milky Way	Formation of tungsten atoms within this period
10.4	4.6	Formation of solar systems	

of such a star, its central region passes through subsequent cycles of nuclear fusion processes. In the first and longest period, called "hydrogen burning," helium is formed by nuclear fusion of hydrogen. When the "fuel" (hydrogen) is consumed, the core contracts further, the temperature and pressure consequently rise, and the next fusion reaction (helium burning) becomes possible. The foregoing burning process moves outward in a radial direction, forming a shell around the core. The different cycles, following one after the other, are summarized in Table 2.2.

This cycling can be exhibited schematically as follows:

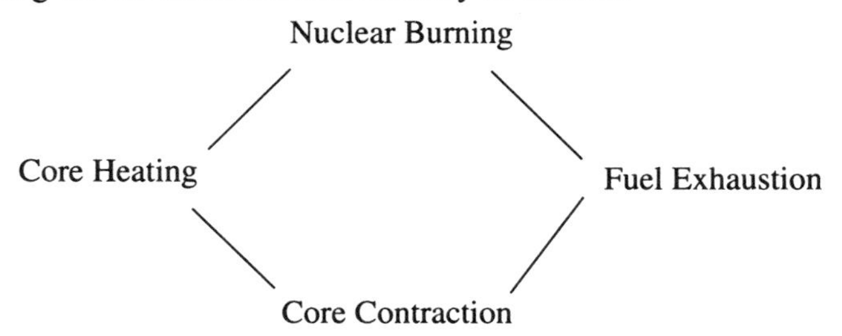

and causes an onion-skin structure of the interior of the star.

At the end of the silicon burning, the core consists of ~1.5 solar masses iron and nickel, while, in the covering shells the foregoing burning processes are still in progress. The life of the star comes to an end with a super nova explosion. Due to the fact that no more fuel is available in the core, it starts to collapse. The temperature will rise over 10^{10} K, and a gas of free nucleons will be formed. The increased pressure finally balances the collapse, and a neutron star of about 1 solar mass is born. Consequently, a shock wave runs through the outer parts of the star combined with high pressure and high temperatures originating in different nuclear reactions. The chemical composition will be changed by this reaction to a certain degree. From the original material of the star, only about 1.5 solar masses remain as a neutron star. The rest is distributed into interstellar space by the blast.

All elements formed within the burning processes by nuclear fusion have even atomic numbers, because in all cases α-particles are involved in these reactions. The nuclei of iron and nickel generated in the last burning process are of optimal composition, which means that the bonding strength of the nucleons is of maximal value. This is why elements having higher atomic numbers than nickel, or cobalt—and this includes tungsten—cannot be formed by nuclear fusion processes, because this would consume energy. All the fusion

TABLE 2.2. Nuclear Fusion Reactions in the Course of the Evolution of a Massive Star

Name	Reaction	log T
H burning	H→He	7
He burning	He→C, O	8.3
C burning	C→Mg, Ne	8.7
N burning	N→O, Mg	8.9
O burning	O→Si, S	9.5
Si burning	Si→Fe, Ni	9.9

reactions described above are exothermic due to a mass defect. Higher elements as well as elements having uneven mass numbers can only be formed by neutron and/or proton absorption reactions.

Neutron absorption processes occur at different times and places in the course of the evolution of massive stars. The *S-process* (slow neutron capture) occurs in the He burning region (state of a red giant). The *R-process* (rapid neutron capture) occurs during the super nova explosion, either within a short distance of the forming neutron star or in the shell where He burning took place prior to the blast when the shock wave hits this area.

S-process. Neutrons freed as a result of the nuclear fusion and decay reactions provide a neutron density of 10^5 cm^{-3}. Higher elements are formed by neutron absorption and subsequent β-decay. Hence S-elements at about the solar abundance are generated. The S-process, depending on the stars mass, lasts between 100 and 100,000 years.

R-process. This occurs during a super nova explosion and lasts only 0.1–10 seconds. Not far from the neutron star at 5×10^9 K, the neutron density is 10^{25} cm^{-3} so enabling the absorption of neutrons by atoms to a high degree. Again higher elements will form by subsequent multiple β-decay. If the pressure wave reaches the shell where He burning took place prior to the blast, a neutron density of 10^{20} cm^{-3} will be generated and the R-process becomes possible.

Proton absorption (P-process). Besides neutron absorption, also protons can be captured; these are generated in a super nova, leading to atoms of higher mass and atomic number. The process is possible in a time interval between 10 and 100 seconds.

It seems likely that during the early life of galaxies, the formation of massive stars and consequently super novae events took place much more often than today. As a result, larger amounts of their "ash" were distributed to interstellar space. During the above-mentioned period of 10 billion years, the ratio of hydrogen to heavier elements in the interstellar medium was continuously decreased, due to star evolutions and explosions.

2.2. HOW TUNGSTEN ATOMS CAME ON EARTH [2.6–2.9]

4.6 billion years ago the solar system was formed by condensation, contraction, and subsequent collapse of an interstellar cloud. More than 99% of the matter of the solar system is concentrated in the sun; the rest is contained in planets, meteorites, and comets. This event stopped the continuous evolutionary change in element concentrations of the system. Besides hydrogen and helium, the cloud contained the elements formed in exploded stars, as described in Section 2.1. One of the heavier elements in this "ash" is tungsten. We do not know to how many star cycles this material, or part of it, was subjected.

The material remaining outside the sun has undergone one or more of the following processes: oxidation, accretion, melting, segregation, and fractional crystallization. Planetesimals were formed out of dust and ice particles; consequently, planets were built up by collisions of planetesimals. Regarding the earth already at a stage of 85% accretion, a separation of the iron–nickel core by segregation was terminated. During the further evolution of the planet, volcanism leads to chemical reactions and differentiations, which are still going on today.

Knowledge about the evolution of the solar system is mainly based on meteorite investigations. With the exception of material brought to earth by spacecrafts from the moon, meteorites represent the only extraterrestrial substances which can be analyzed in laboratories by man. Most meteorites have been formed 4.6 billion years ago like the earth, but have not changed since. Due to the fact that the earth is still an active planet, rocks are continuously changing; older ones are destroyed by melting or corrosion and new ones are formed by igneous processes. The oldest rocks on earth have an age of 4 billion years while the majority are only one billion years old.

Many meteorites have never been subjected to processes of planetary differentiation. These undifferentiated meteorites come from planetesimals that were never molten and resemble the composition of the solar nebula at the time and place of their formation. They exhibit approximately solar system composition, are called chondritic meteorites, and represent one big group of meteorites. The second group consists of the differentiated meteorites which represent pieces of partially or totally molten parent bodies. Examples of this group are meteoritic basalts (eucrites) or iron meteorites. The latter are pieces of the segregated core of the parent body.

Investigations of meteorites in regard to their chemical and mineralogical composition as well as their microstructure not only offer a view into the material composition and properties of the early solar nebula, but also support an understanding of the formation and evolution of the early solar system, as well as the formation and internal creation of the planets and moon and the segregation of the core within planets, etc. (comparison of normalized abundances tell us about depletions, origination, and dictate further the time setting for different events). Tiny fragments within inclusions of carbonaceous chondrites are thought to be of presolar origin and represent the oldest material we have on hand.

Tungsten, in regard to meteorite research, is a very important and interesting element. As a highly refractory element, it condenses under solar nebula conditions in a common alloy of refractory metals (tiny nuggets in chondritic meteorite inclusions). However, the degree of condensation depends on the oxygen concentration and may be incomplete due to the formation of volatile oxides. In the metallic stage tungsten belongs to the siderophile elements, but in the oxidized form it is highly lithophile and incompatible. The relative abundance of tungsten in comparison to refractory elements, which cannot be volatilized by oxidation, allows calculation of the oxygen concentration during condensation. The low abundances of siderophile elements like tungsten in the mantle of the earth are the result of the formation of an iron–nickel core. During the segregation process, most of the metallic elements were removed to the core. The partition coefficient of tungsten between the iron–nickel melt and the silicate phase is strongly influenced by the oxygen concentration. The more oxidizing conditions prevail, the more tungsten remains incompatible in the silicate minerals.

The partly oxidizing conditions during core segregation in the earth evolution are responsible for the fact that we find tungsten deposits in the crust.

2.3. AVERAGE ABUNDANCE [2.2, 2.10, 2.11]

Spectrographic analyses of stars and galaxies as well as of the sun's photosphere reveal that the average abundance of the elements within a certain scatter is comparable throughout the cosmos. As already pointed out, the chondritic meteorites also reflect this general scheme of abundances, but some differences could be found among them originating from the various conditions prevalent during their formation. The biggest anomalies are found in planetesimals, on planets or the moon, where processes like melting, segregation, and differentiated crystallization occurred accompanied by partition, enrichment, or depletion in diverse areas.

The probability of the formation of higher atomic number elements, like tungsten, by absorption of neutrons and/or protons during a star evolution and subsequent super nova blast is quite low. Therefore, the abundances of higher atomic number elements are considerably smaller, as for the elements which were formed by nuclear fusion reactions in an evoluting star.

Figure 2.1 shows the dependence of element abundances on atomic number for the cosmos. In cosmochemistry, such abundances are always presented as a ratio—number of atoms of the respective element/10^6 atoms silicon—in order to eliminate the influence of the strongly varying concentrations of hydrogen and helium. The remarkable drop in abundance from atomic number 47 onward indicates the boundary between elements formed by nuclear fusion and those built up by subsequent neutron absorption reactions. Maxima are always related to especially high nuclear stability of the respective element (a typical example is the iron peak).

Table 2.3 summarizes the tungsten abundances so far known for the cosmos, the solar system, the sun photosphere, different types of meteorites, and the earth crust and mantle. It is interesting to note that the tungsten concentration in the sun photosphere is three times higher than in chondritic meteorites. According to some specialists, it might be a wrong result. By far, the highest tungsten abundances ($\times 7000$) have been found in the aforementioned tiny metallic microinclusions of chondrites.

Accordingly, tungsten can be regarded as a rare element. It is rated 56th of the elements in terms of the earth crust (lithosphere) abundance and ranks point 18 among the metallic elements. Under oxidizing conditions, tungsten forms the hexavalent ion W^{6+}, which is either tetrahedrally or octahedrally coordinated by four resp. six oxygen atoms. In this form, it counts among the lithophile (rock-forming) elements; therefore, we can find it in the rocks of the earth's crust. Under reducing conditions, however, tungsten occurs as a metal and is siderophile. This is the reason why iron–nickel meteorites contain increased concentrations of tungsten. These meteorites belong to the segregated core of their parent bodies. By analogy, and moreover due to the low abundance of tungsten in the earth's mantle, it can be assumed that the earth's core is enriched in tungsten.

2.4. GEOLOGY: FORMATION OF ORE DEPOSITS [2.12–2.15]

The increase in concentration by 3 magnitudes, locally even by 4 magnitudes via igneous processes, is the most important fact in understanding tungsten's geology. As already explained, the average abundance of tungsten is approximately 1 ppm ($\mu g/g$) throughout the solar system, and consequently on earth too. In sharp contrast, one finds concentrations of one-tenth of a percent and sometimes of 1–2% as average contents of tungsten ore deposits. All of them are formed by igneous activities which are best discussed in terms of the magmatic hydrothermal model.

This model (see Fig. 2.2) starts with any magma of changing composition and of different sources. Changes in chemical and physical properties take place. Moving occurs sometimes to higher levels of the earth crust, and consequently temperature as well as pressure gradually drop and differential crystallization starts. The first crystallization step

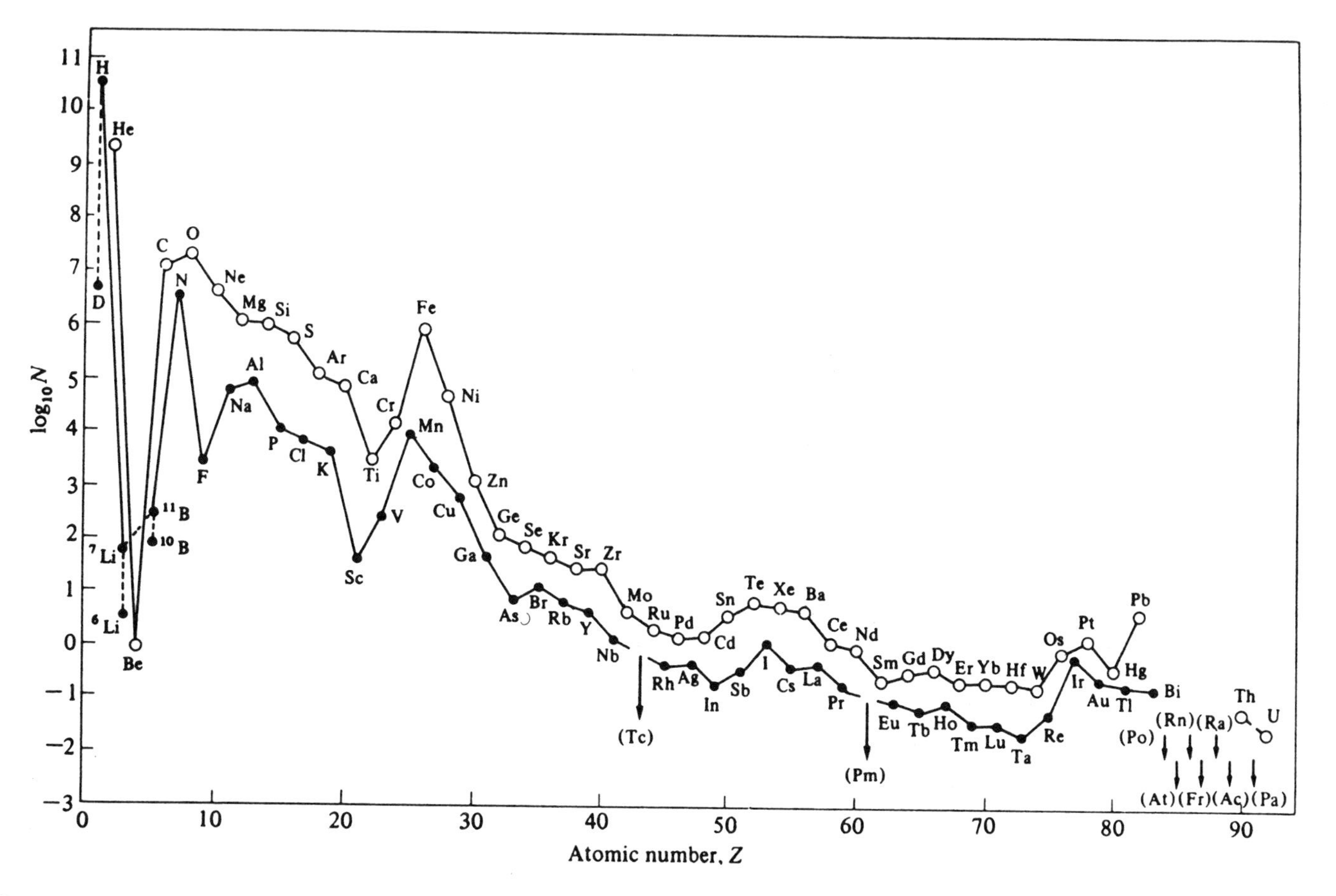

FIGURE 2.1. Cosmic abundance of elements as a function of atomic number Z [2.4]. Abundances are expressed as numbers of atoms per 10^6 atoms of Si and are plotted on a logarithmic scale.

TABLE 2.3. Abundance of Tungsten[a]

	log N_W (log $N_H = 12$)	Atoms W/10^6 atoms Si	μg W/g
Cosmos		0.2–0.3/0.16	
Solar system	0.66	0.127	
Sun photosphere	1.11/0.8	0.36/0.15	
Chondrites	0.68	0.13	
Earth mantle			0.01/0.0164
Earth crust			1.3
Iron meteorites			0.07–5.0
Eucrites			0.024–0.205
Metallic microparticles in chondrites			574
Moon basalt			0.09
Moon highland soil			0.30

[a] Additional literature:
Landolt Börnstein, *Numerical Data and Functional Relationships in Science and Technology*, Group VI, Vol. 2: *Astronomy and Astrophysics*, Subvol. a (K. Schaifers and H. H. Voigt, eds.) Springer-Verlag, Heidelberg (1981).
N. Grevesse and E. Anders, *Solar Element Abundances in Solar Interior and Atmosphere* (A. N. Cox, W. C. Livingston, and M. S. Mathews, eds.), The University of Arizona Press, Tucson (1991).
To convert the Si scale to log *N*, add 1.554 to the log of the Si scale value.

removes large amounts of Fe, Mg, Ca, and Al from the melt in order to form the so called basic rocks (basalt and gabbro). By that, the remaining magma is enriched in silica, alkalis, and other chemical compounds which cannot cocrystallize with the minerals of basic rocks. In other words, they do not fit into their crystal lattices. When crystallization proceeds, the silica content of the minerals formed increases in series, like gabbro, diorite, granodiorite, monzonite, and granite. The last two steps of the model are the pegmatite and the hydrothermal activity. Following the first two main crystallization phases, a fluid remains, consisting of any material which could not be built in during the foregoing steps, such as excessive silica, alkalis, alkaline earth, metal ions, and volatiles. Pegmatite crystallization includes minerals like quartz, feldspars, and micas and may be regarded as the last stripping of the aluminum silicate constituents from the residual magmatic fluid.

The remaining "hydrothermal liquid" is enriched in excessive silica, volatiles, and metals. Metals and volatiles, present in the original magma only in very low concentrations, have now been concentrated as a result of the removal of the main constituents by the prior crystallization phases. Their concentration in the hydrothermal liquid is high enough to form ore minerals in the successive crystallization. Where a crystallizing magma contains tungsten, this metal will be concentrated in the water-rich residual solution (hydrothermal liquid) because of the notable difference in chemical properties between anions of hexavalent tungsten and the major ions of the rock-forming minerals. Hexavalent tungsten has the wrong size and the wrong charge in order to substitute elements like Fe, Mg, Ca, Al, and Si, which are the major rock-forming components. The same is true for the tungstate ion, which cannot substitute silicate or aluminate ions.

It is generally believed that tungsten is born in mineralizing fluids chiefly as tungstate ion, tungstic acid, sodium tungstate, or as heteropoly acid. The relative amount of these compounds depends on temperature, pH, and silica concentration of the solution. Tungsten

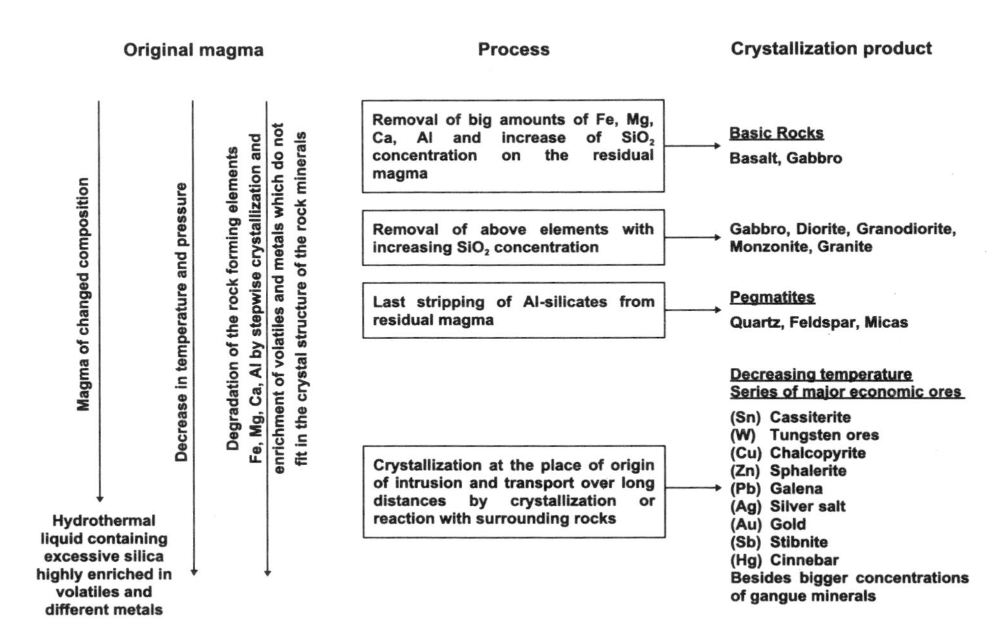

FIGURE 2.2. Magmatic hydrothermal model.

will be deposited as wolframite if the calcium concentration of the environment is low, or as scheelite if it is high (for example, in contact with limestone).

Tungsten ore crystallization can be classified as near-source high-temperature deposits, because wolframite as well as scheelite range on the high-temperature end of the series of minerals crystallizing from hydrothermal liquid. Due to the properties of this liquid, a move to further distant places may also occur.

The physical state of the hydrothermal liquid is a supercritical liquid (around 600 °C and under extremely high pressure). During intrusion of rocks and combined fracturing, pressure decrease leads to boiling. Volatiles and liquid may move over extensive distances from their source. Also, metals may be carried away as volatile compounds. So it is easy to understand that tungsten ores are not only associated with pegmatites, but also due to the above moving processes with other geologic formations. Although the solubility of tungsten compounds is generally low in pure water, it will be enhanced by other constituents of the magmatic fluid to permit the transport of the metal during ore-forming processes and over a wide range of temperatures. A likely form for that transport might be the heteropoly compound, silico-tungstic acid.

Neutralization of a silico-tungstic acid solution in the presence of Ca^{2+}, Fe^{2+}, and Mn^{2+} ions in laboratory experiments yields successive precipitation of Fe, Mn, and Ca tungstate. This order parallels a frequently observed mineral paragenesis, where ferberite, huebnerite, and scheelite are deposited successively.

Tungsten ore deposits have been classified according to their formation, which is closely related to the properties and composition of the residual magma and the hydrothermal liquid as well as to the physical and chemical properties of the surrounding rocks and intrusives. A short summary of the different kinds of tungsten ore deposits is given in Table 2.4.

2.5. MINERALS [2.12, 2.13]

Tungsten minerals can be divided into economically important and minor minerals. The first group, comprising the minerals wolframite, huebnerite, ferberite, and scheelite, occurs in sufficient abundance to guarantee economic mining and processing. The second group is of minor importance due to their small amounts, low concentration, and rare occurrence.

Chemically, nearly all the minerals are tungstates. In the group of the minor minerals, two exceptions exist: (1) tungstenite—which is a sulfide, and (2) tungstite—which is an oxide hydrate.

The Wolframite Group. The wolframite group comprises iron and manganese tungstate minerals of varying composition. The iron tungstate up to a manganese content

TABLE 2.4. Types of Tungsten Deposits

Type of deposit	Tungsten mineral	Formation and typical criteria
Pegmatite	Scheelite Wolframite	Tungsten minerals crystallize together with Pegmatite minerals.
Greisen	Wolframite	Tungsten-bearing hydrothermal fluid stays with the intrusive. Intruded rocks are similar in chemistry to the intrusive rocks. Granite-like rock made up of quartz, mica, topaz, tourmaline, fluorite, cassiterite, wolframite. Typical association of W, Sn, and F.
Contact metasomatic	Scheelite	Hydrothermal liquid invades limestone, causing a violent reaction of the acidic liquid and the limestone. Close proximity with the intrusive. Formation of huge amounts of Ca-silicates (scarn or tactite). Majority of the tungsten deposits.
Vein	Scheelite Wolframite	Intrusion into fractures already present or formed by temperature and pressure, followed by crystallization. Distant from the place of origin.
Pneumatolytic	Scheelite	Transport of volatile tungsten compounds via the gas phase followed by condensation and reaction with the surrounding rocks.
Hot spring		Tungsten compounds in aqueous solution reach the surface. Seldom! Chemism unknown.
Secondary enrichment		Contain tungstite and are very rare. Tungstite is a weathering product of all four W minerals.
Placer	Scheelite	If tungsten deposits are eroded, placers can be formed during the transport of the material by water due to the high gravity and the relative insolubility of the tungsten minerals.

of 20% is called ferberite, the manganese tungstate up to 20% iron is called huebnerite, and the mixed crystals between these limits are named wolframite, containing 20–80% of each, iron and manganese tungstate.

At an elevated temperature, iron and manganese tungstates form a continuous solid solution series. Due to the fact that the temperature during mineral crystallization was sometimes below the limit of miscibility, ferberite and huebnerite can also be found side by side and not as a mixed crystal, as would be the case at correspondingly high temperatures.

The chemical composition and important physical and mineralogical properties are summarized in Table 2.5. Well crystallized mineral specimens are shown in Fig. 2.3.

Scheelite. Chemically, scheelite is calcium tungstate ($CaWO_4$). It is isomorphous to powellite ($CaMoO_4$) with which it forms an incomplete series of mixed crystals. Scheelite contains mostly only minor concentrations of powellite. Its chemical composition as well as physical and mineralogical properties are given in Table 2.6. A large scheelite single crystal is shown in Fig. 2.3.

A very important property of scheelite is its bright bluish-white fluorescence under short-wave UV radiation. The color of the fluorescence radiation is influenced by the mixed crystal formation with powellite. The color changes with increasing concentration of molybdenum from blue to cream, pale yellow, and orange. The fluorescence of scheelite is of great help for prospecting and mining.

Minor Minerals. Chemical as well as physical and mineralogical properties of these minerals are summarized in Table 2.7.

2.6. ORE DEPOSITS AND RESERVES [2.14–2.23]

The map in Fig. 2.4 reveals that known tungsten deposits are scattered over all the continents with the exception of Antarctica. The various types of these deposits can be assigned to three macrotectonic units: Precambrian, Paleozoic orogenic belts, and Mesozoic-Tertiary orogenic belts.

Deposits of the Mesozoic-Tertiary orogenic belts are the most abundant. Eighty-seven percent of the reserves, or 93% of the potential reserves, are located in these areas (orogenic belts of the American Cordillera-Andes, Alpes, Central Asia, Malaysia, and East Asian Islands). These belts can be regarded as the most important areas for tungsten mining and also for further prospects.

The second most abundant deposits can be found in Paleozoic orogenic belts. They represent about 10% of the reserves, or 6% of the potential reserves.

Precambrian deposits are rare. Only 3% of the reserves, or 1% of the potential reserves, belong to them.

Wolframite. About one-third of the knownworld tungsten reserves are hydrothermal wolframite-bearing veins and pegmatites. One may distinguish between three different types of this species:

Plutonic hydrothermal vein deposits, which can be found in China, (Jiangxi; Guangdong; Guangxi and Hunan provinces), Burma, countries of the former Sovient Union, Bolivia, Spain, and Portugal.

Subvolcanic vein deposits,, located in Bolivia, USA, and Japan.

TABLE 2.5. Physical and Chemical Properties of Wolframite Group Minerals

Property	Ferberite	Wolframite	Huebnerite
Formula (pure)	$FeWO_4$	$(Fe, Mn)WO_4$	$MnWO_4$
WO_3 content (%)	76.3	76.5	76.6
Mn content (%)	0–3.6	3.6–14.5	14.5–18.1
Fe content (%)	18.4–14.7	14.7–3.7	3.7–0
Crystal structure	Monoclinic	Monoclinic	Monoclinic
Lattice parameters			
a (Å)	4.71	4.79	4.85
b (Å)	5.70	5.74	5.77
c (Å)	5.94	4.99	4.98
β	$90°$	$90°26'$	$90°53'$
Cleavage	Perfect in one direction	Perfect in one direction	Perfect in one direction
Specific gravity ($g \cdot cm^{-3}$)	7.5	7.1–7.5	7.2–7.3
Color	Black	Dark gray to black	Reddish brown to black
Tenacity	Very brittle	Very brittle	Very brittle
Luster	Submetallic to metallic	Submetallic to metallic	Submetallic adamantine
Fracture	Uneven	Uneven	Uneven
Hardness (mohs)	5	5–5.5	5
Magnetism	Sometimes feebly magnetic	Slightly magnetic	—
Streak	Dark brown	Dark brown	Brownish red to greenish yellow
Diaphaneity	Opaque to translucent in cleavage plates	Opaque	Opaque to translucent
Common form of occurrence	Well-defined crystals, massive crystalline	Irregular masses radiating groups of bladed crystals	Radiating groups of thin-bladed crystals

FIGURE 2.3. Huebnerite, Ferberite, and Scheelite: (a) *Huebnerite*: ruby red, transparent euhedral crystals; length about 6 cm; small white tetragonal scheelite crystals grown on top of them; Pasto Bueno Mine, Ancash, Peru. (b) *Ferberite*: black, plate-like crystals; dimensions 3 × 5 cm; together with pyrite and arsenophrite; Palca XI Mine, Puno, Peru.

FIGURE 2.3. Huebnerite, Ferberite, and Scheelite: (c) *Scheelite*: milky, nearly transparent single crystal; height 5.5 cm; Felbertal, Austria. By courtesy of Dr. Niedermayr, Naturhistorisches Museum Wien.

Pegmaticic-pneumatolytic tin–tungsten deposits, found in Saxonia, Bohemian Erzgebirge, England, Spain, Portugal, Zaire, Australia, and China.

For example, in 1979, half of the world tungsten production came from these three types of deposits.

Scheelite. Approximately two-thirds of the world tungsten reserves consists of scheelite deposits. Distinction can be made between:

TABLE 2.6. Physical and Chemical Properties of Scheelite

Formula (pure)	$CaWO_4$
WO_3 content (%)	80.6
Crystal structure	Tetragonal
Lattice parameters	
a (Å)	5.257
c (Å)	11.373
c/a	2.163
Cleavage	Good in four directions
Specific gravity	5.9–6.1
Color	Pale yellow, brown, commonly white
Tenacity	Very brittle
Luster	Vitreous to resinous
Fracture	Uneven
Hardness (mohs)	4.5–5
Magnetism	Nonmagnetic
Streak	White
Diaphaneity	Transparent to translucent
Common form of occurrence	Massive and in small grains, exist sometimes as pseudomorph after wolframite.

TABLE 2.7. Minor Tungsten Minerals[a,b]

Name	Formula	Crystal structure	Density ($g \cdot cm^{-3}$)		Hardness (mohs)	Streak
Alumotungstite	$AlW_2O_6(OH)_3$	cub	3.39	white		white
Anthoinite	$Al(WO_4)(OH) \cdot H_2O$	tric	4.84	white	1	white
Cerotungstite	$(Ce,Nd)W_2O_6(OH)_3$	mono	6.27	orange yellow	3	
Cuproscheelite	$(Ca,Cu)WO_4$					
Cupritungstite	$CuWO_4$					
Cuprotungstite	$Cu_2WO_4(OH)_2$	tetr	7.06	green, brownish to yellowish		greenish yellow, greenish gray
Farrallonite	$2MgO \cdot W_2O_5 \cdot SiO_2 \cdot nH_2O$					
Ferritungstite	$Ca_2Fe_2^{2+}Fe_2^{3+}(WO_4)_7 \cdot 9H_2O$	tetr	5.2	pale yellow to brownish yellow		
Hydrotungstite	$H_2WO_4 \cdot H_2O$	mono	4.5	pleochroitic	2	
Jixianite	$Pb_9Fe_4(WO_4)_{11}(OH)_8 \cdot 6H_2O$	cub			1	
Kiddcreekite	$Cu_6SnW(S,Se)_8$	cub	4.87			
Mpororoite	$(Al, Fe)[OH\|WO_4] \cdot H_2O$	mono	4.59	greenish yellow		
Meymacite	$WO_3 \cdot 2H_2O$	amorph	4.02	yellow brown		
Phyllotungstite	$H(Ca,Pb)Fe_3(WO_4)_6 \cdot 10H_2O$	ortho	5.26	yellow	2	yellowish
Qitanglinite	$(Fe,Mn)_2(Nb,Ta,Ti)_2WO_{10}$	ortho	6.42	black	5.25	dark brown
Rankachite	$CaFeW_8V_4O_{36} \cdot 12H_2O$	ortho	4.5	dark brown, brownish yellow	2.5	brown
Raspite	$PbWO_4$	mono	8.46	yellowish brown light yellow, gray	2.5–3.0	
Russelite	Bi_2WO_6	tetra	7.35	pale yellow to green	3.5	
Sammartinite	$(Zn,Fe,Ca,Mn)WO_4$	mono	6.70	dark brown to brownish black		reddish brown
Scheteligite	$(Ca,Y,Sb,Mn)_2Ti_2(Ta,Nb,W)O_6(OH)$	cub	4.74	black	5.5	pale yellow to grayish
Stolzite	$PbWO_4$	tetra	8.12	reddish brown, yellow, green, red	2.75	
Thorotungstite	$2W_2O_3 \cdot H_2O+(ThO_2,Ce_2O_3,ZrO_2)+H_2$	ortho				
Tungstenite 2H	WS_2	hexa	7.75	dark lead gray	2.5	lead gray
Tungstenite 3R	WS_2	trig	7.73			
Tungstite	$WO_2 \cdot H_2O$	ortho	5.5	bright yellow	2.5	greenish yellow
Uranotungstite	$(Fe,Ba,Pb)(UO_2)_2(OH)_4(WO_4) \cdot 12H_2O$	ortho	4.27	yellow, orange brownish	2	yellow
Welinite	$Mn_6[(W,Mg,Sb,Fe)O_3][(OH)_2\|O\|(SiO_4)_2]$	trig	4.47	dark red brown reddish black	4	
Wolframoixiolite	$(Fe,Mn)W(Ta,Nb)_2O_8$	mono	6.55	black	5	
Yttrocrrassite	$(Y,Th,U,Ca)_2(Ti,Fe,W)_4O_{11}$	ortho	4.80	black	5.75	

[a] Additional literature: A. R. Hölzel, *Systematics of Minerals*, Alexander R. Hölzel, Mainz (1989).
[b] Some chemical formulas given are putative.

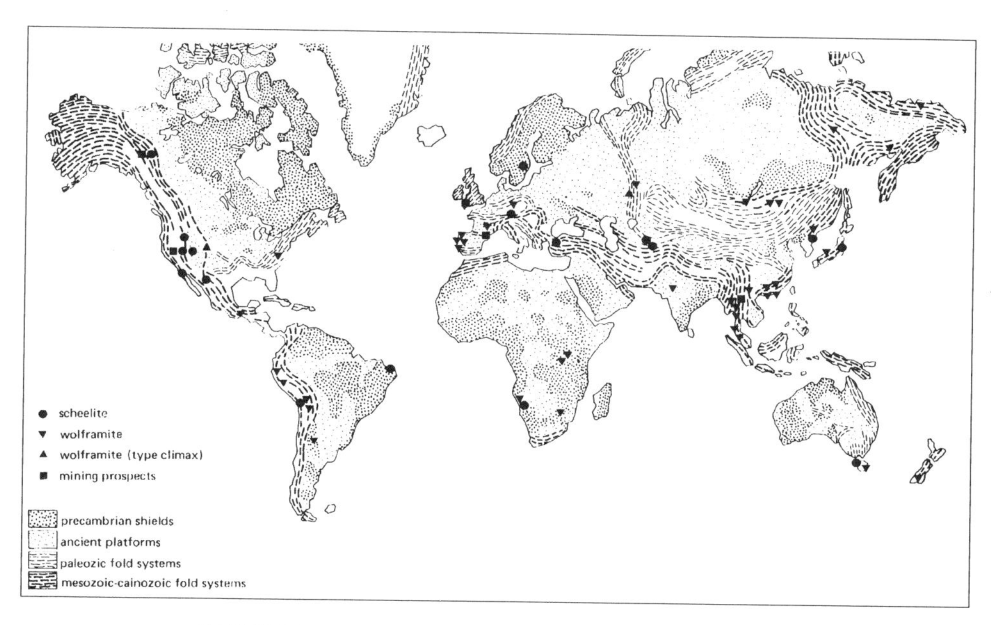

FIGURE 2.4. Principal tungsten deposits and mining disricts [2.19].

contact metamorphic deposits, which can be found in USA, Sweden, Brazil, and the former Soviet Union; and

volcanogenic-sedimentary deposits, located in Austria, South Korea, France, Austria, Australia, and Canada.

In 1979, nearly half of the world production stemmed from scheelite deposits. Around 1980, the ratio of scheelite:wolframite deposits in the western world was 73%:27%.

From time to time, summaries about the tungsten ore reserves are published in the current literature. In principle, it must be taken into account that all those figures are based on estimates. Moreover, they are influenced by the demand of tungsten, the market situation, and consequently by the world market price for tungsten.

Generally, figures for the tungsten reserves are always correlated with production. The higher the demand for tungsten, the bigger the ore reserves are. The reason for this dependence is that high demand generates high prices, and this again intensifies further prospecting activities as well as corrected calculations of the potential reserves. For example, higher prices will bring down the cut-off grade in mining operations. Therefore, figures concerned with ore reserves are only valid for a certain time and a specific situation and will change in the future.

A comparison of the global cumulative production of tungsten, the reserves, and the static lifetime presented in Fig. 2.5 is very informative. From this comparison, it can be seen that the reserves increased from 700.000 t in 1951 to 2.3 million t in 1978, in spite of considerable production growth rates. This increases the static lifetime of the reserves from 38 to 49 years.

Table 2.8 informs about the tungsten ore reserves of countries, continents, and the earth at different times. By far the largest reserves lie in the Peoples Republic of China

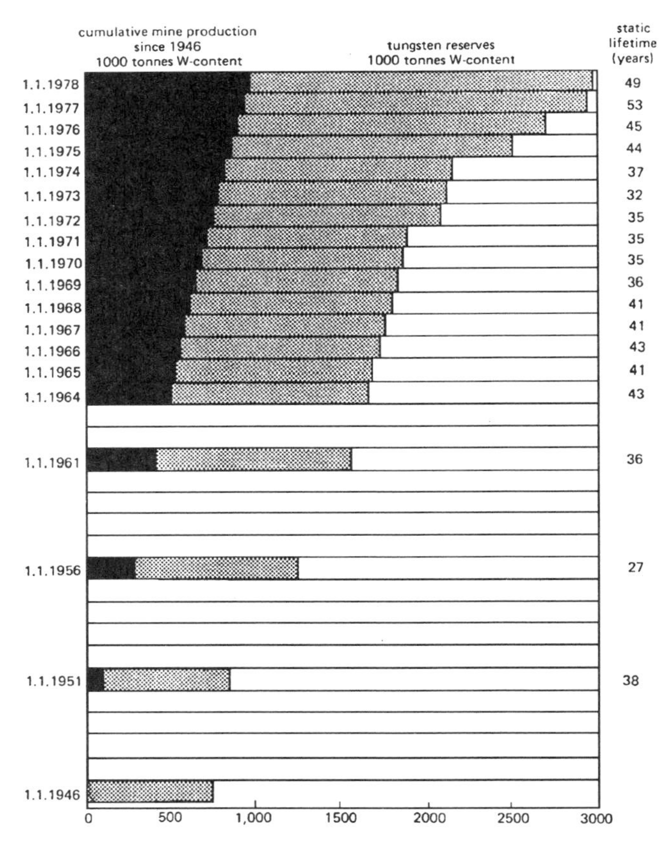

FIGURE 2.5. Cumulative mine production, reserves, and static lifetime of tungsten (1.1.1946–1.1.1978) [2.19].

TABLE 2.8. Reserves of Tungsten Ore at Different Times ([10^3t] tungsten content) [2.23]

	1995		1985		1980	
Country	Reserves	Reserve base	Reserves	Reserve base	Reserve	Reserve base
Australia	5	129	130	140	109	259
Austria	10	15	15	20	18	54
Bolivia	53	105	45	70	40	86
Brazil	20	20	20	20	18	41
Burma	15	34	15	15	32	73
Canada	260	493	480	670	270	318
China	980	1330	1200	1230	1361	2268
France	20	20	20	20	16	2
Korea	58	77	58	60	82	77
Portugal	26	26	40	40	24	27
Russia	250	355	280	490	213	318
Thailand	30	30	30	30	18	18
U.S.A.	140	200	150	290	125	327
Others	347	440	317	405	572	1290
Total	2200	3300	2800	3500	2598	4158

followed by Canada. This area in southern China is a special section of the circum pacific metallogenic belt.

The average tungsten concentration in ore deposits is generally low, which means mostly between 0.4 and 0.8% WO_3 and only seldom between 1 and 2%. An exception is the Cantung Mine in the North West Territories of Canada with about 2.5%.

We refrain from any description of mines in regard to geology, mining, lifetime, etc. This was already done by Yih and Wang [2.12] for some major mines. However, our steadily changing economy has a marked impact on mining operations and their feasibility. A typical situation exists since 1988. All the larger mines of the western world had to close due to extremely low tungsten prices. So, "mothballing" and reopening of tungsten mines occur relatively often.

2.7. EARLY DISCOVERIES OF ORES, COMPOUNDS, AND OF THE ELEMENT [2.24–2.31]

This period can be regarded as 300 years of scattered laboratory experiments in European countries (mainly Sweden and Germany), which elucidated the composition of the ores scheelite, wolframite, the constitution of several tungsten compounds, and a preparation method for the element itself.

Tungsten minerals were already known during the Middle Ages in the tin mines of the Saxony-Bohemian area and in Cornwall, long before the discovery of tungsten itself, because of their negative influence during the tin production. Gregorius Agricola named it "spuma lupi" (latin), "Wolfsschaum" (german), and "Wolfs Foam" (English). In tin melting, the presence of wolframite minerals causes a foam which retards a certain amount of tin, thus decreasing the yield [2.27, 2.30].

Because of their high density, tungsten minerals in Cornwall were called "Call" or "Mock Lead".

Until the end of the 18th century, tungsten minerals were generally named "Wolfram". The name Wolfram can be deduced from the German words "Wolf" (English wolf) and "Rahm" = Geifer (English spittel). The underlying idea is that the wolf eats the tin. Manifold variations of this name can be found in the old literature, such as Wolform, Wolfrumb, Wolfferam, Wolferam, Wolffram, Wolfert, Wolfort, Wolfrig, Wolferan, Wolfish, Woolfram, Wolframit, and Wolframicum. The word Wolfram has remained and today it is the official designation of tungsten in the German language.

1564. The German teacher and priest Johann Mathesius was the first to mention the mineral wolframite in the literature. In his *Sarepa* [2.29], sixteen prayers for miners are summarized. Wolfram is mentioned as *wolform* (1555, third prayer) and later as wolfrumb (1559, ninth prayer); see Fig. 2.6.

1574. The words wolfferam and wolffram were used a few years later by Lazarus Ercker, who assumed the mineral to be an arsenic and iron containing tin ore because of its association with tin.

1757. The mineral Scheelite was first mentioned by A. F. Cronstedt. Also, in regard to the high density ($6 g \cdot cm^{-3}$), he named it *tungsten*, derived from the Swedish tung sten, which means "heavy stone" (*tung stone* in English or *tungstene* in French). At that time, it was regarded as a calcium-containing iron ore.

1781. C. W. Scheele analyzed *tungsten* (scheelite ore) to be a calcium salt of an

Die III. Predig/ von Vrsprung

So behalt nun heut jr jungen bergkleut/das metal dreyerley ding begreiffet/ein Minerische oder bergart von allerley handfarben/die kein euglein metal inn sich helt. Wir nennen solche taube oder lehre arten inn vnserm silber bergkwerck/ glantz oder marchasit/kiß/ cobalt/speyse/ greuß/gilbe/letten/eysenschuß/bleyschweif/quartz oder quatertz/ spate.

Zu Freyberg hat man roten vnd weissen zinck/Inn der Rauriß vnd andern orten/schwillen oder obergel/welches Plinij Sil ist. Inn Hungern bricht ein gelff/stein gallen/stein marck. Auff zihn bergtwerck hat es auch mancherley art/von wolforn/ mißpieckel/farbstein/glaß-kopff/ blutstein.

Die IX. Predig/ Vom zin/ bley/

Wolfrumb/welches die Lateiner Wolffsschaum/etliche wolffshar heyssen/darumb das es schwartz vnd lenglicht ist/bricht auch neben dem zwitter/wie glantz neben dem silber ertz/welcher lind vnnd mild ist/der ist schedig vnnd flüchtig im wasser/den groben vnnd spissigen muß das fewer wegreumen/damit die zin nicht vnscheinlich werden.

Es bricht auch kobelt in zwitter (wie sich im Buchholtz silber vnd zwittergenge mit einander schleppen) der macht die zin auch hart vnd weyßflecket / Spießglaß oder wie andere fürgeben / der gifftige kiß/so beim zwitter bricht/sihet schier dem Wolfrumb ehnlich/ist sehr ein schedlich vnnd giftig metal/ welches durchs fewer vberweltiget/vnnd vom zwitter abgebrent wird/der rauch vnd stanck daruon/verderbet laub/ graß/hoppen vnnd getreyde/vnnd das wasser so von den lauterträgen vnd henden fellet/ist sehr vergifftet/wie viel leut vnnd viehe daruon zu Erbarfdorff gestorben sein.

FIGURE 2.6. Copied sections of the manuscript of J. Mathesius, published in 1564 [2.29]. The third and ninth prayer, mentioning wolform, wolfrumb, wolffshar, wolffsschaum. By courtesy of Dr. V. Dufek, Prague; original available at the Charles University in Prague.

unknown acid. He digested the ore with potassium carbonate and isolated the acid by subsequent precipitation with nitric acid. Moreover, he described the digestion of scheelite by hydrochloric acid, the formation of tungstic acid, and their dissolution in ammonia. In principle, he applied the same chemical reactions to be used in industrial scales 200 years later.

T. Bergmann proposed the preparation of tungsten metal by reduction of tungstic acid with powdered charcoal in 1781.

1783. Don Juan Jose de Elhuyar and his brother Fausto prepared tungsten metal as proposed above and named it *Wolfram*. As already indicated, this name remained only in the German and Swedish language while, according to the official IUPAC designation, all others use *Tungsten*, although the symbol is W.

1785. The de Elhuyar brothers describe the formation of a grayish-white, hard, and brittle material, which forms as a reaction between wolfram and pig iron.

1820. A. Breithaupt adopts the word *Wolframite* for the mineral $(Fe,Mn)WO_4$.

1821. K. C. von Leonhard suggested the name *Scheelite* (in German and Swedish, Scheelit) for the mineral $CaWO_4$ (formerly *tungsten*), in this way honoring the famous Swedish chemist.

1841. R. Oxland was the real founder of systematic tungsten chemistry. He gave procedures for the preparation of sodium tungstate, tungsten trioxide, and tungsten metal. He was the first to propose a method for ferrotungsten production forming the basis for modern high speed steels. Although patents have been granted for those procedures (1847, 1857), an industrial application was not granted mainly due to the enormous price.

2.8. TECHNICALLY IMPORTANT DISCOVERIES [2.24, 2.27, 2.31–2.33]

1855–57. The use of tungsten for special steels was patented by F. Köller (Austria), and later by Mushet.

1868. Mushet started to manufacture a self-hardening V–Mn–W steel (containing 5.5% W) in England.

1890. During experiments to produce diamonds in a self-constructed electric arc furnace, Henry Moissan actually found very hard compounds. None of them were diamonds, but one was tungsten carbide. This fluke was to have an unbelievable impact on the further development of tungsten technology on a long-term basis. Tungsten carbide grew within a period of only 30 years to be the biggest tungsten consumer.

End of the 19th Century. This marked the development of the first high speed steels (produced by quenching the steel just below the liquidus point) by Taylor and White.

1900. At the Paris World Fair, the Bethlehem Steel Company presented the first high speed steel (HSS) cutting tools. For several decades after that, steel was the most prominent tungsten consumer.

1903. According to a patent of A. Just and F. Hanamann (at that time assistant professors at the Technical University of Vienna), the first tungsten filaments for incandescent lamps were produced in Hungary. Tungsten filaments were made by forming a paste from fine tungsten powder after addition of sugar solution and gum. The paste was then squirted through diamond dies and the wire loops or coils were then sintered by electric current in hydrogen [2.33]. The elevated carbon content (reaction with carbon of the binder) was removed during sintering by a moistened hydrogen–nitrogen atmosphere.

1907. Commercial production of incandescent bulbs with these "squirted" tungsten filaments gradually replaced Edison's carbon filament bulbs due to their much better light yield and lower energy consumption.

1909. W. D. Coolidge applied for a patent, describing the powder metallurgical production of ductile tungsten wire. The main features of that patent are still valid for today's technology. Moreover, the procedure gave birth to large-scale powder metallurgy.

1911. This year saw the commercial production of light bulbs containing the new ductile tungsten filaments.

1922. Investigations of F. Skaupy and co-workers at the *Osram Studiengesellschaft* with iron, nickel, or cobalt additions to tungsten carbide were the basis for a patent about production of cemented carbides.

1923-1925. The outstanding results of the above research enabled the application of the famous patent of K. Schröter, DRP 420,689 (WC + 10% Fe, Ni, or Co), followed by DRP 434,527 (WC up to 20% Co). Diamond, at that time very expensive, was in use as drawing die for refractory metal wires. *Osram* substituted it successfully by cemented carbides. But the inventors did not at all realize the real importance of the new material. They even missed the deadlines for patent announcements in several countries and the existing patents were sold to the *Friedrich Krupp Aktiengesellschaft.*

Although K. Schröter is named the inventor in the above patents, it was not at all his sole merit, but the logical consequence of the united efforts of the whole *Osram* research group under F. Skaupy [2.32].

It was *Krupp*, a company experienced in steel technology and machining problems, who really recognized the potential of cemented carbides (hardmetals).

1927. Krupp offered the first cemented carbide (WC-6%Co) at the Leipzig tool fair. The brand name was WIDIA, an abbreviation of Wie Diamant (like diamond in German) and is still known all over the world.

In summary, it can be said that within a short period of less than 40 years at the change from the 19th to the 20th century, the foundations were laid for an avalanche-like growth of tungsten technology.

2.9. INDUSTRIAL EVOLUTION [2.34–2.37]

Before and around 1920, tungsten was only important as a steel alloying element and as a filament in incandescent lamps. At that time, tungsten was added to the steel melt as tungsten powder derived from tungstic acid by hydrogen reduction (low carbon, 96–98% W). In principle, this was complicated and expensive. In 1914, ferrotungsten production was started. The prealloy could be prepared directly from ore concentrates by carbo-, silico-carbo-, or metallothermic reduction, offering a much cheaper possibility of adding tungsten to steel. Since 1940, the prealloy "melting base" has also been in use and is produced by reductive melting of tungsten bearing scrap materials.

In 1935, the first tungsten heavy metal alloys were produced as a material group of growing importance that would continue for the next four decades.

By 1940, tungsten and its alloys and compounds already had widespread application as demonstrated by Wah Chang's Tungsten Tree (Fig. 2.7).

Figure 2.8 illustrates the growth in tungsten demand since 1930. Except for temporary fluctuations, a steady growth due to increasing industrialization in the world can be observed. War times were always related to maxima and economical recessions to minima.

Figure 2.9 demonstrates the shares of the four main tungsten consumers. In slightly more than 60 years, the No. 1 of the past (steel) became No. 2, and the cemented carbides now consume more than half of the total tungsten production.

In the early years of industrial tungsten evolution, around 1930, the largest consumer of the metal was the steel industry, and only small percentages were used for lamp

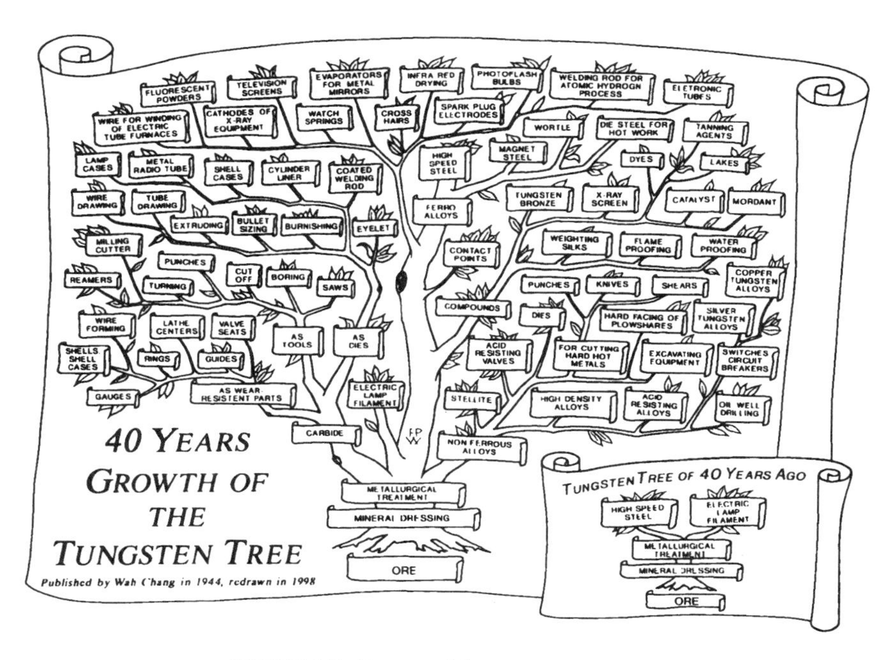

FIGURE 2.7. 40 years of the tungsten tree [2.34].

filaments and the just-born cemented carbides. This marked the starting point of a steady competition between high speed steel and cemented carbides. The share of the latter was steadily growing; thereby, decreasing the percentage of tungsten applied to high speed steel production, a process which still exists today.

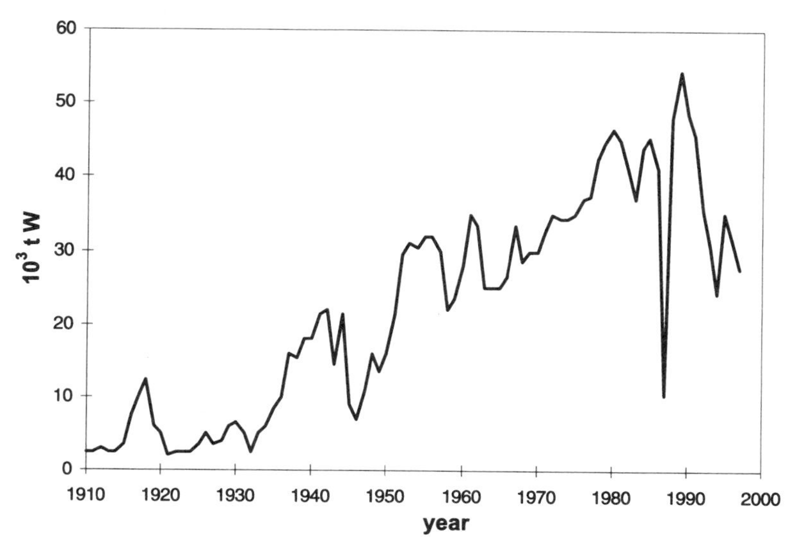

FIGURE 2.8. Tungsten world production 1910–1996. By courtesy of ITIA and A. P. Newey, in: *Proc. 3rd Int. Tungsten Symp. Madrid*, pp. 19–33 (1985).

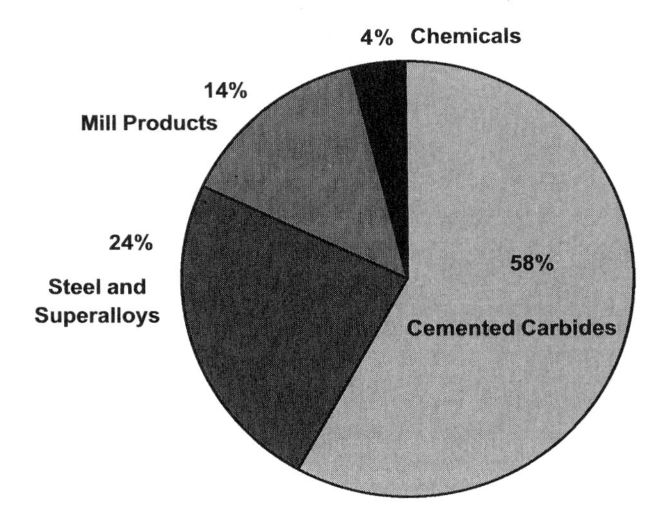

FIGURE 2.9. Tungsten consumption by end use (worldwide).

Since 1930, cemented carbides (also called hardmetals) steadily attained a greater share in tungsten consumption. It is of interest to ask why the demand for cemented carbide grew so rapidly. Table 2.9 shows a chronological table indicating the most important events in cemented carbide research and development, a process which is still under way today. We recognize that what we call cemented carbides or hardmetals are in reality a very wide palette of materials with different properties. Cemented carbide properties can be adjusted by several variations and combinations of the components, as shown in Table 2.10. Hence cemented carbides could be applied widely. Figure 2.10 presents a breakdown of the fields of application of cemented carbides.

The ability to adjust the properties of a tool to the special problem of the material to be worked marks an important advantage over high speed steel, and explains the replacement of the latter in many fields.

TABLE 2.9. Evolution of Cemented Carbides (Chronological Table)

1927	WC-6%Co cutting tool (Krupp)
1929	WC-6%Co fine WC Grain (Krupp)
1930	WC-11%Co (Krupp)
1931	WC-10%TiC-6%Co (Krupp)
	WC-27%TaC-13%Co (Carboloy)
1932	WC-15%Co (Krupp)
	WC-TiC-TaC-CO (Firth Sterling)
1969	Coatings (TiC, TiN, Ti(C,N), Al_2O_3)
1970	Improved micrograin qualities
1973	HIPing
1979	Ni binder
1990	Ultrafine qualities/nanograin powder precursors

TABLE 2.10. Reason for the Expanding Applicability of Cemented Carbide

Widespread property variation and combination through variation and combination of:
—Carbide/binder ratio (influencing hardness and toughness)
—Binder composition (influencing corrosion resistance)
—Carbide phase(s) grain size (influencing hardness, toughness and strength)
—Second carbide phase composition (influencing wear resistance)

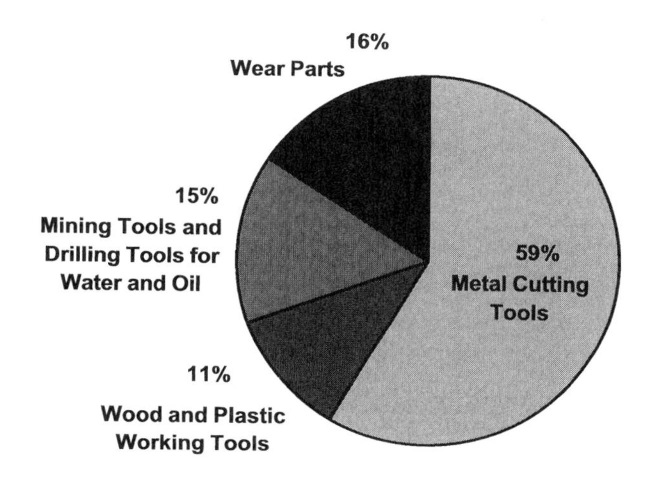

FIGURE 2.10. Field of application of cemented carbides.

Although cemented carbides themselves have been replaced by other materials (ceramics, cermets, diamonds) in special applications, and although the quality improvements and coatings have increased the lifetime of tools, a steady increase in cemented carbide demand is evident.

REFERENCES FOR CHAPTER 2

2.1. R. Kippenhahn and A. Weigert, *Stellar Structure and Evolution*, Springer-Verlag, Berlin, Heidelberg (1990).
2.2. S. Mitton, ed., *Cambridge Enzyklopaedie der Astronomie*, Orbis Verlag, Munchen (1989).
2.3. W. Hillebrandt and W. Ober, *Naturwissenschaften* **69** (1982), 205–211.
2.4. N. N. Greenwood and A. Earnshaw, *The Chemistry of the Elements*, Pergamon Press, Oxford (1984).
2.5. H. Palme, "Cosmetic Chemistry," in: *Reviews in Modern Astronomy* (G. Klare, ed.), Vol. 1, Springer-Verlag, Berlin, Heidelberg (1988).
2.6. E. R. D. Scott, *Earth Planet, Sci. Lett.* **39** (1978), 363–370.
2.7. H. E. Newsom and H. Palme, *Earth Planet. Sci. Lett.* **69** (1984), 354–364.
2.8. H. Palme and F. Wlotzka, *Earth Planet. Sci. Lett.* **33** (1976), 45–60.
2.9. H. Palme and W. Rammensee, *Proc. Lunar Planet. Sci.* **12B** (1981), 949–964.
2.10. H. Palme, "Chemical Abundances in Meteorites," in: *Review of Modern Astronomy* (G. Klare, ed.), Springer-Verlag, Berlin, Heidelberg (1988).
2.11. Landolt Börnstein, *Numerical Data and Functional Relationships in Science and Technology*, Group VI,

Vol. 2, *Astronomy and Astrophysics*, Subvol. a (K. Schaifers and H. M. Voigt, eds.), Springer-Verlag, Heidelberg (1981).

2.12. S. W. H. Yih and C. T. Wang, *Tungsten, Sources, Metallurgy, Properties and Application*, Plenum Press, New York (1979).

2.13. H. J. Rösler, *Lehrbuch der Mineralogie*, VEB Deutscher Verlag für Grundstoffindustrie, Leipzig (1984).

2.14. V. F. Barbanov, *Int. Geol. Rev.* **13** (1971), 332–344.

2.15. K. B. Krauskopf, "Tungsten," in: *Handbook of Geochemistry* (K. H. Wedepohl, ed.), Springer-Verlag, Berlin (1970).

2.16. H. Grundlach and W. Thormann, *Z. Dtsch. Geol. Ges.* **112** (1960) 1–35.

2.17. R. F. Horsnail, "The Geology of Tungsten," in: *Proc. 1st Int. Tungsten Symp., Stockholm*, pp. 18–31, Mining Journal Books Ltd., London (1979).

2.18. A. Smith, in: *Proc. 1st Int. Tungsten Symp., Stockholm*, pp. 32–63, Mining Journal Books Ltd., London (1979).

2.19. F. Bender, in: *Proc. 1st Int. Tungsten Symp., Stockholm*, pp. 2–17, Mining Journal Books Ltd., London (1979).

2.20. Wu Weisun, in: *Proc. 1st Int. Tungsten Symp., Stockholm*, pp. 118–126, Mining Journal Books Ltd., London (1979).

2.21. R. F. Stevens, *Tungsten*, Bull. U.S. Bur. of Mines, Minerals, Facts, Problems p. 1161 (1980).

2.22. R. F. Stevens, *Tungsten*, Bull. U.S. Bur. of Mines, Minerals, Facts, Problems, p. 667 (1975).

2.23. *The Economics of Tungsten*, Roskill Information Services, London (1990).

2.24. E. Lassner, W. D. Schubert, E. Lüdertiz, and H. U. Wolf, "Tungsten, Tungsten Alloys, and Tungsten Compounds," in: *Ullmann's Enclyclopedia of Industrial Chemistry*, Vol. A27, pp. 229–266 (1996).

2.25. S. Engels and R. Stolz, *ABC-Geschichte der Chemie*, VEB Deutscher Verlag für Grundstoffindustrie, Leipzig (1989).

2.26. I. Kappel, *Chem. Ztg.* **50** (1926), 969–971.

2.27. *Gmelins Handbuch der Anorganischen Chemie*, 8. Auflage, Syst. Nr. 54, Verlag Chemie, Berlin (1933).

2.28. V. Dufek, *Lazarus Ercker von Schreckenfels*, manuscript prepared for the Symposium on Science and Technology of Rudolfinian Time, National Technical Museum, Prague (1997).

2.29. J. Mathesius, *Sarepa oder Bergpostill*, Nürnberg (1564).

2.30. J. C. Bailar, Jr., H. J. Emeleus, R. Nyholm, and A. F. Trotman-Dickenson, eds., *Comprehensive Inorganic Chemistry*, Pergamon Press, Oxford (1973).

2.31. H. Pastor, in: *Proc. of the 1996 European Conf. on Advances in Hard Materials Production, Stockholm*, pp. 3–11, EPMA, Shrewsbury (1996).

2.32. H. Kolaska, "The Dawn of the Hard Metal Age," *Powder Metall. Int.* **24** (1992), 311–314.

2.33. J. Gurland and J. D. Knox, "No Child of Chance . . . "—From Sintered Tungsten to Cemented Tungsten to Cemented Tungsten Carbide," in: *Tungsten and Refractory Metals 3* (A. Bose and R. J. Dowding, eds.), pp. 219–227, MPIF, Princeton (1996).

2.34. Wah Changs, "Unique Plant for Concentrating Tungsten," *Eng. Min. J.* (Sept. 1944).

2.35. A. P. Newey, in: *Proc. 3rd Int. Tungsten Symp., Madrid*, pp. 19–33, MPR Publishing Services, Shrewsbury (1985).

2.36. *ITIA Information*; 119–123 Hackford Road, London SW9 0QT, UK.

2.37. P. Borchers and W. Müller, in: *Proc. 5th Int. Tungsten Symp., Budapest*, pp. 40–55, MPR Publishing Services, Shrewsbury (1990).

3

Important Aspects of Tungsten Chemistry

3.1. OXIDATION OF TUNGSTEN METAL BY AIR OR OXYGEN [3.1–3.3]

Tungsten's sensitivity toward oxygen restricts its high-temperature application to protecting atmospheres or vacuum only.

Adsorption. Oxygen is adsorbed on clean tungsten surfaces in a variety of atomic and molecular states. At low temperature ($<0\,°C$), oxygen is adsorbed molecularly, but at room temperature this adsorption is a precursor state to the atomic adsorption. A covered surface shows an ordered oxygen superstructure. If the temperature is increased, a more extensive coverage occurs and oxide-like structures are formed. The surface layer can be described as adsorbed oxide.

The adsorption of oxygen on tungsten surfaces has been investigated most extensively. Relevant data can be found elsewhere [3.1].

Diffusion. Oxygen diffuses from the surface into the crystal lattice forming a bcc (α) solid solution. The solubility is very low. Calculated diffusion coefficients at 1700 °C are given as

$$D = 10^{-7} \text{ and } 3 \times 10^{-8} \text{ to } 5 \times 10^{-8}\ \text{cm}^2 \cdot \text{s}^{-1} \text{ for polycrystalline tungsten [3.4, 3.5]}$$

and

$$D = 7 \times 10^{-8} \text{ to } 5 \times 10^{-8}\ \text{cm}^2 \cdot \text{s}^{-1} \text{ for monocrystalline tungsten [3.4]}$$

Permeation. For the diffusion of oxygen through tungsten in the temperature range 1950–2300 °C, the following relationship was found:

$$P = \begin{cases} 510 \exp(-40{,}000/RT) \text{ [3.5], resp. } 380 \exp(-38,100/RT) \text{ [3.6] for P/M tungsten} \\ 850 \exp(-44{,}000/RT) \text{ [3.5] for arc-cast tungsten} \end{cases}$$

where P = permeation coefficient in ($\text{mm}^3\ O_2 \cdot \text{cm}^{-2} \cdot \text{min}^{-1} \cdot \text{atm}^{-1/2}$ and R = gas constant in $\text{cal} \cdot \text{mol}^{-1} \cdot \text{K}^{-1}$ with T in K.

Oxidation [3.3]. The oxidation of tungsten depends strongly on temperature and on the oxygen partial pressure at elevated temperatures. A clean-looking tungsten surface between room temperature and 370 °C contains oxide.

It was shown by ESCA investigations that the oxidation of tungsten by oxygen or air starts at room temperature. The oxide formed in oxygen as well as in dry or humid air is always tungsten trioxide. The thickness of the oxide layer increases slowly with increasing temperature up to 200 °C, but increases rapidly above 200 °C. Increase in humidity also accelerates the oxidation rate.

From 327–400 °C, a firmly adherent blue and thin oxide film will be formed. It acts as a protective layer. Its formation rate is determined by diffusion and proceeds according to a parabolic rate law. The film composition is given as $WO_{2.75}$, but this is in contradiction to above ESCA findings. The color is not real but is caused by interference.

Above 500 °C the oxide layer cracks, and above 600 °C, WO_3 forms on the blue layer. The WO_3 is permeable to oxygen. The WO_3 formation rate is determined by the tungsten ion transport to the phase boundary $WO_{2.75}/WO_3$. As long as the inner layer has not yet attained its maximum thickness, the growth follows a mixed parabolic linear rate law. When the maximum thickness is reached, it is strictly linear.

Sublimation of tungsten trioxide starts at 750 °C and becomes substantial above 900 °C. Above 1300 °C, the WO_3 sublimation rate corresponds to the oxidation rate, and the surface remains free of oxide.

At elevated temperature, the oxygen pressure strongly influences the reaction rate. The tungsten consumed by oxidation in the temperature range between 700 °C and 1300 °C and oxygen pressures from 0.0013 to 20.8 atm can be calculated as follows [3.7]:

$$-\frac{d(m/A)}{dt} = (5.89 \times 10^6)\exp\left(\frac{12,170}{T}P^{1/2}\right) \qquad \text{resp.}$$

$$-\frac{dx}{dt} = (3.05 \times 10^2)\exp\left(-\frac{12,170}{T}P^{1/2}\right)$$

where m/A is the weight change per unit area, dx/dt the depth of oxidation per unit time, P the pressure in atmospheres, and T the absolute temperature. Units in $mg \cdot cm^{-2} \cdot h^{-1}$ resp. $cm \cdot h^{-1}$

At high temperature (>2100 K) and low oxygen partial pressure, gaseous WO_2 is formed more rapidly than WO_3. Thermal dissociation slows down the oxidation and, above 2500 K, volatilization of the metal occurs.

Oxidation of Tungsten Metal Powder. Tungsten powder, of average grain size >1 μm, reacts like bulk tungsten. Finer powder qualities, depending on grain size and preparation method, can be pyrophoric.

Tungsten powder that is reduced by hydrogen at low temperature from tungstic acid or tungstic-acid-derived WO_3 contains large concentrations of β-W and burns immediately when in contact with air. If such a powder is saturated by an inert gas containing only a low concentration of oxygen, it is no longer pyrophoric. It is assumed that the high density of lattice defects is responsible for the pyrophoricity.

3.2. REACTION OF TUNGSTEN WITH WATER [3.3, 3.8]

This section is closely related to the following one (3.3.) The basic equation for the reactions involved is

$$W + xH_2O \Leftrightarrow WO_x + xH_2$$

This section deals with reactions from left to right, while Section 3.3 deals with reactions in the opposite direction. The hydrogen partial pressure in Section 3.2 is normally low, because hydrogen is a reaction product. In Section 3.3 hydrogen is the reducing agent, which is usually applied in excess, and accordingly the partial pressure is higher.

Water reacts with tungsten below the boiling point. In principle, water acts as an oxidizing agent. According to ESCA investigations WO_2, WO_3, and oxide hydrates are formed [3.3]:

$$W + 2H_2O \rightarrow WO_2 + 2H_2$$

$$W + 3H_2O \rightarrow WO_3 + 3H_2$$

$$WO_3 + xH_2O \rightarrow WO_3 \cdot xH_2O \qquad (x = 1\text{–}2)$$

The reaction with distilled water is very slow at low temperature: the corrosion rate is $3.8\,\mu g\,W \cdot m^{-2} \cdot h^{-1}$ at 38 °C. The reaction is accompanied by the formation of a nonprotective oxide film, which is in equilibrium with 10^{-3}mol/l WO_4^{2-} in solution. At elevated temperature and pressure (150–360 °C and 70–80 atm), the corrosion rate increases to $1.6\,g\ W \cdot m^{-2} \cdot h^{-1}$.

Reaction with water vapor or humid air (60–95% relative humidity) in the temperature range 20–500 °C results in the formation of a WO_3 layer. No lower oxides or hydrates could be found by ESCA. The thickness of the oxide layer increases with increasing humidity. The reaction rate depends on temperature and $[H_2O]/[H_2]$ partial pressure ratio and is more rapid than with liquid water. Water molecules are adsorbed at the tungsten surface and dissociate. The oxygen atoms diffuse into the tungsten metal, forming at first a solid solution and then the oxide compound, while, hydrogen escapes as element.

Depending on temperature and the p_{H_2O}/p_{H_2} ratio, all known oxides can be formed. As long as both parameters are low, the reaction product is WO_2. An increase in both parameters yields in the formation of the higher oxides $WO_{2.72}$, $WO_{2.9}$ and WO_3. Loss by evaporation shows a linear increase with p_{H_2O} between 1050 and 1275 °C. Changes of volatility occur at the equilibrium pressures of the instability points of the oxides formed. For example, the point of coexistence of $W/W_3O/WO_2$ corresponds to 1150 K and a log p_{H_2O}/p_{H_2} ratio of 0.13.

WO_3 evaporation at 1480–1950 K occurs at [3.9]:

$$p_{H_2O} = 6.3\text{–}20 \times 10^{-4}\ \text{Torr}$$

An argon stream saturated with water vapor (21 °C) flowing over tungsten at 1470 K produces red-violet $W_{18}O_{49}$ at the cooler parts of the furnace. At higher humidity, blue $WO_{2.9}$, and yellow WO_3 will be formed.

At temperatures >600 °C, gaseous $WO_3 \cdot H_2O = WO_2(OH)_2$ is the most volatile compound formed in the system W–O–H. It is responsible for all vapor-phase transport processes at temperatures where volatilization of oxides can be neglected:

$$W + 4H_2O \Leftrightarrow WO_2(OH)_2 + 3H_2$$

The logarithm (log) of the equilibrium constant

$$K = (p_{WO_3 \cdot H_2O}) \cdot (p_{H_2})^3/(p_{H_2O})^4$$

plotted versus $1/T(\mathrm{K})$ is linear and given by

$$\log K = \frac{T - 5840\ (\pm 370)}{T} + 1216\ (\pm 0.23)$$

The standard free energy change for the reaction between 1200 and 1500 °C is given by: $\Delta F = 26,700 - 5.56 \cdot T(\pm 500\ \mathrm{cal})$ and the standard entropy of $WO_3 \cdot H_2O$ at 1600 °C as $S^0 = 136.7 \pm 1.05$ eu [3.10].

Tungsten sheet or powder reacts with water vapor between 700 and 1200 °C to form WO_2. Above 900 °C, volatilization via the oxide hydrate occurs. In a moist N_2–H_2 atmosphere, after one hour at 1700 °C, the loss in W is:

$$0.03\ \mathrm{mg/cm^2}\ \text{at}\ p_{H_2O}/p_{H_2} = 0.15 \qquad \text{and} \qquad 10\ \mathrm{mg/cm^2}\ \text{at}\ p_{H_2O}/p_{H_2} = 2.3$$

During the reaction of a tungsten filament with water vapor, the following species can be found at temperatures between 1580 and 3200 K: WO, WO_2, WO_3, $(WO_3)_2$, $(WO_3)_3$, $(WO_3)_4$ [3.11]. From 2250 °C upward, oxygen replaces the water molecule because of its dissociation. The trimeric and quadruple species form between 2200 and 3200 °C:

$$3W + 9H_2O \Leftrightarrow 9H_2 + (WO_3)_3$$
$$4W + 12H_2O \Leftrightarrow 12H_2 + (WO_3)_4$$

Knowledge of these reactions is important in many respects. Tungsten is used for high-temperature engineering components, as heater or cathode in high vacuum techniques, and as filaments in incandescent lamps. Water vapor may be responsible for:

- increased burn-off of cathodes,
- corrosion of contacts,
- water cycle in incandescent lamps,
- destruction of tungsten in contact with alumina (water vapor oxidizes W to WO_2, which reacts with Al_2O_3 to form $AlWO_4$).

3.3. REDUCTION OF TUNGSTEN OXIDES BY HYDROGEN [3.12–3.15]

3.3.1. Introduction

Today's only technically important method for tungsten powder production is the hydrogen reduction of tungsten oxides according to the overall equation

$$WO_3 + 3H_2 \rightarrow W + 3H_2O$$

at temperatures between 600–1100 °C in a streaming hydrogen atmosphere. As starting materials besides the trioxide, tungsten blue oxide (WO_{3-x}) or tungstic acid (H_2WO_4) are in use.

The reduction process is not only a chemical conversion from oxide to metal. It is combined with a chemical vapor transport of tungsten which is responsible for the final powder characteristics, and which is a peculiarity of the W–O–H system. By changing the reduction parameters the powder properties like average grain size, grain size distribution, grain shape, agglomeration, etc. can be regulated within a considerable wide range and this makes the process unique in a certain sense.

Besides chemical conversion and chemical vapor transport, the reduction process is a purification step, too. Trace impurities, always present in the oxide, may evaporate. On the other hand, foreign phases can be incorporated during the CVT growth of tungsten, finally leading to inclusions in the tungsten powder particles. This is of special interest in the production of "non-sag" tungsten wire used for incandescent lamps.

The reduction of WO_3 to W as described by the simple equation above is in reality a rather complex process. Most of the scientific background and understanding is relatively new, although the process has been in use on a technical scale for more than 80 years.

In the past an attempt was made to describe the reduction process in terms of temperature, time, diffusion-path lengths, and oxide bulk density, based on theoretical diffusion equations [3.16], but such models are not used in practice due to the complexity of parameter interactions. Industrial production today is still based on extensive empirical knowledge, supported by in-depth understanding of the basic aspects governing the reduction process, accompanied by a feel for the matter.

The following subsections will present the major aspects of the reduction process in order to understand the chemical and physical processes that arise during the reduction sequence and determine the quality of the final metal powder. The industrial procedure including technical details will be presented in Section 5.4.

3.3.2. *Thermodynamic Considerations*

The W–O–H System. Six condensed phases appear in the W–O–H system: WO_3, $WO_{2.9}$ ($W_{20}O_{58}$), $WO_{2.72}$ ($W_{18}O_{49}$), WO_2, α-W, and β-W. The latter is metastable and was earlier described as W_3O. In regard to their properties we refer to Section 4.2.7. Phase relations were studied by several authors [3.17–3.24], and thermodynamic data were evaluated for both the solid [3.18, 3.22] and gaseous oxides (W_4O_{12}, W_3O_9, W_3O_8, W_2O_6, $WO_2(OH)_2$) [3.10, 3.19, 3.25].

Standard Gibbs functions of various reactions of the W–O system are given below for the temperature range of 700 to 1000 °C [3.23]:

$$W(s) + O_2(g) \leftrightarrow WO_2(s) \qquad \Delta G^0 = -574{,}547 + 170.0 \cdot T$$

$$\frac{2}{0.72} \cdot WO_2(s) + O_2(g) \leftrightarrow \frac{2}{0.72} \cdot WO_{2.72}(s) \qquad \Delta G^0 = -499{,}067 + 125.6 \cdot T$$

$$\frac{2}{0.18} \cdot WO_{2.72}(s) + O_2(g) \leftrightarrow \frac{2}{0.18} \cdot WO_{2.9}(s) \qquad \Delta G^0 = -567{,}852 + 202.6 \cdot T$$

$$20 \cdot WO_{2.9}(s) + O_2(g) \leftrightarrow 20 \cdot WO_3(s) \qquad \Delta G^0 = -558{,}689 + 223.9 \cdot T$$

Based on the Gibbs functions, the temperatures for the invariant three-phase equilibria can be calculated:

620 °C [3.23]/560 °C [3.26] for the equilibrium:

$$5 \cdot WO_{2.72}(s) \leftrightarrow 4 \cdot WO_{2.9}(s) + WO_2(s)$$

280 °C [3.23]/260 °C [3.26] for the equilibrium:

$$10 \cdot WO_{2.9}(s) \leftrightarrow 9 \cdot WO_3(s) + WO_2(s)$$

1480 °C [3.26] for the equilibrium:

$$WO_2(s) \leftrightarrow \frac{2}{2.72} \cdot WO_{2.72}(s) + \frac{0.72}{2.72} \cdot W(s)$$

Vapor pressures above WO_3 and WO_2 are listed as a function of temperature elsewhere [3.19, 3.25].

The stability regions of the oxides are determined by the temperature and the p_{H_2O}/p_{H_2} ratio, as demonstrated in Fig. 3.1 [3.17]. Any of the equilibrium oxides can be produced by annealing metallic tungsten or WO_3 under the respective equilibrium-temperature and humidity conditions.

In one of the older equilibrium studies [3.20] an additional oxide was described (W_3O) to occur at low-temperature/humidity conditions. However, this compound was not confirmed in any of the later investigations [3.22–3.24], and it is well accepted today that it (W_3O/β-W) is a metastable transient and not a stable equilibrium phase.

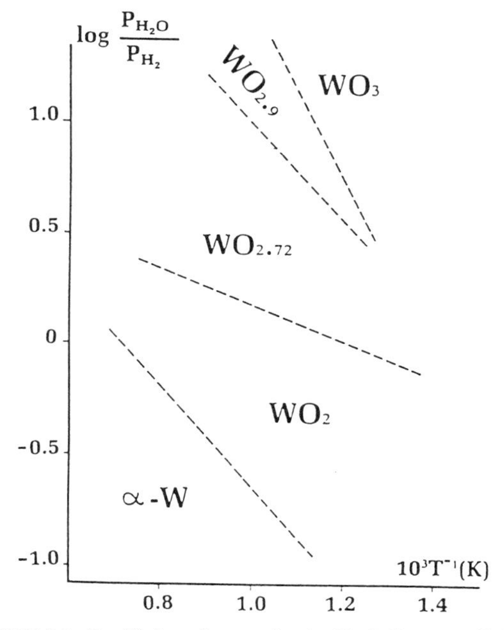

FIGURE 3.1. Equilibrium diagram for the W–O–H system [3.17].

3.3.3. Formation of a Volatile Tungsten Oxide Hydrate [$WO_2(OH)_2$]

Tungsten oxides and tungsten metal exhibit an apparent increase in vapor pressure at elevated temperatures in the presence of water vapor [3.27–3.31]. This increase is due to a reaction between the respective oxide or the metal and the water vapor, to form a volatile tungsten oxide hydrate. The reactions can be described by the following equations:

$$WO_3(s) + H_2O(g) \leftrightarrow WO_2(OH)_2(g)$$

$$\frac{1}{2} \cdot W_{20}O_{58}(s) + 11H_2O(g) \leftrightarrow 10 \cdot WO_2(OH)_2(g) + H_2(g)$$

$$W_{18}O_{49}(s) + 23H_2O(g) \leftrightarrow 18 \cdot WO_2(OH)_2(g) + 5H_2O(g)$$

$$WO_2(s) + 2H_2O(g) \leftrightarrow WO_2(OH)_2(g) + H_2(g)$$

$$W(s) + 4H_2O(g) \leftrightarrow WO_2(OH)_2(g) + 3H_2(g)$$

The presence of the monomere, volatile oxide hydrate [$WO_3 \cdot H_2O$ resp. $WO_2(OH)_2$] was proven by mass spectroscopy [3.28], and thermodynamic data are available for all of the above phase equilibria [3.10, 3.29]. Based on these free energy data as well as on those of the solid oxides [3.23], the equilibrium partial pressure of the volatile compound can be calculated as a function of humidity. The result of such a calculation is shown in Fig. 3.2 for a temperature of 1000 °C [3.32]. In addition, the equilibrium pressures of the other volatile tungsten compounds are also presented. From these calculations it is evident, that the oxide hydrate is by far the most volatile tungsten compound in the W–O–H system.

The oxide hydrate is responsible for the chemical vapor transport of tungsten, which occurs throughout the reduction sequence, and which decisively codetermines the physical and chemical properties of the metal powder. Its actual partial pressure during reduction will depend on both the temperature and the prevailing humidity. It will be lower toward the end of reduction, due to the steady decrease in the oxygen partial pressure (humidity) of the system as the reduction proceeds.

Chemical vapor transport of tungsten also occurs during equilibrium annealing of the oxides with hydrogen–water mixtures (i.e., overall not-reducing conditions), as demonstrated in Fig. 3.3 for the annealing of WO_2 at 800 °C for 1 hour, respectively 6 hours in humid hydrogen (dew point: 72 °C). In such a case a significant particle coarsening occurs via the vapor phase, leading to the evaporation of smaller grains and growth of the larger grains (phenomenologically similar to an Ostwald ripening process).

3.3.4. Kinetic Considerations

Oxide Reduction in a Powder Bed. Industrial reduction of tungsten oxides is carried out either in a static (push-type kilns) or a dynamic (rotary furnace) powder bed. During reduction (accompanied by water formation) this powder bed (by its porosity) exerts considerable diffusion resistance against the removal of water from the layer. The higher the layer, the greater the diffusion resistance and the more slowly the reaction water will be removed. Even increasing the hydrogen flow over the powder layer to very high rates will mainly affect the rate of removal from the uppermost part of the layer and not as much from the interior. The reactions inside the layer remain diffusion-controlled.

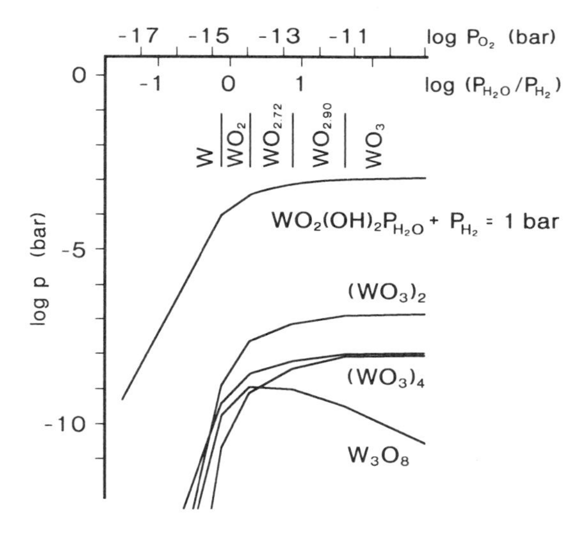

FIGURE 3.2. Equilibrium partial pressure at 100 °C of various tungsten oxide and hydroxide species [3.32]

The influence of the powder layer height and its porosity on the retention of the reaction water is of fundamental importance for the technical process. The diffusion resistance determines a dynamic humidity level within the layer, and hence the nucleation and growth conditions for all phases which form or might form (from a thermodynamic point of view) during reduction. It also determines the reduction path, which can be quite different, depending on the prevailing humidity conditions. Finally, the layer humidity is an important contribution to all tungsten vapor transport reactions which occur during the reduction sequence and which are decisive for the final properties of the tungsten powder, since the total tungsten concentration of the vapor phase (i.e., the sum of volatile tungsten species), and thus the vapor transport capacity, is markedly enhanced by the water vapor (see Fig. 3.2).

The Reduction Sequence. Under industrial conditions (layer heights of several millimeters up to centimeters; temperatures between 700 and 1100 °C) the transport of the water vapor out of the layer is the overall rate-determining step. Any increase in the diffusion resistance, for example by an increase in the layer height, will lead to a significant prolongation in reduction time ("transport controlled reaction"). Under such conditions the humidity in the layer is always high enough to prevent short-cuts in the reduction sequence (for example, $WO_3 \rightarrow W$). The reduction process proceeds in distinct stages, through the formation of all stable oxides:

$$WO_3(\text{yellow}) \rightarrow WO_{2.9}(\text{blue}) \rightarrow WO_{2.72}(\text{violet}) \rightarrow WO_2(\text{brown}) \rightarrow W(\text{gray})$$

An example for this reduction sequence is shown in Fig. 3.4, where the water formation during reduction of pure WO_3 is followed by continuous thermal conductivity measurements of the reaction gas. The reduction water curve can be separated into four segments, according to the stepwise character of the reduction process ($WO_3 \rightarrow WO_{2.9}$, $WO_{2.9} \rightarrow WO_{2.72}$, $WO_{2.72} \rightarrow WO_2$, $WO_2 \rightarrow W$). Furthermore, it is demonstrated that an

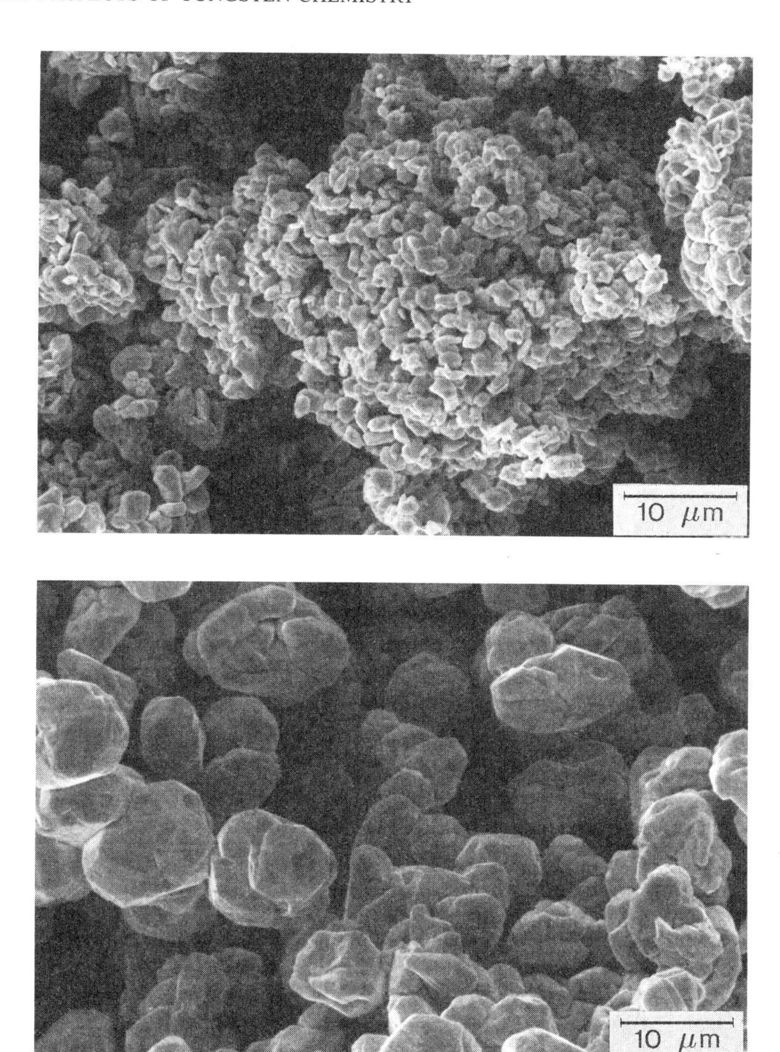

FIGURE 3.3. WO_2 formed at 800 °C and a hydrogen dew point of 72 °C [3.15]: annealing time 1 h (upper image); 6 h (lower image).

increase in the oxide layer height leads to a significant prolongation of the reduction time, as characteristic for transport or mixed chemical and transport-controlled processes [3.14].

Chemical reaction control is observed only at low reduction temperatures, as well as in the case of thin powder layers (< 1 mm) and high hydrogen flow (i.e., optimal material exchange). Under such dry reduction conditions, direct formation of the metal (α-W or β-W) from the higher oxides becomes possible ($WO_{2.9} \rightarrow W$, $WO_{2.72} \rightarrow W$).

At low reduction temperatures (500–750 °C) different transitions can occur simultaneously in the powder layer, indicating a strong influence of local differences in the oxygen

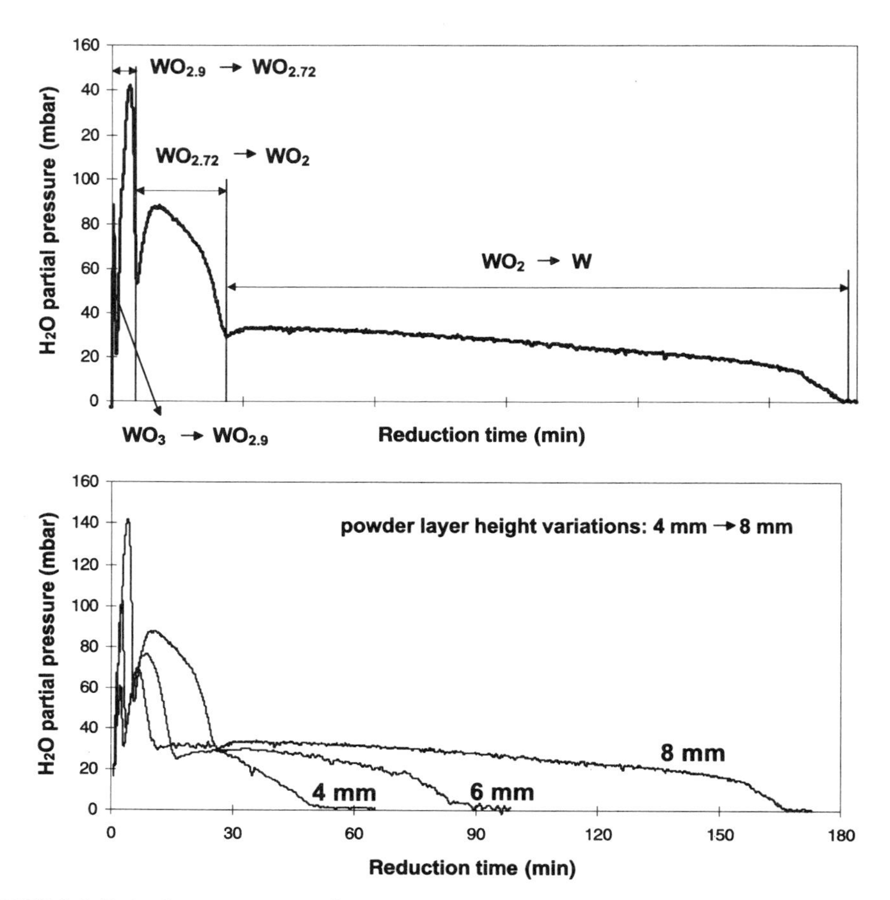

FIGURE 3.4. Reduction water curves of pure WO_3 at 800 °C under near-industrial conditions; influence of powder layer height variations (4 mm–8 mm). Under these "humid" conditions the reduction occurs stepwise ($WO_3 \rightarrow WO_{2.9} \rightarrow WO_{2.72} \rightarrow WO_2 \rightarrow W$) within the powder layer. The dynamic humidity is always high enough to prevent shortcuts in the reduction sequence (e.g., $WO_{2.72} \rightarrow W$).

partial pressure (local humidity) within the layer, or even within a reacting particle, on the nucleation and growth rates of different phases. An example for this reduction sequence is given in Fig. 3.5 (reduction water curve of WO_3 at 650 °C).

An overview of the broad variety of experimentally proven reduction sequences is given in Fig. 3.6. The main reaction paths, as observed under near-industrial reduction conditions, are marked therein by thicker connecting lines. For more information on the individual reactions the reader is referred elsewhere [3.15].

Metastable β-W is always formed out of $WO_{2.9}$ and $WO_{2.72}$ (and from tungsten bronzes), but never out of WO_2, which might indicate that structural relationships between the individual phases are of importance for phase formation, in particular at low reduction temperatures.

Reduction Mechanisms. Based on morphological investigations, two fundamentally different reduction mechanisms can be postulated for the reduction process.

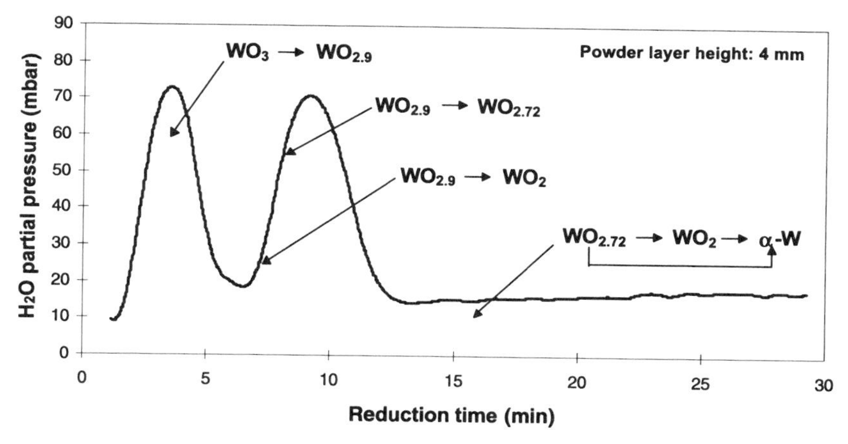

FIGURE 3.5. Reduction water curve of pure WO_3 at 650 °C. Different reactions occur simultaneously within the powder layer or even within an oxide particle, depending on the local oxygen partial pressure.

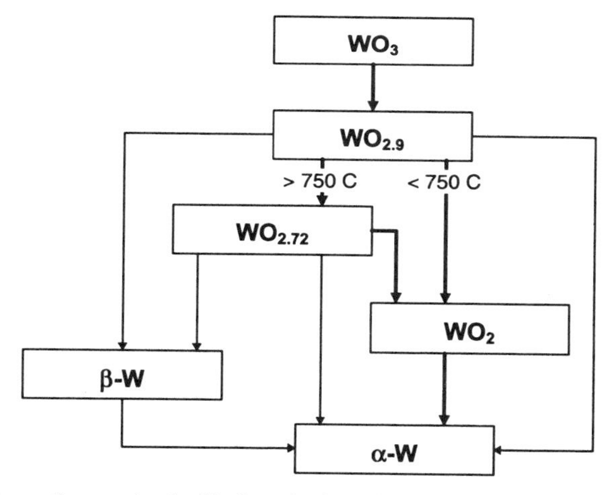

FIGURE 3.6. Reduction paths associated with the reduction of pure WO_3; main transitions are marked with thicker lines [3.15].

(1) *Diffusion in the solid state (oxygen transport)*

The reduction occurs by direct oxygen removal from the solid oxides (solid-state diffusion). The basic underlying mechanism is not known (diffusion of O, OH, H_2O) and is likely to vary for different for different phase transitions. On the final reduction the metal remains pseudomorphous to the starting oxide, forming a polycrystalline metal sponge. Solid-state reactions are characteristic for low reduction temperatures (<750 °C) and the early $WO_3 \rightarrow WO_{2.9}$ transition ("crystallographic shear" transition).

An example for a solid-state transition is the formation of metastable β-W out of $WO_{2.9}$ (Fig. 3.7):

$$WO_{2.9}(s) + 2.9 \cdot H_2(g) \rightarrow \beta W(s) + 2.9 \cdot H_2O(g)$$

FIGURE 3.7. SEM image of β-tungsten formed at 550 °C. The metal is pseudomorphous to the starting oxide (WO_3, APT origin); the pseudomorphs are built-up of nanocrystalline grains.

(2) *Chemical vapor transport (tungsten transport)*
The reduction is combined with a chemical vapor transport of tungsten via the volatile $WO_2(OH)_2$. The morphology of the solid phases changes significantly. The reaction can be separated into three sequential stages:

Stage 1: Formation of the volatile oxide hydrate by a surface reaction of the oxide with water vapor:

$$WO_2(s) + 2H_2O(g) \rightarrow WO_2(OH)_2 + 2H_2 \text{ (oxidation)}$$

Stage 2: Transport of the $WO_2(OH)_2$ from the higher to the lower oxide (resp. to W metal).

Stage 3: Reduction of the volatile $WO_2(OH)_2$ at the surface of the growing oxide resp. tungsten metal:

$$WO_2(OH)_2(g) + 4H_2(g) \rightarrow W(s) + 4H_2O(g)$$

An important example for this mechanism is the formation of small, individual grains of α-W from the WO_2 matrix.

Which of the two mechanisms prevails during the reduction sequence is determined by the interaction among temperature, humidity, and the kinetics of phase formation [3.15]. It is very likely, that solid-state reactions govern all the phase nucleation processes, i.e., also in the case where subsequent growth occurs via chemical vapor transport (CVT).

CVT reactions occur under industrial reduction conditions throughout the reduction sequence, thereby significantly altering the morphology of the intermediate and final reduction products (Fig. 3.8). SEM images of the most important phase transitions are shown in Fig. 3.9. Note that the gap formation between the reacting particles as well as the formation of specific crystal habits for the individual phases ($WO_{2.72}$ needles, WO_2 rosettes, W cubes, icositetrahedrons, etc.) are characteristic for CVT.

During the reduction, the formation of $WO_2(OH)_2$ takes place by the reaction of the oxide with the water which forms during reduction. Above a substance which is stable at the conditions applied (e.g., α-W; see Fig. 3.10), a lower $WO_2(OH)_2$ partial pressure is built up than above a reacting unstable substance (e.g., WO_2). This difference in vapor pressure provides the driving force for a chemical vapor transport of tungsten, which finally leads to a dissolution (oxidation) of the unstable substance through the formation of $WO_2(OH)_2$ out of WO_2 and the growth of the stable substance through the deposition of W by decomposition (reduction) of $WO_2(OH)_2$. This CVT growth is shown schematically in Fig. 3.11 for the $WO_2 \rightarrow W$ transition. It is, however, also valid for the other transitions.

A schematic presentation of the reduction sequence under near-equilibrium conditions is presented in Fig. 3.12, together with the equilibrium vapor pressures of the volatile $WO_2(OH)_2$. Under such conditions the tungsten is transported three times via the vapor phase, as indicated in Fig. 3.13, before it is reduced to the metal. This behavior (which is also valid for humid industrial conditions) is a fascinating peculiarity of the W–O–H reduction system. Only during the early $WO_3 \rightarrow WO_{2.9}$ transition was no CVT observed.

Nucleation and Crystal Growth. When a new phase forms during the reduction process, the first step is phase nucleation. If the structure of the new phase does not match the host matrix, phase nucleation problems can occur (incoherent nucleation). The

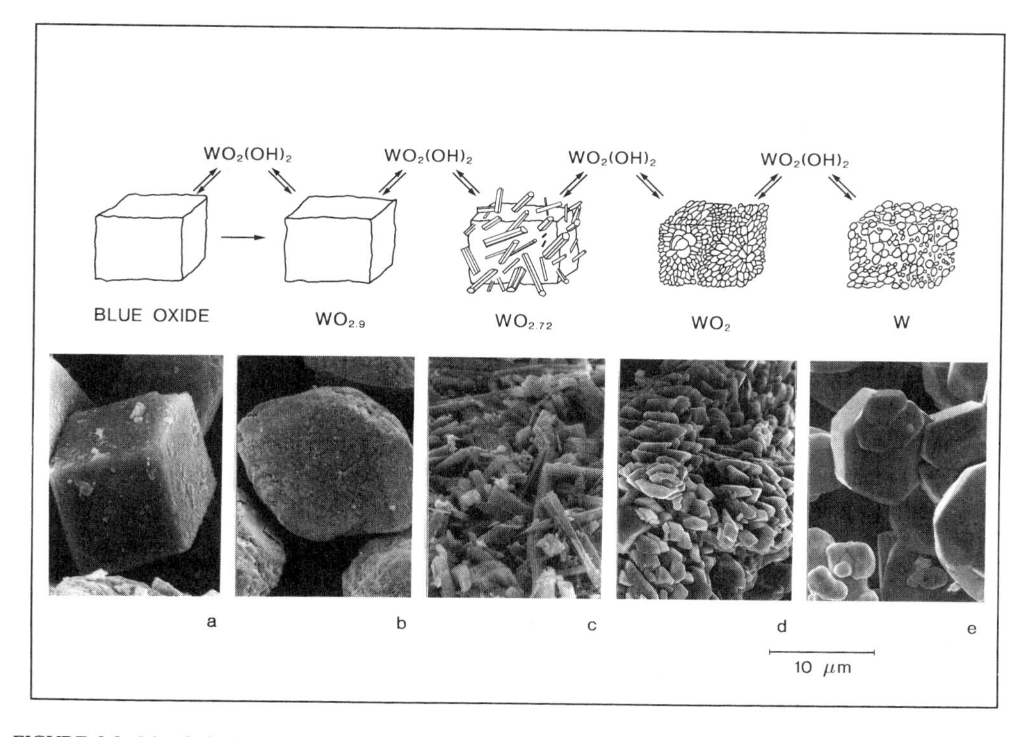

FIGURE 3.8. Morphological changes associated with the hydrogen reduction of tungsten blue oxide under near-industrial conditions (1000 °C); by courtesy of Wolfram Bergbau and Hüttenges.m.b.H.

FIGURE 3.9. Transition $WO_{2.72} \rightarrow WO_2$ (left) and $WO_2 \rightarrow W$ (right). Note the gap which forms between the reacting particles. It is characteristic for the CVT growth of the phases. In case of tungsten, small single crystals are formed.

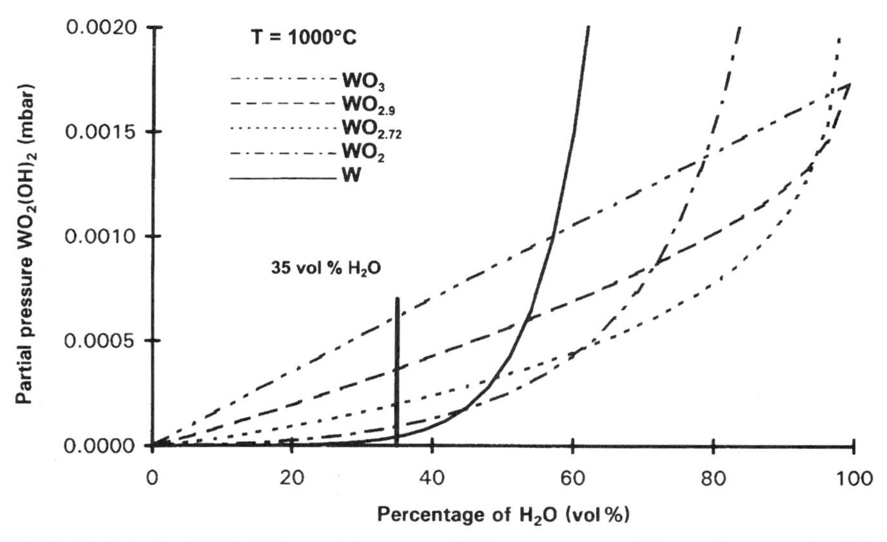

FIGURE 3.10. Calculated $WO_2(OH)_2$ partial pressure of different tungsten oxides at 1000 °C vs. humidity; at a certain humidity (e.g., 35 vol% H_2O) the phase with the lowest $WO_2(OH)_2$ partial pressure is the stable one (e.g., W) [3.38].

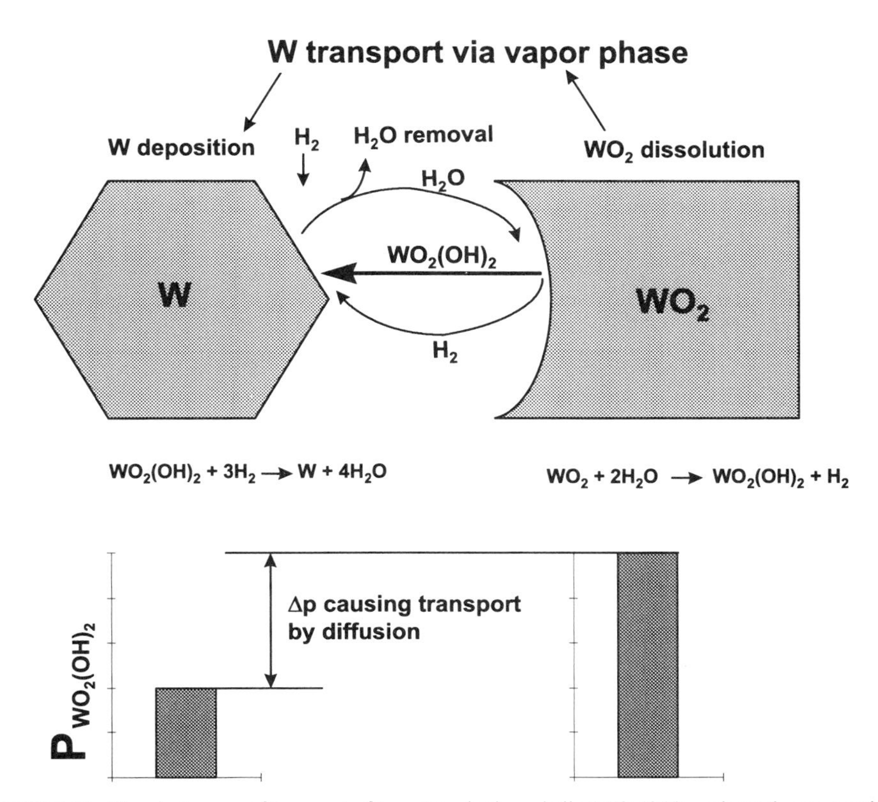

FIGURE 3.11. Chemical vapor of transport of tungsten via the volatile ($WO_2(OH)_2$; schematic presentation.

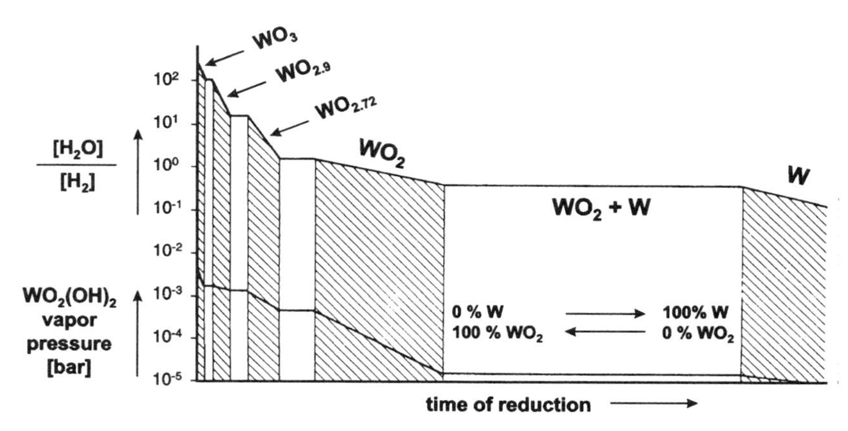

FIGURE 3.12. Changes in the $[H_2O]/[H_2]$ ratio and $WO_2(OH)_2$-vapor pressure during reduction; 3-step chemical vapor transport via $WO_2(OH)_2$; 1000 °C.

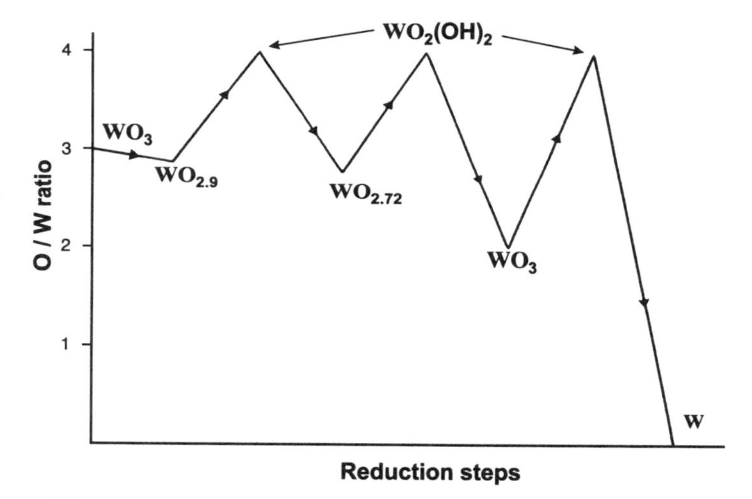

FIGURE 3.13. Changes in O/W ratio during the CVT reduction (schematic).

formation of W from WO_2 is a good example for this aspect. When WO_2 is reduced by wet hydrogen, the reaction comes to a standstill already at comparatively low humidities, although from a thermodynamic point of view the reaction should proceed even at much higher humidities (see Fig. 3.14). Obviously, under these wet conditions no W nucleation takes place. However, tungsten growth by CVT is possible, which can be easily demonstrated by adding nucleation aids, such as metallic tungsten. In this case, a slow further overall reduction occurs, but the W is deposited mainly on already present W particles. Similar reactions take place on the walls of the reduction boats, where W is readily deposited on the wall surface.

The strong influence of the humidity on the nucleation rate of the metal has an important consequence for the technical process. Besides the reduction temperature, it is

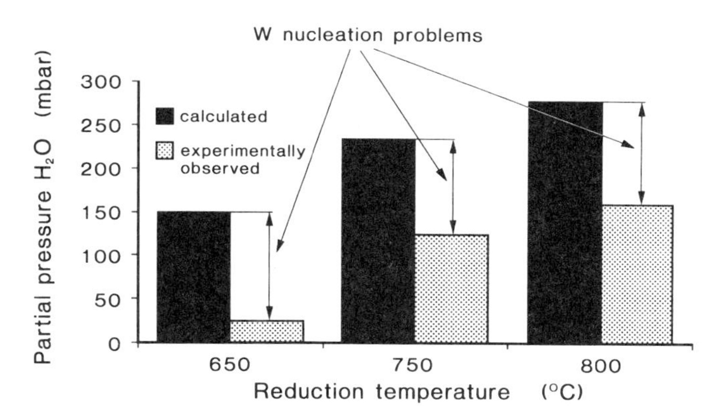

FIGURE 3.14. Humidity limits for the reaction $WO_2 \rightarrow W$, calculated from phase equlibria and experimentally observed [3.38].

the dynamic humidity within the layer which determines the nucleation and growth conditions. The lower the humidity during the $WO_2 \rightarrow W$ transition (i.e., the lower the powder layer and the closer to the top of the layer), the higher the nucleation rate and hence the lower the growth of the individual grains, resulting in smaller average grain sizes. The reverse is true for high humidities, where only few nuclei form, leading to a coarse powder. Due to the varying growth conditions for the individual particles the powder always exhibits a certain grain size distribution. Grain growth of the metal occurs, as described above, by CVT through the formation of $WO_2(OH)_2$.

Mutual Interactions of Oxide Properties and Reduction Conditions. Figure 3.15 [3.13] shows schematically the influence of different reduction parameters on tungsten nucleation and growth. It is divided into three parts. In the first part, the raw material aspects and the more industrial-technical aspects concerning the furnace parameters are considered. The space dimensions range from meters down to centimeters. The second part considers the powder layer itself, where the dimensions range from centimeters downward. Gas permeability of the powder layer and macrogas convection dominate and determine the local events. The third part considers the birth of the individual tungsten grains which happens in the dimensions near 1–100 μm. The microporosity within the oxide agglomerates as well as the local conditions for mass transport (diffusion, microconvection), crystal growth, and nucleation determine the ongoing chemical reactions.

Alternative processes. In the case of a transport-controlled process, the overall reduction rate can be significantly increased if a continuous renewal of the gas in contact with the solid oxide is readily achieved, i.e., when the oxide particles are not in close contact with each other, as, for example, in a fluidized bed, or a laminar flow reactor, where the reaction is virtually instantaneous. Under these conditions, however, the change in the water removal rate will have a strong impact on the powder properties, in particular on the average grain size. It is very unlikely that under such conditions the production of coarse powder (i.e., the formation of small W single crystals of size $\geq$10 μm) can be successful. Thus the inherent ability of the powder bed to retard the water vapor within the layer and to build up locally humid conditions is an important aspect in industrial powder manufacture,

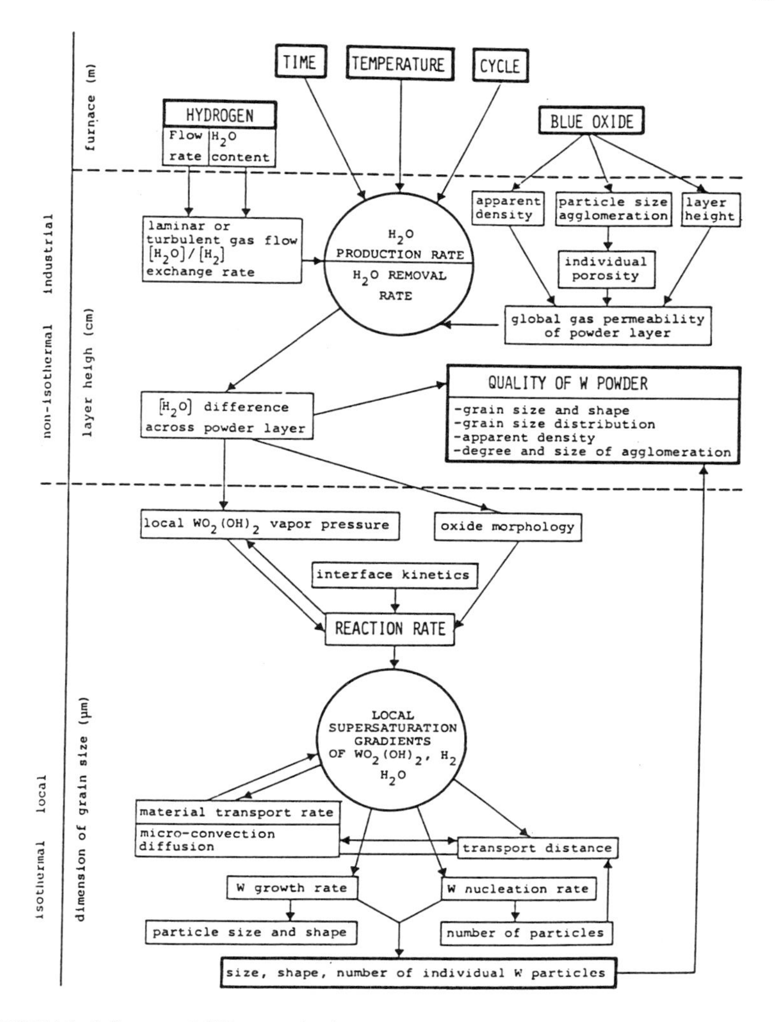

FIGURE 3.15. Influence of different reduction parameters on tungsten nucleation and growth [3.35]

requiring a variety of different powder sizes (ranging from submicron to 60 μm) and specific size distributions. This has to be considered when discussing alternative methods for W powder production.

The importance of the powder layer in the hydrogen reduction of tungsten oxides, although long known, is still today sometimes not altogether understood and therefore its influence is here, once again, briefly summarized.

The capability of retaining water vapor within the layer is to a certain extent the reason why it can be used as a steering parameter for the final tungsten metal powder properties. The water vapor retention capacity of the layer depends on the layer height and its porosity. The layer height can be chosen within limits in connection with the temperature. The porosity depends on the oxide powder properties. The macroporosity is a consequence of the agglomerate size of the oxide powder (which is related to the size of the crystallized APT particles prior to decomposition) and the microporosity is due to the porosity of a single agglomerate (APT pseudomorph). Both parameters combine to yield the overall porosity of the layer. It is noteworthy that due to morphological changes during reduction (formation of different intermediate oxides by CVT) the porosity may also change. The following rules are, however, valid:

- The powder layer height is directly proportional to the water vapor concentration inside the layer, and thus to the average grain size of the W powder (length of diffusion path).
- The layer porosity is inversely proportional to the water vapor concentration and average grain size (diffusion rate).

As regards the influence on other powder properties, the reader is referred to Section 5.4 (industrial W powder manufacture).

In addition, it must be stressed here that this reduction process is also unique in so far as it presents the possibility to produce single crystalline powder of a highly refractory metal at temperatures between 800 and 1100 °C (less than one-third of the melting temperature) by CVT.

3.3.5. *Influence of Foreign Elements* [3.33–3.38]

Traces elements have considerable influence on the quality of the W powder. They originate either from raw materials or are introduced during manufacture. They can, however, also be added intentionally prior to reduction, if specific powder properties are desired.

During reduction the foreign elements are either evaporated, or remain partially or completely in the as-reduced powder. In general, the higher the reduction temperature and the lower the humidity, the stronger is their evaporation. There are, however, exceptions to this rule in case the oxides are volatile (for example, phosphorous). The presence of two or more foreign elements at the same time may change the influence of a single element drastically by forming more stable compounds. Foreign elements can be contained as oxides, tungstates, silicates, phosphates, metals, or intermetallic compounds, either as separate particles or incorporated into the tungsten grain. During reduction they can be solid or liquid. In case of low concentrations, foreign element atoms can also be present in solid solution.

Foreign elements may also influence the reduction process, either by altering the reduction sequence and/or by changing the nucleation and growth conditions for the individual phases. In particular, the latter aspect is of great technical importance. Most interactions known today refer to the $WO_2 \rightarrow W$ transition. Certain trace elements enhance the reduction rate during this stage (Li, Na, K, Mn, Fe, Co, Ni) while others retard it (B, Al, Cr). Crystal growth can be significantly promoted (by alkali metals), but also reduced (Co, Al, V, Cr), or even suppressed (B).

Not only is the grain size affected by foreign element traces, but also size distribution, grain morphology, agglomeration, apparent density, and compactability of the W powder.

Several foreign elements can stabilize the formation of metastable β-W, shifting the transition temperature from about 630 °C to above 800 °C [3.39]. Since all the stabilizers have a very high affinity to oxygen (Be, Al, Th, Sr, Zr, B, P), it was argued, that these elements form two-dimensional oxide compounds on the surface of the β-W, and that the

β-W $\rightarrow$ α-W transformation may occur only after these compounds are decomposed by hydrogen [3.40].

Elements or element combinations of industrial importance are discussed below.

Alkali Metals. Alkali metal compounds significantly enhance crystal growth during reduction at both low (700 °C) and high (1000 °C) temperatures, and increase the reduction rate. The higher the temperature, the more pronounced are these effects. Alkali additions are used on an industrial scale to produce extraordinarily large grained W powder (see Section 5.4).

The metals (Li, Na, K) as well as the tungstates and hydroxides (which form during reduction) are liquid at the usual temperatures. No evaporation occurs below 800 °C but, above 800 °C, a partial or total evaporation takes place, depending on the respective element and the nature of the compound (tungstate, hydroxide, silicate, borate, etc.). Lithium as Li_3BO_3, for example, or sodium as silicate are less volatile than LiOH or NaOH. In contrast to Na and K, Li cannot be volatilized due to its high boiling point.

The effect of the addition of 0.1 mol of alkali metal to tungsten blue oxide is shown in Table 3.1. It can be seen that the influence in grain growth is considerable. Moreover, there are differences between the different metals which are mainly based on the properties of the metal. At the end of the reduction process the respective alkali tungstate is reduced and at 1000 °C the metal is evaporated. The vapor pressure of Li is the lowest of the three metals and therefore the amount remaining is the highest.

The acceleration of the reduction process is attributed to a catalytic action of the liquid alkali compounds on the reaction of WO_2 with water vapor forming the volatile oxide hydrate. This step (evaporation) is said to be the rate-limiting stage in the chemical vapor transport reaction [3.35]. The overall reaction (3) can be described as the sum of reactions (1) and (2), according to the following:

(1) $K_2WO_4\ (l) + 2H_2O(g) \rightarrow WO_2(OH)_2(g) + 2KOH\ (l)$

(2) $WO_2(s) + 2KOH\ (l) \rightarrow K_2WO_4\ (l) + H_2(g)$

(3) $WO_2(s) + 2H_2O(g) \rightarrow WO_2(OH)_2(g) + H_2(g)$

Equimolar concentrations of Li, Na, and K have about the same effect on grain coarsening, which supports this theory. Nevertheless, the presence of a volatile alkali–tungsten–oxygen compound has also been made responsible for grain coarsening [3.41].

TABLE 3.1. Effect of Alkali Compounds on Tungsten Crystal Growth (addition of 0.01 mol alkali hydroxide)

Element	Undoped		Li		Na		K	
Temperature (°C)	750	1000	750	1000	750	1000	750	1000
Deagglomerated W Powder particle Size (μm)	reduction incomplete	11	5.4	46	3.5	25	2.3	24
Concentration of doping elements in % of addition	—	—	84	68	31	8	80	37

Aluminum. Aluminum additions severely disturb the W crystal growth, leading to the build-up of AlO_x phases in the tungsten lattice. The reaction rate is significantly decreased, and consequently the nucleation is favored, resulting in a decreased average grain size. About 50 μg/g are built-in. At higher concentrations, AlO_x phases are deposited onto the crystal surfaces. The growing W crystals are rough-faced during the incorporation period, but then they become smooth and finally have a completely regular shape in the as-reduced condition (Fig. 3.16).

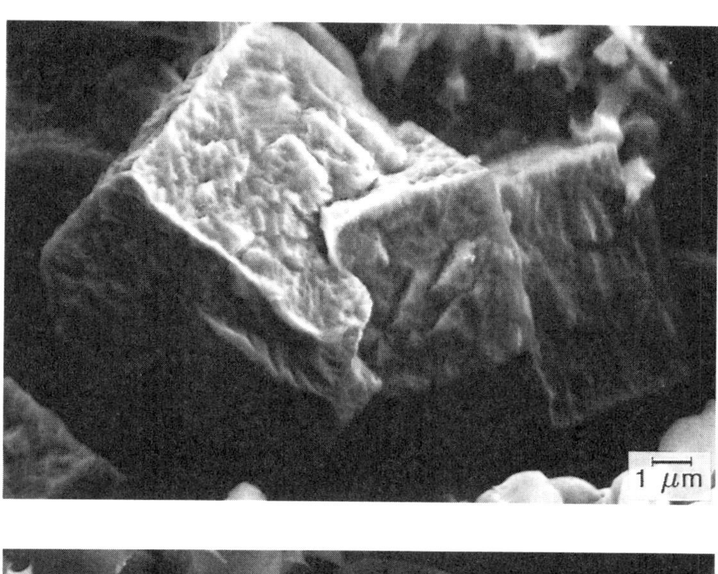

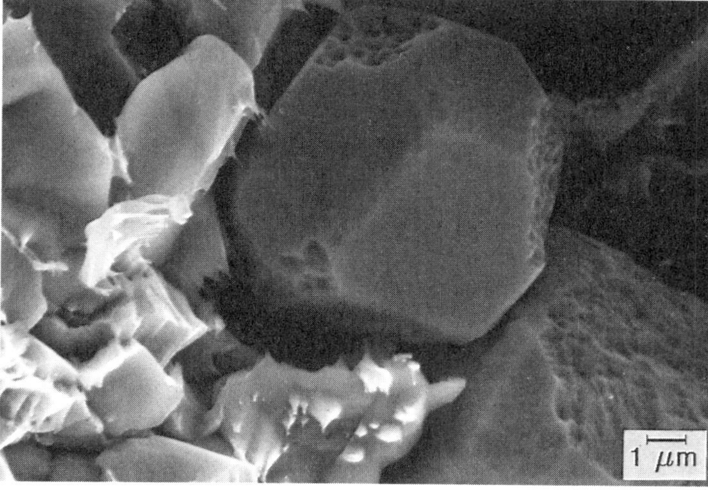

FIGURE 3.16. Influence of aluminum on the morphology of tungsten particles; at first, cubic crystals are formed with rough surfaces (upper figure) which become smooth during further crystal growth (lower figure); 200 ppm Al added to the oxide; 1000 °C.

The influence of Al can be completely suppressed by the addition of sodium compounds. In this case all Al remains outside of the tungsten crystals as an Na–Al phase.

Iron, Nickel, and Cobalt. Traces of iron, nickel, or cobalt may originate from boat and furnace materials. They are also present in large amounts in recycled oxides of heavy metal scrap. High concentrations lead to an activated sintering of the powder and thus significantly alter the powder properties, in particular the agglomeration. During the early part of reduction iron and cobalt form tungstates, which are then reduced to intermetallic phases (Fe_7W_6, Co_7W_6). The latter phase can be incorporated into the tungsten grain. Nickel is present in the metallic state. Small amounts of Ni (200 μg/g) lead to a characteristic whisker growth of W metal (Fig. 3.17). If such whiskers are present in a regular powder charge, they are the first witnesses of an enhanced corrosion of the reduction boat (made of Inconel).

Phosphorus. Phosphorus is a detrimental impurity in most sintered tungsten materials since it commonly occurs in all technical powder grades in the range of 10 to 20 μg/g. It is a strong β-W stabilizer and, as such, responsible for enhancing the agglomeration of fine-grained W powders under dry reduction conditions. Phosphorous additions to the tungsten blue oxide lead to the formation of phosphate "bronzes" of the general formula $P_4O_8(WO_3)_{2m}$, with m between 2 and 16.

AKS or NS Doping. This type of oxide doping is used for the production of non-sag filaments in incandescent lamps. The doping compounds are potassium silicate and aluminum chloride. The term AKS refers to the initials for the German names of the three elements Aluminium, Kalium (potassium), and Silizium (for details see Chapter 5.4). During hydrogen reduction of AKS-doped tungsten oxides, incorporation of the dopants occurs by CVT overgrowth during the $WO_2 \rightarrow W$ transition in the form of aluminosilicates. Less than 10% of the dopants added remain in the W powder after an HCl/HF-acid leaching. On sectioning the W powder particles the dopant phases appear as holes in the W

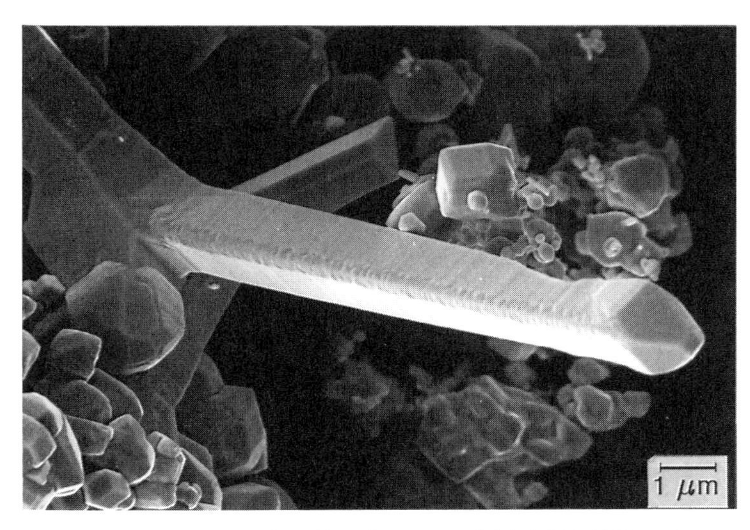

FIGURE 3.17. Tungsten whisker. Whiskers form if small amounts of nickel (200–1000 ppm) are present; higher amounts lead to an activated sintering of the W particles.

matrix (Fig. 3.18). The mechanism behind this incorporation is still under discussion [3.42] but seems to be linked to the intermediate formation and transformation of potassium tungsten bronzes.

3.4. REDUCTION OF TUNGSTEN OXIDES BY CARBON OR CARBON-CONTAINING COMPOUNDS. [3.2, 3.43, 3.44]

The reduction of tungsten oxides by carbon or carbon-containing compounds can be easily performed. Statements about the starting temperature for the reaction between WO_3 and solid carbon (carbon blacks, graphite) vary in the current literature between 655 °C and 783 °C. Differences in WO_3 and C properties (particle size of the powders, preparation history, crystallinity, etc.) as well as in atmospheres may be responsible for that. The temperature range coincides with the beginning of self-conductivity and sublimation of WO_3. Carbon monoxide starts to react with WO_3 at 535 °C (reduction pressure 1 bar, p_{CO_2}/p_{CO} equilibrium ratio 8.52) [3.45].

The reduction/carburization sequence can be assessed from so-called predominance area diagrams (Kellogg diagrams), which can be derived from free energy of formation data of the compounds occurring in the W–C–O system (WO_3, $WO_{2.9}$, $WO_{2.72}$, WO_2, W, W_2C, WC, CO, CO_2) [3.45–3.47]. Such a diagram is shown in Fig. 3.19 for 1100 °C [3.46]. It predicts that at 1100 °C the reduction/carburization proceeds via the oxide phases ($WO_{2.9} \rightarrow WO_{2.72} \rightarrow WO_2 \rightarrow W \rightarrow W_{(2)}C$), with the possibility of a direct carburization of $WO_{2.72}$ and WO_2.

The reduction kinetics are based on an adsorption–autocatalytic reaction. Reduction enthalpy and temperature are linked by a linear relationship. The calculated activation energy is about 75 kJ · mol^{-1}. Carbide formation occurs only when all oxygen is removed.

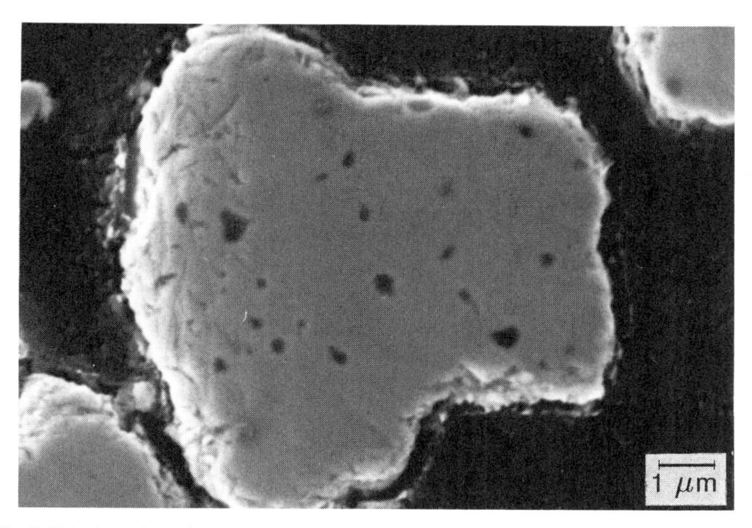

FIGURE 3.18. Polished section of a large tungsten particle showing several holes in the matrix that formed on polishing; these holes were originally filled with potassium aluminosilicates.

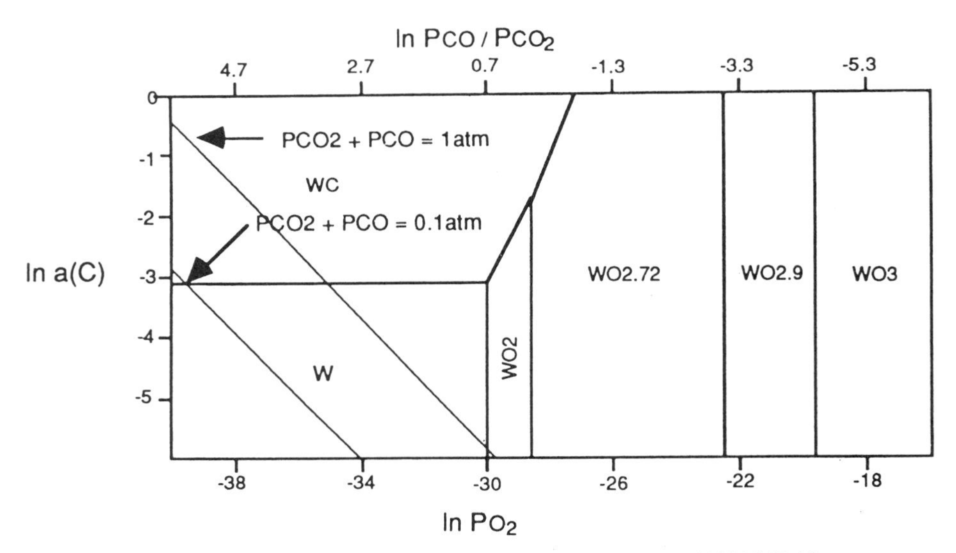

FIGURE 3.19. Kellogg diagram for the W–C–O system at 1100 °C [3.46].

The apparent activation energy at the beginning of the reaction is 121 kJ/mol WO_3 and increases during further progress to 205 $kJ \cdot mol^{-1}$ WO_3, due to the fact that the reaction becomes diffusion-controlled (diffusion of oxygen to the particle surface).

The assumed reaction mechanism is as follows: direct reaction of WO_3 and C at contact zones and, for the main portion, sublimation of WO_3 to the carbon particles, reduction at the surface of the carbon particles, and desorption of CO and CO_2.

Although the reducing agent—carbon—is very cheap, and carbon reduction was the early basis of tungsten powder production, so far none of the numerous carbothermic procedures has been established in the production of pure tungsten.

The reasons for that are:

- The carbon balance is difficult to control; furthermore, the metal forms carbides. Although the theoretical carbon consumption to reduce WO_3 to W is 15.5 wt%, only 12–14 wt% are normally sufficient at 1200 °C to 1400 °C because the reaction does not only proceed via $WO_3 + 3C \rightarrow W + 3CO$ but also to a certain extent via $2WO_3 + 3C \rightarrow W + 3CO_2$. Therefore, in most cases the reduction is either incomplete or the metal powder is contaminated by carbides.
- Carbon is always a source of increased tungsten contamination because it contains impurities like Ca, Si, Fe, S, and P.
- Another disadvantage of the carbothermic reduction is the limited possibility of steering the grain size by varying the reduction conditions as compared to the hydrogen reduction process.

Carbothermic reduction of WO_3 or ore concentrates is of technical importance in melting metallurgy—preparation of ferrotungsten, melting base, and cast carbide (see Chapter 8).

The reduction of WO_3 by carbon in the solid state, however, has gained technical interest and importance in the case when the desired final product is tungsten carbide. This process is named "Direct Carburization" and the basic reaction is described by the following equation:

$$WO_3 + 4C \rightarrow WC + 3CO$$

In contrast to the conventional procedure for WC production—WO_3 reduction by hydrogen followed by W + C mixing and carburization—this method allows the formation of WC in only one step, while, carbon acts as carburizing and as reducing agent.

The above equation is only of theoretical value because, besides the main reaction, the following reactions also occur:

$$WO_3 + \tfrac{5}{2}C \rightarrow WC + \tfrac{3}{2}CO_2$$
$$C + CO_2 \rightarrow 2CO$$

Consequently, in this procedure, an uncertainty in the carbon demand is given. This means that, depending on the operating conditions, the carbon content of the final product may vary significantly.

Under industrial conditions, the process can be performed in two steps, one following the other [3.44, 3.48]. The first step—reduction—affords the complete exclusion of hydrogen and is performed in nitrogen atmosphere. In the presence of hydrogen, water vapor would form which generates tungsten crystal growth and consumption of carbon ($C + H_2O \rightarrow CO + H_2$). This carbon consumption is not only restricted to the carbon black or graphite used as reagent (disturbance of the carbon balance), but also to the furnace carbon tube (shortening its lifetime). The second step—carburization—needs a hydrogen atmosphere, which supports the carbon transport via methane to the tungsten particle surface (see also Section 3.6).

Accordingly, the direct carburization process starts with a pelletized mixture of WO_3 and graphite passing step-by-step two rotary furnaces, the first operating at 950–1250 °C under nitrogen (reduction) and the second at 1400 °C under hydrogen (carburization).

Direct carburization is advantageous mainly in the production of submicron tungsten carbide powders. The idea behind the process is to circumvent the expensive hydrogen reduction of fine tungsten powder by excluding water vapor during the reduction step, which is responsible for grain growth reactions and low capacity. Roughly speaking, the grain size of the final carbides is related to the size of the intermediate oxides [3.48].

Only recently was a new carbothermal reduction process developed in which the WC is synthesized by a rapid carbothermal reduction of tungsten oxides in a vertical graphite transport reactor (RCR entrainment process) [3.49]. Rapid heating of the WO_3/C mixture driven by thermal radiation allows conversion of the mixture into a carbide precursor (WC_{1-x}) within very short reaction times (a matter of seconds). In a second step, additional carbon is added to the carbide precursor to form a mixture, which then undergoes a second heat treatment to convert the precursor into substantially pure WC.

Carbothermal reduction of tungsten oxides with carbon monoxide [3.47], or gas mixtures of CO/CO_2, CO/H_2, CH_4/H_2 [3.50], C_2H_4/H_2, and C_2H_4/H_2 [3.51], as well as by reaction between metal oxide vapor and solid carbon [3.52] have recently attracted attention for producing high surface area tungsten carbides (up to 100 m^2/g), for use as catalyst (see Section 10.4), and for nanophase WC/Co composite powders (see also Section 9.2.1.4) [3.53].

3.5. REDUCTION OF TUNGSTEN HALIDES

3.5.1. Introduction [3.2, 3.54–3.57]

The direct chlorination of tungsten raw materials and subsequent reduction of the chloride to tungsten metal would offer a much simpler and cheaper production path in comparison to the rather complicated one via hydrometallurgy to APT and a two-stage reduction to tungsten metal. Although several promising attempts have been made, it is interesting to note that chlorination chemistry of tungsten has not found its way into large-scale production, unlike other refractory metals such as zirconium, hafnium, titanium, niobium, and tantalum. None of the processes mentioned below survived for long or are in use today.

Chlorination of scheelite or wolframite ore concentrates according to the equations

$$2CaWO_4 + 3C + 6Cl_2 \rightarrow 2CaCl_2 + 2WOCl_4 + 3CO_2$$
$$2FeWO_4 + 3C + 7Cl_2 \rightarrow 2FeCl_3 + 2WOCl_4 + 3CO_2$$

and subsequent treatment of the chlorides offers a selective separation of most other chlorides formed with the exception of molybdenum but bears certain disadvantages:

- It mainly yields in the formation of $WOCl_4$ and not in the desired WCl_6.
- Therefore, an additional conversion step from $WOCl_4$ to WCl_6 is necessary; the separation from other chlorides can be done only in the WCl_6 stage by fractional distillation.
- Calcium as well as iron or manganese form molten chlorides at the reaction temperature, causing difficulties in reactor design.
- Finally, hazards associated with HCl, Cl_2, $COCl_2$, and chlorides as well as environmental considerations are further limiting factors.

In order to circumvent the formation of $WOCl_4$, the following path was described and was even more complicated and more expensive [3.2]. A low melting tungsten silicide was prepared from scrap or ore concentrates in advance by melting with aluminum or silicon in an arc furnace at 1800 to 2000 °C. The alloy settles from the slag and can be separated. During the subsequent chlorination of the alloy WCl_6 and $SiCl_4$ form, which can be easily separated due to their different condensation temperatures.

Based on a patent of Jonsson and Svanstrom [3.56], chlorination of tungsten scrap and subsequent fractional distillation to purify the crude WCl_6 was used by Sandvik Co.

The most advanced installation was made in the USA in the early 1970s. L. Ramquist of the Axel Johnson Institue of Industrial Research in Sweden described an on-line, computer-controlled industrial-scale process for tungsten powder production from ore concentrates and scrap. The process included chlorination, treatment of the chlorides formed to prepare pure WCl_6, and reduction by hydrogen to tungsten powder. Based on this concept, P. R. Mallory and Co. Inc. in the USA installed a new plant, which was, however, closed shortly after start-up.

The main complication in producing larger amounts of tungsten powder by hydrogen reduction of WCl_6 is the fact that the powder surface is covered by chlorine (and HCl), which is gradually desorbed during the further handling. Due to the fine powder grain size, the specific surface is quite high and the adsorbed amount of chlorine is accordingly high too. The consequences were hazard to workers as well as corrosion problems. The chlorine layer at the tungsten particle surface is one of the reasons why these fine powders are not pyrophoric besides their perfect crystal structure.

The two important halide compounds which are technically in use today are WF_6 and WCl_6. They can be easily prepared by direct combination of the elements. High purity is achieved by using pure starting materials. The total exclusion of moisture is very important.

WF_6 is the predominant precursor for the production of tungsten thin layers in microelectronics via CVD. Because of its electromigration resistance and low resistivity of $5.6\,\mu\Omega \cdot cm$, tungsten is and will be the most promising material for integrated circuit metallization.

WCl_6 is the starting material for the preparation of extremely fine grained tungsten and tungsten carbide powders (small scale only), for the deposition of tungsten on substrates to prepare intricate forms by CVD, and for the preparation of high-purity tungsten whiskers of high strength. Furthermore, it is also in use, like WF_6, as a precursor for thin layers in microelectronic devices.

CVD tungsten exhibits several advantageous properties, such as high purity, resistivity to grain growth (low energy grain boundaries and absence of stored strain energy), near-theoretical density, and high ductility. Depending on deposition conditions and substrate properties, different types of layers can be produced, such as single crystal layers, layers of a determined crystal orientation, columnar crystallites rectangular to the substrate surface, and so on. The crystal orientation can be influenced by the WF_6/H_2 ratio and the temperature: low temperature and low ratio favor (100) orientation while high values favor (111) orientation.

3.5.2. *Reduction of WF_6* [3.58]

Today, high-purity WF_6 is available at low cost. This makes it the most important tungsten source for thin layers in microelectronic devices prepared by CVD. Different reduction agents can be used. Hydrogen, silicon, and silane are the most common reducing agents for WF_6 in integrated-circuit metallization processes. Moreover, reduction can be achieved too by Si_2H_6, B_2H_6, PH_3, SiH_2Cl_2, and GeH_4 at much lower deposition temperatures. Reduction with hydrogen or silane allows the deposition of tungsten in the temperature range 300–500 °C. Reduction by silicon is important to deposit tungsten selectively on Si surfaces.

The WF_6–Si System. Depending on temperature, the following two reactions are possible:

$$\text{below } 400\,^{\circ}\text{C:} \quad 2WF_6 + 3Si \rightarrow 2W + 3SiF_4$$

$$\text{above } 400\,^{\circ}\text{C:} \quad WF_6 + 3Si \rightarrow W + 3SiF_2$$

Accordingly, twice as much Si is consumed at higher temperature. Reduction of WF_6 by a silicon substrate allows for selective tungsten deposition in the contact region of highly doped silicon. No deposition occurs on insulator surfaces such as SiO_2 or Si_3N_4. It is a replacement reaction which is highly sensitive to the substrate pretreatment. The reaction is very fast and stops as soon as a critical tungsten thickness is reached (10–15 μm). The thickness depends on the reactor setup, the amount of oxygen on the silicon surface, and pretreatment. It is assumed that a continuous tungsten layer of a certain thickness will inhibit silicon diffusion to the tungsten surface, consequently the penetration of WF_6 to the W/Si interface is stopped.

The WF_6–H_2 System. The reaction $WF_6 + 3H_2 \rightarrow W + 6HF$ is possible from 300 to 800 °C and is also used in thin-film production. There are several disadvantages in comparison to the other reduction methods. The HF formed during the reaction may cause defects, like encroachment or wormholes. The layers show poor adhesion on native SiO_2, which is always present on Si. Therefore, tungsten is not directly deposited on Si but on a bilayer. One layer provides an ohmic contact with Si, and the other acts as an adhesion promotor for W.

The WF_6–SiH_4 System. The advantages over the WF_6–H_2 system are: the deposition rate is higher, the film is smoother, good adhesion prevents encroachment and wormhole formation and suppresses silicon substrate consumption. A disadvantage is the dependence of the deposition on the SiH_4/WF_6 ratio and on the temperature. Depending on these parameters, either W or W silicide can be deposited and blanket or selective deposition on a phase can occur. A ratio up to 1.6 yields in selective W deposition and higher in blanket deposition. As long as the ratio is smaller than 1, tungsten is deposited as α-W and, between 1 and 3, as β-W. The amount of silicon incorporation increases with temperature. The dependence of film properties on deposition conditions and on the ratio is given in Table 3.2.

The reaction rate of the WF_6 with silane is very fast. The hazard of an explosion may be overcome by low total pressure, low silane partial pressure, and fast flow rate. Silane reduction is only used for a short duration to initiate W nucleation and is followed by WF_6/H_2 reduction.

The WF_6–GeH_4 System. Deposition at 300–400 °C yields in layers of higher resistivity (approximately 200 μΩ · cm) which consists of β-W stabilized by a Ge incorporation of 10–15 at%. Above 400 °C, the β-W and Ge concentrations decrease and pure α-W is deposited with less than 1 at% Ge. The layers show good adhesion to Si and SiO_2 with no encroachment and wormholes.

TABLE 3.2. Deposition Conditions and Tungsten Film Properties using WF_6 Precursor and Different Reducing Agents [3.58]

Temperature (°C)	Carrier and reducing agent	Pressure (torr)	Substrate	Film resistivity (μ · cm) and purity
420–800	H_2/Ar, H_2He	1–10, 760	Si, SiO_2, Sapphire	6–15; 100% α-W
260–400	H_2/Ar	0.1–1	Si, SiO_2, $TaSi_2$	10–15
288–403	WF_6:H_2 = 1:15	0.2–10, 760	Si (100)	6
300	SiH_4	UHV	Si (100)	30
250–550	SiH_4	Vacuum	Si (100)	7.5–15; low in Si
500–570	H_2/Ar	3.5	GaAs	50; 2% C, O
350–550	H_2/Ar	0.5–4.5	InP	55
<400	GeH_4	1.5	Si, SiO_2	200; β-W, 10–15% Ge
>500	GeH_4	15	Si, SiO_2	10; α-W, 1% Ge

3.5.3. Reduction of WCl_6 by Hydrogen [3.58–3.64]

The reduction of WCl_6 by hydrogen is the basis for preparing tungsten powder by CVD. The reaction may be performed in low-pressure reaction chambers, in chlorine/hydrogen flames, plasma jets, or fluidized bed reactors. These methods offer the possibility to prepare very fine grained tungsten powders with close grain size distribution and smooth particle surfaces. Grain sizes are in the range of 10–100 nm and can be regulated by the reaction conditions. With the exception of plasma jet originated samples, these powders are not pyrophoric in contrast to those reduced from oxides by hydrogen and having comparable grain size. This can be explained, on the one hand, by the more perfect crystal constitution and smooth surfaces of powders produced by CVD. On the other hand, the adsorbed chlorine at the powder particle surface prevents a rapid reaction with oxygen or water vapor. The fact that nanosized powders can be produced which are not pyrophoric is a big advantage, but, on the other hand, the adsorbed chlorine has until now prevented the application of these production methods on a bigger scale.

A typical procedure using a low-pressure reaction chamber is as follows: WCl_6 is heated to evaporate and is carried to the reactor by a stream of argon. The argon flow rate regulates the WCl_6 input. By varying the WCl_6 feed rate, the hydrogen flow rate, and the temperature (600–920 °C), different average particle sizes can be obtained (16–46 nm). The principle in preparing such small grains is that, by controlling the dilution, only a minimum number of atoms and molecules involved in the reaction may form agglomerates.

The preparation of 30–60 nm tungsten powder by feeding WCl_6 into a hydrogen/chlorine flame burning in a CVD reactor yields also spherical particle shape and close grain size distribution. Variations in flame temperature, flow rate of reactants, and residence time of particles allow grain-size regulation.

Fine CVD-derived tungsten powders have been carburized successfully and further processed to cemented carbides (Axel Johnson Process).

Single crystals can be prepared by reductive dissociation of WCl_6 at a tungsten wire. The same method can also be used to produce bigger tungsten rods (200 mm length and 6 mm diameter).

An advantage of WCl_6 for thin-layer production in microelectronic application is that Cl_2 and HCl cause much weaker encroachment and corrosive effects on Si and semiconductors in contrast to F_2 and HF. A disadvantage is the lower vapor pressure of WCl_6 compared to WF_6.

The WCl_6–H_2 system works at 600–900 °C. Lower temperature, high total pressure, and high WCl_6 concentration favor the process and the selectivity. The chloride process is more affected by the oxide layers on substrates than is the fluoride process (slower etching of SiO_2). This is why thickness reproducibility is worse. Also, film composition shows poor reproducibility (α- and β-tungsten).

3.5.4. Reduction of WCl_6 by Carbon-Containing Reagents [3.65–3.67]

The basic idea in these reactions was always to produce fine and stable WC powder (0.01–0.1 μm) in one step starting from WCl_6, and to circumvent the reduction to W by hydrogen and the subsequent carburization. Chemical vapor reactions are ideal for producing fine powders consisting of discrete particles. Particle size distribution can be kept quite close and agglomeration is low.

For the system WCl_6–CH_4–H_2, the following reactions can occur:

$$
\begin{aligned}
&(1)\ WCl_6 + CH_4 + H_2 \rightarrow WC + 6HCl \\
&(2)\ WCl_6 + \tfrac{1}{2}CH_4 + 2H_2 \rightarrow \tfrac{1}{2}W_2C + 6HCl \\
&(3)\ WCl_6 + 3H_2 \rightarrow W + 6HCl \\
&(4)\ W + CH_4 \rightarrow WC + 2H_2 \\
&(5)\ W + \tfrac{1}{2}CH_4 \rightarrow \tfrac{1}{2}W_2C + H_2 \\
&(6)\ WCl_6 + \tfrac{1}{2}H_2 \rightarrow WCl_5 + HCl \\
&(7)\ WCl_6 + H_2 \rightarrow WCl_4 + 2HCl \\
&(8)\ WCl_6 + 2H_2 \rightarrow WCl_2 + 4HCl \\
&(9)\ WCl_6 + \tfrac{5}{2}H_2 \rightarrow WCl + 5HCl \\
&(10)\ CH_4 \rightarrow C + 2H_2
\end{aligned}
$$

The log K_p values (equilibrium constants) for the direct carburization to W_2C as well as to WC for equations (1) and (2) are quite high (>20). Also high is the log K_p value for the hydrogen reduction of WCl_6 to W (17–20) (equation 3); while the log K_p for the carburization of W by CH_4 (equations 4 and 5) to W_2C or WC, respectively, are low (2–4). All lower tungsten chlorides can be easily reduced by hydrogen. The high equilibrium constants for reactions 1, 2, and 3 are why direct carbide formation and hydrogen reduction take place at the same time. W formed by hydrogen reduction is subsequently carburized by CH_4. This step can be the rate-determining stage in the formation of the carbide, particularly at comparatively low temperatures ($\leq$1400 °C). Moreover, the high equilibrium constants are a prerequisite for the powder formation. At low temperature mainly W_2C is formed and above 1400 °C, WC is the only product. Powder particle size can be influenced by the reaction conditions. Powders have been produced in high-temperature reactors, tube reactors, and hydrogen arc plasmas. Typical grain sizes are between 5 and 100 nm. The main and important drawback of the direct formation of WC via the vapor phase is the difficulty in producing monophase WC without the presence of either carbon black (equation 10), or W_2C, or both of them. This is only possible by proper control of the vapor phase, the temperature, and the exposure time, and so far has not been successfully mastered in large-scale production.

Besides methane, other hydrocarbons were used as reducing and carburizing agents, such as C_3H_8 and C_2H_4. They do not offer any advantage.

3.6. REACTION OF TUNGSTEN WITH CARBON OR CARBON-CONTAINING COMPOUNDS (CARBURIZATION)

This section deals primarily with the carburization of tungsten powder, the most widely used process for producing WC on a technical scale. Also, the carburization performed in melts is discussed briefly owing to historical reasons and the production of hard facing alloys. Finally, the formation of coarse tungsten carbide crystals in auxiliary melts (“Menstruum WC)” is discussed.

3.6.1. The Carburization of Tungsten Powder by Solid Carbon

Tungsten reacts with carbon (carbon black, graphite, CVD-produced carbon layers, etc.), as well as with a variety of carbon-containing compounds to form carbides (W_2C, WC, γ-WC_{1-x}) at elevated temperatures. Indications of a lower temperature limit for these reactions are extremely widespread in the current literature. For example, carbides were detectable after long-time high-energy ball milling of a W + C powder mixture (mechanosynthesis) [3.68]. The overall temperature in that procedure should have been somewhere between room temperature and 100 °C. However, no knowledge is available about the local temperature when a ball hits a W + C increment. Also, at room temperature, W_2C formation was observed in sputter-deposited W/C multilayers [3.69]. Thin layers of W and C form W_2C and WC at 800 °C. Technically, the usual carburization temperatures for W + C powder mixtures range between 1050 °C and 2100 °C, depending on the powder particle size. Bulk tungsten, like sheet or wire, embedded in carbon powder starts carburizing at 1550 °C, while, single-crystal wires react only at 1900 °C [3.70].

This wide range in temperatures can be understood because the carburization of tungsten is influenced by the properties of the starting materials (W and C) as well as several other parameters.

Properties of Tungsten. In the case of bulk tungsten, one has to distinguish between polycrystalline metal, which carburizes more easily than monocrystalline metal. This is because the diffusivity of carbon in tungsten is strongly increased by lattice defects.

For tungsten powder, the grain size is determining. Smaller particles carburize more readily than larger particles, due to their higher surface area. The rate of carburization is inversely proportional to the square of the nominal particle size [3.71]. Since the powders always consist of a distribution of particles, the larger particles determine the time for complete carburization. The coarser the tungsten particles, the higher is the temperature required for complete carburization and the longer is the time required [3.72].

Furthermore, the purity of tungsten has an influence on the carburization. Certain impurities, such as cobalt, accelerate the carburization, and others, like the alkali metals, retard it [3.73].

Properties of Carbon Black/Graphite. The fineness, chemical reactivity, and purity influence the carburization process. The finer the powder, the better the contact with the tungsten particles and the higher the reactivity. Regarding the impurity level, artificially produced graphite will be advantageous owing to its higher purity. But this is only important in regard to harmful impurities like alkali metals, sulfur, calcium, magnesium, etc. In regard to hydrogen, carbon black is much more impure than graphite. Hydrogen in carbon black is contained chemically bound to carbon. These bondings are relicts of the raw material (hydrocarbons), C–H bonds which have not been cracked during the burning process. The hydrogen promotes the carburization, as described below. Moreover, this hydrogen content of carbon black makes it possible to carburize W + C mixtures in big, closed graphite containers by induction heating in an open furnace without any specially provided protective atmosphere.

The Atmosphere. Industrial carburization of tungsten powder is commonly carried out in hydrogen or hydrogen-containing gas mixtures. Hydrogen reacts with the solid carbon in the W + C mixture to form methane (at temperatures $\leq$1600 °C) or acetylene (>1600 °C), which transport the carbon in tungsten–carbon black mixes over larger distances. At the surfaces of the reacting particles (W, W_2C, WC), the hydrocarbons are dissociated (cracked) and the freed carbon atoms diffuse into the particles. The diffusion of the carbon through the carbide phases is the rate-determining step of carburization.

Reactions also occur with the carbon parts of the furnace and the graphite boats, but these are of minor importance. Excessive carbon losses occur in flowing hydrogen if the carburization is carried out in open graphite boats. These losses can be prevented by using graphite covers, thus forming a quasi-stationary atmosphere above the W + C powder mixture.

Carburization in vacuum, argon, or helium is always very slow and incomplete. The carbon transport in this case is only possible via surface diffusion, and this is a very slow process to overcome larger distances [3.71]. In this case, the supply of carbon to the surface of the reacting particles is becoming rate-determining. Carbon monoxide may also act as a carbon carrier.

Availability of Carbon at the Surface of Tungsten Particles. If we consider a tungsten surface covered with carbon particles, it will depend on the particle size of the carbon source, whether the surface is densely or loosely covered. Finer powder will result in a more dense layer. On the other hand, finer powders show a higher tendency to form agglomerates. Therefore, proper mixing of the two components, which guarantees the break-up of these agglomerates, is important (long-time mixing in ball mills). In addition, to ensure a dense and close contact of carbon particles to the tungsten surface, compacting prior to the carburization can be applied.

Upon heating, only the carbon atoms in contact or in the near-neighborhood react with tungsten. All others must to be transported to the tungsten surface, either by surface diffusion (slow) or by chemical vapor transport (CVT), i.e., formation and subsequent dissociation of methane (fast). Thus, during the initial stage of the carburization, the availability of carbon atoms is quite high and becomes less later.

Figure 3.20 shows the progress of carburization under different atmospheres with time for a 10-μm W/carbon black powder mixture (containing 6.2 wt% C) on heating to 1800 °C, and then holding for 1.5 hours [3.74]. Under hydrogen or carbon monoxide

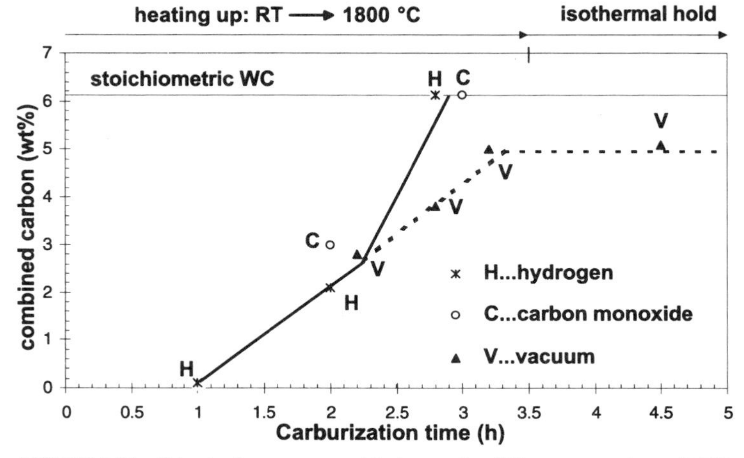

FIGURE 3.20. Caburization progress with time under different atmospheres [3.74].

(50 mbar), the carburization was completed on heating the mixture to the isothermal hold (combined carbon 6.13 wt%), while, under vacuum condition (<0.2 mbar), even after 1 hour at 1800 °C, about 1.3 wt% of the carbon remained unreacted. In the latter case there was evidently no further carbon supply to the particle surfaces.

The Carburization Process. Unfortunately, only scarce data about diffusion of carbon in W, W_2C, and WC are available in the current literature. The diffusion coefficient of carbon in metallic tungsten ranges from $9 \times 10^{-8}\,cm^2 \cdot s^{-1}$ at 1500 °C to $4.5 \times 10^{-6}\,cm^2 \cdot s^{-1}$ at 2400 °C [3.75]. Activation energies are stated between 37.8 and $53.5\,kcal \cdot mol^{-1}$ [3.69, 3.75–3.78], depending on temperature, tungsten quality, and method of measurement. Diffusion coefficients for carbon in W_2C were determined as $8.1 \times 10^{-10}\,cm^2 \cdot s^{-1}$ at 1526 °C, and $5 \times 10^{-6}\,cm^2 \cdot s^{-1}$ at 2400 °C, based on layer-growth measurements [3.79]. The respective activation energy is cited between 100–$111\,kcal \cdot mol^{-1}$ [3.79–3.81]. No diffusion coefficients were found for the diffusion of carbon through WC, but the activation energy is given as $58\,kcal \cdot mol^{-1}$ [3.71] which, however, seems comparatively low.

The diffusion of carbon in tungsten and its carbides proceeds via a vacancy mechanism rather than via the grain boundaries. This is because the carbon atoms are located in the interstices between the metal atoms and diffuse by squeezing between the host atoms to move from one interstitial site to another. In the hexagonal carbide phases of densest packing, the octahedral interstices are either half filled (W_2C) or completely filled (WC). The anomalously low WC layer-growth rate, as compared to W_2C or other transition metal monocarbides, might therefore originate in the fact that WC exhibits a narrow homogeneity range, and all sites on the carbon sublattice are occupied [3.79].

The Carburization Sequence. As a consequence of the high carbon supply in the initial stage of carburization, a thin layer of WC is always formed at first [3.72]. The carbon transport from the outer to the inner surface of that layer is then only possible by diffusion through it. The layer is polycrystalline, even when the tungsten part or particle was a single crystal. The smooth interface indicates that preferentially fast grain-boundary diffusion does not occur. The further carbon supply to the interior is insufficient for WC formation, and consequently W_2C forms. No significant growth of the WC layer occurs, because all the incoming carbon is consumed for W_2C formation, building up a second layer inside the particle. This layer grows in the radial direction and comprises large crystals. The carbon diffusion in W_2C is faster than in WC. Consequently, during this stage, only the W_2C layer is growing while the WC layer thickness remains about constant. After all the tungsten metal in the interior is transformed to W_2C, the WC shell starts growing further in the radial direction until the whole particle consists of WC. Diffusion of carbon through the growing WC shell is the rate-determining stage of the carburization process.

Figure 3.21 shows the change of the product composition with carburization time in hydrogen for a tungsten–carbon black mixture at 1119 °C. After only 10 minutes at this temperature most of the metal was transformed to W_2C, which was then only slowly converted to WC. This result is characteristic for the temperature range of 1050 to 1850 °C and W particle sizes of 1.3 to 20 μm [3.71]. It is remarkable that W_2C forms during carburization even at 900 °C, which is well below its eutectoid decomposition temperature of 1250 °C.

Simultaneous with the carburization reaction, a diffusion of impurity element traces takes place. It can be assumed that these atoms migrate preferably along grain boundaries

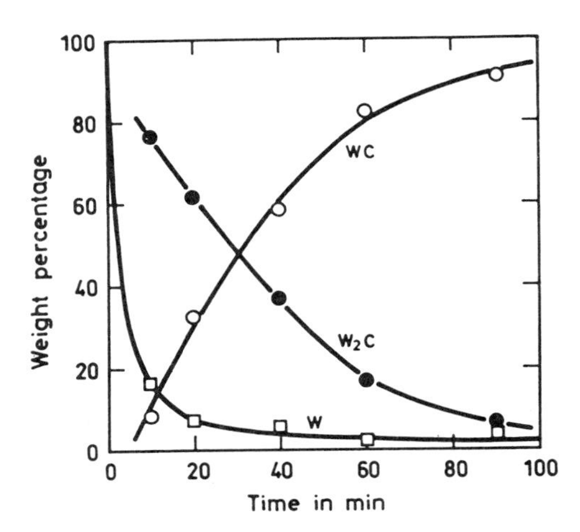

FIGURE 3.21. Reaction products from nominal 1.30 μm tungsten in W–C powder mixtures heated in H_2 at 1119 °C for different periods of time [3.69, 3.71].

to the surface, where they evaporate. The driving force for that process is the concentration gradient between the clean surface and the impure interior. Hence purification of the grain boundaries takes place, enabling their migration. The WC crystals grow and, depending on temperature and time, finally single-crystal WC particles or particles consisting of some very large WC crystals form by coalescence.

The average grain size of tungsten powder to be carburized for cemented carbide production varies between submicron and 50–100 μm. Time and temperature, according to the diffusion processes discussed above, must be adjusted in relation to the grain size. Finer powders up to 2 μm can be carburized quantitatively at 1400 °C and heating periods of 1–2 hours. In order to keep the heating time at about the same length also for coarser powders up to 10 μm, temperatures between 1600 °C and 1700 °C must be applied, and for very coarse tungsten powders (>20 μm) 1800–2000 °C is necessary.

3.6.2. Formation of Tungsten Carbides in Melts

This classical production method, first used by H. Moissan in 1896, is a fusion process in electric furnaces. Either W_2C or WC–W_2C mixtures can be produced. The latter containing 3.5–4.0% C are technically important as castings and hard facing alloys and are produced in large-scale operations (see Chapter 9). Mixtures of tungsten powder and carbon are heated in graphite tube furnaces up to 3000–3200 °C. The melt is cast either directly into wear parts or onto cold plates. In the latter case, the material is then ground to powder. The usual composition is 70% W_2C and 30% WC.

3.6.3. Preparation of Coarse Tungsten Carbide Powder in an Auxiliary Melt

The solubility of WC in iron metal melts (Fe, Ni, Co) is relatively low. Supersaturation of a respective melt by W and C results in crystallization of WC. On slow cooling, well faceted, big crystals (100 μm up to several mm) are formed. In comparison to the above-described WC particles formed by solid–solid or solid–gas reaction, these crystals are by far more perfect. In regard to the technical importance of the process, see Section 9.2.1.4.

3.6.4. Preparation of Tungsten Carbide in Salt Melts [3.82]

Tungsten dissolved in a NaCl/NaF halide melt, which is used to extract tungsten from ore concentrates, can be crystallized as WC by sparkling the melt with carbon monoxide. The advantage in comparison to the above-described method is that no acid is necessary to dissolve the auxiliary "bath" material. Only water is needed.

3.7. CHEMISTRY OF AQUEOUS TUNGSTEN SOLUTIONS [3.83–3.88]

A characteristic feature of the tungstate ion which only occurs monomerically in alkaline or neutral solutions is its tendency to form condensed, complexed isopolytungstate ions in slightly acidic medium. The ability of polycondensation is common for group 5 and 6 elements and is more pronounced for elements of higher atomic mass within the same group. The dimeric dichromate ion is a typical example.

Different techniques have been applied to elucidate the tungstate species forming in neutral and slightly acidic solutions, such as polarography, paper chromatography, light absorption, turbidimetry, molecular weight, ultracentrifugation, Raman, infrared and absorption spectroscopy, radiochemistry, precipitation, and solvent extraction.

Nevertheless, the present knowledge of tungstate species in aqueous solutions is still rather limited, because what can be determined is the charge of the ion present at a certain pH and its condensation grade. Proposed structures in the literature are based exclusively on investigations of solid crystalline compounds. Deducting the structure of a dissolved ion species from that of the corresponding water soluble, solid, crystallized compound is not possible, because of a lack in knowledge about the hydration of tungstate ions in aqueous solutions. The situation is described typically by the following two examples.

Paratungstate A ion is written in the literature sometimes as $[HW_6O_{20}(OH)_2]^{5-}$, or as $[HW_6O_{21}]^{5-} \cdot H_2O$ and $[HW_6O_{21}]^{5-}$.

At pH $\geq$6.2, the monomeric tungstate ion exists, commonly written in the form WO_4^{2-}. The assumption derived from the solid tungstates is that the tungsten ion is coordinated tetrahedrally by the four oxygen ions. Already in the crystallized stage, the ions are hydrated as in the case of $Na_2WO_4 \cdot 2H_2O$. It can be assumed that in aqueous solutions with their large excess of water molecules the degree of hydration might be higher. The hydration by two water molecules in case of monotungstate leads to the proposed octahedral structure $[WO_2(OH)_4]^{2-}$, but more likely seems to be the tetrahydrate with eight equivalent hydroxo ligands $[W(OH)_8]^{2-}$.

As already explained, the monomeric tungstate ion is only stable above pH 6.2 as far as solutions are considered which are free of complexing agents. At lower pH values the tungstate ions start to condense to polytungstate species. The condensation grade increases with decreasing pH until finally, at pH approximately 1, insoluble, highly polymeric, amorphous tungstic acid is precipitated.

Normally, polycondensation is defined by the reaction of anionic species and protons, such as

$$2WO_4^{2-} + 2H^+ \rightarrow W_2O_7^{2-} + H_2O$$

However, it seems unlikely that two negatively charged particles can react with each other.

Therefore, polycondensation was also explained as follows: in analogy to molybdenum it is assumed that also cationic tungsten entities are in equilibrium with the anionic tungstates, although their concentration is low. On acidification, their concentration will increase because hydroxo ligands of a portion of the anionic tungstate undergo protonation, in which case cationic entities result. These species reacted instantaneously with some of the remaining anionic hydroxo ions to form polycondensed tungstates. Accordingly, the reaction of positively and negatively charged tungsten ions is the basic mechanism of the polycondensation, as will be shown in the following simplified equations (each in a hydrated as well as dehydrated version):

Protonation of the hydroxo complex (tungstate anion):

$$[W(OH)_8]^{2-} + 4H^+ \rightarrow [W(OH)_4(H_2O)_4]^{2+}$$
$$[WO_4]^{2-} + 4H^+ \rightarrow [WO_2]^{2+} + 2H_2O$$

Condensation: aquo ligands of the cationic species are substituted by commonly shared hydroxo groups of the anionic entity (aquo ligands are more loosely bound as hydroxo ligands):

$$2[W(OH)_8]^{2-} + [W(OH)_4(H_2O)_4]^{2+} \rightarrow \left[(OH)_6W \begin{smallmatrix} \text{H} & & \text{H} \\ \text{O} & (OH)_2 & \text{O} \\ & \text{W} & \\ \text{O} & (OH)_2 & \text{O} \\ \text{H} & & \text{H} \end{smallmatrix} W(OH)_6\right]^{2-}$$

$$2[WO_4]^{2-} + [WO_2]^{2+} \rightarrow [O_3W\text{–}O\text{–}WO_2\text{–}O\text{–}WO_3]^{2-}$$

Not only can polycondensation be explained by the existence of positively charged tungsten entities, but also the reaction of tungstate containing solutions with anions like fluoride or anions of chelating agents like polycarboxylic acid, etc., always keeping in mind that a reaction of two negatively charged particles is not possible. Some chelate complexes are formed with dimeric tungsten species, and this is the proof for a dimeric cationic tungsten entity which is also analogous with molybdenum.

In principle, the polycondensation may also be regarded as a dehydration process caused by the increase in proton concentration. In the starting species of the polycondensation process, the fully hydrated monotungstate, the W:H_2O ratio is 1:4 (pH $\geq$ 6.2). In the various polytungstates prevailing between pH 6.2 and 1, the ratio is somewhere between 1:4 and 1:2, and attains exactly 1:2 in the precipitated tungstic acid at pH 1.

Considering polytungstates as being formed by addition of protons to tungstate ions, a general, overall equation for their formation can be written in the form

$$A \cdot [H_3O]^+ + 12 \cdot [WO_4]^{2-} \rightarrow B \cdot [W_xO_yH_z]^n + C \cdot H_2O$$

The respective formulas and corresponding values of A, B, and C are given for different polytungstate species in Table 3.3.

TABLE 3.3. Isopolytungstate Formulas and Corresponding Coefficients [3.84]

Coefficient			
A	*B*	*C*	Formula
4	1	6	$[W_{12}O_{46}]^{20-}$
8	1	12	$[W_3O_{11}]^{4-}$
	4	4	$[H_4W_3O_{13}]^{4-}$
14	1	16	$[H_{10}W_{12}O_{46}]^{10-}$
	2	20	$[HW_6O_{21}]^{5-}$
16	1	24	$[W_{12}O_{40}]^{8-}$
18	1	26	$[H_2W_{12}O_{40}]^{6-}$
	2	24	$[H_3W_6O_{21}]^{3-}$
24	12	24	$[WO_3 \cdot 2H_2O]$

It can be seen that preferential condensation grades exist, the most important of them being W_6 and W_{12}. In a neighborhood close to the starting pH for the condensation, a trimeric ion also exists and a dimeric is postulated.

The reason why investigations in this field are difficult and in most cases not very precise is the existence of equilibria between the different species, and, as was described above, they do not differ in condensation grade alone but can also be positively or negatively charged. Another complicating fact is that equilibration rates are fast in some cases and slow or even very slow in others. This means that although the equilibria are pH-dependent, there will always be more than one species present in solution at the same time due to the equilibrium constants. In addition, the equilibria are influenced by the temperature, the concentration, and the presence of a third ion or compounds. Higher concentration shifts the equilibrium to higher condensation grades.

Figure 3.22 presents a scheme of the species in equilibrium between pH 6 and 1. The number of tungstates is surely confusing, but certainly still incomplete. Once more, it must be stated that the indices of H, OH, and O are always based on assumptions and are therefore of less importance.

The hexameric para(A)- and the pseudo-meta-tungstate as well as tungstic acid are formed instantaneously if the pH is lowered to 6 or 4 or 1, respectively. The reactions to the dodecameric ions para(B)- or meta-tungstate are slow or very slow, and the equilibration may last for days up to one month. In the literature this is often called the "aging" of tungsten solutions—meaning the change in properties with time.

Here is an example: A sodium tungstate solution acidified to pH 2.5 behaves differently during further processing via solvent extraction in comparison to the same solution after some days.

The influence of foreign ions on these equilibria can be different. A variety of ions form heteropolytungstates. Others shift the polycondensation to the alkaline region like ammonium. Complexing ions may depress the polycondensation.

Anions like silicate, phosphate, arsenate, etc. form heteropolytungstates. For information about their structure see Section 4.2.7.3.3. The heteropolytungstates are always higher condensed species in comparison to the isopolytungstates existing in solution under

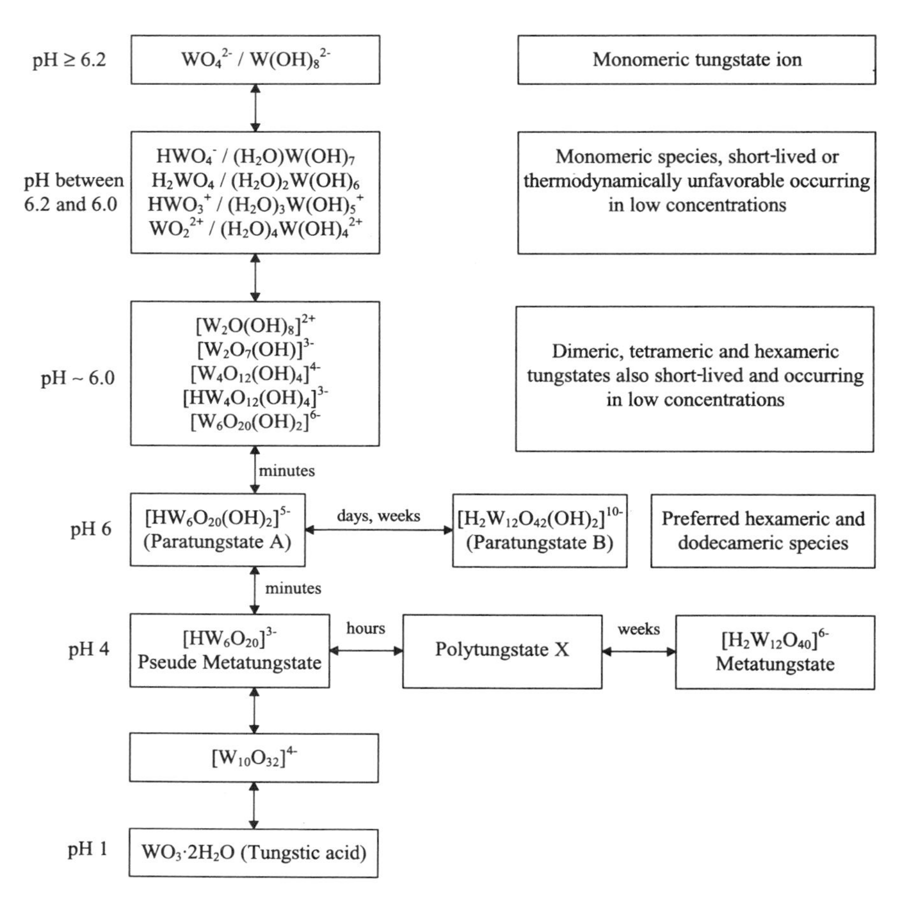

FIGURE 3.22. Scheme of tungsten ions in aqueous solution; updated version of an original scheme published by Van Put [3.86].

the same conditions. This is often combined with a decrease in solubility or with a change to colored solutions.

Ammonium ions cause a shift of the condensation pH limit to higher values. This means that in ammoniacal solutions, polytungstate species are also stable at pH values where normally only the monotungstate ion exists. For this reason no monomeric ammonium tungstate can be obtained by crystallization from aqueous solutions, only polytungstates.

The OH ions bound to the central tungsten ion can be easily substituted by ligands which form bondings of higher strength mainly by forming more stable chelate ring structures. The intermediate formation of cationic tungsten species is a prerequisite for reactions of this type. Hence the ability to condense will be suppressed.

The simplest and best known ion in this respect is fluoride. Therefore, tungsten solutions containing fluoride are stable also in strongly acidic solution, without any precipitation of tungstic acid.

Another well-known but bidentate ion is the peroxo group. Both fluoride and peroxo ions can occupy up to 8 coordination centers at the tungsten ion. Hydrofluoric acid as complexing agent in combination with an oxidizing acid or hydrogen peroxide (which represents both properties) are very important solvents for analytical samples of tungsten metal and several of its compounds.

Tungsten species in aqueous solutions combine with a variety of chelating agents offering oxygen, sulfur, selenium, and nitrogen as donor atoms. With tungsten as central ion in a chelate ring, the following systems have been reported:

$$
\begin{array}{c}
-\mathrm{O}-\mathrm{W}-\mathrm{O}- \\
-\mathrm{O}-\mathrm{W}-\mathrm{S}- \\
-\mathrm{O}-\mathrm{W}-\mathrm{N}- \\
-\mathrm{S}-\mathrm{W}-\mathrm{S}- \\
-\mathrm{S}-\mathrm{W}-\mathrm{N}-
\end{array}
$$

Important examples are chelates formed with polycarboxylic acids or polyhydroxy–polycarboxylic acids like oxalic, tartaric, or citric acid, which can be used to stabilize acidic tungsten solutions (to prevent polycondensation), or chromogenic agents which react to colored chelates, like f.e. toluene-3,4-dithiol. Several chelates play an important role in the analytical chemistry of tungsten.

3.8. ELECTROCHEMISTRY OF TUNGSTEN

3.8.1. Introduction [3.89]

The electrochemical behavior of tungsten in aqueous solutions is very closely linked to two facts: the high affinity of tungsten to oxygen and the complexity of tungsten species in aqueous solution.

There are an immense number of publications dealing with the electrochemistry of tungsten, based on practical investigations (a complete compilation is presented elsewhere [3.90]). The interpretations of the results are often contradictory. Their hypothetical nature is associated with a general lack in knowledge of tungsten chemistry in aqueous solutions (see Section 3.7). For example, it is never mentioned that for the deposition of tungsten at the cathode the existence of a positively charged particle is a prerequisite.

In aqueous solution, no oxygen-free tungsten cation exists but only monomeric or dimeric species like WO_2^{2+}. The discharge of such a species at the cathode yields an oxidic deposit.

Tungsten, due to its great affinity to oxygen, may never form a metal/metal ion electrode system but always a metal/metal oxide/metal ion system, which is quite irreversible. The oxide coating of the metallic tungsten in its chemical composition, properties, etc. will be strongly influenced by the ions and equilibria present in solution, which themselves are closely associated with pH, temperature, concentration, presence of complexing agents or oxidants, etc. Therefore, not only the chemistry of tungsten in solution but also the oxide layers at the W electrode are quite complex and not always understood. For these reasons, no experimental values for standard potentials are available.

The steady-state potential measured against a saturated calomel electrode (SCE) is negative in alkaline and positive in acidic solution, and even higher positive in the presence of oxidizing agents. Calculated standard potentials from thermodynamic data are around -0.1 V in acidic and around -0.9 V in alkaline solution against NHE.

The pH dependence (0–10) is linear (about 50 mV/pH). The potential becomes more negative with increasing temperature due to a partial transformation of a metal oxide to a metal electrode (the electrode properties shift from semiconductive to metal conductive).

In aqueous solution, the following reactions and corresponding potentials are assumed:

$$WO_2 + 4H^+ + 4e^- \rightarrow W + 2H_2O \quad E = -0.119 - 0.0591 \cdot pH \tag{1}$$
$$W_2O_5 + 2H^+ + 2e^- \rightarrow 2WO_2 + H_2O \quad E = -0.031 - 0.0591 \cdot pH \tag{2}$$
$$2WO_3 + 2H^+ + 2e^- \; W_2O_5 + H_2O \quad E = -0.029 - 0.0591 \cdot pH \tag{3}$$
$$WO_4^{2-} + 2H^+ \rightarrow WO_3 + H_2O \quad \log(WO_4^{2-}) = -14.05 + \cdot 2\,pH \tag{4}$$
$$WO_4^{2-} + 8H^+ + 6e^- \rightarrow W + 4H_2O \quad E = 0.049 - 0.0788\,pH + 0.098(\log WO_4^{2-}) \tag{5}$$
$$WO_4^{2-} + 4H^+ + 2e^- \rightarrow WO_2 + 2H_2O \quad E = 0.386 - 0.1182\,pH + 0.0295(\log WO_4^{2-}) \tag{6}$$
$$2WO_4^{2-} + 6H^+ + 2e^- \rightarrow W_2O_5 + 2H_2O \quad E = 0.801 - 0.1773\,pH + 0.0591(\log WO_4^{2-}) \tag{7}$$

Areas of passivation, corrosion, and immunity in aqueous systems can be seen in Fig. 3.23. Accordingly, electrodeposition of tungsten is only possible when the hydrogen formation is blocked. However, this will not only be influenced in practice by thermodynamic but also by kinetic properties. Therefore, it is only possible to deposit tungsten

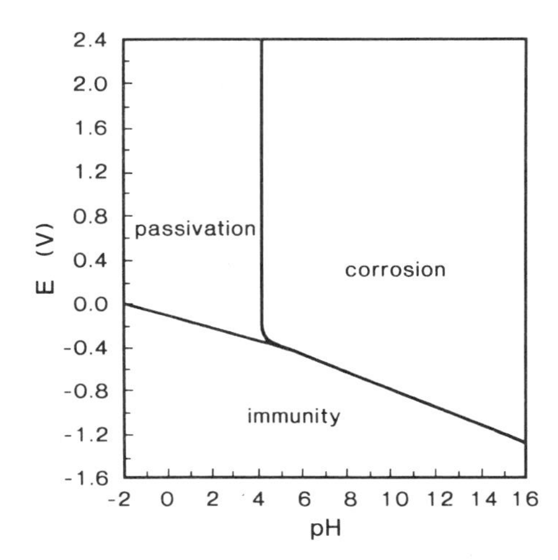

FIGURE 3.23. Theoretical regions of passivation, corrosion, and immunity of tungsten at 25 °C [3.89].

from aqueous solutions together with alloying elements. Another possibility is deposition from salt melts where water is completely excluded.

3.8.2. Cathode Reactions [3.89]

An oxide layer is immediately formed during electrolysis of a tungstate solution at a tungsten cathode. The deposits are high molecular oxides, which vary in degree of reduction, depending on the conditions of the electrolyte solution. The acidity and presence of other constituents determine the deposition rate and consequently the layer thickness.

WO_3 is a n-type semiconductor and is chemically reduced at the cathode. Therefore, a high potential at the metal–metal oxide interface is initiated, and hence reduction.

WO_3 has a chain structure. The partially reduced ion generated at the metal–metal oxide interface imparts high conductivity to the whole chain. Consequently, the charge produced at one end of the chain can be found at its other end (the whole system is an electrical field). The increase in conductivity of WO_3 is generated by changes in the electronic structure of the W atoms. Progress in reduction corresponds to decrease in oxygen concentration, a rising percentage of d^5-configuration atoms, increase in percentage of metallic bonds, and higher conductivity.

The partial W reduction is not only obtained at the metal–metal oxide or at the metal oxide–electrolyte interface, but the oxide turns blue immediately throughout the whole layer. This proves the above statement concerning the macromolecular film structure.

When the film thickness has attained a certain thickness, its further reduction is suppressed, and the generation of hydrogen starts by the discharge of protons. Partially reduced tungsten oxides may act as catalysts for that reaction. Film thickness and a start of hydrogen generation depend on the overvoltage of hydrogen, which itself is influenced by pH, temperature, other ions, or compounds in solution, solvents, etc.

From aqueous solutions, tungsten can only be deposited electrolytically together with other metals like Co, Fe, Ni, Cu, etc. Hypotheses dealing with the respective deposition mechanisms are referred to elsewhere [3.89].

3.8.3. Anodic Processes [3.90]

Metallic tungsten as an anode in basic solutions is dissolved electrolytically (oxidized to the hexavalent state). The six-electron stoichiometry of the overall reaction

$$W + 8OH^- \rightarrow WO_4^{2-} + 6e^- + 4H_2O$$

masks a very complex mechanism, which can be described by the following equations (s stands for solid and aqu for dissolved):

$$W(s) + 2OH^- \rightarrow WO^+(s) + 3e^- + H_2O$$
$$WO^+(s) + 2OH^- \rightarrow WO_2(s) + e^- + H_2O$$
$$WO_2(s) + OH^- \rightarrow WO_3H(s) + e^- \text{ (rate det. step)}$$
$$WO_3H(s) + OH^- \rightarrow WO_3(s) + e^- + H_2O \text{ (fast)}$$
$$WO_3(s) + OH^- \rightarrow HWO^{4-}(aqu)$$
$$HWO^{4-}(aqu) \rightarrow WO_4^{2-}(aqu) + H_2O$$

This corresponds to a stepwise oxidation from the metallic (elemental) state, via W^{3+}, W^{4+}, W^{5+} to W^{6+}.

3.8.4. *Molten Salts* [3.89]

As in aqueous solution also, in molten salts, too, tungsten behaves in an irreversible manner. The electrochemical activity of tungsten in melts depends on its position in the electromotive series, but this position changes with solvent and temperature. In NaOH, $NaPO_3$ and $Na_4P_2O_7$, it is electronegative, and in $Na_2B_4O_7$ and Na_3AlF_6 less so. Corresponding examples are given in Table 3.4.

In contrast to aqueous solutions, tungsten can be deposited from melts as metal at the cathode.

TABLE 3.4. Position of Tungsten in the Electromotive Series of Metals in Different Molten Salts

Solvent	Temp (°C)	Series
NaOH	340	W, Mo, Cd, Pb, Bi, Cu, Ag, Au, Pt, Fe, Ni, Ta, Nb, Zr
$NaPO_3$	720	Na, W, Zn, Cd, Fe, Pb, Co, Ni, Cu, Mo, Bi, Sb
$Na_2B_4O_7$	840	Na, Cr, Mo, Fe, Mn, W, Zn, Sb, Ni, Sn, Cd, Bi, Ti, Co, Pb, Cu, Ag
$Na_4P_2O_7$	1000	Na, Zn, W, Cd, Co, Ni, Cu, Pb, Fe, Mo, Sb, Nb
Na_3AlF_6	1000	Al, Mn, Cr, Nb, W, Fe, Co, Mo, Ni

3.8.5. *Practical Applications of Electrochemical Processes* [3.89]

pH-Sensitive Electrodes. The utilization of tungsten or tungsten bronze electrodes in pH measurement is based on the constant slope of the pH–potential dependency (50 mV/pH for tungsten and 58 mV/pH for tungsten bronze).

Electrochemical Treatment of Tungsten and Tungsten Alloys. Electrolytic etching and polishing is applied in industry to clean tungsten surfaces, or to precondition surfaces before coating or metallographic examination. Electrolytes and conditions for tungsten electropolishing are shown in Table 3.5.

Electroforming. The electroforming of W–Co or W–Co–Ni parts, which can be used in space vehicles and aircraft, is rather complicated and comprises the following steps: mechanical processing of an Al matrix, degreasing, zincate treatment, Ag plating, electrodeposition of the alloy, chemical removal of the matrix, and heat treatment [3.91].

Electrolytic Dissolution. The electrolytic dissolution of tungsten and tungsten alloy scrap in alkaline media (NaOH, Na_2CO_3, NH_3, $(NH_4)_2CO_3$) is used technically to recycle tungsten into the APT production (see Section 5.2.3.6.).

Production and Refining. The production of tungsten or its refining via electrolysis of molten salts, although technically possible, has never gained greater importance.

3.8.6. *Electrochromism* [3.92]

Electrochromism is the property of a material or a system to change its optical properties (color) reversibly if an external potential is applied. It is associated with ion insertion/extraction processes, which can be presented schematically by:

$$MeO_n + xI^+ + xe^- \rightarrow I_xMeO_n$$

TABLE 3.5. Electrolytes and Conditions for Tungsten Electropolishing [3.89]

Electrolyte	Anodic current density ($A \cdot dm^{-2}$)	Voltage (V)	Temperature (°C)	Time (s)
NaOH ($100\ g \cdot l^{-1}$)	66–300	—	25	—
	3–6	6	20	—
NaOH ($40\ g \cdot l^{-1}$) Na_2WO_4 ($40\ g \cdot l^{-1}$) $K_2Cr_2O_7$ ($0.5\ g \cdot l^{-1}$)	10–15	7	50	—
NaOH($5\ g \cdot l^{-1}$) Na_2WO_4 ($30\ g \cdot l^{-1}$)	400	21	40–60	—
KOH ($50-100\ g \cdot l^{-1}$)	50–1000	0.5–20	25	—
Na_2CO_3 ($150\ g \cdot l^{-1}$)	9	6–10	44–45	10
($50-150\ g \cdot l^{-1}$)	39	6–12	40–46	5–7
	50–100	0.5–20	25	—
Na_3PO_4 ($50-100\ g \cdot l^{-1}$)	10–200	0.5–20	25	—
($50-150\ g \cdot l^{-1}$)	30–1000	0.5–20	50	—
NaCN ($70\ g \cdot l^{-1}$) Ethanol ($300\ g \cdot l^{-1}$)	2	50	35–45	—
Na_2S ($50\ g \cdot l^{-1}$)	50–1000	0.5–20	25	—
Methanol (93.75 vol%) Hydrofluoric acid (1.25 vol%) Sulfuric acid (5.00 vol%)	—	15–20	—	15–20
Methanol (590 ml) Sulfuric acid (25 ml)	62	—	—	15–20
Methanol (590 ml) 2-Butoxyethanol (350 ml) Perchloric acid (60 ml)	—	15–20	—	15–20

where Me = metal atom, I = single charged small ion, e = electron, and n depends on the oxide type.

Electrochromism is in principle a device property, although the optical function can sometimes be caused by a single layer. The basic design of an electrochromic device, presented in Fig. 3.24, consists of several layers. The substrate (mostly glass) is covered by a transparent, conducting film in contact with a film of the electrochromic substance. These films are followed by a layer of a fast ion conductor (electrolyte), an ion storage film, and another transparent conductor. The electrochromic and ion storage layers are conductors for ions and electrons while, the ion conductor has zero conductance for electrons.

WO_3 was the first discovered compound with electrochromism [3.93], and today it remains the most viable option for the respective devices [3.94–3.97]. The basic reaction can be presented by the following equation:

$$\underset{\text{colorless}}{WO_3} + xI^+ + xe^- \rightarrow \underset{\text{blue}}{I_xWO_3}$$

where I stands for H^+ or alkali and x is about 0.3.

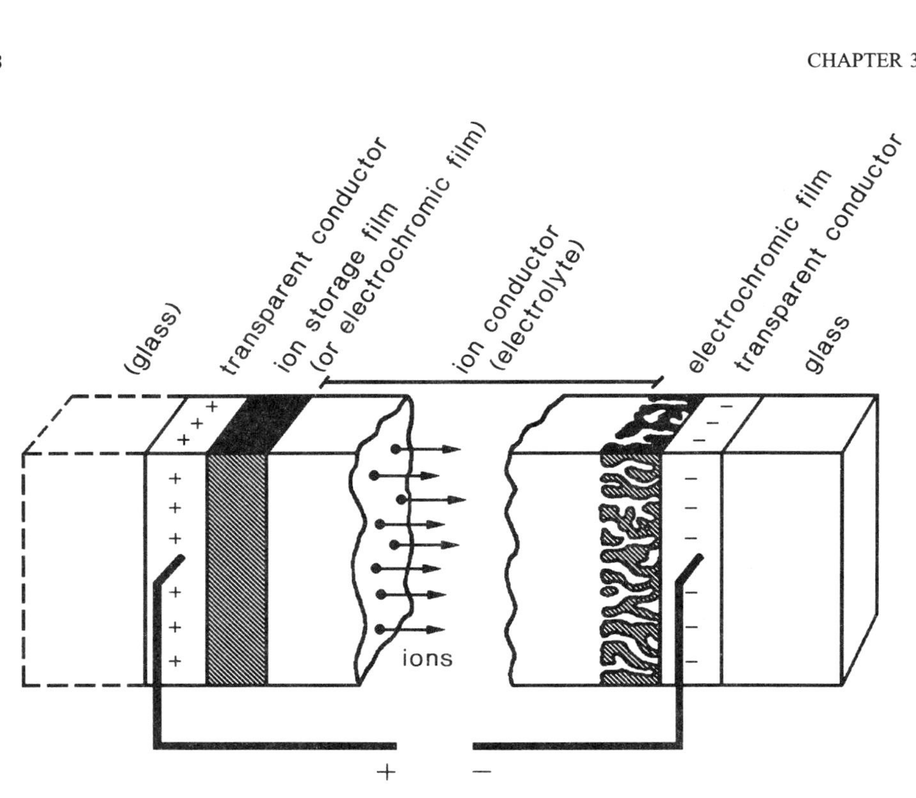

FIGURE 3.24. Basic design of an electrochromic device, indicating transport of positive ions under the action of an electric field [3.94].

In comparison to bulk WO_3, thin films always exhibit a high amount of microstructural disorder. This can be further influenced by the preparation mode.

Disordered WO_3 films transform from an optically transparent to an absorbing state under ion insertion (intercalation) (cathodic coloration). The octahedral WO_6 units are favorable for both ion as well as electron transport. The pertinent crystal structure allows long-range diffusion through tunnels or between layers (about WO_3 structures, see Section 4.2.7). Thin films show a cluster-type microstructure and a column-type macrostructure. This type of coordination leads to electronic bands, responsible for the electrochromic properties.

The electron insertion/extraction occurring jointly with the ionic movement is responsible for the actual electrochromism. In highly disordered WO_3, the optical effects are associated with the amount of W^{6+} transformed to W^{5+}.

In these thin films, the electrochromic processes are entirely reversible.

WO_3 films can be prepared by evaporation, sputtering, CVD, plasma-enhanced CVD, anodization, and the sol-gel technique. As already mentioned, the film properties depend strongly on the preparation mode and parameters.

In devices, diverse types of electrolytes have been used, such as liquid, solid inorganic bulk type, or thin film and solid organic polymer electrolytes.

Electrochromic devices are potential candidates for numerous applications:

(i) Modulation of the diffuse reflectance can be used in nonemissive display devices.

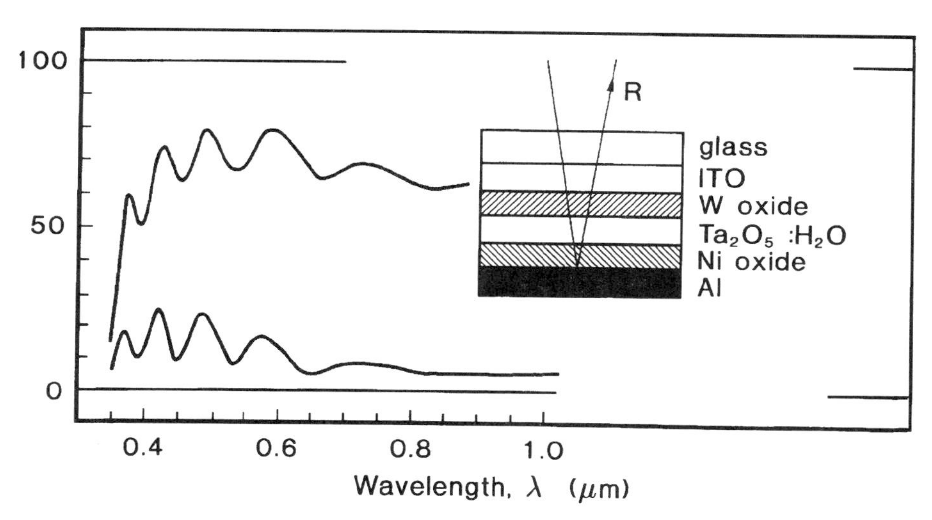

FIGURE 3.25. Spectral reflectance in colored and bleached states for an asymmetric electrochromic device with hydrated dielectric thin-film ion conductors; the design is sketched in the inset [3.94].

(ii) Modulation of the specular reflectance can be applied in antidazzling rear-view mirrors in automobiles.
(iii) Modulation of the luminous transmittance could be used in future buildings for superior daylight (optimal working conditions and saving of electrical lighting, as well as in sunglasses and windows of cars, trucks, and ships).
(iv) Modulation of solar energy transmittance for energy-efficient architecture and diminished need for airconditioning (smart windows).

An “all solid state” electrochromic device as used in antidazzling mirrors is shown schematically in Fig. 3.25. It consists of a glass plate coated with a transparent electrode, covered with the electrochromic WO_3 film, and attached to a solid electrolyte, a reflector, and a counterelectrode. An applied voltage drives the charge into the electrochromic layer, thus causing the change in absorption. Depending on voltage, electrons migrate from the counterelectrode to the working electrode (absorbing), or vice versa (transparent).

REFERENCES FOR CHAPTER 3

3.1. *Gmelin Handbuch der anorganischen Chemie*, 8th ed., Syst. No. 54, *Tungsten*, Suppl. Vol. B1, Springer-Verlag, Berlin (1978).
3.2. S. W. H. Yih and C. T. Wang, *Tungsten, Sources, Metallurgy, Properties and Application*, Plenum Press, New York (1979).
3.3. A Warren, A. Nylund, and I. Olefjord, *Int. J. Refract. Met. Hard Matter.* **14** (1996), 345–353.
3.4. C. H. Lee, *Nature* **203** (1964), 1163.
3.5. E. A. Aitken, H. C. Brassfield, P. K. Conn, E. C. Duderstadt, and R. E. Fryxell, *Trans. AIME* **239** (1967), 1565–1574.
3.6. R. E. Fryxell, *GEMP* **439** (1996), 1–13.
3.7. J. N. Ong and W. M. Fassel, *Corrosion*, **18** (1962) 382T–389T.
3.8. *Gmelin Handbuch der anorganischen Chemie*, 8th ed., Syst. No. 54, Tungsten Suppl. Vol. A7, Springer-Verlag, Heidelberg (1987).

3.9. T. Hamamura, *Bull. Chem. Soc. Jpn.* **32** (1959), 1180–1204.
3.10. G. R. Belton and R. L. McCarron, *J. Phys. Chem.* **68** (1964), 1852–1856.
3.11. Yu. G. Goryachkovskii, V. I. Kostikov, and G. A. Solodkin, *Zh. Fiz. Khim.* **50** (1976), 1959–1962.
3.12. R. Haubner, W. D. Schubert, E. Lassner, M. Schreiner, and B. Lux, *J. Refract. Met. Hard Mater.* **2** (1983), 108–115.
3.13. R. Haubner, W. D. Schubert, H., Hellmer, E. Lassner, and B. Lux, *J. Refract. Met. Hard Mater.* **2** 1983), 156–162.
3.14. H. Hellmer, W. D. Schubert, E. Lassner, and B. Lux, in: *Proc. 11th Int. Plansee Seminar* (H. Bildstein and H. Ortner, eds.), Vol. 3, pp. 43–86, Metallwerk Plansee GhmbH, Reutte, Austria (1985).
3.15. W. D. Schubert, *J. Refract. Met. Hard Mater.* **4** (1990), 178–191.
3.16. D. Parson, *Electrochem. Technol.* **3** (1965), 280–283.
3.17. C. Choain and F. Marion, *Compt. Rend.* **252** (1961), 3258–3260.
3.18. G. R. St. Pierre, W. T. Ebihara, M. J. Pool and R. Speiser, *Trans. TMS-AIME* **224** (1962), 259–264.
3.19. R. J. Ackermann and E. G. Rauh, *J. Phys. Chem.* **67** (1963), 2596–2601.
3.20. J. Bousquet and G. Perachon, *Comp. Rend.* **8** (1964), 934–936.
3.21. J. Bousquet and G. Perachon, *Comp. Rend.* **258** (1964), 3869–3871.
3.22. B. Phillips and L. Chang, *Trans. TMS-AIME* **230** (1964), 1203–1206.
3.23. F. E. Rizzo, L. R. Bidwell, and D. F. Frank, *Trans. TMS-AIME* **39** (1967), 1901–1905.
3.24. L. L. Y. Chang and B. Phillips, *J. Am. Ceram.* **52** (1969), 527–533.
3.25. P. E. Blackburn, M. Hoch, and H. L. Johnston, *J. Phys. Chem.* **62** (1958), 769–773.
3.26. P. Taskinen, P. Hytoenen, and M. H. Tikkanen, *Scand. J. Met.* **6** (1977), 228–232.
3.27. T. Millner and J. Neugebauer, *Nature* **163** (1949) 601–602.
3.28. O. Glemser and H. Ackermann, *Z. Anorg. Allg. Chem.* **325** (1963), 281–286.
3.29. O. Glemser and H. G. Voltz, *Naturwissenschaften* **43** (1956), 33.
3.30. O. Glemser and R. Haeseler, *Z. Anorg. Allg. Chem.* **316** (1962), 168.
3.31. J. Neugebauer, *Acta Chim. Acad. Sci. Hung.* (1963), 247–249.
3.32. W. Sahle and S. Berglund, *J. Less-Common Met.* **79** (1981), 271–280.
3.33. H. Mayer, E. Lassner, M. Schreiner, and B. Lux, *High Temp. High Pressures* **13** (1981), 529–539.
3.34. E. Lassner, M. Schreiner, and B. Lux, *J. Refract. Met. Hard Mater.* **1** (1982), 52–97.
3.35. R. Haubner, W. D. Schubert, E. Lassner, and B. Lux, in: *Proc. 11th Plansee Seminar*, Vol. 2 (H. Bildstein and H. Ortner, eds.), pp. 69–97, Verlagsanstalt Tyrolia, Innsbruck, Austria (1985).
3.36. R. Haubner, W. D. Schubert, E. Lassner, and B. Lux, *J. Refract. Met. Hard Mater.* **6** (1987), 40–45, 111–116, 161–167.
3.37. R. Haubner, W. D. Schubert, E. Lassner, and B. Lux, *J. Refract. Met. Hard Mater.* **7** (1987), 47–56.
3.38. W. D. Schubert, B. Lux, and B. Zeiler, in: *The Chemistry of Non Sag Tungsten* (L. Bartha, E. Lassner, W. D. Schubert, and B. Lux, eds.), Elsevier, Oxford (1995).
3.39. A. J. Hegedüs, T. Millner, J. Neugebauer, and K. Sasvari, *Z. Anorg. Allg. Chem.* **281** (1955), 64–82.
3.40. W. R. Morcom, W. L. Worrell, H. G. Sell, and H. I. Kaplan, *Metall. Trans.* **5** (1974), 155–161.
3.41. J. Qvick, *Theoretical and Experimental Studies of Trace Elements in Tungsten and Cemented Carbides*, Doctoral Dissertation, Uppsala University, Sweden (1987).
3.42. L. Bartha, E. Lassner, W. D. Schubert, and B. Lux, eds., *The Chemistry of Non Sag Tungsten*, Elsevier, Oxford (1995).
3.43. *Gmelin Handbuch der anorg. Chemie*, 8th ed., Syst. No. 54, *Tungsten*, Suppl. B2, Springer-Verlag, Heidelberg (1979).
3.44. M. Miyake, A. Hara, T. Sho, and Y. Kawabata, in: *Proc. 5th Eur. Symp. Powder Met.*, Vol. 2, pp. 94–98 (1978).
3.45. W. L. Worrell, *Trans. Met. Soc. AIME* **233** (1965), 1173–1177.
3.46. J. K. Fisher and D. R. Moyle, in: *Proc. 13th Int. Plansee Seminar* (H. Bildstein and R. Eck, eds.), Vol. 2, pp. 425–439, Plansee Metall AG, Reutte, Austria (1993).
3.47. A. K. Basu and F. R. Sale, *Trans. Met. Soc. AIME* **9B** (1978), 603–613.
3.48. Y. Yamamoto, a. Matsumoto, and Y. Doi, in: *Proc. 14th Int. Plansee Seminar* (G. Kneringer, P. Roedhammer, and P. Wilharitz, eds.), Vol. 2, pp. 596–608, Plansee AG, Reutte, Austria (1997).
3.49. S. D. Dunmead, W. G. Moore, A. W. Weimer, G. A. Eisman, and J. P. Henley, US Patent No. 5,380,688 (1995).
3.50. F. H. Ribeiro, R. A. Dalla Betta, M. Boudart, J. Baumgartner, and E. Iglesia, *J. Catal.* **130** (1991), 86–105.

3.51. S. Decker, A. Loefberg, J.-M. Bastin, and A. Frennet, *Catal. Lett.* **44** (1997), 229–239.
3.52. S. T. Oyama, in: *The Chemistry of Transition Metal Carbides and Nitrides* (S. T. Oyama, ed.), pp. 1–27, Chapman & Hall, Glasgow (1996).
3.53. L. E. McCandlish and P. Seegopaul, in: *Proc. Europ. Conf. on Advances in Hard Mater. Production, Stockholm*, pp. 93–100, EPMA, Shrewsbury, UK (1996).
3.54. *Gmelin Handbuch der anorganischen Chemie*, 8th ed., Syst. No. 54, *Tungsten*, Suppl. Vol. A1, Springer-Verlag, Heidelberg (1979).
3.55. Mallory Tungsten Powder Brochure.
3.56. K. A. Jonsson and E. K. A. Svanstrom, German Patent 1,945.154 (1970).
3.57. R. Huenert, G. Winter, W. Kiliani, and D. Greifendorf, *J. Refract. Met. Hard Mater.* **11** (1992), 331–335.
3.58. A. A. Zinn, "Chemical Vapor Deposition of Tungsten," in: *The Chemistry of Metal CVD* (T. T. Kodas and M. J. Hampden-Smith, eds.), VCH, Weinheim (1994).
3.59. Union Carbide Corp., J. B. Culbertson, H. Lamprey, and R. L. Ripley, U.S. Patent 3.062.638 (1961, 1962).
3.60. J. E. Tress, T. T. Campbell, and F. E. Block, *U.S., Bur. Mines, Rep. Invest.* No. 6835 (1966).
3.61. L. Ramqvist, in: *Modern Developments in Powder Metallurgy* (H. H. Hausner, ed.), Vol. 4, pp. 75–84, Plenum Press, New York (1971).
3.62. G. Y. Zhao, V. V. Revankar, and V. Hlavacek, *J. Less-Common Met.* **163** (1990), 269–280.
3.63. V. Revankar, G. Y. Zhao, and V. Hlavacek, *Ind. Eng. Res.* **30** (1991), 2344–2349.
3.64. H. Ishikawa, S. Watabiki, S. Takajo, and K. Otsuka, *Solid State Phenomena* **25–26** (1992), 175–178.
3.65. J. Hojo, T. Out, and A. Kato, *J. Less-Common Met.* **59** (1978), 85–95.
3.66. Soo-Sik Kim and Han-Sam Kim, *J. Korean Inst. Met. Mat.* **29** (1991), 640–648, 651–658, 659–664.
3.67. E. Neuschwander, *J. Less-Common Met.* **11** (1966), 365–375.
3.68. P. Matteazzi and G. Le Caer, *J. Am. Ceram. Soc.* **74** (1991), 1382–1390.
3.69. *Gmelin Handbook of Inorganic Chemistry*, 8th ed., Syst. No. 54, *Tungsten*, Suppl. Vol. A5b, pp. 131–153, Springer-Verlag, Heidelberg (1993).
3.70. W. Geiss and J. A. M. van Liempt, *Z. Metallkd.* **16** (1924), 317–318.
3.71. L. V. McCarthy, R. Donelson, and R. F. Hehemann, *Metall. Trans.* **18A** (1987), 969–974.
3.72. Tao Zhengji, *J. Refract. Met. Hard Mater.* **6** (1987), 221–225.
3.73. E. Lassner, M. Schreiner, and B. Lux, *Proc. 10th Plansee Seminar* (H. Bildstein and H. Ortner, eds.), Vol. 2, pp. 761–793, Metallwerk Plansee GmbH, Reutte, Austria (1982).
3.74. A. Tschinkowitz, *Diploma-Thesis*, Vienna University of Technology (1989).
3.75. I. I. Kovenskii, *Diffusion in Body-Centered Cubic Metals*, (1965) A.S.M., Metals Park, Novelty, Ohio, p. 283.
3.76. A. Shepela, *J. Less-Common Met.* **26** (1972), 33–43.
3.77. L. N. Aleksandrov and V. Y. Schelkonogov, *Poroshk. Metall. Akad. Nauk Ukr. SSR* **22** (1964), 28.
3.78. A. I. Nakonechnikov, L. V. Paulinov, and V. N. Bykov, *Fiz. Met. Metalloved.* **22** (1966), 234.
3.79. R. J. Fries, J. E. Cummings, C. G. Hoffman, and S. A. Daily, *Proc. 6th Plansee Seminar*, pp. 568–607, Reutte, Austria (1968).
3.80. M. Andrews and S. Dushman, *J. Phys. Chem.* **29** (1925), 462–471.
3.81. G. S. Kreimer, L. D. Efros, and E. A. Voronkova, *Zh. Tekhn. Fiz.* **22** (1952), 858.
3.82. J. M. Gomes, A. E. Raddatz, and T. G. Carnahan, *Proc. 3rd Int. Tungsten Symposium, Madrid*, pp. 96–109, MPR Publishing Services Ltd., Shrewsbury, England (1985).
3.83. N. N. Greenwood and A. Earnshaw, *The Chemistry of the Elements*, Pergamon Press, Oxford (1984).
3.84. T. K. Kim, R. W. Mooney, and V. Chiola, *Sep. Sci.* **3** (1968), 467–478.
3.85. V. Cordis, K. H. Tytko, and O. Glemser, *Z. Naturforsch.* **306** (1975), 834–841.
3.86. J. W. Van Put, "Crystallization and Processing of Ammonium Paratungstate," in: *The Chemistry of Non Sag Tungsten* (L. Bartha, E. Lassner, W. D. Schubert, and B. Lux eds.), Elsevier Science, Oxford (1995).
3.87. R. Püschel and E. Lassner, in: *Chelates in Analytical Chemistry* (H. A. Flaschka and A. J. Barnard Jr. eds.), Marcel Dekker, New York (1967).
3.88. O. Glemser, N. Holznagel, W. Holtze, and E. Schwarzmann, *Z. Naturforsch.* **206** (1965), 192–196.
3.89. A. T. Vas'ko, in: *Encyclopedia of Electrochemistry of the Elements* (A. J. Bard, ed.), pp. 69–126, Marcel Dekker, New York (1967).
3.90. *Gmelin Handbook of Inorganic Chemistry*, 8th ed., Syst. No. 54, *Tungsten*, Suppl. Vol. A5a, Springer-Verlag, Heidelberg (1990).
3.91. M. E. Browning and E. W. Turn, in: *Symposium on Electroforming Application; Uses and Properties of Electroformed Metals*, p. 107, Dallas, Texas (1962).

3.92. C. G. Granquist, *Handbook of Inorganic Electrochromic Materials*, Elsevier, Amsterdam (1995).
3.93. T. Kraus, *Laboratory Report*, Balzers, AG, Liechtenstein, 30.07.1953.
3.94. C. G. Granquist, "Electrochromic Tungsten-Oxide-Based Thin Films: Properties, Chemistry, and Technology, in: *Physics of Thin Films* (M. Francombe and C. Vossen, eds.), Vol. 17, pp. 301–370, Academic Press, San Diego (1993).
3.95. C. G. Granquist, *Sol. Energy Mater. Sol. Cells* **32** (1994), 369–382.
3.96. C. G. Granquist, *Solid State Ionics* **70/71** (1994), 678–685.
3.97. A. Azens, A. Hjelm, D. de Bellac, C. G. Granquist, J. Barczynska, E. Pentjuss, J. Gabrusenoks, and J. M. Wills, *Solid State Ionics* **86–88** (1996), 943–948.

4

Tungsten Compounds and Their Application

The chapter starts with a summary of the tungsten compounds with metallic elements (intermetallic compounds), and is followed by a more or less detailed description (depending on the importance of the respective compound) of the tungsten compounds with nonmetallic elements in sequence from the 14th to the 17th group of the Periodic Table. Because of the extremely large number of corresponding compounds known, mixed ligand compounds and coordination compounds as well as organometallics are treated only in general in the last two sections.

A short, general discussion at the beginning emphasizes that the chemistry of tungsten compounds is rather complicated and complex. The following important facts must be taken into account.

- Tungsten forms compounds in the valence states -2, -1, 0, $+2$, $+3$, $+4$, $+5$, and $+6$. The most stable compounds belong to the higher valence states $+5$ and $+6$. The stability of lower valence states is less and, in particular, the $+2$ state can be stabilized only by a high degree of metal-to-metal multiple bonds.
- Normally the maximal coordination number is 9, but in combination with peroxo ions or σ-bound aromatic ring systems it may attain 13.
- The hexavalent oxygen compounds have the tendency to form condensed compounds with preferential condensation grades of W_6 and W_{12}. The tungsten atoms in those compounds are connected via oxygen bridges. Higher condensed compounds form in the presence of heteroions, like, for example, silicate or phosphate. Several isopolytungstate ions surround the central heteroion. Typical condensation grades are 6, 9, and 12.
- Lower valence compounds ($+2$, $+3$, and partially $+4$) are not monomeric but consist of clusters having a high degree of W–W bondings.

Examples of compounds for different valence states are given in Table 4.1 [4.1]. The following abbreviations are used throughout the chapter:

TABLE 4.1. Typical Examples of Tungsten Compounds for the Different Valence States [4.1].

Valence state	Number of electrons in 5d	Coordination number	Structure	Formula
−2	8	5	Trigonal bipyramid	$[W(CO)_5]^{2-}$
−1	7	6	Octahedral	$[W_2(CO)_{10}]^{2-}$
0	6	6	Octahedral	$W(CO)_6$
+2	4	5	Square bipyramid	$[W_2(CH_3)_8]^{4-}$
		6	Octahedral	$[W_2(\text{diarsine})_2I_2]$
		9		$[W(\eta^5\text{-}C_5H_5)(CO)_3ClW_6Cl_{12}]$Cluster
+3	3	4	Tetrahedral	$[(R_2N)_3W{\equiv}W(NR_2)_3]$
		6	Octahedral	$[W_2Cl_9]^{3-}$
+4	2	6	Octahedral	$[WCl_6]^{2-}$
		6	Trigonal prismatic	WS_2
		8	Dodecahedral	$[W(CN)_9]^{4-}$
		8	Antiprismatic	$[W(\text{Picolinate})_4]$
		12		$[W(\eta^5\text{-}C_5H_5)_2X_2]$
+5	1	6	Octahedral	$[WF_6]^-$
		8	Dodecahedral	$[W(CN)_8]^{3-}$
		13		$[W(\eta^5\text{-}C_5H_5)_2H_3]$
+6	0	4	Tetrahedral	$[WO_4]^{2-}$
		5		$[WOX_4]$
		5	Square pyramid	$[W{\equiv}(CCMe_3)({=}CHCMe_3)(CH_2CMe_3)\{(PMe_2CH_2)_2\}]$
		6	Octahedral	$[WO_6]$ in polytungstates
		6	Trigonal prismatic	$[W(W_2S_2H_2)_3]$
		7	Pentagonal bipyramid	$[WOCl_4(\text{diarsine})]$
		8		$[WF_8]^{2-}$
		9	Tricapped trigonal	$[WH_6(Pr_2PPh)_3]$

PR	Preparation	BP	Boiling point
PP	Physical properties	ΔH°_{298}	Enthalpy of formation per mol (298.15 K)
D	Density	CP	Chemical properties
MH	Microhardness	HR	Homogeneity range
CO	Color	STR	Stereochemical structure
CR	Crystal structure: Lattice parameters	I	Importance
MP	Melting point	A	Application

Thermodynamic parameters (with exception of the enthalpy of formation) are not considered in this book. They can, however, be easily obtained from the relevant standard compilations [4.2, 4.3]. Data for the following compounds are available: WBr (g), WBr_5 (s, l, g), WBr_6 (s, g), WC (s), W_2C (s), $W(CO)_6$ (s, g), WCl (g), WCl_2 (s, g), WCl_4 (s, g), WCl_5 (s, l, g), WCl_6 (s, l, g), W_2Cl_{10} (g), WF (g), WF_6 (g), WO (g), WO_2 (s, g), $WO_{2.72}$ (s), $WO_{2.9}$ (s), WO_3 (s, l, g), W_2O_6 (g), W_3O_8 (g), W_3O_9 (g), W_4O_{12} (g), H_2WO_4 (s),

H_2WO_4 [$WO_2(OH)_2$] (g), $WOCl_4$ (s, l, g), WO_2Cl_2 (s, g), WOF_4 (s, l, g), WO_2I_2 (g), and WS_2 (s).

4.1. TUNGSTEN AND METALS: INTERMETALLIC COMPOUNDS AND PHASES [4.4–4.8]

Tungsten forms intermetallic compounds and phases with the following elements: Be, Al, Ca, Ce, Ge, Zr, Hf, Tc, Re, Fe, Ni, Co, Ru, Rh, Os, Ir, and Pt.

In most cases such binary phases are not of importance for the production of alloys, because they exhibit completely different physical properties in comparison to the materials of the solid solution regions (higher hardness, brittleness, etc.) which are in use as alloys. A typical example are σ-phase inclusions in powder metallurgically produced W-10Re alloy, which lead to a pronounced embrittlement. They are the consequence of improper mixing of the two components prior to the sintering process. Re-enriched areas cannot be completely equalized by diffusion during sintering.

In principle, one could imagine that in an alloy a dispersion-strengthening influence could be achieved by a finely distributed intermetallic compound. Complex intermetallic phases account for the precipitation hardening in low carbon high-speed tool steels (see Section 8.1.3) and in superalloys (see Section 8.2). Furthermore, intermetallic phases can appear in heat-treated WC–Co alloys, increasing the hardness but significantly lowering the toughness.

Table 4.2 summarizes these compounds and details some properties.

TABLE 4.2. Tungsten Intermetallic Compounds

Group[a]	Element	Formula	Properties
2	Be	WBe_{22}	4.3 at% W CR: Cubic $Zn_{22}Zr$-type $a = 11.61–11.64$ Å
		WBe_{12}	D: 4.25 $g \cdot cm^{-3}$ 7.7 at% W CR: Tetragonal $D2_b$ $Mn_{12}Th$-type $a = 7.234–7.362$ Å; $c = 4.216$ Å
		WBe_2	HR: 28–36 at% W CR: Hexagonal C_{14} $MgZn_2$-type $a = 4.446–4.578$ Å; $c = 7.289–7.429$ Å
4	Zr	W_2Zr	66.7 at% W CR: Cubic C_{15} Cu_2Mg-type $a = 7.612–7.6195$ Å
	Hf	W_2Hf	66.7 at% W CR: Cubic C_{15} Cu_2Mg-type $a = 7.5825–7.591$ Å
7	Tc	WTc_3	σ-Phase HR: 25–28 at% W (1800 °C) CR: Tetragonal $D8_b$ $a = 9.479–9.507$ Å; $c = 4.952–5.166$ Å

(continued)

TABLE 4.2. (*continued*)

Group[a]	Element	Formula	Properties
	Re	κ-Phase	26.5 at% W CR: Cubic A12 α-Mn-type $a = 9.54–9.588$ Å σ-Phase HR: 29.0–56.5 at% W CR: Tetragonal $D8_b$ σ-CrFe-type $a = 9.55–9.645$ Å; $c = 4.98–5.038$ Å
8	Fe	WFe_2	λ-Phase metastable 33.3 at% W CR: Hexagonal C_{14} $MgZn_2$-type $a = 4.727–4.745$ Å; $c = 7.694–7.726$ Å
		W_6Fe_7	μ-Phase 42.1 at% W CR: Hexagonal $D8_5$ Fe_7W_6-type $a = 4.714–4.764$ Å; $c = 25.728–25.85$ Å
		WFe	δ-Phase 50.0 at% W CR: Orthorhombic MoNi-type $a = 7.76$ Å; $b = 12.48$ Å; $c = 7.10$ Å
	Ru	W_3Ru_2	ρ-Phase HR: 59.5–66.5 at% W CR: Tetragonal $D8_b$ σ-CrFe-type $a = 9.55–9.561$ Å; $c = 4.96–4.973$ Å
	Os	W_3Os	σ-Phase 66.7 at% W CR: Tetragonal $D8_b$ σ-CrFe-type $a = 9.6343–9.659$ Å; $c = 4.9810–5.001$ Å
		WOs_3	CR: Hexagonal (existence uncertain) $a = 2.75$ Å; $c = 4.37$ Å
9	Co	WCo_3	χ-Phase 25 at% W CR: Hexagonal $D0_{19}$ Ni_3Sn-type $a = 5.126–5.130$ Å; $c = 4.125–4.128$ Å
		W_6Co_7	γ-Phase 46 at% W CR: Rhombic $D8_5$ Fe_7W_6-type $a = 8.957$ Å; $\alpha = 30.66°$
	Rh	WRh	ε-Phase HR: 20–52 at% W CR: HCP A3 Mg-type $a = 2.708$ Å; $c = 4.328$ Å
		WRh_3	ε′-Phase HR: 25 at% W CR: Hexagonal $D0_{19}$ Ni_3Sn-type $a = 5.453$ Å; $c = 4.350$ Å
	Ir	W_3Ir	σ-Phase 75 at% W CR: Tetragonal $D8_b$ σ-CrFe-type $a = 9.672–9.700$ Å; $c = 4.99–5.010$ Å
		WIr	ε′-Phase CR: Orthorhombic B_{19} β-AuCd-type $a = 4.452$ Å; $b = 2.760$ Å; $c = 4.811$ Å
		WIr_3	ε′-Phase HR: 22–66 at% W CR: Hexagonal close-packed A3 Mg-type $a = 2.74–2.76$ Å; $c = 4.389–4.56$ Å
10	Ni	WNi_4	β-Phase 20.0 at% W CR: Tetragonal Dl_a $MoNi_4$-type $a = 5.730$ Å; $c = 3.533$ Å

(*continued*)

TABLE 4.2. (*continued*)

Group[a]	Element	Formula	Properties
		WNi	δ-Phase 50.0 at% W CR: Orthorhombic MoNi-type $a = 7.76$ Å; $b = 12.48$ Å; $c = 7.10$ Å
		W_2Ni	γ-Phase 66 at% W CR: Body-centered tetragonal $a = 10.40$ Å; $c = 10.90$ Å
	Pt		γ-Phase HR: 33–37 at% W CR: Tetragonally distorted fcc structure $a = 3.895$ Å; $c = 3.943$ Å ε-Phase HR: 45–52 at% W CR: Hexagonal close-packed Pt_3Mo_2-type $a = 2.796$ Å; $c = 4.493$ Å
13	Al	WAl_{12}	D: 3.88 $g \cdot cm^{-3}$ measured HR: 7.5–8.0 at% W CR: Cubic phase $a = 7.580$ Å
		WAl_5	HR: 16.7–18.0 at% W D: 5.50 $g \cdot cm^{-3}$ measured 5.71 $g \cdot cm^{-3}$ X-ray CR: Hexagonal δ-phase $MoAl_5$-type $a = 4.902$ Å; $c = 8.857$–8.860 Å
		WAl_4	HR: 19.0–21.5 at% W D: 6.60 $g \cdot cm^{-3}$ measured 6.70 $g \cdot cm^{-3}$ X-ray CR: Monoclinic ε-phase $a = 5.272$ Å; $b = 17.771$ Å; $c = 5.218$ Å; $\beta = 100.20°$
	Ga		
	No intermediates at atmospheric pressure but at 77 Kbar		
		W_2Ga_5 (400–800 °C)	31 at% W CR: Tetragonal Hg_5Mn_2-type $a = 8.948$ Å; $c = 2.674$ Å
		W_2Ga_3 (>800 °C)	HR: 39–47 at% W CR: Hexagonal $a = 3.006$ Å; $c = 4.738$ Å
14	Ge		
	No intermediates at atmospheric pressure but at ≥77 Kbar		
		β-WGe_2	66.7 at% W CR: $C11_b$ $MoSi_2$-type $a = 3.320$ Å; $c = 8.192$ Å
		α-WGe_2	66.7 at% W CR: C23 $PbCl_2$-type $a = 6.399$ Å; $c = 8.544$ Å
		W_5Ge_3	40 at% W CR: $D8_m$ W_5Si_3-type $a = 9.81$ Å; $c = 4.92$ Å
		W_5Ge_3	40 at% W CR: $D8_l$ Cr_5B_3-type $a = 6.25$ Å; $c = 11.72$ Å
L[b]	Ce	W_2Ce	No details available, existence uncertain

[a] New IUPAC proposal 1985.
[b] Lanthanoid series.

4.2. TUNGSTEN AND NONMETALS

4.2.1. Tungsten and Boron [4.4, 4.7–4.12]

The following stable intermediates exist: W_2B, WB, W_2B_{5-x}, and $W_{1-x}B_3$ (earlier WB_4, WB_{12}, $W_{2-x}B_9$). For more information on the tungsten–boron system and recent thermochemical data, see elsewhere [4.10].

PR: By solid state reaction between W and amorphous B powder pressed in compacts at elevated temperature; pretreatment at 500 °C in hydrogen for 1 hour followed by 800–1200 °C in argon for 2 hours. The formation of the individual boride depends on the W/B ratio.
By Al reduction of WO_2–B_2O_3 mixtures.
By salt melt electrolysis.
By vapor deposition.

PP: They all exhibit high hardness (>9 Mohs), brittleness, and metal-like electrical conductivity.

Ditungsten Boride: W_2B

PP: D: 16.72 g · cm^{-3}; MP: 2670–2780 °C; MH: 2350 kg · mm^{-2} (100 g); $-\Delta H°_{298} = 65.3$ kJ · mol^{-1}; CR: tetragonal *C*16, $CuAl_2$-type; $a = 5.564$–5.5673 Å, $c = 4.735$–4.748 Å; HR: narrow; transforms to WB when heated with carbon at 1900 °C.

Tungsten Boride: WB

PP: Occurs in a low-temperature (α) and high-temperature modification (β). Values for the transition temperature scatter in the literature are between 1950 and 2400 °C.

β-WB

PP: D: 16.00 g · cm^{-3}; CO: gray; MP: 2665–2920 °C; MH: 2600 kg · mm^{-2} (50 g); **modulus of elasticity**: 608 GPa; **electrical resistivity**: 25 μΩ · cm; $-\Delta H°_{298} = 54.5$ kJ · mol^{-1}; CR: orthorhombic B_f; CrB-type; $a = 3.124$–3.19 Å, $b = 8.4$–8.445 Å, $c = 3.06$–3.07 Å; HR: 48–52 at% B.

α-WB

PP: D: 16.00 g · cm^{-3}; CO: gray; $-\Delta H°_{298} = 63$ kJ · mol^{-1}; CR: tetragonal B_g, αMoB-type; $a = 3.101$–3.115 Å, $c = 16.91$–16.95 Å; HR: 49.0–51.5 at% B.

Ditungsten Pentaboride: W_2B_{5-x}

PP: D: 13.1 g · cm^{-3}; MP: 2200–2980 °C; MH: 2700 kg · mm^{-2} (100 g); **modulus of elasticity**: 755 GPa; **electrical resistivity**: 19 μΩ · cm; $-\Delta H°_{298} = 192.5$ kJ · mol^{-1}; CR: hexagonal $D8_h$, Mo_2B_5-type; $a = 2.982$–2.986 Å, $c = 13.81$–13.89 Å; HR: 68–69.3 at% B at 2170 °C.

Tungsten Tetraboride: $W_{1-x}B_3$

PP: D: 8.40 g · cm^{-3}; MP: 2020–2200 °C; $-\Delta H°_{298} = 82.5$ kJ · mol^{-1}; CR: tetragonal, $Mo_{1-x}B_3$-type; $a = 5.189$–5.200 Å, $c = 6.332$–6.340 Å.

4.2.2. Tungsten and Carbon [4.4, 4.7, 4.11, 4.13–4.16]

The binary tungsten–carbon system (Fig. 4.1) [4.13–4.15] is of high technical importance. It contains three intermediate compounds: $W_2C(\beta)$, $WC_{1-x}(\gamma)$, and $WC(\delta)$, the latter being the main constituent in most of the commercial cemented carbides (hardmetals; see Chapter 9). Besides, ternary and even more complex compounds exist which are of interest in alloyed steels and cemented carbides. Tungsten hexacarbonyl is an important precursor for organic synthesis.

A thermodynamic evaluation of the W–C system has been presented [4.16].

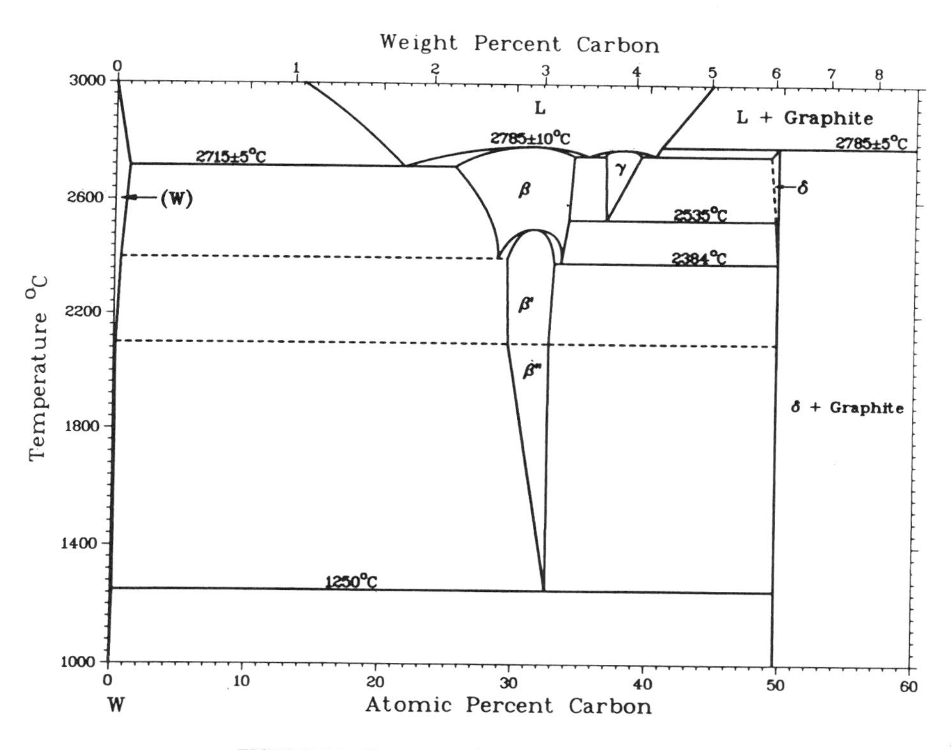

FIGURE 4.1. Tungsten–carbon phase diagram [4.15].

Ditungsten Carbide: W_2C

PR: It can be produced in pure state by mixing the appropriate amounts of tungsten powder and carbon black or graphite and subsequent heating in hydrogen atmosphere to at least 1600 °C; alternatively, it can be easily produced by melting; compositions below $W_2C_{0.82}$ cannot be retained at room temperature, even by quenching in liquid tin [4.17].

PP: D: measured 17.2 $g \cdot cm^{-3}$; X-ray density: 17.34 $g \cdot cm^{-3}$; MP: 2785 ± 10 °C; MH: 1990 $kg \cdot mm^{-2}$ (50 g); **modulus of elasticity**: 420 GPa; **linear coefficient of expansion**: $1.2 \times 10^{-6} \cdot K^{-1}$ (*a*-axis), $11.4 \times 10^{-6} \cdot K^{-1}$ (*c*-axis); **electrical resistivity**: 76–80 $\mu\Omega \cdot cm$ (room temp.), 125 $\mu\Omega \cdot cm$ (2000 °C); **thermal coefficient of**

electrical resistivity: $1.95 \times 10^{-3} \cdot K^{-1}$; **Richardson constant** (thermionic emission): 190 $A/cm^{-2} \cdot K^{-2}$; **superconductivity critical temperature**: 2.74 K; $-\Delta H^{\circ}{}_{298} = 26.4\ kJ \cdot mol^{-1}$.

CR: The crystal structure of W_2C can be described as a slightly distorted hexagonal close-packing of tungsten atoms. The carbon atoms occupy half of the octahedral interstices. They may be distributed in an ordered manner, the type and degree of ordering depending on temperature. There exist 3 temperature dependent modifications: β, β′, and β″. The transition temperatures are:

$\beta' \rightarrow \beta''$	2100–2140 °C
$\beta' \rightarrow \beta$	2490 °C

β (earlier γW_2C)
hexagonal $P6_3/mmc$; , $a = 2.985$–3.000 Å, $c = 4.716$–4.730 Å; HR: 25.6–34.5 at% C; statistical occupation of the interstices.
β′ (earlier βW_2C)
orthorhombic *Pbcn*, ζ-Fe_2-N-type; $a = 4.728$ Å, $b = 6.009$ Å, $c = 5.193$ Å; HR: 29.6–32.7 at% C.
β″ (earlier εW_2C)
hexagonal; $P\bar{3}m1$; ε-Fe_2N-type; $a = 5.184$ Å, $c = 4.721$ Å [4.18]; HR: 29.6–32.5 at% C.

CP: W_2C decomposes at 1250 °C into W and WC. It is resistant to many acids, but will be dissolved by hot, concentrated HNO_3 or a mixture of HNO_3 and HF (1:4). F_2 attacks at room temperature and Cl_2 at 200 °C. It is oxidized by air or oxygen at 500 °C.

A: W_2C is a component of the cast carbide (W_2C–WC mixture), which is in use as a hard facing alloy.

WC_{1-x}

Cubic high-temperature modification; it undergoes a eutectoid decomposition at 2530–2535 °C into β(W_2C) + δ(WC). At room temperature it can only be obtained by rapid quenching in liquid tin.

CR: fcc *B*1, NaCl-type; $a = 4.220$ Å (at 38 at% C).

Tungsten Monocarbide: WC

It is the only stable tungsten carbide at room temperature.

PR: Reaction of tungsten trioxide with carbon in an inert atmosphere.
Vapor phase deposition.
Electrolysis of fused salts.
Direct combination of the elements (about minimum reaction temperatures, depending on the starting materials; see Section 3.6.1).
Due to the fact that WC melts incongruently, it cannot be produced by melting.
The direct combination of the elements is by far the most widely applied method for the industrial production of WC, followed by the reaction of WO_3 with carbon (see Chapter 9).

PP: D: measured 15.7 $g \cdot cm^{-3}$; X-ray density: 15.77 $g \cdot cm^{-3}$; **hardness**: inherent to its hexagonal structure, *WC single crystals* exhibit a pronounced hardness anisotropy; the room-temperature Knoop hardness of the basal plane (0001) varies from 2200 to 2500, depending on the direction of the long diagonal axis of the indenter, and of the prism planes between only 1000 in the [0001] direction to 2500 perpendicular to this direction [4.19]; the hardness anisotropy prevails at least up to 900 °C [4.20]; for *polycrystalline* WC the hardness increases with decreasing grain size (Hall–Petch relationship); for fine-grained WC it is in the range of 2400–2800 $kg \cdot mm^{-2}$ (HV30); **modulus of elasticity**: 670–707 GPa; **shear modulus**: 262–298 GPa; **compressive modulus**: 384 GPa; **plasticity** [4.21]: WC exhibits a pronounced plasticity; the formation of slip bands can be observed in the vicinity of hardness indentations or on compressive testing; the prism planes are the slip planes (1010), and the slip directions are both ⟨1120⟩ and ⟨0001⟩; the plasticity of WC is an important aspect for its use as hard component in hardmetals; WC is significantly tougher than other, covalent hard compounds (fracture toughness: 6–7 $MN \cdot m^{-3/2}$) [4.22], which contributes to the comparatively high fracture toughness of the WC–Co composite; **surface energy**: 1.7 $J \cdot m^{-2}$; $-\Delta H°_{298} = 40.5$ $kJ \cdot mol^{-1}$; **linear coefficient of thermal expansion**: $5.2 \times 10^{-6} \cdot K^{-1}$ (*a*-axis); $7.3 \times 10^{-6} \cdot K^{-1}$ (*c*-axis); **electrical resistivity**: 17–25 $\mu\Omega \cdot cm$; **thermal conductivity**: 1.2 $J \cdot cm^{-1} \cdot s^{-1} \cdot K^{-1}$; **magnetic susceptibility**: $+0.005 \times 10^{-6}$; **superconductivity critical temperature**: 10 K.

The extremely high modulus of elasticity of WC (only exceeded by diamond and W_2B_5), and the high electrical and thermal conductivity are further important criteria for its use in hardmetals. The latter two properties also reflect the strong metallic component of the mixed covalent (W5d-C2p) metallic bonds in the carbide. Fermi surface properties of WC and electronic band structure calculations can be found elsewhere [4.23, 4.24].

CR: Hexagonal *Bh*, WC-type; $a = 2.906$–2.9066 Å, $c = 2.8364$–2.8374 Å. The trigonal prismatic interstices of the parent metal lattice are completely filled; HR: WC exhibits a very narrow homogeneity range and is practically stoichiometric. This means that there are neither tungsten nor carbon vacancies or interstitials other than those created by thermal activation.

CP: It is highly resistant to many acids. Unlike W_2C, it is not attacked by a mixture of HNO_3 and HF (1:4) at room temperature, but dissolves on heating. Table 4.3 presents data on the resistance to chemical attack in various aqueous solutions. WC is resistant against Cl_2 up to 400 °C, but is attacked by fluorine at room temperature. Fine WC powders are readily oxidized in air or oxygen at $T > 300$–500 °C, depending on the particle size. They are also decarburized in wet hydrogen.

A: WC is the most important tungsten compound in regard to the percentage of total tungsten consumption. More than 60% of the tungsten produced (i.e., 20,000–24,000 t/year) are used for the cemented carbide production via WC. It is also used as a catalyst. For more details the reader is referred to Chapters 9 and 10.

Ternary Compounds: eta-carbides

Me_6W_6C, Me_2W_4C–Me_3W_3C (Me = Co, Fe, Ni).

TABLE 4.3. Resistance to Chemical Attack of Tungsten Carbide in Various Media

	Insoluble residue (%)	
	24 h at 20–25 °C	2 h at boiling of the medium
HCl (density 1.19)	97	48
HCl (1:1)	96	92
H_2SO_4 (density 1.84)	91	1
H_2SO_4 (1:4)	96	95
HNO_3 (density 1.43)	63	1
HNO_3 (1:1)	72	10
H_3PO_4 (density 1.7)	91	93
H_3PO_4 (1:3)	96	90
$HClO_4$	98	40
$HClO_4$ (1:3)	98	93
$H_2C_2O_4$ (saturated solution)	95	95
HCl + HNO_3 (3:1)	28	3
H_2SO_4 + HNO_3 (1:1)	92	42
HNO_3 + HF (4:1)	0	0
H_2SO_4 + $H_2C_2O_4$ (1:1)	95	70
H_2SO_4 + H_3PO_4 (1:1)	96	93
NaOH		
20% solution	98	98
10% solution	97	98
20% + bromine water (4:1)	70	60
20% + H_2O_2 (4:1)	88	87
20% + $K_3[Fe(CN)_6]$ (10% solution)	68	68

Corresponding Co compounds may form during liquid-phase sintering of cemented carbides if the carbon balance is low; analogous Fe compounds (containing also Mo, V, and Cr) occur in tungsten alloyed steel.

Tungsten Hexacarbonyl: $W(CO)_6$ [4.25]

PR: By direct reaction of tungsten and carbon monoxide under high pressure.
By Al reduction of WCl_6 dissolved in anhydrous ether at a CO pressure of 1 bar at 70 °C with a yield of 90%.
Purification by sublimation in vacuum or by steam distillation.

PP: D: 2.65 $g \cdot cm^{-3}$, white crystalline solid with low vapor pressure (13.3 Pa at 20 °C and 160 Pa at 67 °C); it sublimes in vacuo at 60 to 70 °C and decomposes at ~150 °C.

CP: Tungsten hexacarbonyl is a zero-valent monomeric compound. The bonding can be understood as follows: between each carbon atom and the central tungsten a weak σ-bond is formed by donation of the free pair of electrons of the carbon atom. In addition, a π-bonding is formed by back donation of electrons from the 5d level of the tungsten atom. This explains why the W–C bonding distance (2.06 Å) is shorter than a normal single bond. It is slightly soluble in organic solvents like CCl_4, and insoluble in water. It is stable in air, water, and acids, but will be decomposed by strong bases and attacked by halogens.

A: It is used as a catalyst in organic synthesis and plays a central role as a precursor to synthesize a vast variety of coordination as well as organometallic tungsten compounds. The special importance lays in the possibility to synthesize low-valence and zero-valent compounds.

4.2.3. *Tungsten and Silicon* [4.4, 4.7, 4.8, 4.11, 4.26]

Two compounds are known in the tungsten–silicon system: W_5Si_3 and WSi_2.

PR: By reaction of the elements in powder form in a protective atmosphere, like argon or vacuum at elevated temperature. Starting materials, their size, crystallinity, presence of defects, oxide films, and experimental conditions determine the reaction product, mechanism, and kinetics.
By silicothermic reduction of tungsten oxides.
By chemical or physical vapor deposition.

A: Tungsten silicides form a protective layer at the surface of tungsten and prevent destructive oxidation at elevated temperatures. The layer fails to protect at lower temperature (disilicide pest). Tungsten silicide thin films play an important role in microelectronics (see Section 5.7.5).

Pentatungsten Trisilicide: W_5Si_3

PP: D: measured 12.2 $g \cdot cm^{-3}$; X-ray 14.56 $g \cdot cm^{-3}$; MP: 2370 °C; $-\Delta H°_{298} = 125$ $kJ \cdot mol^{-1}$; CR: tetragonal $D8_m$, W_5Si_3-type; $a = 9.601$ Å, $c = 4.972$ Å.

Tungsten Disilicide: WSi_2

PP: D: measured 9.30 $g \cdot cm^{-3}$; X-ray 9.75 $g \cdot cm^{-3}$; MP: 2150–2182 °C; MH: 1200 $kg \cdot mm^{-2}$ (100 g); **modulus of elasticity**: 430 GPa; **electrical resistivity**: 12 $\mu\Omega \cdot cm$; CO: blue gray; $-\Delta H°_{298} = 92$ $kJ \cdot mol^{-1}$; CR: tetragonal CII_b, $MoSi_2$-type; $a = 3.211$ Å, $c = 7.868$ Å.

CP: Insoluble in water. Will be attacked by fluorine, chlorine, fused alkalis, and HNO_3–HF mixtures. Stable in air up to 900 °C.

A: Can be used in high-temperature thermocouples in combination with $MoSi_2$ in oxydizing atmosphere.

4.2.4. *Tungsten and Nitrogen* [4.11, 4.27, 4.28]

There are three nitrides which are considered to be equilibrium phases: W_2N, WN, and WN_2. However, no reversible equilibrium transformation has been reported for any of these phases [4.27]. In addition, twelve other phases have been quoted. Thermodynamic properties of the W–N system are reported elsewhere [4.28].

Nitrides usually form by reaction of tungsten with dry ammonia (formation of atomar nitrogen; "in-situ" nitrogen pressures in the range of 10^4–10^5 bar). Exceptions are reactions in a nitrogen jet stream or chemical and physical vapor deposition (plasma-aided formation).

The solubility of nitrogen in solid tungsten is very low even at high temperature and high pressure. Nitrogen is interstitially dissolved in the tungsten lattice.

Ditungsten Nitride: β-W_2N ($WN_{0.67}$–$WN_{1.0}$)

PR: Tungsten powder reacts with ammonia gas at 825–875 °C.
PP: $-\Delta H°_{298} = 12.6$ kJ · mol^{-1}; CR: fcc *B*1, NaCl-type; $a = 4.12$–4.14 Å.

Tungsten Nitride: δ-*WN*

PR: Tungsten powder reacts at 800 °C in flowing dry ammonia within a few hours.
PP: $-\Delta H°_{298} = 24.5$ kJ · mol^{-1}; CR: hexagonal B_h, WC-type; $a = 2.893$ Å, $c = 2.826$ Å.

Tungsten Dinitride: $\delta_R{}^V$ WN_2

PR: Formation during direct heating of tungsten wire in pure nitrogen at 2500 °C. Evaporating tungsten combines with nitrogen. The nitride deposits on the enclosure wall.
PP: MP: 400 °C (decomposition); CO: brownish black; CR: rhombic R3m, $\delta_R{}^V$-type; $a = 2.89$ Å, $c = 16.40$ Å.

4.2.5. Tungsten and Phosphorus [4.4, 4.8, 4.11]

Three line compounds occur in the W–P system: W_3P, WP, and WP_2.

Tritungsten Phosphide: W_3P

PR: Observed in arc-melted samples; disappeared on annealing at 800 to 1000 °C; probably a metastable compound.
PP: CR: tetragonal; $a = 9.858$ Å, $c = 4.800$ Å.

Tungsten Phosphide: WP

PR: By dissociation of WP_2 or direct synthesis.
PP: D: 12.3 g · cm^{-3}; MP: 1450 °C (decomposition); CO: gray; CR: orthorhombic *B*31, MnP-type: $a = 5.734$ Å, $b = 3.249$ Å, $c = 6.222$ Å.
CP: Insoluble in water, HCl, and alkalis and soluble in a mixture of HF and aqua regia.

Tungsten Diphosphide: WP_2

PR: By heating of tungsten powder with red phosphorus in vacuo at 700–950 °C. By reduction of tungsten trioxide with phosphorus at 550 °C.
PP: D: 9.17 g · cm^{-3}; CO: black; CR: low-temperature modification α-WP_2: monoclinic, $NbAs_2$-type; $a = 8.501$–8.502 Å, $b = 3.167$–3.1695 Å, $c = 7.466$–7.441 Å, $\beta = 119.33°$–$119.36°$; high-temperature modification β-WP_2: orthorhombic, MoP_2-type; $a = 3.166$ Å, $b = 11.161$ Å, $c = 4.973$ Å.
CP: Insoluble in water, acids, alcohol, and ether.

4.2.6. Tungsten and Arsenic [4.4, 4.8, 4.11]

Two compounds occur in the W–As system: WAs_2 and W_2As_3.

Ditungsten Triarsenide: W_2As_3

PR: Heating stoichiometric amounts of the elements in an evacuated, sealed silica tube at 600–1100 °C.

PP: CR: monoclinic; $a = 13.346$ Å, $b = 3.278$ Å, $c = 9.599$ Å; $\beta = 124.80°$.

Tungsten Diarsenide: WAs_2

PR: Heating of an element mixture at 620 °C.
By reaction of arsine with tungsten hexachloride.
Heating the stoichiometric amounts of the elements in an evacuated, sealed silica tube at 600–1100 °C.

PP: D: 6.9 $g \cdot cm^{-3}$; CO: black; CR: monoclinic, $MoAs_2$-type; $a = 9.078$–9.085 Å, $b = 3.3177$–3.318 Å, $c = 7.686$–7.690 Å; $\beta = 119.42°$–$119.51°$.

CP: The compound is stable in air, reacts with water above 400 °C and with chlorine at 500 °C. It is insoluble in HF, HCl, and aqueous alkali solutions. It is attacked by HF + aqua regia mixtures, molten alkalis, concentrated HNO_3, and sulfuric acid.

4.2.7. Tungsten and Oxygen [4.4, 4.7, 4.29–4.31]

The tungsten–oxygen system is rather complex. Besides the stable stoichiometric binary oxides (WO_3, $WO_{2.9}$, $WO_{2.72}$, and WO_2), and the stoichiometric tungstates and acids, a variety of nonstoichiometric, fully oxidized and reduced compounds exists, according to the scheme in Fig. 4.2.

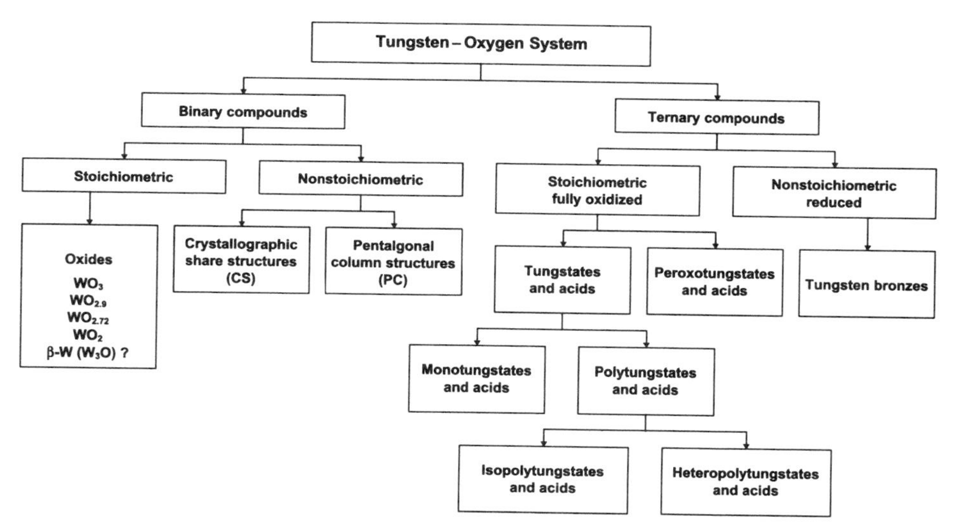

FIGURE 4.2. Survey of tungsten–oxygen compounds.

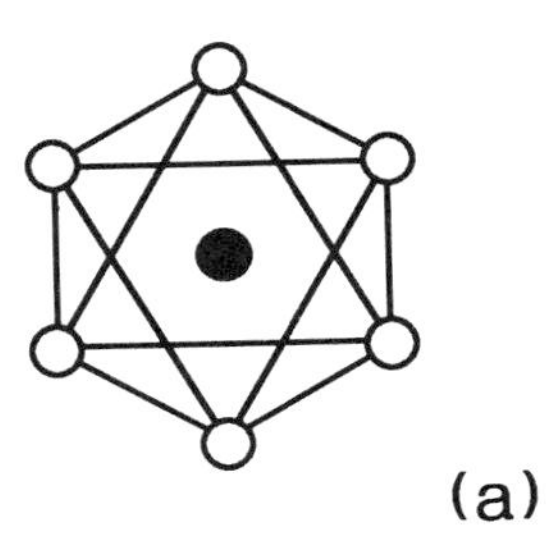

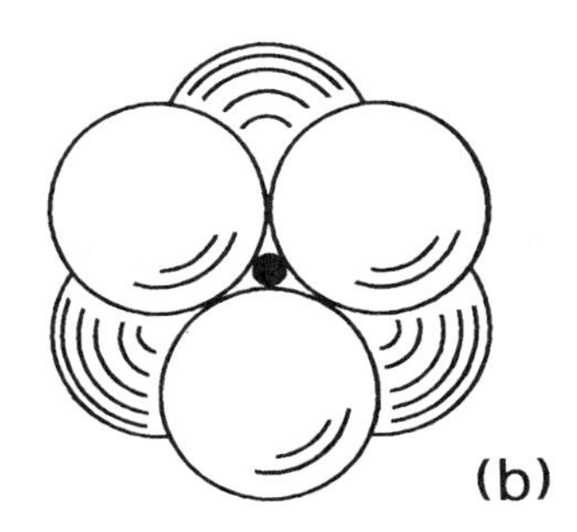

FIGURE 4.3. WO_6 octahedron; (a) Locations of the centers of the atoms; the black circle is tungsten, the white circles oxygen. (b) Atoms shown in full size; the tungsten atom is the small circle.

In the fully oxidized state (W^{6+}) the octahedral coordination dominates (WO_6). The tungsten atoms are situated in the center of octahedrons and are surrounded by six oxygen atoms arranged at the corners. As long as all tungsten ions are hexavalent, these octahedrons are only corner-sharing (Fig. 4.3). However, as the oxygen-to-tungsten ratio decreases, which occurs on reduction (transition to lower oxides), the octahedra become more and more intricately joined in combinations of corner-, edge-, and face-sharing.

By the loss of each oxygen atom from the oxide lattice two electrons are added to the conduction band. Reduced compounds are therefore either semiconducting or conducting.

Besides the tungsten–oxygen octahedra, WO_4 tetrahedra can also be found in fully oxidized compounds (tungstates), as well as WO_7 pentagonal bipyramids in reduced compounds.

Several hundreds or even thousands of ternary and quaternary tungsten oxides are known today. They are only partly described in the literature and can be found in the relevant compilations [4.32–4.35]. They are, however, the top of an iceberg of a much larger number of compounds which might form, in principle, with a large number of elements and/or element combinations. In this sense, the crystal chemistry of tungsten oxides can be regarded as one of the most complex and richest fields in the structural chemistry of the elements.

4.2.7.1. Stoichiometric Binary Oxides

Tungsten Trioxide: α-WO_3

PR: By calcination of tungstic acid or APT in air or oxygen:
$H_2WO_4 \rightarrow WO_3 + H_2O$ (750–800 °C),
$(NH_4)_{10}[H_2W_{12}O_{42}) \cdot 4H_2O) \rightarrow 12WO_3 + 10NH_3 + 11H_2O$
By oxidation of tungsten powder (for stoichiometric WO_3 the oxygen partial pressure has to be higher than in air).

PP: D: 7.21–7.30 $g \cdot cm^{-3}$ (measured), 7.27 $g \cdot cm^{-3}$ (X-ray); MP: 1472 °C; significant sublimation starts already far below the melting point at >750 °C. The presence of water vapor enhances the volatility considerably; BP: 1837 °C (at this temperature the sum of the volatile partial pressures corresponds to 1 bar). The volatile species are only polymeric molecules (in mol%: 23.7 W_2O_6, 0.57 W_3O_8, 58.2 W_3O_9, 17.53 W_4O_{11}); MH: 83–163 $kg \cdot mm^{-2}$ (50 g); CO: yellow (smallest diminution of oxygen changes the color to different types of green); at lower temperature (-27 to -50 °C) it is bluish white and at <-50 °C white. At elevated temperature the color changes to brownish yellow; $-\Delta H^{\circ}_{298} = 832-853$ kJ · mol^{-1}; **electrical resistivity**: 0.14–0.18 Ω · cm; HR: $WO_{2.9873+0.0003}$.

CR: Several allotropic modifications exist:

>740 °C	tetragonal; P4/nmm- $D^7{}_{4h}$ $a=5.272$ Å, $c=3.920$ Å (at 950 °C)
330 to 740 °C	orthorhombic; Pmnb-$D^{16}{}_{2h}$ $a=7.340$ Å, $b=7.546$ Å, $c=7.728$ Å.
+17 to 330 °C	monoclinic γ-WO_3; WO_3 (I); $P2_1/n$ $a=7.302$–7.306 Å, $b=7.530$–7.541 Å, $c=7.690$–7.692 Å, $\beta=90.83$–90.88°
−50 to +17 °C	triclinic β-WO_3, WO_3 (II); P1-$C^1{}_i$ $a=7.30$ Å, $b=7.52$ Å, $c=7.69$ Å, $\alpha=88.85°$, $\beta=90.92°$, $\gamma=90.95°$ (at 10 °C).
−143 to −50 °C	monoclinic α-WO_3, Pc-$C^2{}_s$ $a=5.275$ Å, $b=5.155$ Å, $c=7.672$ Å, $\beta=91.7°$ (at −70 °C).

Commercially available WO_3 consists almost entirely of γ-WO_3.

The reason for the numerous modifications originates in the structure, which is a three-dimensional array of corner-sharing metal–oxygen octahedra (Fig. 4.3). The idealized structure (cubic symmetry; ReO_3-type) is drawn in Fig. 4.4, showing the octahedra as squares in projection. The real structure is distorted. This means that the relatively small tungsten atoms tend to be displaced from the octahedral center and these displacements are temperature-dependent. Moreover, the symmetry can be influenced by small amounts of impurities.

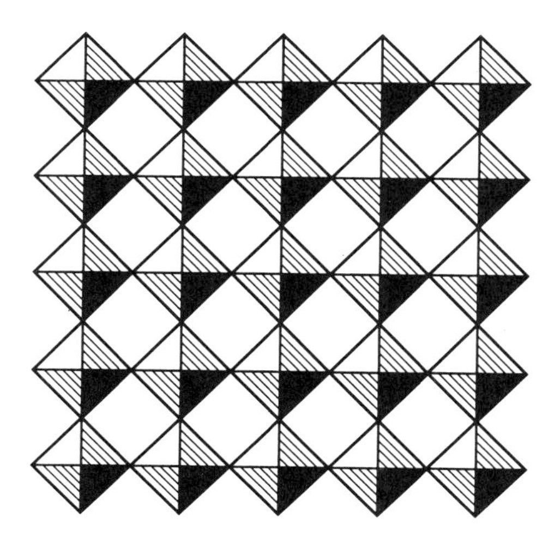

FIGURE 4.4. The idealized WO_3 structure represented as an infinite array of corner-shared octahedra; in the real structure small distortions of the octahedra reduce the symmetry from cubic to monoclinic at room temperature [4.30].

While still maintaining that every octahedron shares every corner with another, other arrays than the chess-board-like arrangement are also possible. Low-temperature dehydration of $WO_3 \cdot \frac{1}{3}H_2O$ yields a metastable oxide (hexagonal WO_3, or h-WO_3), with a structure as shown in Fig. 4.5. Hexagonal WO_3 transforms to monoclinic WO_3 above 400 °C. Another metastable modification is the pyrochlore WO_3, or p-WO_3. It can be prepared from ammonium tungstate by removal of NH_4 ions via acid washing. The flexibility of the WO_6 corner-linked octahedral framework allows a cubic pyrochlore structure, too (the pyrochlore structure will be presented under tungsten bronzes). The material is white and transforms to monoclinic WO_3 above 400 °C.

CP: WO_3 is insoluble in acids with the exception of HF, but dissolves easily in aqueous alkali hydroxide solutions or melts of alkali hydroxides or carbonates. See Chapter 3 for information about reactions of WO_3 with hydrogen, water, and carbon.

A: WO_3 is one of the most important, highly pure intermediates for the production of other tungsten compounds and tungsten metal powder. In the latter application it was substituted to a large extent by tungsten blue oxide (see Chapter 5.3).
Because of its bright yellow color it is used as a pigment in oil and water colors.
It is employed in a wide variety of catalysts, most recently in catalysts for the control of air pollution and industrial hygiene (see Chapter 10).
WO_3 is also used as an addition to VO_2 in "smart windows." Thermochromism [4.36] is the ability of a material to be transparent for infrared rays below a certain transition temperature and to reflect above, due to a crystallographic phase transition. The process is strictly reversible and there is almost no influence on visible light. VO_2 has a transition temperature at 68 °C, which can be lowered by doping with WO_3. Glass coated with a thin film of the above mixed oxide is called a "smart window" and is a promising candidate for use in automobiles and buildings for heat protection.

FIGURE 4.5. The hexagonal framework of octahedra in h-WO_3 [4.30].

Solid state devices use the photochromism of WO_3 thin films in antidazzling mirrors (for more details of WO_3 photochromic properties, see Section 3.8.6).

β-Tungsten Oxide: $WO_{2.9}$ ($W_{20}O_{58}$)

PR: Forms intermediately during the hydrogen reduction of WO_3 to W. In this process it can be obtained in pure state, for example at 500–550 °C and a water vapor partial pressure of 0.6 bar.
By heating corresponding mixtures of W and WO_3. The compacts are heated 6 hours at 800 °C in argon, then powdered and tempered under the same conditions for 24 hours.

PP: D: 7.15 $g \cdot cm^{-3}$ (measured), 7.16 $g \cdot cm^{-3}$ (X-ray); CO: blue to deep blue; **electrical resistivity**: $5 \times 10^{-3}\ \Omega \cdot cm$; $-\Delta H°_{298} = 800-816\ kJ \cdot mol^{-1}$; CR: monoclinic P2/m-$C^1_{2h}$; $a = 12.05$ Å, $b = 3.767$ Å, $c = 3.59$ Å, $\beta = 94.72°$; HR: $WO_{2.8975-2.9014}$ (1000 °C).
$WO_{2.9}$ belongs to the homologous series W_nO_{3n-2}, which is characterized by a crystallographic share (CS) structure. The shear planes lie upon {103} planes with respect to the WO_3 parent structure. An idealized structure of $WO_{2.9}$ is shown in Fig. 4.7.
It crystallizes mostly in needle shape (the monoclinic *b*-axis is parallel to the needle axis).

CP: About its reactions with hydrogen and water, see Chapter 3.

A: It is partially contained in the technical blue oxide and acts in pure form as an important catalyst in industrial chemical syntheses (see Chapter 10). Similar to $WO_{2.72}$, it catalyzes the decomposition of N_2O and the selective oxidation of propene oxygen mixtures to acroleine.

γ-Tungsten Oxide: $WO_{2.72}$ ($W_{18}O_{49}$)

PR: It also forms during the hydrogen reduction of WO_3 to W and can be obtained in pure state, for example, in hydrogen atmosphere at 900 °C and a water vapor partial pressure of 0.8 bar.
In addition, it can be prepared by heating corresponding powder mixtures of W and WO_3 in sealed vessels or in argon atmosphere and subsequent tempering.

PP: D: 7.724–7.989 $g \cdot cm^{-3}$ (measured), 7.78 $g \cdot cm^{-3}$ (X-ray); CO: reddish violet, in whiskers brownish red; **electrical resistivity**: $(2-3) \times 10^{-4}\ \Omega \cdot cm$; $-\Delta H°_{298} = 743-780\ kJ \cdot mol^{-1}$; CR: monclinic, P2/m-$C^1_{2h}$; $a = 18.28$ Å, $b = 3.775$ Å, $c = 13.98$ Å, $\beta = 115.14°$.
The crystals are needle-shaped and the monoclinic *b*-axis is parallel to the needle axis; HR: $WO_{2.7190-2.7224}$ at 1000 °C.
$WO_{2.72}$ is not included in the homologous series of CS structures, because besides edge- and corner-sharing octahedra also pentagonal columns exist (as described under tungsten bronzes).

CP: About behavior with respect to hydrogen, see Chapter 3. At elevated temperature it decomposes. It is soluble in alkaline solutions. Propene oxygen mixtures are oxidized by $WO_{2.72}$ to acroleine.

A: It is partially contained in the technical blue oxide.

Tungsten Dioxide: WO_2

PR: It forms intermediately during the hydrogen reduction of higher tungsten oxides. In a pure state it can be obtained, for example, at 900 °C in hydrogen atmosphere and a water vapor partial pressure of 0.5 bar. In addition, it can be prepared by reduction of WO_3 with W. The direct oxidation of tungsten by oxygen is not a suitable method to synthesize WO_2, since higher oxides are formed, too.

PP: D: 10.82–11.05 $g \cdot cm^{-3}$ (measured); 10.82 $g \cdot cm^{-3}$ (X-ray); MP: At 1530 °C WO_2 disproportionates to W and $W_{18}O_{49}$ prior to melting. All indications of melting points in the older literature are not relevant; CO: brown to violet brown, single crystals are bronze colored and show metallic luster; **electrical resistivity**: $2.9 \times 10^{-3}\ \Omega \cdot cm$; $-\Delta H^\circ_{298} = 571-595\ kJ \cdot mol^{-1}1$; HR: at room temperature quasi-stoichiometric; $WO_{1.97-2.0}$ at 1357 °C; CR: monoclinic, $P2_1/c$ MoO_2-type; $a = 5.550$–5.62 Å, $b = 4.89$–4.96 Å, $c = 5.571$–5.73 Å, $\beta = 118.93$–122.1°; WO_2 exhibits a characteristic band structure, in which the WO_6 octahedra are edge-sharing. Neighboring bands are corner-sharing. The W atoms within the bands have alternating shorter (2.53 Å) and larger (3.13 Å) bonding distances, which indicates a cation–cation bond interaction and the formation of W–W double bonds between some of the W atoms. WO_2 is slightly paramagnetic and is described as both electrically conducting and semiconducting.

CP: WO_2 dissolves in warm, concentrated mineral acids and in boiling, concentrated alkali hydroxide solutions. It dissolves rapidly in diluted and conc. H_2O_2. Nitric acid oxidizes it to WO_3. About the behavior with respect to hydrogen, see Chapter 3. Higher oxides are formed with oxygen at elevated temperature. Reaction with chlorine leads to the formation of $WOCl_4$ and WO_2Cl_2, while with bromine and iodine only the dioxodihalides are formed.

A: In a two-step reduction procedure mainly applied in earlier times in NS tungsten powder production, it was the end product of the first reduction step and the starting material for the second.

β-*Tungsten: W_3O resp. W_3W* [4.37–4.39]

It was regarded as a suboxide and also as metal modification. Today it is described as a metastable metal phase, which is stabilized by small amounts of oxygen. A further stabilization of β-W occurs by the presence of different foreign elements, such as K, Be, P, B, As, Ce, Th, and Al. All these elements have in common that they form rather stable ternary and quaternary tungsten oxides.

PR: By electrolysis of fused mixtures of WO_3 and alkali metal phosphates.
Forms partially during the hydrogen reduction of doped tungsten oxide (NS–W production).
X-ray pure β-W forms at 500–600 °C by the reduction of $W_{20}O_{58}$ with dry hydrogen in very thin powder layers.

PP: D: 19.0–19.1 $g \cdot cm^{-3}$ (measured), 18.94 $g \cdot cm^{-3}$ (X-ray; calculated as W_3W); **transition temperature**: β-W → α-W occurs at 530–800 °C, depending on the presence of third elements; it occurs without any significant weight change; **electrical resistivity**: 200 $\mu\Omega \cdot cm$, CO: gray or black; CR: cubic A15; β-W type $Pm3n\text{-}O^3{}_h$; $a = 5.0512$ Å [4.39].

CP: It easily dissolves in diluted or conc. H_2O_2. The residual oxygen is either present on the surfaces or grain boundaries of the commonly nanocrystalline metal powder, or randomly distributed over all eight positions of the A15 W_3W lattice.

4.2.7.2. Nonstoichiometric Binary Compounds

Crystallographic Shear Structures (CS Structures) [4.30]

STR: When WO_3 is slightly reduced at approximately 730 °C either in sealed tubes or in hydrogen atmosphere, the oxygen deficit is taken up by CS planes. In principle, CS planes can be built up of edge- and plane-sharing structures. This corresponds to a partial replacement of W^{+6} by W^{+5} and W^{+4} ions.

At the lowest degree of reduction, blocks of four edge-sharing octahedra are present (Fig. 4.6.) The CS planes in this case are disordered. If the degree of reduction gets higher, more and more ordered structures exist (bands of parallel CS planes). For equispaced CS planes the composition corresponds to W_nO_{3n-1}. These ordered structures have a composition range between $WO_{2.96-2.933}$ ($W_{25}O_{74}$–$W_{15}O_{44}$).

Below $WO_{2.92}$, the structure changes to six edge-sharing octahedra (Fig. 4.7.) according to the general formula W_nO_{3n-2} where $n = 25$–18, and this corresponds to a range $WO_{2.920-2.889}$ ($\cong W_{20}O_{58}$).

CS structures have the ability to build-in foreign metal ions, especially in the reduced form. This seems to play an important role in the dopant uptake of tungsten blue oxide in *non-sag tungsten* production (for more information see Section 5.4.5).

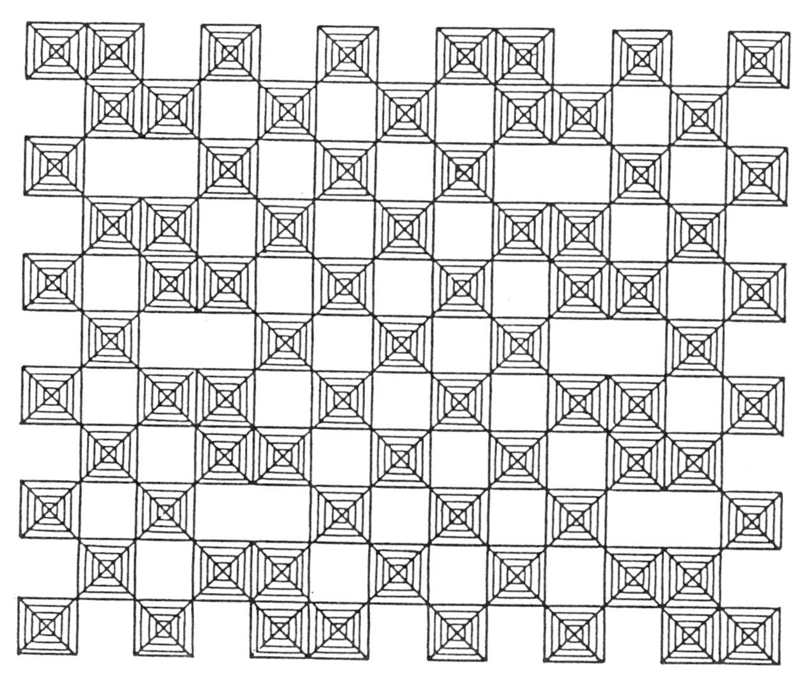

FIGURE 4.6. Idealized crystallographic shear (CS) planes in WO_{3-x}; blocks of four edge-sharing octahedra [4.30].

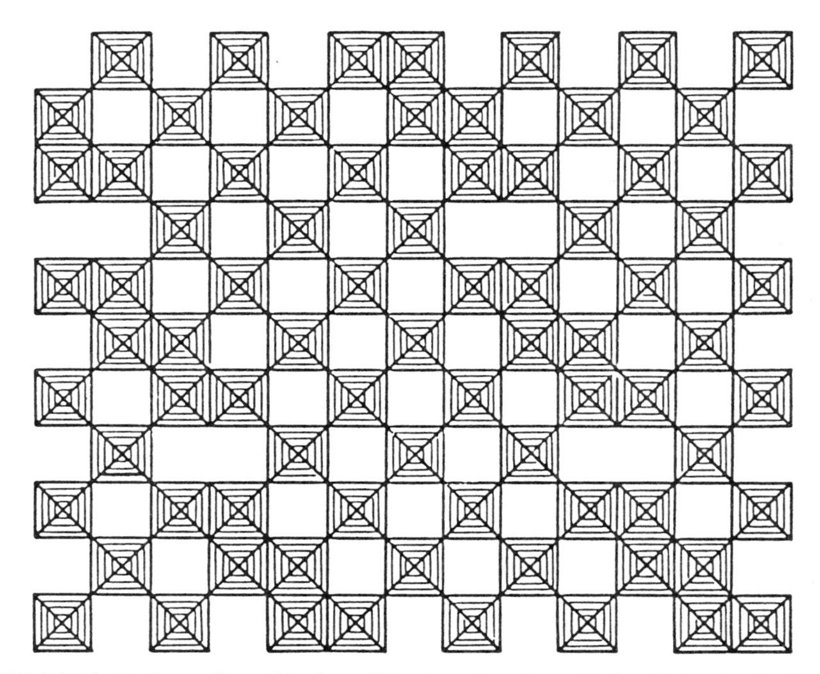

FIGURE 4.7. Idealized crystallographic shear (CS) planes in WO_{3-x}; blocks of six edge-sharing octahedra [4.30].

4.2.7.3. Stoichiometric Ternary Compounds (Oxidized)

(a) Tungsten Trioxide Hydrates and Monotungstates

Tungsten Trioxide Monohydrate: $WO_3 \cdot H_2O$ $(WO_2(OH)_2)$. It is a monomeric, gaseous compound which forms above 600 °C by reaction of water vapor and tungsten oxides. It is responsible for the Chemical Vapor Transport (CVT) of tungsten during the reduction process and hence for the grain size of the finally formed tungsten powder. Another important aspect of the CVT of tungsten is the overgrowth of dopant compounds by tungsten during the powder production of non-sag tungsten (see also Sections 3.3.5 and 5.4.5).

Tungstic Acids. In contrast to the compound of the same overall composition mentioned above, tungstic acids are highly condensed, almost insoluble and amorphous species with decreasing water content:

$WO_3 \cdot 2H_2O$	(W : H_2O ratio 1 : 2)
$WO_3 \cdot H_2O$	(W : H_2O ratio 1 : 1)
$2WO_3 \cdot H_2O$	(W : H_2O ratio 2 : 1)

Within this group dehydration can be achieved by heating. Heating to approximately 500 °C leads to the final stage of dehydration, the water-free, crystallized tungsten trioxide.

Tungstic Acid: H_2WO_4 $(WO_3 \cdot H_2O)$

PR: By reaction of hot alkali tungstate solutions with strong mineral acids, preferably HCl.
APT is reacted with strong acids (preferably HCl). This product is more pure than the above, because of the higher purity of APT.

PP: Amorphous, yellow powder.

CP: Nearly insoluble in water or acid solutions, but freshly prepared tungstic acid forms colloidal solutions when treated with water. It dissolves in strongly alkaline media, ammonia, and HF. It is stable up to 150 °C.

A: Tungstic acid in former times was the most important, highly pure intermediate in the technical process from ore to tungsten powder. It was in most applications substituted by APT. Today, it serves as starting material for the production of submicron tungsten powder and of several tungsten compounds. Tungstic acid is also used as a pigment for bright yellow glazes.

White Tungstic Acid: $WO_3 \cdot xH_2O$ $(x = \sim 2)$

PR: Precipitation by acidifying tungstate solutions in the cold.
Addition of sodium tungstate solution to 1 molar nitric acid until pH 1 is reached.

PP: It is a white, amorphous solid.

CP: Converts to tungstic acid on boiling; it is stable up to 55 °C and loses water in the temperature range 55 °C–75 °C; it seems to be a mixture of hexa- and decatungstate.

Ditungstentrioxide Monohydrate: $2WO_3 \cdot H_2O$

PR: By drying tungstic acid not over 300 °C.

PP: Orange yellow colored solid.

Monotungstates: M^{+1}, M^{+2}, M^{+3}, M^{+4}, M^{+5}, M^{+6}*-tungstates* [4.31]

STR: In the tungstates, tungsten is found either in octahedral or in tetrahedral coordination. If the radius of the ternary cation is plotted against electronegativity, one may observe the following properties (Fig. 4.8.):
With high electronegativity and small ionic radius (0.7 → 0.95 Å) octahedral coordination prevails (*wolframite group*).
With low electronegativity and large ionic radius tetrahedral coordination is observed (*scheelite group*).
Metals at the border line may show polymorphism or other structural complexity (for example, under high pressure, $CaWO_4$ exhibits wolframite structure).

PR: Nonhydrated *sodium tungstate* can be prepared by fusion reactions, such as:

$$WO_3 + 2NaOH \rightarrow Na_2WO_4 + H_2O \quad \text{or} \quad WO_3 + Na_2CO_3 \rightarrow Na_2WO_4 + CO_2.$$

If it is prepared by crystallization of an aqueous solution, the dihydrate always forms $Na_2WO_4 \cdot 2H_2O$.
Potassium tungstate can be formed under the same conditions.
Ammonium tungstate is produced by reaction of liquid ammonia with tungstic acid in a nonhydrated form. A hydrated form does not exist, because ammonium ions

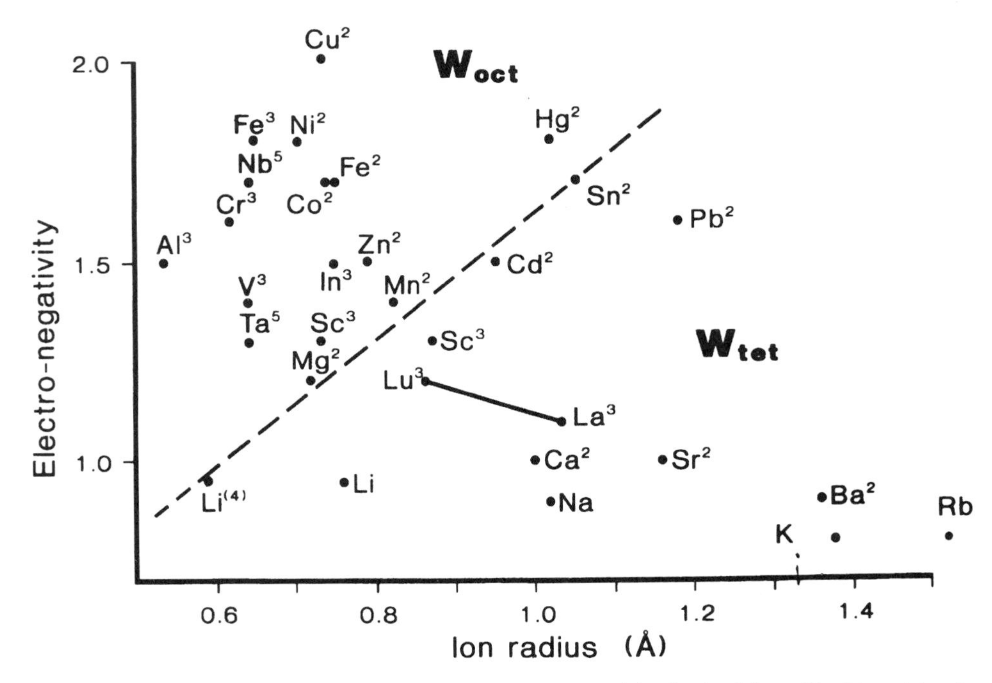

FIGURE 4.8. Plot of the octahedral ionic radius versus electronegativity for the fully oxidized tungstates; the dotted line represents the approximate boundary between structures in which W is tetrahedrally or octahedrally coordinated [4.31].

shift the condensation tendency of tungstate ions to slightly alkaline conditions. This means that only polytungstates can be obtained from ammonia-containing solutions by crystallization.

The preparation of insoluble tungstates (divalent or trivalent metal ions with the exception of Mg) is carried out by reacting aqueous sodium tungstate solutions with the corresponding soluble metal salt.

PP: Alkali tungstates are white, hygroscopic substances. Some further properties of various tungstates are listed in Table 4.4.

CP: Alkali tungstates are very soluble in water and crystallize in hydrated form. They lose the water of hydration nearly quantitatively at 100 °C. All other tungstates with the exception of Mg are sparingly soluble in water. They decompose by addition of hot mineral acids (except phosphoric acid) to tungstic acid and the corresponding metal salt.

Calcium Tungstate: $CaWO_4$

PR: Calcium chloride is added in excess to a sodium tungstate melt.

Addition of calcium chloride solution to sodium tungstate solution. This reaction was used on a large scale in former times as purification step (precipitation of artificial scheelite).

PP: CR: tetragonal Dy5Pd2-type; $a = 5.25$ Å, $c = 11.38$ Å; calcium tungstate is fluorescent when exposed to short-wavelength UV radiation.

TABLE 4.4. Physical Properties of Monotungstates

	D ($g \cdot cm^{-3}$)				
Formula	Measured	X-ray	MP (°C)	CO	CR
Li_2WO_4	3.17		738	white	Rhombic $a = 8.888$ Å; $\beta = 107.78°$
K_2WO_4	3.12		928	white	Monocline
$SrWO_4$	6.184	6.372		white	Tetragonal $a = 5.405$ Å; $c = 11.90$ Å
$CdWO_4$					Tetragonal
$BaWO_4$	6.35	6.00	1490	white	Tetragonal $a = 5.64$ Å; $c = 12.689$ Å
$PbWO_4$			1130	white	Tetragonal $a = 5.44$ Å; $c = 12.01$ Å
$FeWO_4$	7.391	7.647			Moncline $\beta = 90°$ $a = 4.70$ Å; $b = 5.69$ Å; $c = 4.93$ Å
$MnWO_4$	7.135	7.261			Monoclinic $\beta = 89.17°$ $a = 4.84$ Å; $b = 5.76$ Å; $c = 4.97$ Å
$ZnWO_4$					Monoclinic $\beta = 89.30°$ $a = 4.68$ Å; $b = 5.73$ Å; $c = 4.95$ Å
$CoWO_4$					Monoclinic $\beta = 90°$ $a = 4.66$ Å; $b = 5.69$ Å; $c = 4.98$ Å
$MgWO_4$					Monoclinic $\beta = 89.7°$ $a = 4.68$ Å; $b = 5.66$ Å; $c = 4.92$ Å
$NiWO_4$					Monoclinic $\beta = 89.40°$ $a = 4.69$ Å; $b = 5.66$ Å; $c = 4.93$ Å
$Bi_2(WO_4)_3$			842	gray-green	Tetragonal
$Eu_2(WO_4)_3$			1130	white	Monoclinic $\beta = 109.63°$ $a = 7.68$ Å; $b = 11.46$ Å; $c = 11.40$ Å
$Ce_2(WO_4)_3$					

CP: Only sparingly soluble in water.

A: It is one of the two most important minerals: *scheelite*.
As artificial scheelite it was an important intermediate.
For the production of phosphores for lasers, fluorescence lamps, oscilloscopes, luminescence colors, X-ray observation screens, luminous paints, and scintillation counters. The property of fluorescence is very important in prospecting and mining of scheelite ore.

Zirconium Tungstate: ZrW_2O_8 [4.40]

Zirconium tungstate and the isostructural hafnium tungstate (HfW_2O_8) show strong negative thermal expansion from 0.3 K up to their decomposition temperature of ~1050 K.

PP: Thermal expansion: $-8.9 \times 10^{-6} \cdot K^{-1}$ for ZrW_2O_8; decomposes at ~1050 into tungsten and zirkonuim oxide; CR: cubic, $P2_13$; $a = 9.1575$ Å; order–disorder transition at 428 K.

A: For composite materials with an overall zero bulk expansion (electronic devices).

The importance of tungstates is manifold. We find nearly all naturally occurring tungsten minerals in the group of monotungstates. Barium and zinc tungstates are used for bright white pigments. Sodium tungstate in aqueous solution is the most important intermediate between ore and APT or tungstic acid in the technical production.

(b) Isopolytungstates and their Acids. The transition of monotungstate ions to tungstic acid by acidification was already discussed earlier. It is accompanied by the formation of several species of preferential condensation grades. Evidently due to structural reasons, they show a high degree of stability at some pH value. Some important and typical examples are given in Table 4.5. The respective salts of these polyions are not easy to synthesize, because in aqueous solutions equilibria always exist between the different species and, moreover, kinetic effects of variously rapid reactions make the situation even worse.

The stability and optimal H_3O^+/WO_4^{2-} ratio of certain species can be further influenced by other ions, such as ammonium which shifts the polycondensation to higher pH values.

The most important isopolytungstates are listed below.

Ammonium Metatungstate: $[(NH_4)_6(H_2W_{12}O_{40}) \cdot 3H_2O]$

PR: Ion exchange procedure.
Electrodialysis.

CP: The advantage of AMT is its high solubility in water, giving the possibility to prepare highly concentrated, alkali-free tungsten solutions.

A: AMT solutions are in use to treat supports (like alumnia pellets) for catalyst production. AMT is further used in nuclear shielding, as corrosion inhibitor, and as a precursor for other tungsten chemicals.

Sodium Metatungstate: $[Na_6(H_2W_{12}O_{40}) \cdot 3H_2O]$

PR: Formation of the monotungstate and subsequent removal of the sodium ions from the solution by ion exchange to form a dilute sodium metatungstate solution, and subsequent concentration of the solution.

PP: High concentrations of sodium metatungstate provide solutions with densities up to $3.1\ g \cdot cm^{-1}$; viscosity: > 40 cP ($1\ cP = 10^{-3}\ N \cdot s \cdot m^{-2}$) at a specific gravity of $3\ g \cdot ml^{-1}$.

A: It is used as high-density liquid in crude oil drilling, as important intermediate for the production of heteropolytungstates, as a fuel cell electrode material, in cigarette filters, and for fireproofing textiles.

TABLE 4.5. Some Typical Polytungstate Ions

H_3O^+/WO_4^{2-}	Formula	Name
0.333	$W_{12}O_{46}^{20-}$	Para Z
0.667	$W_3O_{11}^{4-}$	Tritungstate
	$H_4W_3O_{13}^{4-}$	
1.167	$H_{10}W_{12}O_{46}^{10-}$	Para B
	$HW_6O_{21}^{5-}$	Para A
1.33	$W_{12}O_{40}^{8-}$	
1.50	$H_2W_{12}O_{40}^{6-}$	Meta
	$H_3W_6O_{21}^{3-}$	Pseudo-meta
2.00	$WO_3 \cdot H_2O$	Tungstic acid

Lithium Metatungstate: $[Li_6(H_2W_{12}O_{40}) \cdot 3H_2O]$ [4.41]

PR: See above.

PP: Solutions of lithium tungstate are even more dense than sodium tungstate solutions, although lithium is the least dense alkali metal; maximum density is about 3.5 $g \cdot cm^{-1}$; viscosity: ~30 cP at a specific gravity of 3 $g \cdot ml^{-1}$.

Ammonium Paratungstate: $(NH_4)_{10}H_2W_{12}O_{42} \cdot 4H_2O$ [4.42]

PR: Its preparation will be treated in detail in Chapter 5.

PP: D: 4.61 $g \cdot cm^{-1}$ (measured), 4.639 $g \cdot cm^{-1}$ (X-ray); CR: monoclinic, $P2_1/n$; besides the technically important tetrahydrate, depending on crystallization conditions, a triclinic hexahydrate and an orthorhombic decahydrate can be formed.

A: It is today the most common, highly pure intermediate for most tungsten products (see Chapter 5.3).

Ion associate complexes of isopolytungstates with secondary and tertiary alkyl amines play an important role in the technical solvent extraction process (see Section 5.3.)

(c) Heteropolytungstates and their Acids [4.43–4.44]. Heteropoly tungstates may be classified according to the ratio of hetero to tungsten atoms. Some typical examples are given in Table 4.6. More than 30 elements are known to function as heteroatom in those compounds with sometimes varying stoichiometric ratios of hetero to tungsten atoms (Table 4.7).

TABLE 4.6. Some Typical Species of Heteropolytungstates

Atom ratio (hetero:W)	Central atom	Typical formula
1:12	P^{5+}, As^{5+}, Si^{4+}, Ge^{4+}, Ti^{4+}, Co^{3+}, Fe^{3+}, Al^{3+}, Cr^{3+}, Ga^{3+}, Te^{4+}, B^{3+}	$[X^{n+}(W_{12}O_{40})]^{(8-n)-}$
1:10	Si^{4+}, Pt^{4+}	$[X^{n+}(W_{10}O_x)]^{(2x-60-n)-}$
1:9	Be^{2+}	$[X^{2+}(W_9O_{31})]^{6-}$
1:6	Series A, Te^{6+}, I^{7+}	$[X^{n+}(W_6O_{24})]^{(12-n)-}$
	Series B, Ni^{2+}, Ga^{3+}	$[X^{n+}(W_6O_{24}H_6)]^{(6-n)-}$
2:18	P^{5+}, As^{5+}	$[X_2{}^{n+}(W_{18}O_{62})]^{(12-2)-}$
2:17	P^{5+}, As^{5+}	$[X_2{}^{n+}(W_{17}O_x)]^{(2x-102-2n)-}$

TABLE 4.7. Ions Capable of Acting as Central Ion in Tungsten Heteropoly Compounds

Group	New IUPAC designation 1985	Ion
I	1 + 11	H^+, Cu^{2+}
II	2 + 12	Be^{2+}, Zn^{2+}
III	13	B^{3+}, Al^{3+}, Ga^{3+}
IV	14	Si^{4+}, Ge^{4+}, Sn^{4+}
V	5 + 15	N^{5+}, P^{3+}, P^{5+}, As^{3+}, As^{5+}, Sb^{3+}, V^{4+}, V^{5+}
VI	6 + 16	Cr^{3+}, S^{4+}, Se^{4+}, Te^{4+}, Te^{6+}
VII	7 + 17	Mn^{2+}, Mn^{4+}, I^{7+}
VIII	8, 9, 10	Fe^{3+}, Co^{2+}, Co^{3+}, Ni^{2+}, Ni^{4+}, Rh^{3+}, Pt^{4+}
Lanthanoid series		Ce^{3+}, Ce^{4+}

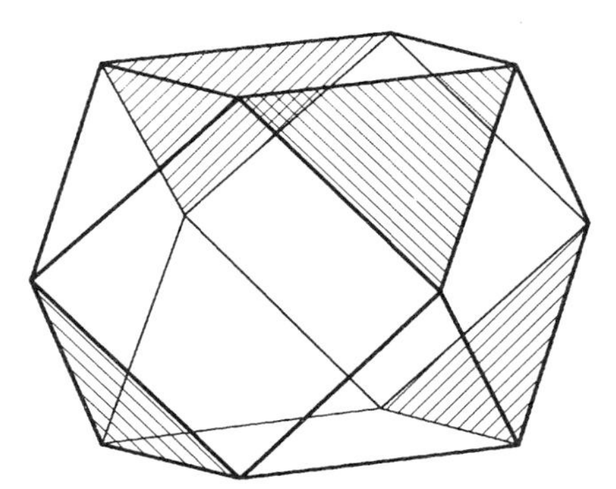

FIGURE 4.9. Cubooctahedron diagram of the $[PW_{12}O_{40}]^{3-}$ anion. The W atoms are at the corners of the polyhedron and the oxygen atoms at the midpoints of the edges.

STR: WO_6 octahedra sharing corners and edges surround the tetrahedrally or octahedrally coordinated heteroatom. Also, the central polyhedron shares corners and edges (or both) with the WO_6 octahedra. Thus each of the W atoms is connected to the central heteroatom via an oxygen bridge. Schematic structures are given in Fig. 4.9. The actual structures are frequently distorted.
In crystals, the relatively large heteropoly anions cause interstices between each other which are occupied by cations or by water molecules. Therefore, there exist no strong bondings between the individual heteropoly tungstate ions. Instead, they are joined by hydrogen bondings through water of hydration molecules.
In case of large cations, they frequently crystallize as acid salt even although an excess of the cation is present in the mother liquor. There is not so much space available which would be required for the cations to form a neutral salt. Those salts are often less hydrated than those of smaller cations. The large cations take so much space that there is not so much left for the water molecules.

CP: Heteropolytungstates can be characterized as follows:
High molecular weight <3000.
High degree of hydration.
Unusually high solubility in water and also in some organic solvents.
Strong oxidizing action in aqueous solution.
Strong acidity in free acid form.
Decomposition into monotungstates in strongly basic aqueous solution.
Highly colored anions or colored reaction products.

A: They were produced in large quantities for:
Precipitants for basic dyes forming lakes and toners.
Catalysts.
Passivation of steel.

Ion associate complexes of (Si, P) heteropolytungstates with alkylamines may disturb the technical solvent extraction process. In particular, if tertiary amines are in use, the complexes are insoluble in the solvent and form third phases (see Section 5.3).

Sodium-12-Tungstophosphate

A: This is a very important compound and is employed for many applications:
Catalyst in petroleum and organic chemical industry (isomerization, polymerzation,

oil bodying, nitrile synthesis, aromatization, desulfurization, dehydrochlorination, dehydration, ring closure, keton synthesis).
Manufacture of organic pigments (precipitating agent for synthetic dye stuffs, forms brilliant colored pigments, highly stable to heat, light and moisture; printing inks, paper tinting, paper coloring, nontoxic paints, wax pigmentation.
Carroting (surface treating of furs).
Antistatic agent for treatment of acryl-based synthetic fibers.
The oxidizing power makes it useful in various photographic processes, manufacture of photosensitive systems, image producing, or printout systems.
Leather tanning, waterproofing of plastic films.
Additive to chrome-plating baths.
In cements and adhesives to impart water resistivity.

(d) Peroxotungstates and Acids

PR: Peroxotungstic acid forms when WO_3 is reacted with hydrogen peroxide.
Peroxotungstic acid is also generated when tungsten powder is dissolved in hydrogen peroxide.
Reaction of tungstates with hydrogen peroxide yields in peroxotungstates.

PP: CO: bright yellow.

CP: Hydrogen peroxide may be regarded as a bidentate ligand which substitutes the oxygen ligands at the tungstate ion. The 3-membered ring configuration $\overset{O}{\underset{O}{|}}\!\!>\!W$ is thermodynamically more stable than the $W\overset{H_2O\downarrow}{=}O$ or $W\!<\!\begin{smallmatrix}OH\\OH\end{smallmatrix}$ bonding, presuming the hydrogen peroxide concentration is high enough.
Depending on the pH value, peroxotungstate species with different $W{:}O_2$ ratio are formed. A survey is presented in Table 4.8. In basic solutions the peroxo groups occupy all eight coordination positions at the W ion. By lowering the pH the $W{:}O_2$ ratio decreases. As soon as oxygen ligands are present at the W ion dimerization becomes possible, as is shown in the following equation:

$$2[W(O_2)4]^{2-} + 2H^+ + 3H_2O \Rightarrow [(O_2)_2OWOWO(O_2)_2]^{2-} + 4H_2O_2$$

This species is sometimes written in the hydrated form

$$(H_2O)(O)(O_2)_2WOW(O_2)_2(O)(H_2O)$$

TABLE 4.8. Peroxotungstate Species

W:O ratio	Formula		Formation
1:4	$WO_8{}^{2-}$	$[W(O_2)_4]^{2-}$	In basic solutions by an excess of hydrogen
1:3	$WO_7{}^{2-}$	$[WO(O_2)_3]^{2-}$	Thermal decomposition of 1:4 peroxotungstates
1:2	$WO_6{}^{2-}$	$[WO_2(O_2)_2]^{2-}$	By slight acidification of alkalitungstate solutions containing excess of hydrogen peroxide
	$W_2O_{11}{}^{2-}$	$[(O_2)_2OWOWO(O_2)_2]^{2-}$	
1:1	$WO_5{}^{2-}$	$[WO_3(O_2)]^{2-}$	By neutralization of peroxotungstic acid solutions increase

and this corresponds to

$$H_4W_2O_{13} \text{ or } H_2W_2O_{11} \cdot H_2O.$$

In general, the presence of hydrogen peroxide prevents the polycondensation tendency of the tungstate ions. All peroxo tungsten compounds are unstable. The degree of instability increases with increasing $W:O_2$ ratio. The 1:4 compound even may explode.

Peroxotungstic acids like H_2WO_6 or $H_2W_2O_{11}$ can be extracted from nitrate solutions into trialkyl-(C7-C9)-benzyl ammonium nitrate dissolved in trichloroethe-lyne.

Peroxotungstates of the following cations have been prepared: Na, K, Ca, Sr, Ba, Co, Sc, and Y.

A: Hydrogen peroxide is used to dissolve tungsten metal powder samples for analytical purposes. The advantage over other solvents is that no further anion is present which may disturb in subsequent reactions, such as fluoride. In this procedure hydrogen peroxide not only functions as solvent, but also as chelating agent preventing the tungsten from precipitating.

4.2.7.4. Nonstoichiometric Ternary Compounds (Reduced); Tungsten Bronzes [4.30, 4.45]. Tungsten bronzes (TBs) are well-defined nonstoichiometric compounds of the general formula [4.45]

$$M^{(z)}xWO_{3-y+z\cdot x/2}$$

where y can be less than or greater than $z \cdot x/2$; z is the valency of the bronze-forming atom.

The ideal composition is $y = z \cdot x/2$ ("true" tungsten bronzes; M_xWO_3).

Not only y but also x influences the physical as well as the chemical properties. For example,

$x < 0.3$: bronzes are semiconducting;

$x > 0.3$: bronzes are conducting (metallic bonding), because x has an influence on the Fermi levels.

The name bronzes derives from the metallic luster and the intense color. The stability of the TBs depends on the size of "M". A decrease in the ionic radius lowers the stability. A survey of TB-forming elements is given in Table 4.9. Very small ions, like Si or Ge, do not form bronzes. TBs shows a high degree of chemical inertness, especially toward nonoxidizing acids (including HF).

According to the above general formula, a phase diagram could be constructed describing the regions for the existence of the following groups of compounds (Fig. 4.10):

monotungstates and polytungstates;
oxides and bronzes of ideal composition;
oxygen-deficient oxides;
reduced polytungstates;
bronzes with cation excess;
bronzes with cation deficiency;
oxygen-deficient oxides with cation contamination.

TABLE 4.9. Tungsten Bronze Forming Elements [4.30]

Group[a]	Element	x	Structure[b]
1	Li	0–0.5	P
	Na	0–0.11	P
		0.26–0.38	T
		0.41–0.95	P
	Rb	0.06–0.10	I_D
		0.19–0.33	H
	Cs	0.06–0.10	I_D
		0.19–0.33	H
2	Ca	0–0.13	P
	Ba	0.03–0.05	I_S
		0.14–0.16	PC
		0.20–0.21	T
4	Zr	0–0.08	P
7	Mn	0–0.01	P
8	Fe	0–0.03	P
9	Co	0–0.04	P
10	Ni	0–0.02	P
12	Cd	0–0.12	P
13	In	0–0.03	P
		0.20–0.33	H
	Tl	?–0.10	I_S
		0.13–0.33	H
14	Sn	0.04–0.06	I_S
		0.15–0.18	I_D
		0.21–0.29	T
	Pb	0.03–0.05	I_S
		0.18–0.35	T
15	As	0–0.07	P
	Sb	0–0.07	P
		0.12–0.20	I_S
	Bi	0–0.05	P
		0.05–0.07	I_S
Lanthanoid series	La	0.02–0.19	
Actinoid series	U	0.07–?	P
	Th	0.07–?	P

[a] New IUPAC proposal 1985.

[b] *Abbreviations*: P, perovskite; T, tetragonal; H, hexagonal; I, intergrowth; I_S, single row of tunnels in parent WO_3 matrix; I_D, double tunnels; PC, pentagonal column bronzes related to tetragonal TB type.

STR: A variety of different structures are known:
The most frequent is the cubic *perovskite* structure. The ternary metal atoms occupy the cage sites present in the WO_3 host framework (Fig. 4.11).
Tetragonal tungsten bronzes are constructed of pentagonal bipyramides forming pentagonal columns (PCs); they can be described as an ordered arrangement of PCs within the WO_3 parent structure. The foreign metal atoms are located in the pentagonal tunnels and sometimes also in the square tunnels (Fig. 4.12).

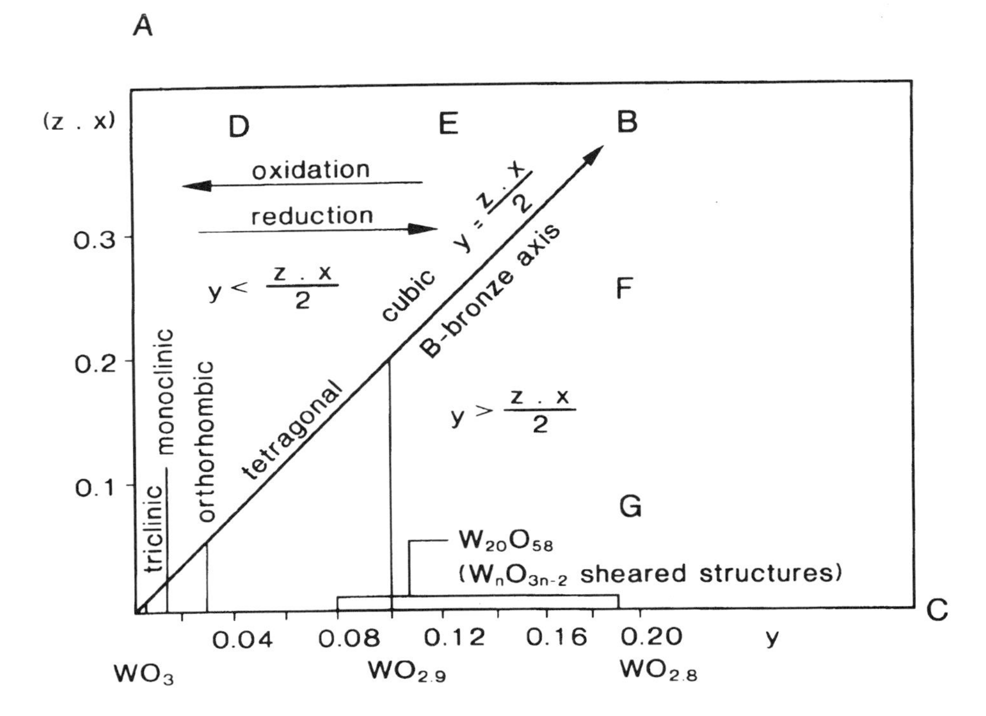

FIGURE 4.10. Idealized phase diagram of the M–W–O system [4.45]: (A) monotungstates and polytungstates; (B) oxide and bronzes of ideal composition; (C) oxygen-deficient oxides; (D) reduced polytungstates; (E) bronzes with cation excess; (F) bronzes with cation deficiency; (G) oxygen-deficient oxides with cation contamination.

In hexagonal tungsten bronzes corner-shared WO_6 octahedra are arranged to form hexagonal tunnels (Fig. 4.13.). The metal atoms are interpolated into the tunnels. Their composition range varies between $x = 0.15$ to 0.33. They form when larger metal atoms are reacted with WO_3 at high temperature.

Intergrowth tungsten bronzes exist when hexagonal tungsten bronzes coherently intergrow with the parent WO_3 structure (Figs. 4.14 and 4.15).

PR: Chemical reduction in acidic solution, either by metal amalgams or electrolytically (mainly for hydrogen bronzes).
Thermal reduction during many different reactions (solid/solid, molten/molten, solid/gaseous state reaction, etc.).
Electrochemical reduction of molten mixtures.

Sodium Tungsten Bronzes

They are the most extensively investigated tungsten bronzes.

PR: Electrolytic reduction.
Vapor-phase deposition.
Fusion.
Solid state reaction at 500–850 °C in vacuum.

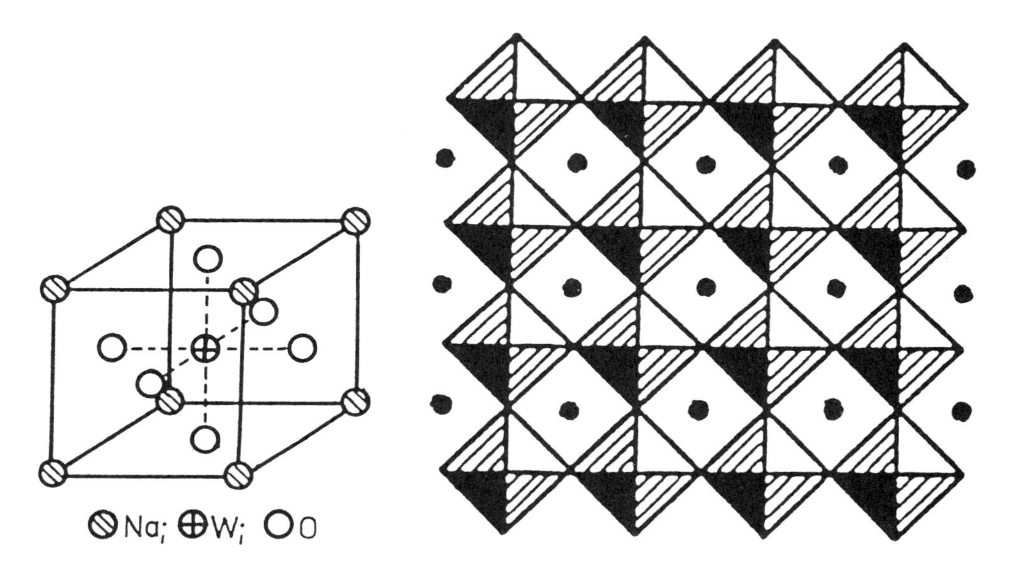

FIGURE 4.11. Perovskite structure: (a) Single cell; $NaWO_3$. (b) Idealized: the large metal atoms are interpolated into the cage sites in the WO_3 parent matrix of WO_3 octahedra; the perovskite bronzes have a partial filling of the structure to produce a nonstoichiometric phase [4.30].

FIGURE 4.12. Tetragonal tungsten bronze structure: the circles represent possible metal atom sites and the shaded squares WO_6 octahedra [4.30].

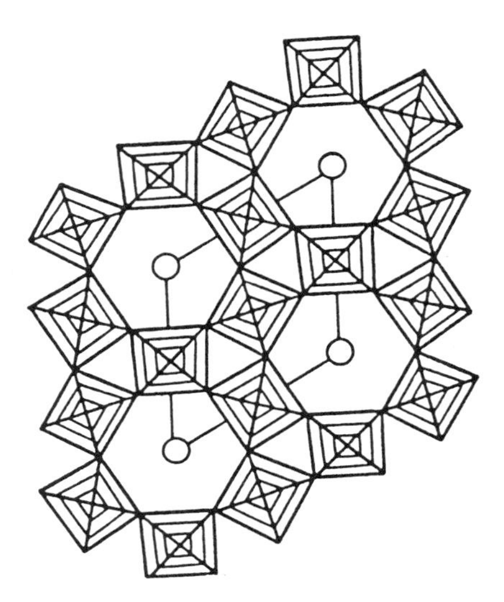

FIGURE 4.13. Hexagonal tungsten bronze structure: the circles represent the positions occupied by a variable population of interpolated metal atoms [4.30].

PP: CO: intensely colored from golden yellow to bluish black, depending on x, metallic sheen; CR: cubic $E2_1$, perovskite type.

x	a (Å)	Color [4.4]
0.93	3.850	Golden-yellow
0.64	3.834	Red-yellow
0.46	3.825	Red-violet
0.32	3.813	Blue-violet

temperature coefficient of resistivity: $x > 0.3 \rightarrow$ positive; $x < 0.3 \rightarrow$ negative.

A: Promotes the catalytic oxidation of CO or reformer gas in fuel cells.

Ammonium and Hydrogen Tungsten Bronzes

Besides the above-mentioned elements, hydronium and ammonium ions also form TB compounds.

Ammonium tungsten bronzes correspond to $x = 0.06$–0.33; $y = 0.030$–0.165.

With the exception of the lowest x and y: $(NH_4)_{0.06}WO_3 \cdot (H_2O)_{0.11}$, which is tetragonal, they belong to the hexagonal system. The dominance of the hexagonal symmetry can be understood as a consequence of the large size of the ammonium ion. They are partially contained in the industrial tungsten blue oxide, especially at low APT decomposition temperatures or short exposure times.

In hydrogen tungsten bronzes x values between 0.03 and 0.53 and y values from 0.015 to 0.3 can be found. They have quite different structures (tetragonal, orthorhombic, monocline, hexagonal, and cubic) and can also be detected in tungsten blue oxides.

We refer the reader to a complete literature survey on tungsten bronzes presented elsewhere [4.32].

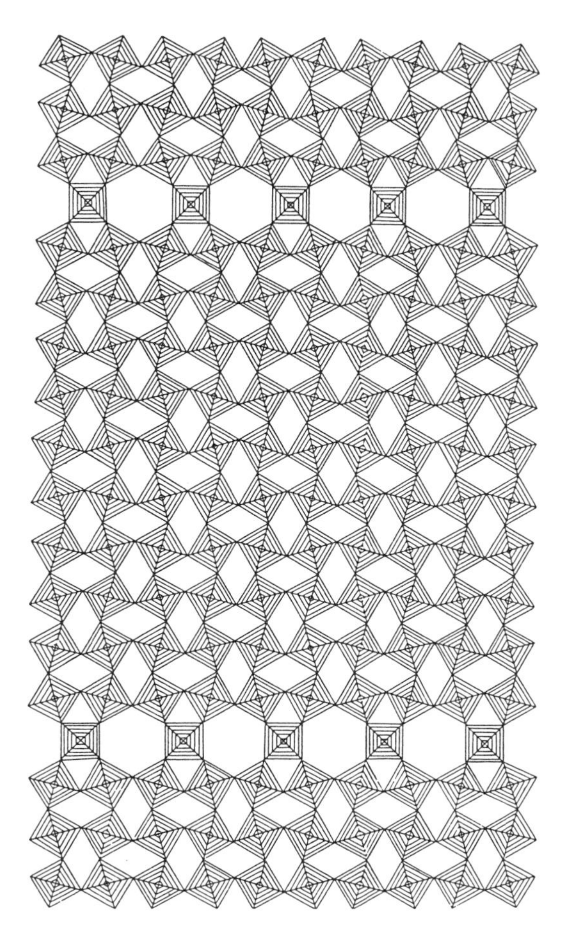

FIGURE 4.14. Intergrowth tungsten bronze structure of the pyrochlore type (single tunnel) [4.30].

4.2.8. *Tungsten and Sulfur* [4.4, 4.7, 4.8, 4.11, 4.25, 4.46]

The binary compounds WS_2 and WS_3 exist.

Tungsten Disulfide: WS_2

PR: Direct synthesis of W and S at 800–900 °C in pure nitrogen atmosphere (β-WS_2).
Sulfur vapor over WO_3 at 1400 °C; $2WO_3 + 4S \rightarrow 2WS_2 + 3O_2$ (α-WS_2).
Thermal decomposition of WS_3.

PP: D: 7.73 $g \cdot cm^{-3}$; MP: 1250 °C (decomposition); CO: black.

CR: β-WS_2: hexagonal C_7; MoS_2(h)-type, $a = 3.145$–3.165 Å, $c = 12.25$–12.35 Å; α-WS_2, rhombohedral, MoS_2(r)-type; $a = 3.162$ Å, $c = 18.50$ Å.
WS_2 has a layer structure, each metal atom being surrounded by a trigonal, prismatic array of S atoms; WS_2 is a diamagnetic semiconductor.

CP: It is insoluble in water, diluted nitric and sulfuric acid. It dissolves in HF–aqua regia mixture, molten alkalis, and alkali carbonates. It decomposes by heating in air, but its thermal stability is 90 °C higher than that of MoS_2.

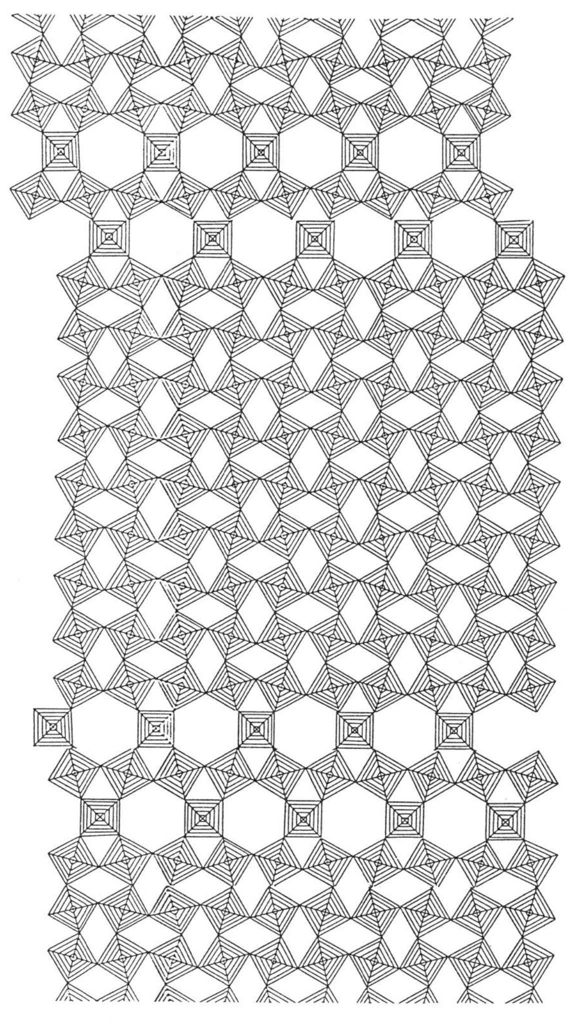

FIGURE 4.15. Intergrowth tungsten bronze structure of the pyrochlore type (double tunnel) [4.30].

A: Forms adherent, soft, continuous films on different surfaces and exhibits good lubricant properties, similar to MoS_2. Is used as lubricant.

Tunsten Trisulfide: WS_3

PR: It can only be produced indirectly by decomposing sodium thiotungstate with hot sulfuric or hydrochloric acid.

$$WS_4^{2-} + 2H^+ \Rightarrow WS_3 + H_2S$$

PP: Reddish brown, amorphous powder; diamagnetic and semiconducting.

CP: It is slightly soluble in cold water and colloidal in hot. It dissolves in alkali carbonates or hydroxides. Decomposition to WS_2 and W occurs at 170 °C.

4.2.9. Tungsten and Selenium [4.4, 4.7, 4.8, 4.11, 4.25, 4.46]

As with sulfur, tungsten forms two binary compounds: WSe_2 and WSe_3.

Tungsten Diselenide: WSe_2

PR: By direct reaction of the elements at elevated temperature in an inert atmosphere. By thermal dissociation of WSe_3.

PP: D: 9.22–9.32 $g \cdot cm^{-3}$; MP: 800 °C (decomposition); CO: black; CR: hexagonal C_7, $MoSe_2$-type; $a = 3.28$ Å, $c = 12.95$ Å.

CP: Only barely dissociated between 700 °C and 800 °C in vacuo, but at higher temperature rapid dissociation.

A: It is in use as lubricant, like the analogous sulfur compound.

Tungsten Triselenide: WSe_3

PR: Sodium tungstate solution is saturated with H_2Se, whereby selenotungstate ions $[WSe_4]^{2-}$ are formed which are subsequently decomposed to WSe_3 by acidification.

PP: Black, amorphous powder.

CP: It is soluble in concentrated HCl and decomposes at 220 °C to WSe_2.

4.2.10. Tungsten and Tellurium [4.4, 4.8, 4.11]

The only known binary compound is WTe_2.

Tungsten Ditelluride: WTe_2

PR: By direct combination of the elements below 750 °C under vacuum.

PP: D: 9.49 $g \cdot cm^{-3}$; MP: 972 °C (decomposition); CO: yellow, crystalline needles.

CP: It is insoluble in water, ammonia, hydrochloric, and sulfuric acid, but is decomposed by nitric acid. Reaction with oxygen occurs at 650–700 °C and thermal decomposition in vacuum starts at 600–620 °C.

4.2.11. Tungsten and Fluorine [4.1, 4.4, 4.7, 4.11, 4.25, 4.46]

Tungsten halide compounds in general and the fluorides and chlorides in particular are the basis for sytheses of a vast variety of substitution, mixed ligands, and coordination compounds, because they are stable compounds and, especially, halide atoms are easy to substitute. Halide atoms also may be substituted by each other, such as in fluoro chloro compounds of the general formula WF_nCl_{6-n}. Besides the hexavalent species WX_6, WOX_4, WO_2X_2, the lower state compounds are also of interest as precursors. The special importance of the oxo and dioxo halides is that they are coordinatively unsaturated. Therefore, they easily form coordination compounds by addition. Analogously to the oxohalides, thio-, seleno-, and nitridohalides are known.

Tungsten forms with fluorine the binary compounds WF_6, WF_5, and WF_4 while the following ternary compounds are formed with fluorine and oxygen: WOF_4, WO_2F_2, and WOF_2.

Tungsten Hexafluoride: WF_6

PR: The direct florination of tungsten in a flowing system at 350–400 °C is the most convenient procedure: $W + 3F_2 \rightarrow WF_6$.
By reaction of tungsten hexachloride with HF: $WCl_6 + 6HF \rightarrow WF_6 + 6HCl$; or with arsenic trifluoride: $WCl_6 + 2AsF_3 \rightarrow WF_6 + 2AsCl_3$; or with antimony pentafluoride: $WCl_6 + 3SbF_5 \rightarrow WF_6 + 3SbF_3Cl_2$.
Tungsten also reacts with ClF, ClF_3, and BrF_3 to form WF_6. WO_3 reacts with HF, BrF_3, and SF_4 to WF_6.

PP: $\leq$2.0 °C white solid; CR: 0 °C cubic, $a = 6.28$ Å; -30 °C orthorhombic; $a = 2.68$ Å, $b = 18.81$ Å, $c = 5.09$ Å; between 2.0 °C and 19.9 °C pale yellow liquid; specific gravity: 3.44 $g \cdot cm^{-1}$ (at 15 °C); $\geq$17.5 °C, colorless gas; D: related to hydrogen 149; related to air 12.9; it is one of the heaviest gases known.

CP: It is unstable in the presence of moisture: $WF_6 + 4H_2O \rightarrow H_2WO_4 + 6HF$.
Soluble in benzene and cyclohexane (bright red), in dioxane (pale red), and in ether (violet-brown). It is very reactive, attacks most metals, erodes glass slowly in the absence of moisture, but rapidly in the presence of moisture.

A: WF_6 is today widely used by the semiconductor industry in the manufacture of integrated circuits as metallization source (chemical vapor deposition of W; see Sections 5.6.3 and 5.7.5). Due to the "explosion" of the electronics industry the demand of WF_6 during the past decades has been rapidly growing and remains presently at approximately 200 t/year worldwide [4.47]. This corresponds to 127 t of W or 0.3–0.4% of the world total production. WF_6 for that purpose must be of highest purity ranging from 99.9995 to 99.98%, depending on application.

Tungsten Pentafluoride: WF_5

PR: By reduction of WF_6 on a hot tungsten filament.

STR It has a tetrameric structure $[(WF_5)_4]$; the four W atoms form a square plane bridged by four F atoms; each W atom is octahedrally coordinated to four terminal fluorides.

CO: Yellow.

Tungsten Tetrafluoride: WF_4

PR: Reduction of WF_6 with PF_3 in presence of liquid anhydrous H_2F_2.

PP: Nonvolatile, hygroscopic, reddish brown solid; CR: orthorhombic.

Tungsten Oxytetrafluoride: WOF_4

PR: By sustitution of Cl: $WOCl_4 + 4HF \rightarrow WOF_4 + 4HCl$.
By reaction of WO_3 with lead fluoride at 700 °C: $2PbF_2 + WO_3 \rightarrow WOF_4 + 2PbO$.
A mixture of oxygen and fluorine reacts with W at elevated temperature to WOF_4.

PP: It is a crystalline, white substance; MP: 110–119 °C; BP: 187.5–190 °C.

STR: The structure is tetrameric $[(WOF_4)_4]$ with W–F...W bridges and terminal W=O groups.

CP: It is extremely hygroscopic (decomposition to H_2WO_4 and HF) and it is soluble in $CHCl_3$ and slightly soluble in CS_2.

Tungsten Dioxydifluoride: WO_2F_2
PR: Hydrolysis of WOF_4.
PP: White solid.

Tungsten Oxydifluoride: WOF_2
PR: Reaction of tungsten dioxide and HF at 600 °C: $WO_2 + 2HF \rightarrow WOF_2 + H_2O$.
PP: Inert, gray solid substance; polymeric with O and F bridges.

4.2.12. Tungsten and Chlorine [4.1, 4.4, 4.7, 4.11, 4.25, 4.46]

The binary compounds WCl_6, WCl_5, WCl_4, and WCl_2 and the ternary oxychlorides $WOCl_4$, WO_2Cl_2, $WOCl_3$, and $WOCl_2$ exist.

Tungsten Hexachloride: WCl_6
PR: By direct chlorination of tungsten in a flowing system at 600 °C. On cooling between 168 °C and 170 °C a violent, explosion-like expansion of the solid occurs due to an α_2–α_1 transition.
By reaction of compounds like S_2Cl_2, PCl_5, HCl, $COCl_2$, or CCl_4 with tungsten or tungsten trioxide.
PP: D: 3.52 $g \cdot cm^{-3}$; MP: 275 °C; BP: 346.7 °C; CO: blue black solid; CR: hexagonal; $a = 6.100$ Å, $c = 16.70$ Å.
CP: It is very soluble in CS_2, alcohol, CCl_3, CCl_4, ether, benzene, acetone, and ammonia. Decomposition with water starts above 60 °C. Gaseous WCl_6 will be stepwise reduced by hydrogen to lower chlorides and finally to W.
A: It is commercially available and is used as starting material for the chemical vapor deposition of tungsten (see Sections 5.6.3. and 5.7.5) and for the preparation of metathesis catalysts which form double and triple bonds with carbon (see Chapter 10).

Tungsten Pentachloride: WCl_5
PR: Reduction of WCl_6 by red phosphorus at 280 °C or by hydrogen at 400 °C.
Reaction of WCl_6 with tetrachlorethene at 100 °C: $2WCl_6 + C_2Cl_4 \rightarrow 2WCl_5 + C_2Cl_6$.
Disproportion of WCl_4.
PP: Crystalline, deliquescent solid; D: 3.875 $g \cdot cm^{-3}$; MP: 243–248 °C; BP: 276 °C; CO: black or dark green.
CP: Hygroscopic, decomposes in water to $W_{20}O_{58}$, slightly soluble in CS_2, disproportionates into WCl_4 and WCl_6. It is dimeric $(WCl_5)_2$. The long W–W distance of 3.814 Å and a magnetic momentum of 1.0 μB (per atom at room temperature) indicate metal bonding.

Tungsten Tetrachloride: WCl_4
PR: Reaction of WCl_6 with $W(CO)_4$ or P_4 in a sealed tube.
Thermal gradient reduction of WCl_6 by Al or hydrogen.

PP: Nonvolatile, crystalline, deliquescent solid; diamagnetic; D: 4.624 $g \cdot cm^{-3}$; CO: black; CR: orthorhombic; $a = 8.07$ Å, $b = 8.89$ Å, $c = 6.85$ Å.

CP: It is hygroscopic and hydrolyzes, sublimes, and decomposes on heating. Consists of infinite linear chains of octahedra sharing a common edge. The alternating short and long metal distances imply metal–metal bonding.

A: It is a useful precursor for many coordination compounds.

Tungsten Trichloride: WCl_3

PR: By oxidation of $[(W_6Cl_8)Cl_4]$ with chlorine.

CP: Consists of clusters $[(W_6\{\mu_2\text{-}Cl\}_{12}]Cl_6$; CO: red.

Tungsten Dichloride: WCl_2

PR: Reduction of WCl_6 with Al in a $NaAlCl_4$ melt or with sodium amalgam.
Disproportionation of WCl_4.

PP: Not volatile, not fusible, amorphous powder; D: 5.436 $g \cdot cm^{-3}$; CO: gray or yellow.

CP: It is a cluster compound containing the face-bridged octahedral core tetracation $[W_6\{\mu_3\text{-}Cl\}_8]^{4+}$; $[W_6Cl_8]Cl_4$.

Tungsten Oxytetrachloride: $WOCl_4$

PR: Refluxing $SOCl_2$ on WO_3 (purification by sublimation).
Direct reaction of chlorine with W powder in presence of oxygen or water vapor.
Reaction of WCl_6 and WO_3: $WO_3 + 2WCl_6 \rightarrow 3WOCl_4$.
$2CCl_2F_2 + 2WO_2 \rightarrow WOCl_4 + WOF_4 + CO$.

PP: Crystalline solid; D: 11.92 $g \cdot cm^{-3}$; MP: 209–211 °C; BP: 327 °C; CO: red.

CP: Soluble in CS_2, benzene, sulfur chloride; reacts with water to form tungstic acid; carbon $\rightarrow WCl_6$; oxygen $\rightarrow WO_2Cl_2$; dry $NH_3 \rightarrow$ W nitrides; $PCl_5 \rightarrow WCl_6$; HF $\rightarrow WOF_4$; it is a polymer containing slightly distorted $OWOCl_4$ octahedra connected via $Cl_4W{=}O \rightarrow (W{=}O)Cl_4$ bonding.

Tungsten Dioxydichloride: WO_2Cl_2

PR: Reaction of WO_2 and CCl_4 in a bomb.
Chlorination of WO_3 by chlorine at 900 °C or by HCl at 300 °C.

PP: Crystalline solid; MP: 266 °C (sublimes); CO: pale yellow.

CP: Soluble in cold water and in alkaline solutions; decomposes in hot water and is insoluble in common organic solvents. It is polymeric and consists of $[\{WO_2Cl_2\}_n]$ with W–O. . .W bridges.

Tungsten Oxytrichloride: $WOCl_3$

PR: Heating of a mixture of WCl_5 and Sb_2O_3 at 100 °C up to 170 °C.
$3WCl_5 + Sb_2O_3 \rightarrow 3WOCl_3 + 2SbCl_3$.
Reduction of $WOCl_4$ by Al in sealed tubes at 100–140 °C.

PP: Crystalline solid; CO: green.

CP: Infinite W–O–W chains.

Tungsten Oxydichloride: $WOCl_2$

PR: Heating of a mixture of W, WO_3, and WCl_6 at 250–450 °C.
$W + WO_3 + WCl_6 \rightarrow 3WOCl_2$.

Thermal decomposition of $WOCl_3$ results in a blue-black polymorph.
A golden brown polymorph forms by reduction of $WOCl_4$ with $SnCl_2$.

PP: Needle-like crystals with copper luster; D: 5.92 $g \cdot cm^{-3}$; CR: monoclinic; $a = 12.87$ Å, $b = 3.76$ Å, $c = 6.46$ Å, $\beta = 104.2°$.

CP: On heating in air it forms WO_2Cl_2 and finally WO_3. It does not react with acids at room temperature, but is soluble in hot nitric acid and hydrogen peroxide. It is a polymer with O and Cl bridges.

4.2.13. Tungsten and Bromine [4.1, 4.4, 4.7, 4.11, 4.25, 4.46]

As in the case of chlorine, there exist five binary compounds from WBr_6 to WBr_2, and two ternary compounds with bromine and oxygen, $WOBr_4$ and WO_2Br_2.

Tungsten Hexabromide: WBr_6

PR: Metathetical exchange reaction of BBr_3 with WCl_6.
Reaction of bromine vapor and tungsten powder.

PP: Crystallized substance; MP: 232 °C; CO: bluish black.

CP: Decomposes on heating, hydrolyzes easily, and dissolves in ammonia.

Tungsten Pentabromide: WBr_5

PR: Reaction of bromine vapor with tungsten powder at 450–500 °C.
Reduction of WCl_6 with HBr.

PP: Crystallizes in needles; MP: 276 °C; BP: 333 °C; CO: brownish black to violet brown.

CP: Extremely sensitive to moisture, unstable in air, soluble in ether, CCl_4, $CHCl_3$, $CHBr_3$, and benzene.

Tungsten Tetrabromide: WBr_4

PR: Reduction of WBr_5 by Al or W.
Decomposition of WBr_5 at 180 °C in vacuum.

PP: CR: orthorhombic; $a = 8.49$ Å, $b = 9.29$ Å, $c = 7.25$ Å; CO: black.

CP: Decomposes at 400 °C; has the same structure as WCl_4.

Tungsten Tribromide: WBr_3

PR: WBr_2 will be soaked in liquid bromine at 50 °C.

PP: Black powder.

CP: Disproportionates at 450–500 °C into WBr_5 and WB_2, and is soluble in water.

Tungsten Dibromide: WBr_2

PR: Reduction of WBr_5 by hydrogen in presence of $ZnCl_2$ at 400 °C.
Disproportion of WCl_4 at 450 °C: $3WBr_4 \rightarrow WBr_2 + 2WBr_5$.

PP: Black solid.

CP: Decomposes at 400 °C.

Tungsten Oxytribromide: $WOBr_3$

PR: Heating a mixture of WBr_5 and Sb_2O_3 under vacuum at 100 °C for 1 hour and at 150 °C for 24 hours: $3WBr_5 + Sb_2O_3 \rightarrow 3WOBr_3 + 2SbBr_3$.

PP: Brown fibrous structure.
CP: Decomposes at 250 °C: $3WOBr_3 \rightarrow WOBr_4 + WO_2Br_2 + WBr_3$.
STR: It consists of infinite W–O–W chains.

Tungsten Oxytetrabromide: $WOBr_4$
PR: By reaction of bromine vapor with a mixture of WO_3, C and W.
By reaction of CBr_4 with WO_2 at 250 °C.
$2W + WO_3 + 6Br_2 \rightarrow 3WOBr_4$.
PP: Deliquescent needles; MP: 277 °C; BP: 327 °C; CO: black.
CP: Very hygroscopic, decomposes rapidly in air, the polymeric structure is analogous to $WOCl_4$.

Tungsten Dioxydibromide: WO_2Br_2
PR: Passing a mixture of oxygen and bromine over tungsten at 300 °C or higher.
PP: Infusible red or brown substance.
CP: Less volatile than $WOCl_4$, decomposes on heating: $2WO_2Br_2 \rightarrow WO_3 + WOBr_4$; the polymeric structure is analogous to WO_2Cl_2.

Bromine additions play an important role in tungsten–halogen lamps (see Chapter 7). It is commonly added as alkylbromides.

4.2.14. Tungsten and Iodine [4.1, 4.4, 4.7, 4.11, 4.25, 4.46]

The following binary tungsten–iodine and ternary tungsten–oxygen–iodine compounds are known: WI_4, WI_3, $WI_2(W_6I_{12})$, W_6I_{15}, WOI_2, WO_2I_2, WO_2I, and WOI_4.

Tungsten Tetraiodide: WI_4
PR: Reaction of concentrated hydroiodic acid with WCl_6 at 100 °C.
WCl_6 with HI gas at 110 °C.
WO_2 and AlI_3.
PP: Black crystallized powder; D: 5.2 $g \cdot cm^{-3}$.
CP: Will be decomposed by air to WI_2 and I_2, is soluble in water, but hydrolyzes; insoluble in ether and $CHCl_3$, slowly soluble in cold alcohol; chlorine replaces iodine at 18 °C, and bromine replaces it at 100 °C. Dissolution and iodine release occurs with KOH solutions, molten alkali carbonates, molten potassium bisulfide, and liquid potassium.

Tungsten Triiodide: WI_3
PR: Reaction of iodine gas and $W(CO)_6$ in sealed tubes at 120 °C.
Is also one of the products of the reaction of W_6I_{12} with iodine gas.
PP: Gray needles.

Tungsten Diiodide: WI_2

PR: Passing iodine gas over tungsten powder at 800 °C.
Reaction of iodine and $W(CO)_6$ in nitrogen.
PP: Brownish crystalline substance; hexamer structure: $[W_6I_8]I_2I_{4/2}$.

Hexatungsten Pentadecaniodide: W_6I_{15}

PR: By reaction of iodine gas and W_6I_{12} at 350 °C in vacuum (coexists with WI_3).
PP: Brown.
CP: Stable in air, decomposes in KOH solutions and in concentrated sulfuric acid, and also has a hexameric structure $[W_6I_8]I_4[I_3]_{2/2}$.

Tungsten Oxydiodide: WOI_2

PR: By heating a mixture of tungsten, tungsten trioxide, and iodine powder under vacuum in a two-zone furnace (500 °C and 700 °C): $2W + WO_3 + 3I_2 \rightarrow 3WOI_2$ followed by purification via benzene extraction.
PP: Diamagnetic glittering plates; CO: black; CR: $a = 18.01$ Å, $c = 7.58$ Å.
CP: Slightly soluble in water and organic solvents and very soluble in NaOH and KOH solutions.

Tungsten Dioxydiiodide: WO_2I_2

PR: Direct reaction of WO_2 and iodine gas.
PP: D: 6.39 $g \cdot cm^{-3}$; MP: 6.40 °C; BP: 200 °C; CO: dark brown; CR: monocline, $a = 17.095$ Å, $b = 3.899$ Å, $c = 7.492$ Å, $\beta = 102.66°$.
CP: Decomposes at 277 °C to WO_2I and iodine. As a side product also WOI_3 is formed.
A: Formation of gaseous WO_2I_2 is considered to effect the removal of tungsten from lamp walls in halogene lamps (Chapter 7).

4.3. MIXED LIGANDS AND COORDINATION COMPOUNDS OF TUNGSTEN [4.1, 4.25, 4.46]

4.3.1. Mixed Ligand Compounds

Some compounds belonging to this group have already been treated in the foregoing subsection, namely, the oxide hydrates and oxohalides. Mixed ligand compounds may contain only inorganic, or mixed inorganic and organic, or merely organic ligands. A large number of these compounds are coordination compounds (complexes) and a high percentage are organometallics. Therefore, it is not easy to draw a borderline between these groups and in various cases some overlapping occurs. For example, carbonyl, thiocarbonyl, cyanide, and isocyanide ligands are counted with the organometallics in some publications while others regard them as pure inorganic ligands.

Some typical mixed ligand compounds are substituted carbonyl or halide compounds of the general formula:

$$WX_{6-n}L_n \qquad \begin{array}{l} X = \text{halide or carbonyl} \\ L = \text{different ligands} \end{array}$$

Typical examples are $[W(CO)_5Cl]$, $[W(CO)_4Br_2]$, $[W(CO)_3Cl(\eta^5\text{-}C_5H_5)]$, $[W(CO)_2Cl_3(\eta^5\text{-}C_5H_5)]$, and $[W(CO)(\eta^5\text{-}C_5H_5)]$.

4.3.2. *Coordination Compounds*

The chemistry of tungsten coordination compounds is quite complicated. The reasons are:

- Tungsten forms complexes in the valence states -2, -1, 0, $+2$, $+3$, $+4$, $+5$, $+6$.
- The coordination number is variable (max 13).
- Tungsten has the tendency to form clusters and polynuclear complexes with varying number of atoms. In these compounds the tungsten-to-tungsten bond varies between single and quadruple. The lower the valence state the higher the degree of W–W bonding.

A short survey of the possible donor atoms and corresponding ligand ions or groups is given in Table 4.10.

Hexavalent tungsten can be found in the form of hexahalides, which give rise to a variety of substitution products containing the structural unit W^{6+}. Hexavalent tungsten in addition shows a strong tendency to form bonds of higher order than one with donor atoms like O, S, Se, or N. Consequently, one must distinguish between complexes of the structural units W^{6+}, WO^{4+}, and $WO_2{}^{2+}$ and the analogous complexes which contain =S, =Se, or, =NR instead of oxygen. Thiotungstate ions like $WS_4{}^{2-}$, $WOS_3{}^{2-}$, and $WO_2S_2{}^{2-}$ act as bidentate ligands for other metal ions.

In pentavalent tungsten complexes the units W^{5+}, WO^{3+}, and WO^{2+} also exist, but in addition dimeric structures can be found containing the $W_2O_4{}^{2+}$ unit. They are not as common as in pentavalent molybdenum chemistry.

Tetravalent tungsten complexes contain the structural units W^{4+}, WO^{2+}, or dimeric configuration having W=W double bonds, as well as trinuclear clusters with three W atoms bound together in a triangular configuration.

Most complexes of trivalent tungsten are dinuclear such as, for example, in the $[W_2Cl_9]_6{}^{3-}$ ion with some of them having a triple bond between the two tungsten atoms.

In divalent tungsten complexes besides monomer configurations, also dimeric configurations having a quadruple W–W bond are known.

TABLE 4.10. Donor Atoms and Corresponding Ions or Groups in Tungsten Coordination Compounds

Donor atom	Ion/group
C	Alkylidine, carbonyl, isocyanide, cyanide
N	Nitrido, imido, amino, thiocyanate, nitrilo, pyridine, dinitrogen, diazenido, hydrazido, diazene, nitrene, hydroxylamino
P	Phosphine derivatives
As	Arsine derivatives
O	Aqua, oxaqua, oxo, peroxo, alkoxide, aryloxide, carboxylate, β-diketonato
S	Sulfido, persulfide, thiolato, dithiocarbamate, dithiol, xanthate
Se	Selenido
Te	Tellurido
F	Fluoride
Cl	Chloride
Br	Bromide

As already mentioned above, 0, −1, and −2 valent tungsten complexes also exist.

The ligands can be inorganic or/and organic ions and groups, and can be mono- or polydentate. A huge number of mixed ligand complexes as well as chelate complexes have been prepared. Further details can be found in other compilations [4.25, 4.46, 4.50, 4.52].

Some typical examples are shown below.

TABLE 4.11. Typical Examples for Organometallic Tungsten Compounds

Designation	Interacting atoms	Ligand type	Typical examples
η^1	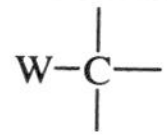 W–C—	Alkyl	$(CH_3)WCl_5$
	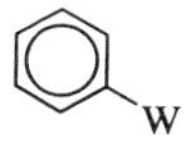 W (phenyl)	Aryl	$(C_6H_5)WCl_3$
	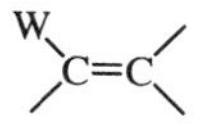 W–C=C	Alkene-(1)yl	$(\eta^5\text{-}C_5H_5)(CO)_3W\text{–}CH\text{=}C(L)(R)$
	W–C≡C—	Alkyne-(1)yl	$K_3[W(\text{–}C{\equiv}CH)_3(CO)_3]$
	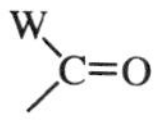 W–C=O	Acyl	$(\eta^5\text{-}C_5H_5)(CO)_2LW\text{–}C(CH_3)(\text{=}O)$
	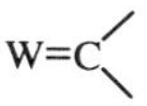 W=C	Alkylidene	$(CO)_5W\text{=}C(R)(OR)$
	W≡C—	Alkylidyne	$Cl(CO)_3W{\equiv}C\text{–}R$
η^2	 C=C / W	Alkene	$[W(\eta^2\text{-ethylene})_2(CO)_4]$
	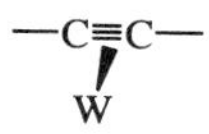 —C≡C— / W	Alkyne	$[W(\eta^2\text{-alkyne})_3CO]$
η^3	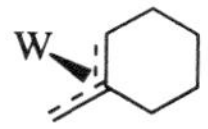 W	Allyl	$[W(\eta^3\text{-allyl})X(CO)_4]$
η^4	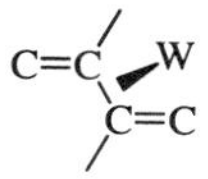 C=C, W, C=C	Diene	$[W(\eta^4\text{-diene})_2(CO)_2]$
η^5	 W	Cyclopentadienyl	$[W(\eta^5\text{-}C_5H_5)_2Cl_2]$
η^6	 W	Arene	$[W(\eta^6\text{-arene})(CO)_3]$
η^7	 W	Cycloheptatrienyl	$[W(\eta^7\text{-}C_7H_7)(CO)_3]^+$

Carbonyl complexes and derivatives: $[W(CO)_4]^{4-}$, $[W(CO)_5]^{2-}$, $[W_2(CO)_{10}]^{2-}$, $[W(CO)_4Br]^-$, $[W(CO)_3(\eta^5\text{-}C_5H_5)]^-$, $[W(CO)_4(\eta^5\text{-}C_5H_5)]^+$.

Cyanide and Isocyanide complexes and derivatives: $[W(CN)_8]^{4-}$, $[W(CN)_8]^{3-}$, $[W(CN)_6]^{4-}$, $[W(CNR)_7]^{2-}$, $[W(CNR)_3(\eta^5\text{-}C_5H_5)]$.

Thiocarbonyl complexes: $[W(CS)(CO)_4I]$, $[W(CS)(CO)_2(\eta^5\text{-}C_5H_5)]$.

4.4. ORGANOMETALLIC TUNGSTEN COMPOUNDS [4.1, 4.48, 4.49]

Organometallic tungsten compounds can be characterized by having one or several bondings between a tungsten atom and one or more carbon atoms belonging to one organic ligand. The bonding between the tungsten and one carbon atom can be single, double, or triple.

The number of compounds synthesized so far is extremely large. The organic ligands can be classified by the number of carbon atoms (1–7) interacting with tungsten, as will be shown in Table 4.11.

The most important precursors for the preparation of organometallic tungsten compounds are the halides and oxohalides, as well as hexacarbonyl.

In lower valence states stabilization is gained by π-bonding. The preferential oxidation state is 2+, but also lower and higher valent compounds are known. The most important class of these compounds is σ-organyl-tricarbonyl-(cyclopentadienyl)-tungsten and their derivatives (especially the pentamethyl-cyclopentadienyl) or analogous compounds having six-membered ring systems. The general formula corresponds to $W(CO)_3(\eta^5\text{-}C_5H_5)R$ where R can be alkyl, acyl, alkenyl, or aryl. These compounds are used in catalytic systems of diverse organic syntheses (see Chapter 10). For more details reference is made to the compilations [4.48, 4.49, and 4.51].

REFERENCES FOR CHAPTER 4

4.1. N. N. Greenwood and A. Earnshaw, *The Chemistry of the Elements*, p. 1291, Pergamon Press, Oxford (1984).

4.2. I. Barin, *Thermochemical Data of Pure Substances*, 3rd edition, Vol. II, VCH, Weinheim (1995).

4.3. O. Knacke, O. Kubaschewski, and K. Hesselmann, eds., *Thermochemical Properties of Inorganic Substances II*,, 2nd edition, Springer-Verlag, Heidelberg (1991).

4.4. S. W. H. Yih and C. T. Wang, *Tungsten*, Plenum Press, New York (1979).

4.5. *Gmelin Handbook of Inorganic Chemistry*, 8th ed., Syst. No. 54, *Tungsten*, Suppl. Vol. A6a, Springer-Verlag, Heidelberg (1991).

4.6. *Gmelin Handbook of Inorganic Chemistry*, 8th ed., Syst. No. 54, *Tungsten*, Suppl. Vol. A6b, Springer-Verlag, Heidelberg (1988).

4.7. J. A. Mullendore, in: Kirk-Otmer: *Encyclopedia of Chemical Technology*, Vol. 23, 3rd ed., *Tungsten and Tungsten Alloys*, Wiley, Chichester (1983).

4.8. S. V. Nagender Naidu and P. Rama Rao, *Phase Diagrams of Binary Tungsten Alloys*, Indian Institute of Metals, Calcutta (1991).

4.9. E. Rudy and S. Windisch, *Ternary Phase Equilibria in Transition Metal-B-C-Si Systems*, Part I, Vol. 3, Technical Report AFML-TR-65-2 (1965).

4.10. H. Duschanek and P. Rogl, *J. Phase Equilibria* **16** (1995), 150–161.

4.11. *Gmelin Handbook of Inorganic Chemistry*, 8th ed., Syst. No. 54, *Tungsten*, Suppl. Vol. A5b, Springer-Verlag, Heidelberg (1993).

4.12. F. Binder, *Radex Rundsch.* **4** (1975), 531-557.
4.13. E. Rudy, *Report AFML-TR-65-2*, Part V, Aerojet-General Corp., Sacramento, California (1969).
4.14. E. Rudy, *Report AFML-TR-2*, Part II, Vol. 18, Wright-Patterson Airforce Base, Ohio (1968).
4.15. T. B. Massalski, *Binary Alloy Phase Diagrams*, 2nd edition, Vol. 1, p. 896, ASM International (1990).
4.16. P. Gustafson, *Mater. Sci. Technol.* **2** (1986), 653–658.
4.17. E. Rudy and J. R. Hoffman, *Planseeber. Pulvermetall.* **15** (1967), 174–178.
4.18. K. Yvon, H. Nowotny, and F. Benesovsky, *Monatsh. Chem.* **99** (1968), 726–729.
4.19. D. N. French and D. A. Thomas, *Trans. AIME* **233** (1965), 950.
4.20. M. Lee, *Metall. Trans.* **14A** (1983), 1625.
4.21. L. Pons, in: *Anisotropy in Single-Crystal Refractory Compounds* (F. V. Vahldiek and J. Mersol, eds.), Vol. 2, pp. 393–444, Plenum Press, New York (1968).
4.22. B. Roebuck, *Int. J. Refract. Met. Hard Mater.* **13** (1995), 265–279.
4.23. Y. Ishizawa and T. Tanaka, *Inst. Phys. Conf. Ser.* No. 75, pp. 29–43, Adam Hilger Ltd., Bristol and Boston (1986).
4.24. L. F. Mattheis and D. R. Hamann, *Phys. Rev. B* **30** (1984), 1731–1738.
4.25. J. A. McCleverty, "Tungsten; Inorganic and Coordination Chemistry," in: *Enyclopedia of Inorganic Chemistry* (R. B. King, ed.), Vol. 8, pp. 4240–4268, Wiley, Chichester (1994).
4.26. V. A. Maksimov and P. T. Shamrai, *Inorg. Mater.* **5** (1969), 965–966.
4.27. H. A. Wriedt, *Bull. Alloy Phase Diagrams* **10** (1989), 358–367.
4.28. A. F. Guillermet and S. Jonsson, *Z. Metallkd.* **84** (1993), 106-117.
4.29. *Gmelin Handbuch der Anorganischen Chemie*, 8th ed., Syst. No. 54, *Tungsten*, Suppl. Vol. B2, Springer-Verlag, Heidelberg (1979).
4.30. R. J. D. Tilley, *Int. J. Refract. Met. Hard Mater.* **13** (1995), 93–109.
4.31. T. Ekström and R. J. D. Tilley, *Chem. Scr.* **16** (1980), 1–23.
4.32. *Gmelin Handbuch der Anorganischen Chemie*, 8th ed., Syst. No. 54, *Tungsten*, Suppl. Vol. B3, Springer-Verlag, Heidelberg (1979).
4.33. *Gmelin Handuch der Anorganischen Chemie*, 8th ed., Syst. No. 54, *Tungsten*, Suppl. Vol. B4, Springer-Verlag, Heidelberg (1980).
4.34. *Gmelin Handbook of Inorganic Chemistry*, 8th ed., Syst. No. 54, *Tungsten*, Suppl. Vol. B5, Springer-Verlag, Heidelberg (1984).
4.35. *Gmelin Handbook of Inorganic Chemistry*, 8th ed., Syst. No. 54, *Tungsten*, Suppl. Vol. B6, Springer-Verlag, Heidelberg (1984).
4.36. M. A. Sobhan, R. T. Kivaisi, B. Stjerna, and C. G. Granqvist, *Sol. Energy Mater. Sol. Cells* **44** (1996), 451–455.
4.37. T. Millner, A. J. Hegedüs, K. Sasvari, and J. Neugebauer, *Z. Anorg. Allg. Chem.* **289** (1957), 288–312.
4.38. W. R. Morcom, W. L. Worrell, H. G. Sell, and H. I. Kaplan, *Metall. Trans.* **5** (1974), 155–161.
4.39. A. Bartl, *Doctoral Thesis*, TU Vienna, Austria (1997).
4.40. J. S. O. Evans, T. A. Mary, T. Vogt, M. A. Subramanian, and A. W. Sleight, *Chem. Mater.* **8** (1996), 2809–2823.
4.41. W. P. C. Duyvesteyn, H. Liu, N. L. Labao, and P. L. Shrestha, US Patent 5,178,848 (1993).
4.42. J. W. van Put, *Doctoral Thesis*, Delft University, The Netherlands, Delft University Press (1991).
4.43. Sylvania GTE, *Tech. Inf. Bull. CM-9003*-(9/80).
4.44. G. A. Tsigdinos, "Heteropoly Compounds of Molybdenum and Tungsten," *Climax Molybdenum Co. Bull.* Cdb-12a (Sept. 1966).
4.45. L. Bartha, A. B. Kiss, and T. Szalay, *Int. J. Refract. Met. Hard Mater.* **13** (1995), 77–91.
4.46. Z. Dori, *Prog. Inorg. Chem.* **28** (1981), 239–299.
4.47. Company brochure, Air Products and Chemicals, Inc., Allentown; PA, USA (1994).
4.48. A. Segnitz: "Organo-Wolframverbindungen," in: Houben-Weyl; *Methoden der Organischen Chemie* (E. Müller, O. Bayer, H. Meerwein, und K. Ziegler, eds.), Bd. 13/7, pp. 489–520, Georg Thieme, Stuttgart (1975).
4.49. A. Mayr, "Tungsten; Organometallic Chemistry," in: *Encyclopedia of Inorganic Chemistry* (R. B. King, ed.), Vol. 8, pp. 4268–4284, Wiley, Chichester (1994).
4.50. R. Püschel and E. Lassner, "Chelates and Chelating Agents in the Analytical Chemistry of Molybdenum and Tungsten," in: *Chelates in Analytical Chemistry*, Vol. 1 (H. A. Flaschka and A. J. Barnard, eds.), Marcel Dekker, New York (1967).
4.51. G. Wilkinson, F. G. A. Stone, and E. W. Abel, eds., *Comprehensive Organometallic Chemistry*, Pergamon Press, Oxford (1981).
4.52. G. Wilkinson, D. Gillard, and J. A. McCleverty, eds., *Comprhensive Coordination Chemistry*, Pergamon Press, Oxford (1987).

5

Industrial Production

5.1. MINING AND ORE BENEFICIATION [5.1–5.5]

5.1.1. Mining

Tungsten mines are relatively small and rarely produce more than 2000 t of ore per day. Mining is mainly limited by the size of the ore bodies, which are not very large. Open pit-mining is the exception. These mines are short-lived and soon convert to underground operations.

Mining methods for tungsten ore are not at all exceptional and usually are adapted to the geology of the ore deposit.

5.1.2. Ore Beneficiation

5.1.2.1. General. The majority of tungsten deposits only contains some tenths of a percent of WO_3. On the other hand, ore concentrates in international trading require 65–75% WO_3. Therefore, a very high amount of gangue material must be separated. This is why ore dressing plants are always located near the mine (to save transportation costs). Companies which process their own concentrates produce low-grade concentrates (6–40% WO_3) in order to minimize the loss of tungsten minerals which increases with increasing concentration grade.

Another important aspect of the beneficiation process today is the disposal of the separated gangue materials, which in case of flotation also contain chemicals. Especially in areas with rigid environmental restrictions, a deposition near the plant is sometimes not possible and transportation over longer distances is therefore necessary. Depending on the mine conditions, a total or partial refill of the tailings into the mine is possible.

In regard to the beneficiation of ore, the *positive* properties of the tungsten minerals are the high specific gravity (scheelite and wolframite) and the ferromagnetism (wolframite). A *negative* property is their brittleness, leading to a partial loss by too fine particles formed during the disintegration steps.

In principle, two important properties of the ore determine the flow sheet of an ore dressing plant:

1. The particle size of the tungsten ore which determines the degree of disintegration necessary to liberate the tungsten mineral (liberation size).
2. The type and concentration of the accompanying minerals (gangue) which have to be separated dictate the mode and number of separation steps.

Ore beneficiation consists of two main steps: comminution and concentration. Comminution is first performed by crushing. Equipment in use comprises jaw, cone, or impact crushers working mostly in closed circuits with vibratory screens. The second step in comminution is grinding, which is undertaken in rod or ball mills working in closed circuits with classifiers.

For concentration (separation of gangue minerals) several methods can be applied, depending mainly on the composition of the ore. They include ore sorting, gravity methods, flotation, magnetic, and electrostatic separation.

Beneficiation of tungsten ores by gravity was the classical method, followed by a "cleaning" step (Fig. 5.1). The recovery depends on the ore characteristics (mainly liberation size) and ranges typically between 60 and 85%. The main loss is in the slimes, because the tungsten minerals are the most friable ones present in ores.

As regards "cleaning," for example, a roast process can be applied to convert pyrite to a magnetic form, followed by its magnetic separation together with garnet and pyroxene. Another cleaning step in the presence of sulfide minerals would be a sulfide flotation.

Gravity methods can also be applied for both scheelite and wolframite. The usual equipment consists of spirals, cones, tables, and a sink-float.

In order to optimize the yield, modern technology includes the following additional steps or combinations:

Preconcentration. This could be accomplished by sorting, use of jigs, or heavy media separation.

Whole Flotation. If tungsten ore mineralization is too fine, the total amount of mined ore can be subject to flotation. A corresponding flow sheet of the Mittersill ore dressing plant (Wolfram Bergbau und Hüttengesellschaft m.b.H.) is presented in Fig. 5.2.

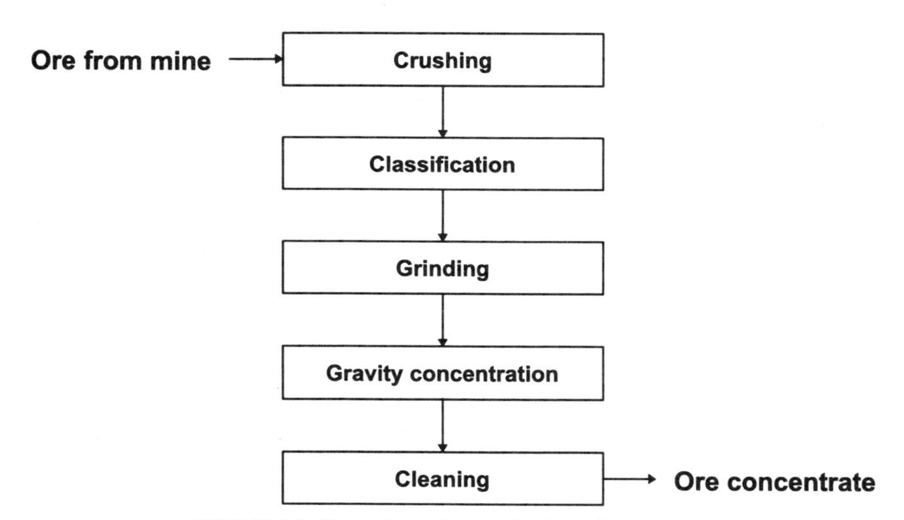

FIGURE 5.1. Classical gravity beneficiation of tungsten ore.

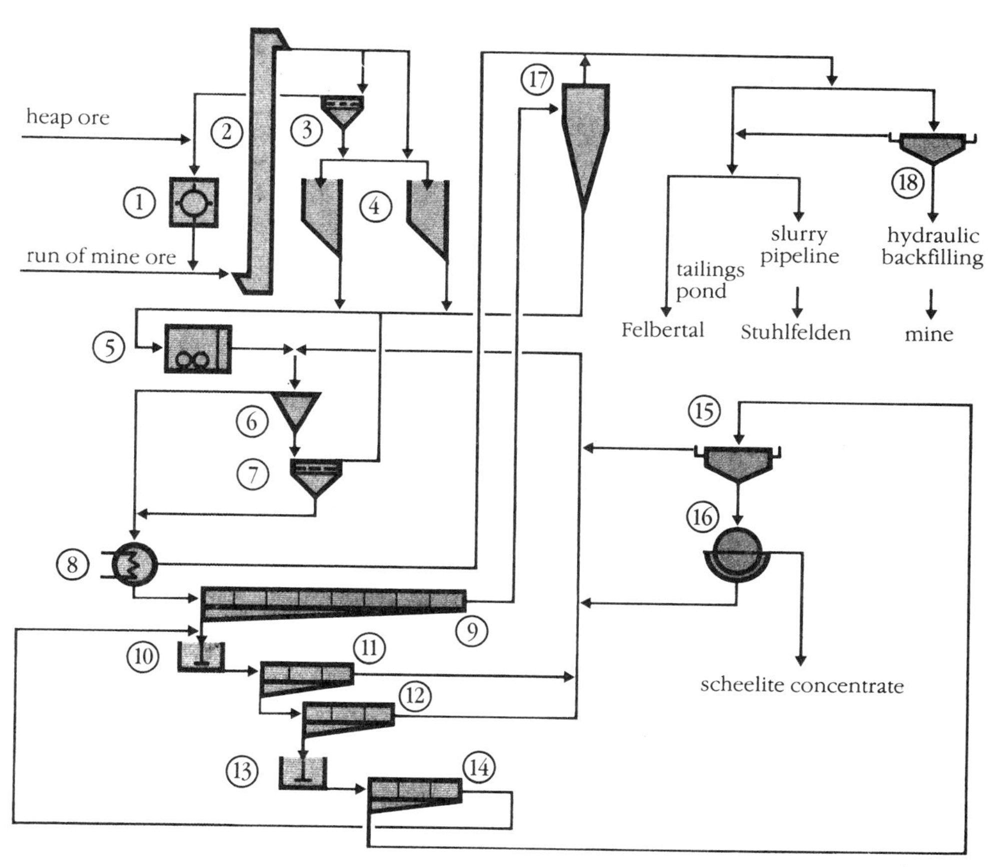

FIGURE 5.2. Mittersill ore dressing plant (Courtesy of Wolfram Bergau- und Hüttenges. m.b.H., Austria): (1) impact crusher, (2) elevator, (3) screen, (4) ore bins, (5) ball mill, (6) cone classifier, (7) dewatering screen, (8) low intensity magnetic separator, (9) rough flotation, (10) conditioner, (11) 1. Cleaner, (12) 2. Cleaner, (13) conditioner, (14) 3. Cleaner, (15) concentrate thickener, (16) drum filter, (17) hydrocyclone, (18) backfilling thickener.

Scavenging Circuits. These are combinations of gravity separation and flotation to recover the loss via slimes occurring in the classical procedure. A simplified flow sheet of the Cantung Mine beneficiation plant (Canada Tungsten Mining Corporation) is given in Fig. 5.3.

A flow sheet can be quite complicated especially in the case of complex ores containing wolframite and scheelite besides other valuable minerals. As an example, ore dressing at Xihuashan Mine is presented in Fig. 5.4.

5.1.2.2. Ore Sorting

Ore sorting is a modern preconcentrating stage ahead of the other conventional dressing methods. In former times, sorting was done by hand. Today, mechanical sorting is possible.

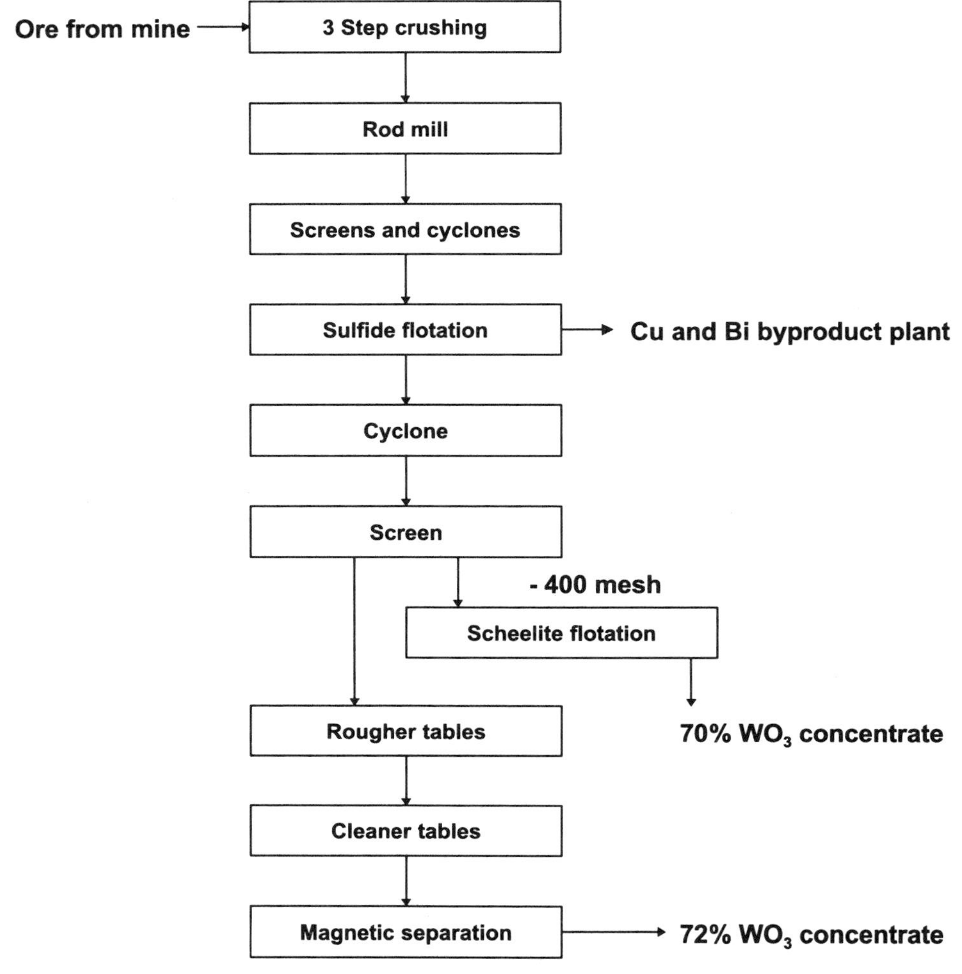

FIGURE 5.3. Simplified flow sheet of a combined gravity–flotation procedure as applied at Canada Tungsten Mining Corporation (before closing down) [5.1].

Successful application of an ore sorter depends on two important criteria:

The valuable fraction must be liberated from gangue at a size range which can be handled by the machine.

The valuable fraction must be identifiable within the time available for examination by the machine.

The principle of an ore sorting process is shown in Fig. 5.5. It consists of 4 main parts which are presentation, sensing, air blast, and data processing. In the presentation step, a monolayer of particles spread from each other is achieved. In the sensing stage, the location of valuables is determined, and in the separation step the valuable particles are blown off, leaving the gangue on the conveyer belt.

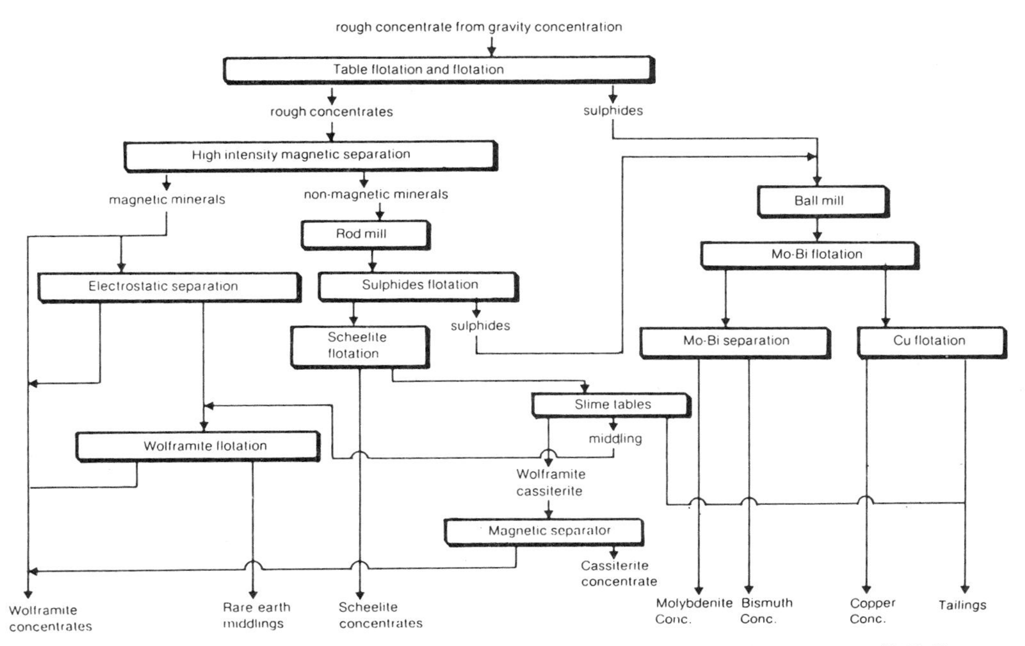

FIGURE 5.4. Flow sheet of the cleaning and comprehensive utilization of Xihuashan tungsten mill [5.4].

Photometric ore sorting can be applied to wolframite. It can also be used for scheelite, if contained in white quarzite veins. Fluorescence ore sorting is a suitable method for scheelite.

5.1.2.3. Scheelite Flotation. The existing literature on scheelite flotation is quite extensive, because the gangue minerals vary in quality and quantity not only from mine to mine, but also within the same deposit with progress in mining. In principle, each ore dressing plant has its own recipe, and this will always be adjusted to the variations in ore constitution. Although patents have been granted, the detailed procedures are kept secret by most companies.

The following chemicals and conditions are generally in use:

Collectors. Oleic acid, mixtures of oleic and linolic acid, sodium oleat, tall oil, saponified tall oil and mixtures of these substances, coriander oil, rape oil, mustard oil, pyridine base, water-soluble carbohydrates of the green oil type, also in saponated form, fatty acids and their salts, fish oil, hydroxy-ethylene-cellulose, carboxy-methyl-cellulose.

Frothers. Alcohols, cresylic acid, aerosol, oil of terpentine, ketones.

Depressants. Quebracho or tannin for calcite, formic acid for apatite, lactic acid for mica, and sodium salts of hydrolyzed polyacrylnitrile.

The optimal pH value is between 9.0 and 10.5 and must be controlled to a tolerance of ± 0.1 units, depending on the ore composition and the reagents in use. Sodium carbonate or sodium hydroxide solutions are used for pH regulation.

Sodium silicate serves as a dispersant for calcite. The addition of polyacrylamide or flocculents increases the flotation yield.

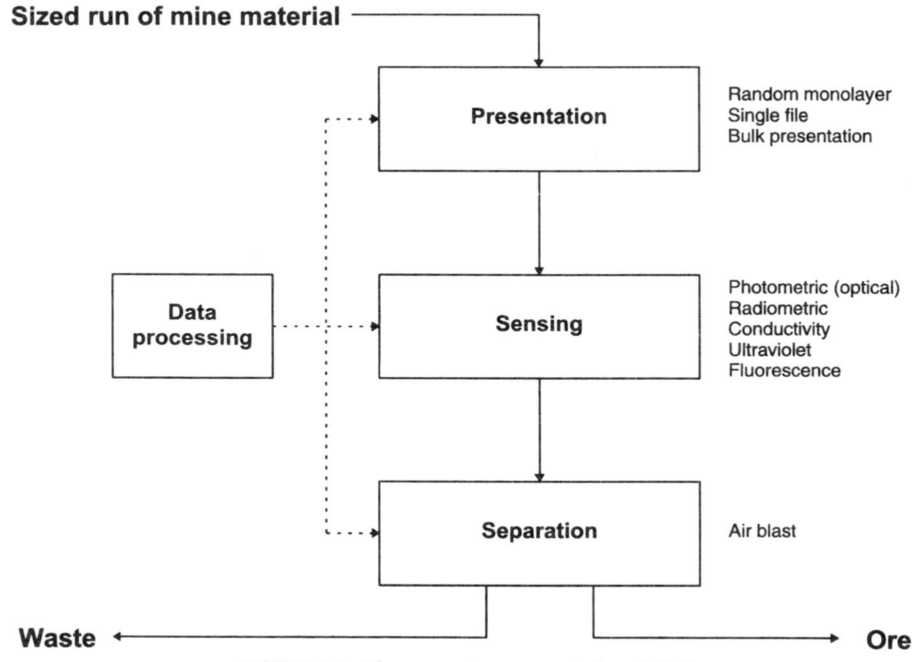

FIGURE 5.5. The ore sorting process (scheme) [5.5].

5.1.2.4. Wolframite Flotation. Wolframite flotation is performed similarly to the above-described scheelite flotation, but is not pH sensitive and can therefore be undertaken in both acidic and alkaline solutions. Generally, flotation is rarely applied to wolframite. The reason is that wolframite occurs mainly in vein-type deposits whose mineralization is much coarser than to most scheelite ores. Therefore, gravity and magnetic methods are more desirable than flotation.

5.1.2.5. Magnetic Separation. Magnetic separation can be applied for two possible reasons: to clean scheelite by removing magnetic minerals like garnet, magnetite, and roasted pyrite; to concentrate wolframite by separation from nonmagnetic minerals like cassiterite or unroasted pyrite. Dry and wet magnetic separators are employed.

5.1.2.6. High Tension Separation. Electrodynamic or electrostatic separators are used only for scheelite–cassiterite mixtures. In contrast to scheelite, cassiterite is conducting.

5.2. HYDROMETALLURGY

5.2.1. Introduction including Ecological and Economical Considerations [5.6]

This section deals exclusively with the modern process of tungsten hydrometallurgy. In this introduction the classical and modern concepts are compared, outlining the differences as well as the advantages and disadvantages. In Figs. 5.6 and 5.7, simplified flow sheets of the classical and modern processes are given.

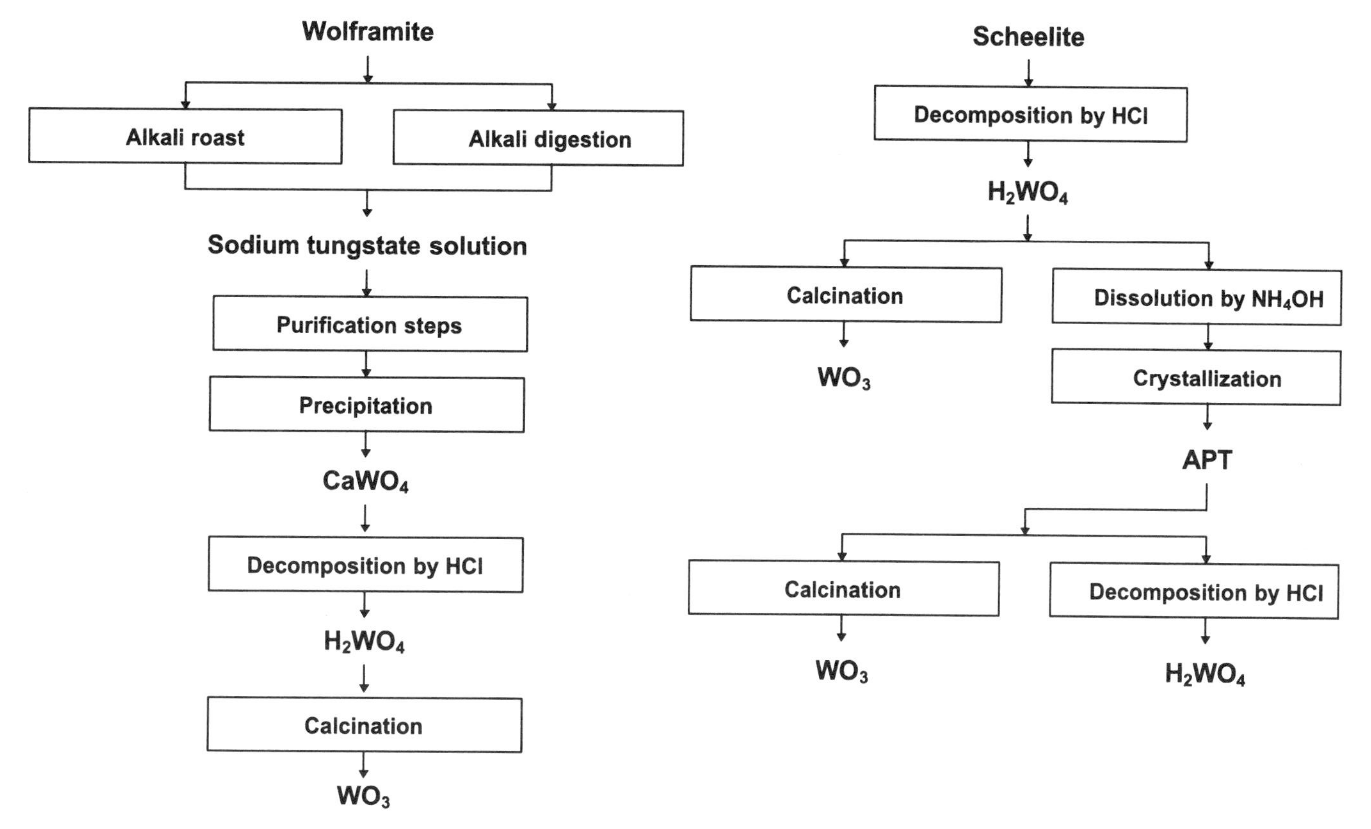

FIGURE 5.6. Tungsten hydrometallurgy—classical methods.

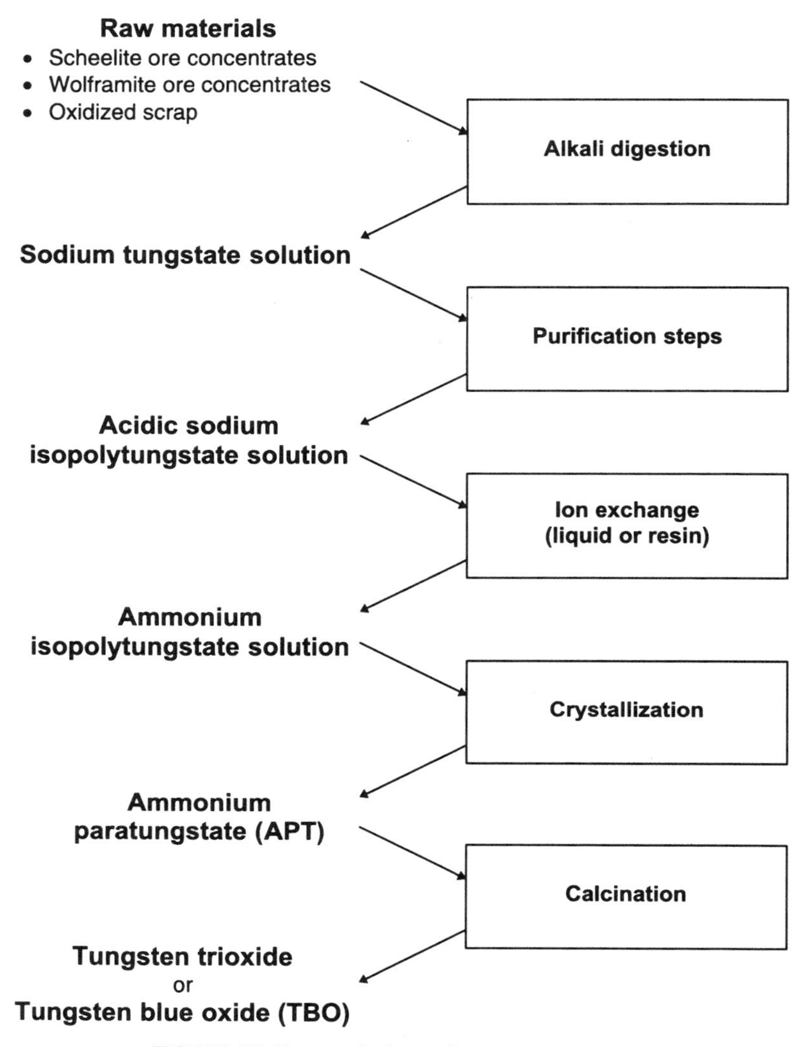

FIGURE 5.7. Tungsten hydrometallurgy—Modern Process.

In the classical procedure, scheelite was mainly processed by acid digestion while wolframite was dissolved by an alkaline digestion. The important solid, pure intermediate was, in both cases, tungstic acid, derived either directly or after several precipitation steps. Tungstic acid is formed by a precipitation, which is a more or less instantaneous process. Foreign ions present in solution during the precipitation are partly entrapped or coprecipitated and contaminate the tungstic acid.

In the modern process, APT is the intermediate which, in contrast to tungstic acid, is gained by crystallization. Crystallization compared to precipitation is a much slower process, consequently less of the impurities present in the mother liquor will be contained in the crystallized product.

The second important difference between classical and modern technology is the method used to separate the large amount of sodium ions. The old method used a calcium tungstate precipitation and subsequent hydrochloric acid treatment for that purpose, while in the modern process solvent extraction or ion exchange converts sodium into ammonium isopolytungstate solution.

The modern technology offers the following further advantages:

- Different raw materials can be handled in the same equipment,
- The energy efficiency is better,
- The labor force is less,
- The process can be easily monitored,
- The efficiency is high,
- The final product APT is of high purity and consistency.

Although recycling of chemicals is performed in many cases, such as sulfide recycling in molybdenum precipitation, ammonia recycling in APT crystallization, and solvent circulation in solvent extraction, still a certain waste of chemicals and also of energy exists which add to the high cost and to environmental pollution. The main example is the sodium carbonate pressure leaching process to digest scheelite. This procedure affords at least a threefold stoichiometric excess of sodium carbonate, and in addition, the corresponding amount of sulfuric acid in the subsequent process to neutralize it. Finally, the equivalent tonnage of sodium sulfate leaves the plant in the solvent extraction raffinate. Depending in which country the plant is located and at which site, the following possibilities exist to dispose of this undesired by-product:

- Disposal to a nearby river is allowed (lowest cost, but environmental pollution).
- Disposal at a distant place (enhanced cost by preconcentration and transport, environmental pollution).
- Disposal not allowed at all (highest cost mainly caused by crystallization, which is an energy-intensive operation, profitable sale of the product not possible, but saving of the environment).
- Research tendencies to solve this and similar problems can be found in the literature and will also be discussed in this section.

A simplified flow sheet of a conversion plant which can deal with all types of raw materials is presented in Fig. 5.8.

5.2.2. Raw Materials and Their Treatment

5.2.2.1. Raw Materials [5.1, 5.2, 5.6]. There exist three different types of tungsten-bearing materials which are currently in use as raw materials for APT production: scheelite ore concentrates and wolframite ore concentrates are the primary sources, and various tungsten-containing scrap materials act as secondary sources.

Primary Tungsten Sources. Marketable ore concentrates of both types usually contain between 65 and 75% WO_3. This guarantees a low percentage of foreign material in contrast to low-grade concentrates, which are in use in fully integrated companies (from mine to APT). Those contain typically between 6 and 40% WO_3 and a correspondingly

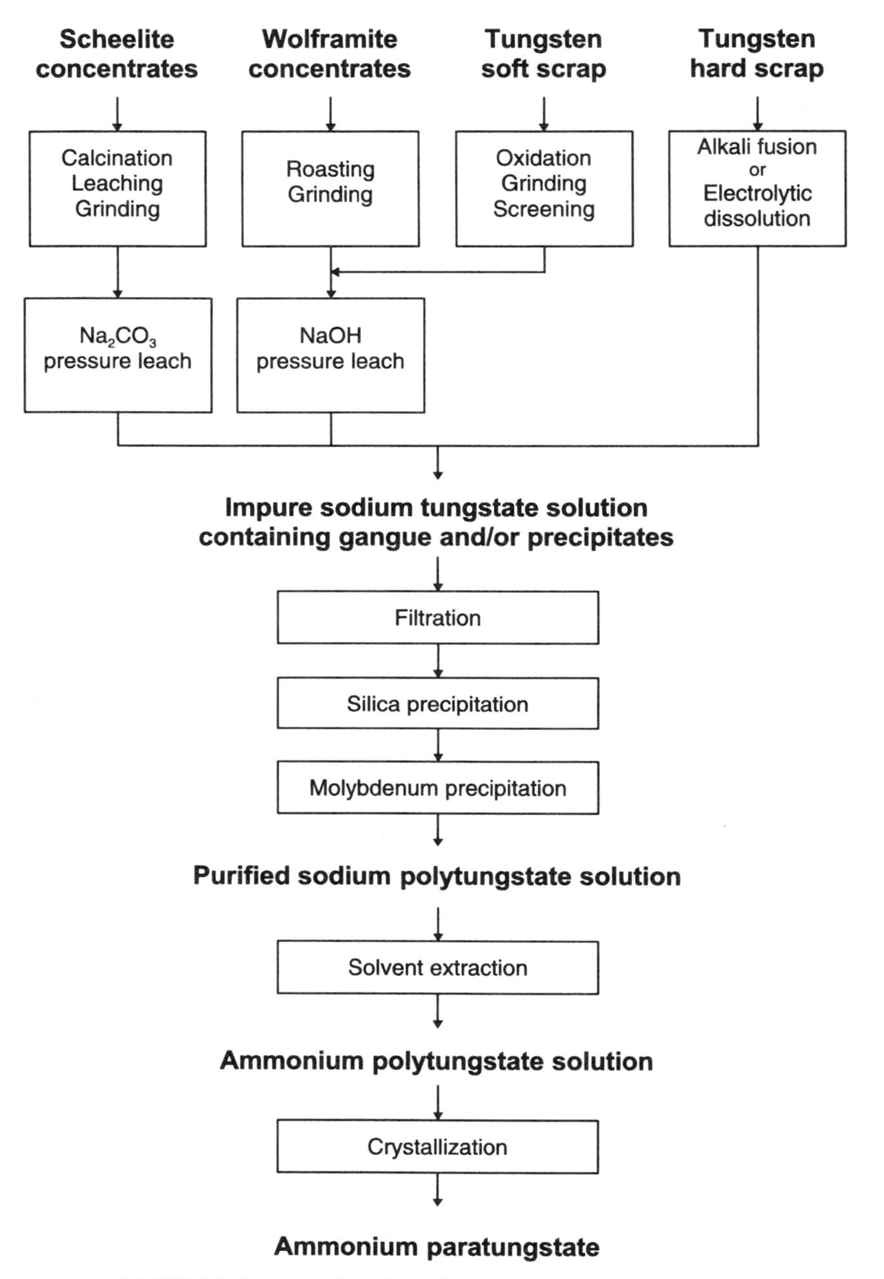

FIGURE 5.8. Simplified flow sheet of a modern tungsten conversion plant.

high percentage of gangue minerals, which sometimes create difficulties in processing when compared to high-grade concentrates. Typical ranges for contents of the main components and of impurities in high-grade scheelite and wolframite concentrates are given in Table 5.1.

Secondary Tungsten Sources [5.7]. Tungsten scrap materials typically range from 40 to 95% W, but can reach up to nearly 100%. Therefore, they are high-grade tungsten raw materials of high value and have been in use as tungsten sources for many decades. One distinguishes between “hard” scrap and “soft” scrap (cutting or grinding sludges, sweepings, powders, turnings, etc.), which consists of more or less fine particles.

Tungsten scrap is not only recycled via the chemical conversion, as will be described here, but also by other methods. Therefore a special chapter is devoted to these methods (see Chapter 11).

Tungsten scrap with only few exceptions is easier to convert to APT than ore concentrates, because usually it does not contain the “harmful" elements like phosphorus, arsenic, or silicon.

5.2.2.2. Pretreatment of Raw Materials. The following reasons may necessitate pretreatment steps:

- The separation of elements or compounds which have not been removed by the concentration and which disturb the further processing. This can be done either by selective dissolution or by roasting processes. In the latter, the disturbing elements are at least partially volatilized.
- The decrease in particle size enabling a quantitative dissolution during digestion. This is especially necessary for concentrates gained by gravity methods.
- The oxidation of tungsten in scrap material to the hexavalent state by heating in air or oxygen-enriched air, because only hexavalent tungsten dissolves in an alkaline leach process.

TABLE 5.1. Composition of High-grade Tungsten Ore Concentrates [5.6]

Constituent	Scheelite (%)	Wolframite (%)
WO_3	66–72 (theor. 80.52)	60.5–65.8 (theor. 76.53 Fe, 76.57 Mn)
CaO	10.9–20.3 (theor. 19.48)	2.14–2.30
Mg	0.03–0.84	0.04
B	<0.001–<0.01	0.005
Al	0.22–2.0	0.40–1.17
Si	0.77–2.94	1.36–2.73
Sn	<0.01–3.0	1.00
Pb	0.025–0.25	0.066
P	0.008–0.17	0.014–0.22
As	0.0066–0.17	0.058–1.00
Bi	0.006–0.10	0.045–0.20
S	0.056–0.41	0.50–1.10
Ti	<0.001–0.075	0.23
V	<0.004–0.01	0.015
Cr	<0.001–0.008	<0.005–0.008
Mo	<0.01–2.0	<0.003
MnO	0.052–0.654	2.53–10.8 (theor. 23.43)
FeO	0.28–3.74	13.2–17.0 (theor. 23.66)
Co	<0.005	0.05
Cu	<0.025–0.10	0.039–0.75
Ni	0.003–0.01	0.01

Leaching of Scheelite with Hydrochloric Acid [5.1]. The process must be performed at room temperature, otherwise scheelite can be dissolved. It reduces phosphorus, arsenic, and sulfur values to less than 200 μg/g P, 100 μg/g As, and 1% S by dissolution of apatite, sulfide, and arsenide minerals.

Roasting [5.1, 5.2, 5.8, 5.9]. This process can be applied to scheelite and wolframite. Treatment of ore concentrates at 660 to 1000 °C in air in rotary furnaces has been proposed to reduce the content of sulfides and/or arsenides by volatilization as sulfur dioxide and As_2O_3, respectively. The addition of powdered coke leads to the evaporation of tin as SnO (900–1000 °C). If CO–CO_2 or water vapor is added to the roasting atmosphere, arsenic and antimony volatilize.

Calcination is also applied to mixtures of concentrates rich in sulfides and those rich in calcite. The SO_2 and SO_3 evolved react with the calcite to form $CaSO_4$.

The roast treatment not only has a positive aspect but, especially at higher temperature, can bear negative consequences. For example, a change in solubility of gangue minerals may occur. MoS_2, which is insoluble during digestion, will be converted at higher temperature to easily soluble MoO_3; also, insoluble arsenides are oxidized to As_2O_3 and in case the temperature is not high enough for its sublimation, it remains and will be dissolved.

Calcination of flotation concentrates at 500–600 °C for a short time is used to oxidize organic reagents stemming from the flotation process. This is important, because as surface-active substances they cause excessive foaming during the further processing steps and interfere in solvent extraction.

A general procedure cannot be given. The individual composition of the ore concentrate together with environmental considerations as well as economical calculations dictate the optimal direction.

Grinding of Ore Concentrates [5.6]. Flotation-derived concentrates in most cases are fine enough to provide a complete dissolution during digestion. In contrast, concentrates prepared by gravity procedures are much too coarse and must be disintegrated by grinding in ball mills mostly working in closed circuits with a screen. Nearly all wolframite and a certain percentage of scheelite concentrates have to be ground.

Particle size plays a crucial role, especially in wolframite digestion. Mostly <325 mesh (44 μm) is recommended and seldom <200 mesh. For scheelite, <200 mesh is sufficient.

Oxidation of Soft Scrap [5.7]. By heating tungsten bearing fine grained scrap in air or oxygen-enriched air, the tungsten is oxidized to the hexavalent state. Other constituents of the scrap, like Co, TiC, TaC, NbC from cemented carbides or Fe, Ni from heavy metal alloys, are oxidized to the respective oxides. WO_3 partly reacts with these oxides to tungstates. Alumina or silicon carbide remain unattacked during the oxidation process while diamond is burned off. These materials are contained in grinding or cutting sludges. In addition to WO_3, the tungstates contained in the oxidized scrap dissolve easily during the alkaline digestion.

Scrap oxidation is performed either in push-type or in rotary furnaces. Also, multihearth roasting furnaces are employed.

The oxidation is followed by grinding and screening. Coarser particles are recycled to the oxidation.

5.2.3. Digestion–Dissolution

The aim of this step is to dissolve the WO_3 or the tungstates and convert them into water-soluble sodium tungstate. Besides, a partial or total separation of gangue minerals can be achieved, depending on the type of compounds present. Raw materials consisting of fine grains can be treated by aqueous solutions of sodium hydroxide or sodium carbonate. The latter is more selective (it does not dissolve so much of gangue minerals) compared to sodium hydroxide. Compact, bigger pieces of scrap, which cannot be oxidized in a reasonable time, can be dissolved either in an oxidizing, alkaline melt (sodium carbonate with additions of sodium nitrate or sodium nitrite), or by electrolysis.

The usual equipment for pressure digestion consists of horizontal roll autoclaves or vertical autoclaves equipped with a stirrer. For NaOH leaching, Inconel-lined vessels are necessary, while mild steel can be used in case of sodium carbonate. Heating is performed by direct steam injection.

Digestion of Wolframite [5.1, 5.10–5.15]. Conventionally, wolframite is attacked by sodium hydroxide either under atmospheric pressure in higher concentrated solution or under elevated pressure in a more diluted solution. Today, the less concentrated solution is preferred, because it dissolves less gangue minerals. Table 5.2 describes digest conditions.

$$(Fe, Mn)WO_4 + 2NaOH \rightarrow Na_2WO_4 + (Fe, Mn)(OH)_2$$

Concentrates low in manganese may be dissolved also by a pressure leach as used for scheelite (Table 5.2).

Usual yields range between 97 and 99%.

Digestion of Scheelite [5.1, 5.16–5.22, 5.24–5.26]. The method mostly applied today in western world countries is dissolution with sodium carbonate under increased pressure. Costs are higher compared to the above-described sodium hydroxide digestion of wolframite due to the fact that a high stoichiometric excess of the reagent must be applied.

$$CaWO_4 + Na_2CO_3 \rightarrow Na_2WO_4 + CaCO_3$$

The equilibrium constant $K = [WO_4^{2-}]/[CO_3^{2-}]$ is increased by temperature. Therefore, the digest is performed at elevated pressure and temperature. An increase in carbonate concentration decreases the equilibrium constant, and when it exceeds a certain limit the formation of an insoluble scale at the surface of the scheelite crystals slows down the further dissolution. The scale consists of $Na_2CO_3 \cdot CaCO_3$ and/or $Na_2CO_3 \cdot 2CaCO_3$. Accordingly, maintaining the sodium carbonate concentration within a certain limit is crucial for a good yield, which should be around 99% or better. There are proposals in the literature to improve the process:

- Countercurrent leaching scheme to save chemicals.
- Continuous process with feedback control.
- Use of ball mill autoclave.

Within the last ten years, in many plants in China, a change was made from sodium carbonate to sodium hydroxide leaching with simultaneous mechanical activation. By mechanical activation the density of lattice defects is increased, and correspondingly the activity and decomposition rate is also increased. Mechanical activation favors the decomposition not only kinetically but also thermodynamically. The advantages of this

TABLE 5.2. Digest Conditions [5.6]

Raw material	Agent concentration	Molar ratio WO_3:agent	Temperature time	Pressure (bar)	Remarks
Wolframite	NaOH 40–50% 15 M	1:1.0–1.5	100–150 °C 0.5–4 h	Atmospheric	After cooling, addition of water to 4.5–5.5 M NaOH
Wolframite	NaOH 7–10%	1:1.05	175–190 °C 4 h	10–12	200–250 g solids/l
Wolframite low in Mn	Na_2CO_3 16.5%	1:3.3	210 °C 3 h	22	
Scheelite and scheelite/wolframite mixtures	Na_2CO_3 10–18%	1:2.5–5.0	190–235 °C 1.5–4 h	12–26	The lower the WO_3 conc., the higher the stoichiometric excess of Na_2CO_3 must be
Scheelite/Wolframite mixtures	Na_2CO_3 12–18% + NaOH	Weight ratio Na_2CO_3/WO_3 1:1.0–1.6	200–250 °C 4 h	15–40	NaOH addition stoichiometric to Fe + Mn content
Scheelite and scheelite/wolframite mixtures	NaOH	1:3–7	180 °C 1.5 h	10	Mechanical activation in a ball mill autoclave
Oxidized scrap	NaOH 20%	1:1.4	150–200 °C 2–4 h	10–12	

method are: less chemical consumption (0.7–0.8 kg NaOH/kg WO_3 compared to 1.35–1.40 kg Na_2CO_3/kg WO_3); very good yield (>98%); and flexibility in regard to the raw material. The method can be applied to high and low scheelite concentrates as well as to all kinds of scheelite and wolframite mixed ore concentrates.

Table 5.2 describes digest conditions.

Digestion of Wolframite–Scheelite Mixtures [5.1, 5.23–5.26]. As long as the wolframite percentage is low, sodium carbonate pressure leach as applied to scheelite can be successful. Another method uses mixtures of sodium carbonate and sodium hydroxide because, as opposed to the digestion of scheelite, $NaHCO_3$ is formed in wolframite digestion by hydrolysis of $FeCO_3$.

$$FeWO_4 + Na_2CO_3 \rightarrow Na_2WO_4 + FeCO_3$$
$$FeCO_3 + Na_2CO_3 + 2H_2O \rightarrow Fe(OH)_2 + 2NaHCO_3$$

This means that an increased sodium carbonate consumption is slowing down the dissolution of the tungstates. The addition of sodium hydroxide converts the primary to secondary sodium carbonate:

$$NaHCO_3 + NaOH \rightarrow Na_2CO_3 + H_2O$$

Finally wolframite–scheelite mixtures can be successfully digested by a pressure leach with sodium hydroxide under instantaneous mechanical activation (ball mill autoclave). As already pointed out, under scheelite digestion this method is frequently applied in China.

Digestion of Oxidized Scrap [5.6]. Dissolution of these materials can be achieved by sodium hydroxide pressure digestion, similar to that used for wolframite (Table 5.2). Depending on the scrap type, yields are between 97 and 99%.

Fusion or Roast Technique for Hard Scrap [5.27–5.29]. Fusion is performed at high temperature with sodium nitrite or sodium nitrate as oxidizing agent and sodium carbonate as dilutant and must be carefully controlled. Roasting is done in rotary kilns at 800 °C using a 1:3 mixture of oxidizer and diluent. After cooling, the cake must be crushed prior to dissolution in water.

Electrolysis [5.30]. The principle of the method is to have the tungsten scrap as anode in an electrolyte like sodium hydroxide or sodium carbonate aqueous solution. The tungsten atoms are oxidized electrolytically to the hexavalent state. The resulting solution contains sodium tungstate besides insoluble oxides or oxide hydrates of the other constituents. Rotating drum or disk electrolytic cells are the applied equipment.

During rotation, the scrap pieces are constantly rearranged, enabling scale and metal skeleton removal. The surface is continuously renewed (cleaned) during the operation. Scale at the surface is formed by oxidation of other scrap constituents to insoluble oxides. Metal skeleton is formed if metallic constituents of the scrap are not oxidized, as in the case of heavy metal. The Ni–Fe binder phase is not attacked; only the tungsten grains dissolve. Both scale and skeleton are diffusion barriers and slow down the process.

Solutions containing 100 g WO_3/l can be produced by the process. Typical parameters are given in Table 5.3.

TABLE 5.3. Typical Energetic and Dissolution Parameters of the Electrolytic Dissolution Process for Different Tungsten Scraps [5.30]

Scrap type	Specific energy consumption ($W \cdot h \cdot g^{-1}$)	Dissolution rate ($g \cdot A^{-1} \cdot h^{-1}$)
W rods or plates	11	1.12
W–Cu rods and plates	6.0	0.86
Heavy metal pieces		
low iron	3.8	0.81
higher iron	4.3	0.83

5.2.4. Purification Steps

These includes the following main steps: separation of the undissolved digestion residue by filtration, and separation of the impurities dissolved in the sodium tungstate solution. Sometimes, the first step can be combined with the first precipitation step (silica precipitation).

Depending on the raw material composition and digestion conditions, the sodium tungstate solution after filtration may contain more or less impurity ions in variable concentrations. Typical contaminating ions are silicate, phosphate, arsenate, molybdate, fluoride, lead, bismuth, and aluminum. Their main origins are the gangue minerals (aluminum may be contained as $[AlF_6]^{3-}$ ion as a consequence of a fluoride content after digestion and subsequent silica precipitation by addition of aluminum sulfate). If their concentrations are too high, they cause disturbances during further processing, or they contaminate the final product APT.

The first separation step is called silica precipitation but, besides silica, other ions like phosphate and fluoride are at least partially separated.

The second step is a precipitation of sulfide-forming cations mainly applied to separate molybdenum, but others like arsenic, antimony, bismuth, cobalt, etc. are precipitated, too.

The silica precipitation is performed in alkaline solution, while the sulfide precipitation is effective at pH 2.5–3.0. This change in pH is usually achieved by addition of sulfuric acid. An interruption at pH 6 offers the possibility to separate the above-mentioned $[AlF_6]^{3-}$ ion (minimum solubility of $Na_3[AlF_3]$). The acidification transforms all excess Na^+ into soluble sodium sulfate.

Acidification can also be achieved by electrodialysis and offers in addition the possibility to recover NaOH solution, which can be used for digestion. This is a modern process until now used only in a pilot plant scale. However, economic as well as ecologic considerations will dictate such processes in the future.

It is interesting to note that the majority of APT plants, including those which have been installed recently, still make use of these precipitation reactions, although in the literature other methods have also been proposed. Basically, separations of this type always involve an increased loss of the valuable metal with difficulties in filtration and washing.

The only exception in this regard is the PR of China. For many years more modern processes have been applied in several plants and are still being improved, based on ion exchange and adsorption. One of the main advantages of the ion exchange process is the fact that separation of disturbing ions like silicate, arsenate, phosphate, and molybdate can be combined with the step consisting of conversion of sodium to ammonium tungstate, so reducing considerably investment as well as working costs.

5.2.4.1. Separation of Undissolved Residues. The residue in the case of ore concentrates consists of undissolved gangue minerals and insoluble reaction compounds such as $CaCO_3$ or $(Fe,Mn)(OH)_2$, or, in the case of grinding scrap, insoluble metal oxides. The latter contain valuable components like Co, Ta, or Ni and can be subject to metal recovery.

The equipment used in this step can be a disk, plate, and frame, multistage-drum or belt filters with counterflow washing with water. Additional repulping and application of a high-pressure filter press reduces the loss of dissolved tungsten.

5.2.4.2. Silica Precipitation [5.1, 5.31, 5.32]. Silicate ions may be precipitated by addition of aluminum sulfate or magnesium sulfate solution to the sodium tungstate solution. Also, mixtures of both reagents are used. The addition is made to the slightly alkaline or neutral solution. The respective chemical reactions are complex but can be described in a simplified manner by the following equations:

$$2Na_2SiO_3 + Al_2(SO_4)_3 + Na_2CO_3 \rightarrow Na_2O \cdot Al_2O_3 \cdot 2SiO_2 + 2CO_2 + 3Na_2SO_4$$
$$Na_2SiO_3 + 2MgSO_4 + 2NaSO_4 \rightarrow 2MgO \cdot SiO_2 + 2Na_2SO_4 + 2CO_2$$

Several procedures were proposed. Especially, in the case of higher silica concentrations, a double precipitation is recommended. Typically 0.08 kg $Al_2(SO_4) \cdot 18H_2O$ and 0.03 kg $MgSO_4 \cdot 7H_2O$ per kg WO_3 are added to the 70–80 °C hot solution after digestion. The pH value should be between 9.0 and 9.5. Filtration can be started after 1 hour.

Another procedure is stepwise addition of the two chemicals to precipitate silicate, phosphate, and fluoride ions:

Step 1: addition of $MgSO_4$ to a final concentration of 0.5–2.0 g Mg/l at 60–110 °C and pH 10–11, followed by filtration.

Step 2: addition of $Al_2(SO_4)_3$ to a final concentration of 0.2–3.0 g Al/l at 20–60 °C and pH 7–8.

In this way, original concentrations of 20–10,000 mg Si/l, 10–120 mg P/l, and 500–4000 mg F/l can be reduced to $\leq$ 10 mg Si/l, $\leq$ 5 mg P/l, and $\leq$ 200 mg F/l.

The precipitates have very poor filtration behavior. Therefore, cellulose is often added as a filter aid.

5.2.4.3. Combination of Gangue Removal and Silica Precipitation [5.1, 5.6, 5.31]. Because of the aforementioned bad filtration behavior of the silica precipitate, the undissolved gangue itself will be used as a filter aid. This comprises the addition of the precipitating agents, as described above, to the digested slurry. In case of higher silica concentration (several g/l) a second precipitation must be applied, reducing the level to $\leq$ 20 mg/l.

5.2.4.4. Molybdenum Separation [5.1, 5.31, 5.33, 5.35]. If sodium sulfide is added to the neutral or slightly alkaline sodium tungstate solution, the molybdate ion is converted to thiomolybdate:

$$MoO_4^{2-} + 4S^{2-} + 4H_2O \rightarrow MoS_4^{2-} + 8OH^-$$

The reaction is completed after one hour at pH 10 and a temperature of 80–85 °C. The addition of sulfuric acid to change the pH to 2.5–3.0 converts the thiomolybdate to insoluble molybdenum trisulfide.

$$MoS_4^{2-} + 2H_3O^+ \rightarrow MoS_3 + H_2O + H_2S$$

The stoichiometric excess for a complete reaction depends on the Mo–W ratio and should be at least 3. Lower Mo concentration affords a higher excess. The precipitate is filtered off (mostly by plate and frame filter), while the remaining molybdenum concentration in solution is ≤ 10 mg/l. As already mentioned, other insoluble sulfide-forming cations are coprecipitated.

The H_2S evolved during acidification has to be scrubbed by NaOH. In case of a solution derived from a sodium carbonate leach during the acidification, a high amount of CO_2 is generated due to the neutralization by sulfuric acid up to pH about 6. This CO_2 is vented, but should not pass the scrubber so as not to consume NaOH (H_2S is only generated at lower pH).

A continuous operation using a multistage stirred-tank reactor system and hydrogen sulfide recycling by absorption in the sodium tungstate solution substantially reduces the amount of sulfuric acid.

5.2.4.5. Molybdenum Removal by Adsorption on Activated Carbon [5.34]. The principle of the method is the selective adsorption of thiomolybdate by activated carbon while tungstate will not be retained.

The neutral sodium tungstate solution (pH 7.2–7.4) is treated with sodium sulfide to convert the molybdate to thiomolybdate and then adjusted to pH 8.2–8.4. Activated carbon is added and mixed for about 30 minutes. The loss of tungsten by coadsorption is less than 0.1%. Specific adsorption of Mo varies between 4.5 and 6.1 mg Mo/g carbon, depending on the carbon quality. A multistage crossflow adsorption is proposed for industrial application, leading to higher specific adsorption in the earlier stages.

5.2.4.6. Electrodialysis [5.30]. The sodium tungstate solution after silica precipitation contains sodium tungstate as well as an excess of sodium ions stemming from the surplus necessary for an optimal yield in digestion (1.5 times excess of NaOH for wolframite and 3.3 and more times excess of Na_2CO_3 for scheelite digest). The electrodialysis offers the possibility to recover the sodium ions as NaOH. The solution is fed into the anode compartment of the electrodialyzer. In the electric field, the Na ions pass through the cation exchange membrane into the cation compartment (NaOH). An equivalent amount of H^+ ions is generated in the tungstate solution. Hence the sodium ion concentration is decreased in the tungstate solution and at the same time increased in the NaOH solution. When the decrease exceeds the stoichiometric ratio for Na_2WO_4, polycondensation starts and isopolytungstates form. The acidification of the tungstate solution by electrolytically generated protons cannot be compared with the usual acidification in the technical process, which is performed by addition of concentrated sulfuric acid. The electrodialysis protons are homogeneously generated in the solution and the pH gradually decreases. During the classical process, locally and intermediately higher sulfuric acid concentrations arise and were equalized briefly by stirring. However, already-formed higher condensed tungstate species will not be dissociated completely. Therefore, the condensation grade of the tungstate species prepared by electrodialysis is more homogeneous, and solutions containing 200 g WO_3/l are stable at pH 1.7–1.8, which is not the case for solutions derived from the classical procedure. A further reason for the different properties is the absence of excessive sodium sulfate in the electrodialyzed solution. High sodium sulfate concentration also influences the condensation. Those solutions behave differently in the next step, namely, solvent extraction.

5.2.4.7. Ion Exchange. As already pointed out in the Introduction, the methods have been evaluated in China and are in industrial large-scale use. The ion exchange method combines the separation of impurities with the sodium-to-ammonium tungstate conversion. The sodium ions pass the anion exchanger. Owing to the chemical similarity of tungstate and molybdate, the latter must be converted first to thiomolybdate, which has a quite different desorption behavior. Silicate, phosphate, and arsenate are also not retained by the resin while tungstate and thiomolybdate are adsorbed. Selective desorption of tungstate is achieved by eluation with a mixture of aqueous ammonia and ammonium chloride and, subsequently, thiomolybdate is eluated by a basic hypochlorite solution. For more details see Section 5.2.5.2.

5.2.5. Conversion of Sodium into Ammonium Isopolytungstate Solution

The goal of this step—in the classical process achieved by tungstic acid precipitation and successive dissolution by ammonia—is almost total separation of sodium ions, which have been necessary for dissolution but which interfere in the further production steps to tungsten metal powder. Their concentration will be decreased quite effectively from approx. 70 g/l to less than 10 mg/l.

Today, two industrial methods are applied to meet this target: *solvent extraction* (liquid ion exchange) and *ion exchange* by resins. Plants using the latter process can be found exclusively in China (smaller works). All others are equipped with liquid ion exchange facilities.

5.2.5.1. Solvent Extraction [5.9, 5.15, 5.36–5.44]. Generally, SX in hydrometallurgy serves to concentrate diluted solutions, to separate, and to purify. In tungsten hydrometallurgy, as already explained, separation is the main target and, to a certain extent, purification from ions which do not condense with tungstate and from those which are not extracted by the solvent can be achieved.

The basic principle is to extract isopolytungstate species from the slightly acidic aqueous solution by the reagent dissolved in the organic phase, thus separating the tungstate from sodium ions. Re-extraction (stripping) from the organic into the aqueous phase is performed by aqueous ammonia solution. The organic solvent circulates from the extraction to scrubbing, stripping, regeneration, and back to extraction. Losses are only due to the very low solubility of the organic compounds in aqueous solutions.

A simplified flow sheet is presented in Fig. 5.9. One distinguishes between 4 steps: extraction, scrubbing, stripping, and solvent regeneration. The last step in some plants is a separated one, as indicated in Fig. 5.9, but may be combined with the extrac-

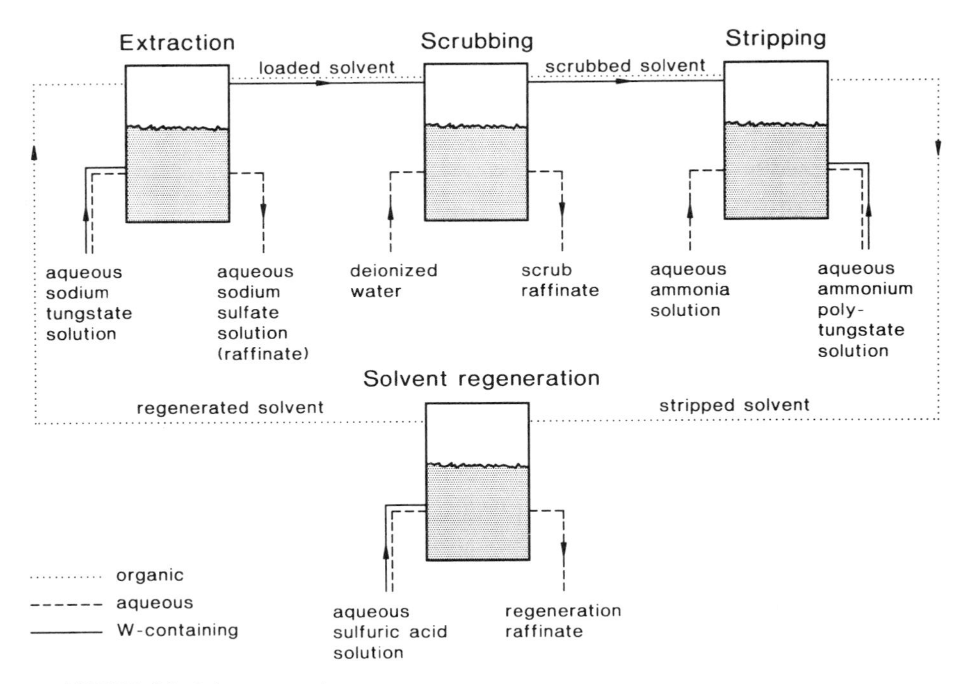

FIGURE 5.9. Solvent extraction process to convert Na in NH_4-tungstate (simplified flow sheet).

tion step in others, by feeding an appropriate amount of sulfuric acid to the aqueous phase.

The equipment used for the different steps consists of mixer-settlers in a two-stage or multistage mode comprising countercurrent flow of aqueous and organic phases, and with partial internal circulation in each step to achieve the retention time required. Also, static mixers combined with mixing pumps and settlers can be used in extraction and scrubbing, where only short retention time is necessary.

Organic Solvent. The organic phase consists of two or three components: *reagent, modifier, and solvent.* A basic study of possible chemicals for these components dates back to 1960 [5.33].

As reagent, long-chained primary, secondary, tertiary amines, and quaternary ammonium compounds (chain length C_8 to C_{10}) like trioctylamine or tridecylamine are used. Commercial products are Primene JMC (primary amine), Amberlite LA-1 or Adogen 283 (secondary amine), Trioctylamine (tertiary amine), and Aliquat 336 (quaternary ammonium compound). Depending on the process conditions, either secondary or tertiary amines are preferred, the latter more seldom. Primary amine is advantageous for higher condensed tungstate solutions as derived from the electrodialytic process. The concentration of the reagent varies in the literature between 1 and 22 vol%.

As solvent, either kerosene or mixtures of alkyl benzenes like toluene or xylene have been proposed. Due to the toxicity of the latter, kerosene is widely preferred.

The modifier is only necessary in case of kerosene as solvent, because of the low solubility of the isopolytungstate–amine complex which is hence increased. Compounds used as modifiers are isodecanol (5%) or tributyl-phosphate (12%).

Extraction. The slightly acidic polytungstate solution is contacted with the regenerated organic phase, whereby the isopolytungstate reacts with the protonated cationic amine to form an ion associate complex. In the following equation, a tertiary amine and, as the tungsten species, a paratungstate A is assumed:

$$5(R_3NH)^+ + \tfrac{5}{2}SO_4^{2-} + HW_6O_{21}^{5-} + 5Na^+ \rightarrow (R_3NH)_5HW_6O_{21} + \tfrac{5}{2}SO_4^{2-} + 5Na^+$$

The equation is idealized, because in the slightly acidic solution a variety of condensed tungsten species can be contained depending mainly on pH and time of standing after acidification [5.34]. Due to equilibria at the same pH, several polytungstates are present at the same time. Therefore, not only one species is extracted. Higher condensed polytungstates should be avoided if tertiary amine is the reagent because of the formation of the third phase. The complex of higher condensed species and tertiary amines is much less soluble and "precipitates" cause worse disengagement. In this case, primary amines are advantageous.

In regard to the throughput of heteropolytungstates, secondary and tertiary amines behave differently. Secondary amine allows tungstophosphate and silicophosphate to pass solvent extraction, while in the case of tertiary amine a honey-like tertiary phase is formed which, due to its high density, settles to the bottom. In both cases, severe disturbances result. If heteropolytungstates pass the solvent extraction in the following step (APT crystallization), the tungsten solubility in the mother liquor will be enhanced (see Section 5.2.6). The formation of the third phase not only consumes amine and separates it from the process, but may also lead to a plug-up of pipings, etc. Therefore, proper separation of Si and P impurities in advance is important.

The equivalent amounts of sulfate and sodium ions, as well as any surplus of these ions applied in excess during digestion and acidification, remain in the aqueous phase and leave the extraction as raffinate. Depending on conditions, it can be discarded or must be treated further (see the introduction).

Indications about extraction conditions in the literature vary due to different amines and differences in concentration, loading, and pH values:

- WO_3 concentration of the aqueous phase, 10–180 g/l.
- pH of aqueous phase, 1.8–2.6.
- WO_3 concentration of the loaded organic phase, 25–120 g/l (usually 30–60 g/l).
- organic:aqueous ratio mainly 1:1.
- Temperature, approximately 50 °C.

Scrubbing. In a technical scale, some entrainment of the organic phase after extraction can not be avoided. In scrubbing, the main portion will be removed. The scrubbing raffinate contains low concentrations of sodium sulfate. A typical aqueous: organic ratio is 1:20. A combination of coalescer and settler can be used for entrainment removal besides mixer settlers.

Stripping. In this stage, the loaded organic is contacted with aqueous ammonia solution. The ion associate complex is decomposed, and the isopolytungstate is re-extracted into the aqueous phase as ammonium salt according to the following idealized equation:

$$(R_3NH)_5HW_6O_{21} + 5OH^- \rightarrow HW_6O_{21}^{5-} + 5(R_3NH)OH$$

Conditions for this step also differ in the literature:

- pH 8–13.
- Ammonia concentration 1.2–10%.
- Retention time of some minutes, due to slow reaction rate.
- Aqueous:organic ratio must be adjusted to the concentration of the loaded organic, between 1:1 and 1:5. High concentrated ammonium isopolytungstate solutions are desirable. The higher the concentration, the less water has to be evaporated during the subsequent production step (crystallization) and that means a saving of energy and a reduction in cost.

The entrainment of the ammonium tungstate solution can be minimized by applying a direct current voltage of 15 V to promote coalescence [5.39].

Solvent Regeneration. In this stage, the stripped organic phase is contacted with an aqueous sulfuric acid solution. Hence the amine is again protonated and at the same time the organic phase is cleaned from the entrainment of the ammoniacal solution. As already mentioned, this stage is substituted in some plants by combining protonation with extraction. Entrainment of the unloaded organic phase can be decreased by contacting it with activated carbon [5.38].

5.2.5.2. Ion Exchange [5.45–5.47]. The process comprises the following steps:

Preconditioning of the sodium tungstate solution. The pH is adjusted to 7.2–7.4 by electrodialysis (see Section 5.2.4.7.). Electrodialysis is preferred to the addition of acids, because additional anions affect the sorption capacity of the resin and furthermore result in stable tungstate solutions. Subsequently, molybdate is converted into thiomolybdate by addition of sodium sulfide solution.

Sorption on the anion exchange resin.

$$RCl_2 + Na_2WO_4 \rightarrow RWO_4 + 2NaCl$$
$$RCl_2 + Na_2MoS_4 \rightarrow RMoS_4 + 2NaCl$$

The resin in the chloride form exchanges chloride against tungstate and thiomolybdate. In the case of tungstate the equation is idealized, because in reality isopolytungstates are present. Sodium ions as well as the anions silicate, phosphate, and arsenate are not retained and pass the column. The pH is too high for heteropolytungstate formation with these anions.

China-made anion exchange resins 930, 931, 934, and 937 are employed.

Desorption. Tungstate is first removed from the resin by a mixture of aqueous ammonia solution and ammonium chloride and exchanged by chloride.

$$RWO_4 + 2NH_4Cl \rightarrow (NH_4)_2WO_4 + RCl_2$$

The equation is also idealized. Thiomolybdate remains combined with the resin and can be subsequently stripped by a basic hypochlorite solution.

In a series column operation, thiomolybdate is sorbed in the first column (resin 930) and tungstate in the second (resin 937). The molybdenum removal is better than 99%. The impurity to WO_3 ratios are:

Mo	$1.2-1.4 \times 10^{-5}$	P	$1.2-3.0 \times 10^{-4}$
As	$0.8-1.5 \times 10^{-5}$	S	$2.5-2.7 \times 10^{-5}$

As already pointed out, the majority of the pertinent literature is in Chinese while the rare contributions in English are not at all detailed. Therefore, we have included in the next subsection (5.2.6) a summary of modern Chinese APT technology written by an expert from China.

5.2.6. *Modern Methods in Chinese Plants* [5.48–5.50].

Ion Exchange Technology in Processing Impure Wolframite Concentrates to APT, by Zhao Quinsheng (Central South University of Technology, Changsha, Hunan, 410083, China).

Hot sodium hydroxide digestion is widely used in China for treating wolframite concentrates. Recent developments show that NaOH digestion can also be used for treating impure wolframite concentrates such as scheelite–wolframite mixed ore concentrates or low-grade wolframite concentrates containing unusually high contents of P, As, Si, and Ca by adding some additives to accelerate the decomposition process of the tungsten minerals and to restrict the dissolution of some impurities. As a result of NaOH digestion, the raw sodium tungstate solution is formed. Impurity elements like Mo, P, As, and Si are also partly dissolved by NaOH, which must be removed. The classical process combines precipitation reactions with liquid solvent extraction and is adopted by most manufacturers worldwide at present.

The ion exchange technology for purifying tungstate solution has been used in China for nearly 15 years [5.48]. More than 50% of the APT produced in China is manufactured by using ion exchange technology. Figure 5.10 shows a pertinent simplified flow sheet.

Normally, the raw sodium tungstate solution after NaOH digestion contains 200 g WO_4^{2-}/l, which is too high for the ion exchange process. The ion exchange capacity decreases with increasing WO_3 concentration, as is shown in Fig. 5.11.

In the ion exchange process, ions with higher affinity to the resin can replace those with lower affinity. A strong basic anion exchange resin (type 201 × 7) is used. There are several anions like WO_4^{2-}, MoO_4^{2-}, $HAsO_4^{2-}$, HPO_4^{2-}, SiO_3^{2-}, OH^-, etc. in the sodium tungstate solution. Beside OH^-, all others have higher affinity than Cl^- and can be adsorbed by the resin as follows:

$$2R_4NCl + \text{Anion}^{2-} \rightarrow (R_4N)_2\text{Anion} + 2Cl^-$$

The affinities of these ions can be arranged in a decreasing series:

$$WO_4^{2-} > MoO_4^{2-} > HAsO_4^{2-} > HPO_4^{2-} > SiO_3^{2-} > Cl^- > OH^-$$

During the early stage of adsorption, all ions with higher affinities than that of Cl^- can be adsorbed on the resin. However, when the diluted tungstate solution continuously passes through the ion exchange column, anions with smaller affinities, adsorbed already on the resin, are replaced by tungstate and removed by the effluent solution, as described by the following equation:

$$(R_4N)_2\text{Anion} + WO_4^{2-} \rightarrow (R_4N)_2WO_4 + \text{Anion}^{2-}$$

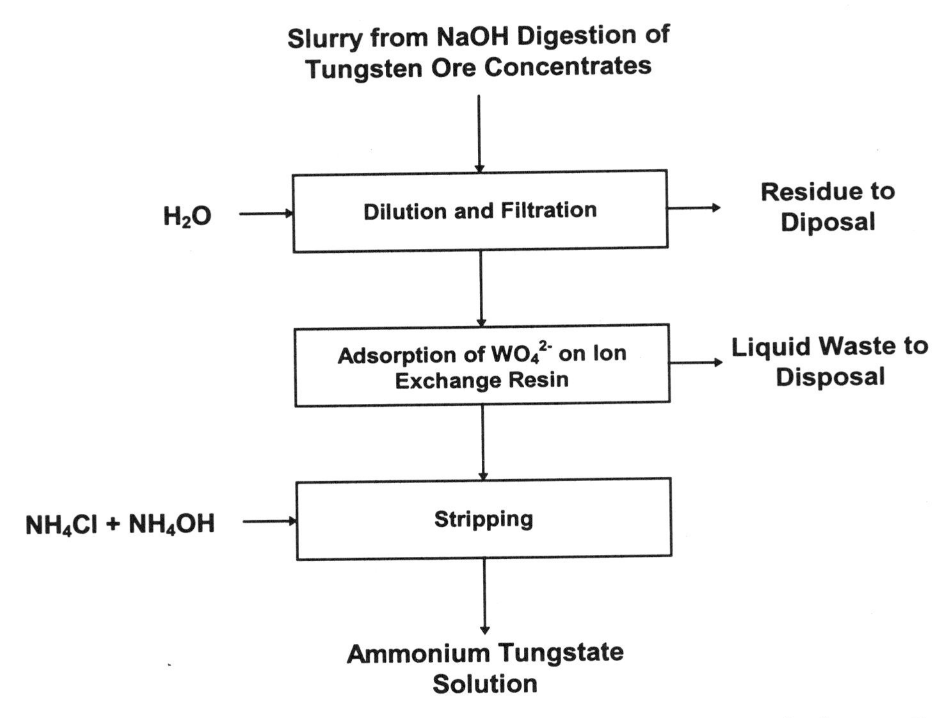

FIGURE 5.10. Flow sheet of the single ion exchange process for purifying and converting impure sodium tungstate solution.

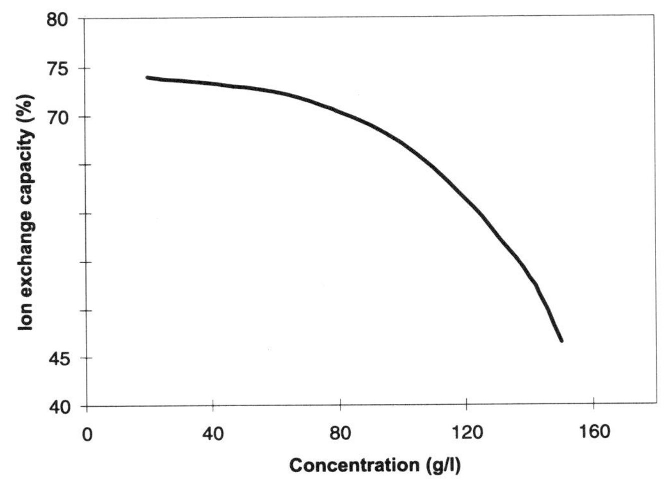

FIGURE 5.11. Relationship between WO_4 concentration and ion exchange capacity [5.48].

By this substitution reaction, removal of P, As, and Si up to 90% is possible; however, the difference in affinities of molybdate and tungstate is too small to separate them. Obviously, all cation impurities can be removed by this process.

The ion exchange technology is not only used for purification of the tungstate solution, but also for converting it into ammonium tungstate solution. The conversion is done while stripping the tungstate from the loaded resin, using ammonium salts as stripping agents. The stripping effects of various ammonium salts is compared in Fig. 5.12. It is seen that ammonium chloride is the most suitable agent. In addition, it regenerates the resin to the desired form:

$$(R_4N)_2WO_4 + 2NH_4Cl \rightarrow 2R_4NCl + (NH_4)_2WO_4$$

In order to avoid precipitation of APT crystals during stripping, an overstoichiometric amount of NH_4OH is applied.

Wolframite concentrates, generally used as raw material in a single ion exchange processing, are first-grade No. 2 quality according to Chinese National Standard GB-2825-81. The corresponding analysis is: $WO_3 > 65\%$ and impurities (%) < 0.7 S, 0.1 P, 0.1 As, 0.05 Mo, 3.0 Ca, 0.25 Cu, 0.2 Sn, and 5.0 Si.

The effect of a single ion exchange treatment of raw sodium tungstate solutions is shown in Table 5.4. As indicated before, the single ion exchange technology is not suitable for removing molybdenum. In order to treat impure wolframite concentrates and low-grade wolframite–scheelite concentrates containing more than 0.1% Mo and higher concentrations of other disturbing impurities, a double ion exchange technology was recently

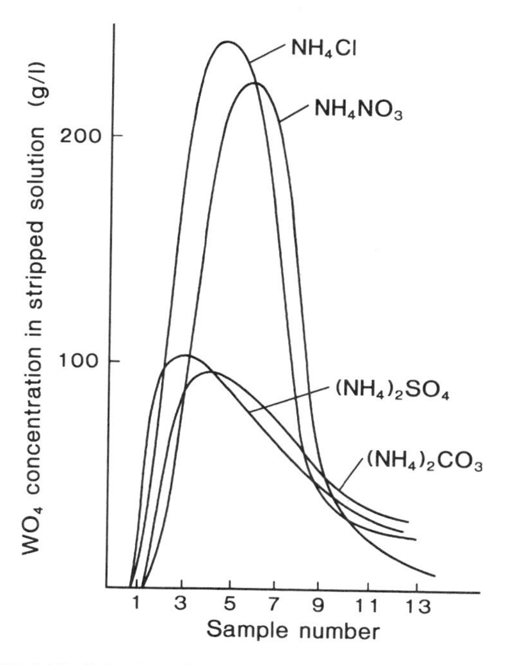

FIGURE 5.12. Stripping effect of various ammonium salts [5.48].

TABLE 5.4. Composition of Solutions of a Single Ion Exchange Process [5.49]

Solution	WO_3 ($g \cdot l^{-1}$)	P	As	Si	Mo
		($\mu g \cdot g^{-1}$ WO_3)			
Raw sodium tungstate	20.08	37.4	104.6	1245	1594
Adsorption effluent	0.1	68	190	2400	1750
Ammonium tungstate	203.2	3.5	10.3	73.8	738

developed and put into production in the Four Rings Metallurgical Factory, Yunan, Guangdong Province.

A process for separating molybdate from tungstate using ion exchange was patented by Chen Zhouxi *et al.* in China [5.50]. The principal flow sheet of the technology is presented in Fig. 5.13.

The feed solution can be either a sodium tungstate solution containing Mo, P, As, and Si or an ammonium tungstate solution containing Mo as the main impurity element stemming from a single exchange technique. After pH adjustment, NaHS or $(NH_4)_2S$ is added to convert molybdate into thiomolybdate. The solution passes a column of strongly basic anion exchange resin (Type D-201). Thiomolybdate is adsorbed and the effluent contains the tungstate purified from Mo. In order to regenerate the anion exchange resin, a solution of an oxidizer is used to desorbe the thiomolybdate.

The original Mo/WO_3 concentration ratio in the feed solution of 0.06–0.84% decreases to 0.001–0.015% in the effluent. The recovery of the operation is better than 99%. It was used in 1992 commercially to produce sodium and ammonium tungstate of high purity. The purification effect is demonstrated by the analyses in Table 5.5.

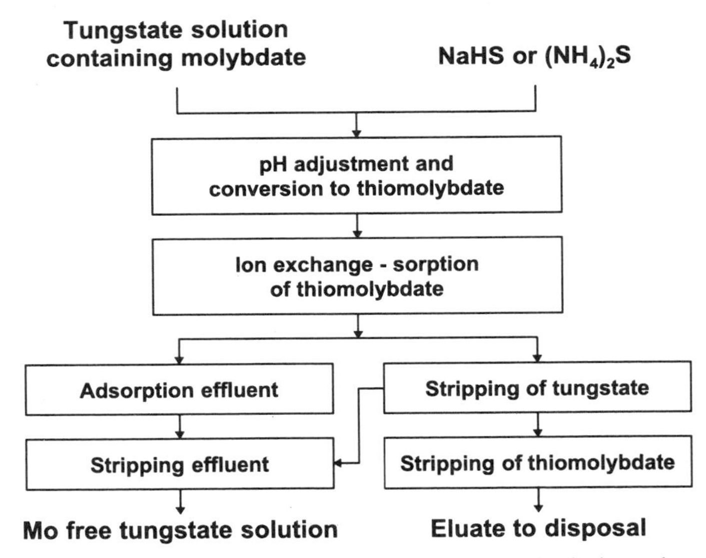

FIGURE 5.13. Flow sheet of molybdate removal from tungstate solution by ion exchange [5.50].

TABLE 5.5. Composition of Solutions of a Double Ion Exchange Process [5.49]

Solution	WO_3 ($g \cdot l^{-1}$)	Mo ($g \cdot l^{-1}$)	Mo/WO_3 (%)
Ammonium tungstate feed	200.5	0.592	0.283
Adsorptioin effluent	195.6	0.005	0.0025
W stripping effluent	53.5	0.04	0.8
Mo stripping effluent	4.4	4.07	92

If the ammonium tungstate solution after the Mo removal is evaporated for APT crystallization, a 98.4% evaporation can be applied. At 92% evaporation the APT contains $< 20\ \mu g/g$ Mo, and at 98.4% evaporation 20–80 $\mu g/g$ Mo.

With this basic knowledge, a new double ion exchange technology for processing impure wolframite concentrates was created by combining the above-described double ion exchange processes. The resulting APT is of utmost purity also in regard to Mo. The complete flow sheet for treating primary tungsten raw materials using the double ion exchange technology for commercial APT production is given in Fig. 5.14. Typical impurity concentrations ($\mu g/g$) in the resulting APT are: Al 2, As 1, Ca 5, Cr 5, Fe 5, K 10, Mg 3, Mn 1, Mo 7, Na 9, Ni 5, P 7, Si 5, Sn 2.

The APT quality is very consistent and either comparable or even better than that of other producers. The main feature of the technology is the lower production cost and the ability to produce APT according to the customers' requirements (e.g., Mo content).

5.2.7. *Ammonium Paratungstate Crystallization* [5.51–5.58]

The ammonium isopolytungstate solution from solvent extraction is evaporated, whereby ammonia and water are volatilized. The ammonia concentration in the mother liquor decreases (pH drop) and at the same time the WO_3 concentration increases. Para tungstate B $[H_2W_{12}O_{42}]^{10-}$ is formed whose ammonium salt has low solubility. Crystallization of $(NH_4)_{10}H_2W_{12}O_{42} \cdot 4H_2O$ is the result. The degree of evaporation depends on the purity of the feed solution and the required purity of the APT. Usually, if proper purification was performed in the foregoing steps it ranges somewhere between 90 and 99%.

The physical properties of the crystallized APT can be influenced by crystallization conditions to some extent. They are of importance for the further processing to metal powder by hydrogen reduction.

The crystallization is not only a conversion of the dissolved to the solid salt, but the last purification step in the hydrometallurgical process. Most of the impurity elements still present in the process solution have much higher solubility than APT and are consequently enriched in the mother liquor.

Impurity concentrations of the feed solution to the crystallizer, of the mother liquor, and of the crystallized APT are related. For a given required APT, purity limits in impurity concentrations of the mother liquor are allowed, and these limits in combination with the impurity level of the feed solution determine the degree of evaporation. Upper impurity

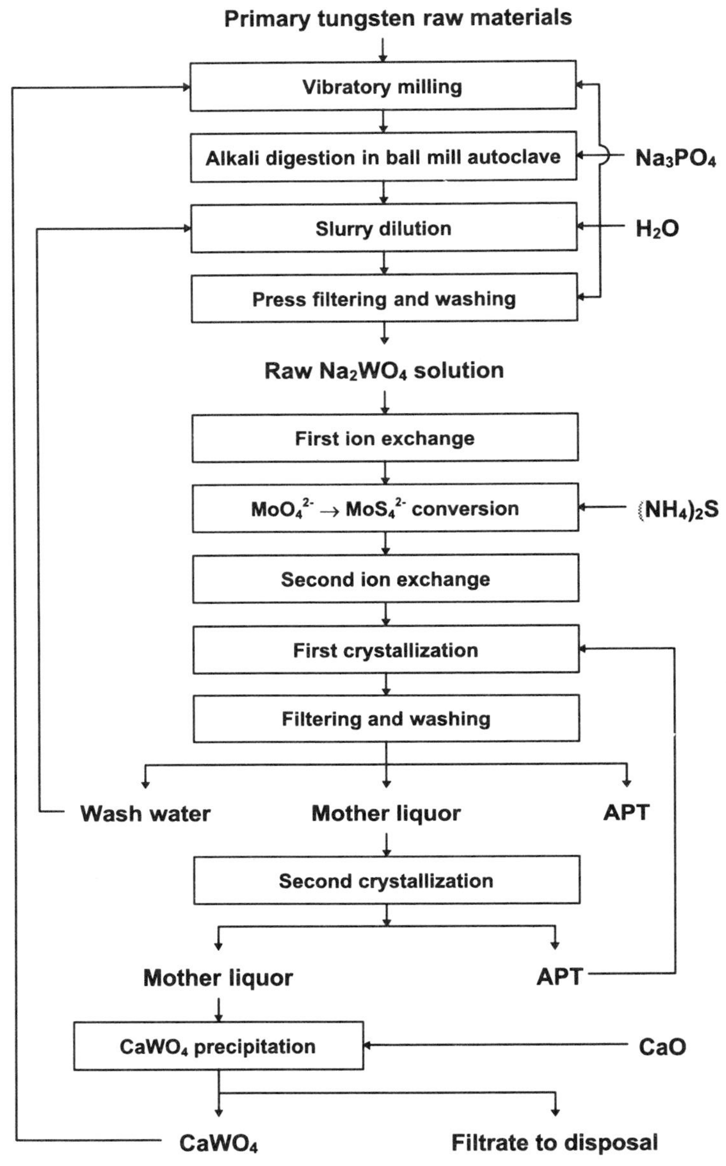

FIGURE 5.14. Flow sheet for treating primary tungsten raw materials by using double ion exchange technology. By courtesy of Four Rings Metallurgical Factory, Yunan, Guangdong, China.

levels in the feed solution and corresponding concentrations in the mother liquor and the crystallized APT are given in Table 5.6.

The crystal slurry is separated from the mother liquor by filtration and washing with deionized water. Filtrate and wash water are recycled and, finally, the APT is dried.

On a technical scale, batch and continuous operation are used. Batch crystallization leads to better consistency of the product, but the continuous process is more economic.

TABLE 5.6. APT crystallization: Impurity Concentrations [5.6]

Element	Feed solution ($g \cdot l^{-1}$)	Concentration of mother liquor ($g \cdot l^{-1}$)	APT ($\mu g \cdot g^{-1}$)
Al	<10	<100[a] <3000[b]	<10
As	<50	<2000[b]	<20
F	<250	<3000	<10
Fe	<10	<200	<10
Mo	<10	<60	<20
Na	<10	<100	<10
P	<50	<400	<20
Si	<10	<200	<20
V	<100	<1200	<20

[a] In the absence of fluoride.
[b] At fluoride concentrations of about 4000 mg/l.

The presence of foreign element impurities not only affects the impurity level of APT, but also in some cases the APT crystal size. The higher the concentration of these impurities, the smaller the average crystal size. Heteropolytungstate-forming ions like silicate, phosphate, and arsenate lead to a considerable increase in the solubility of tungsten in the mother liquor. The consequences are worse crystallization yield and higher cost due to increased recycling of tungsten. A precipitation by aluminum salt solution to separate phosphorus and silicon prior to crystallization is proposed. However, it is preferable to make all advanced purification steps more effective.

Ammonia Recovery. The ammonia evaporating together with water during crystallization is condensed and can be reused in the stripping stage of solvent extraction. Sometimes, a concentration by distillation is applied to maintain the water balance.

Mother Liquor Recycling. In order to maintain the impurity concentration in the APT low and consistent, the evaporation can never be 100%. This means that a certain amount of mother liquor must be recycled to the former steps of the conversion plant. Usually, an amount of mother liquor corresponding to 5–10% of the volume fed into the crystallizer is pumped back (90–95% evaporation). This can be done continuously or batchwise. There exist several possibilities to treat the recycled mother liquor, which is high in impurities.

In some factories it is added to the feed make-up prior to digestion. The disadvantage of this method is that, during heating, an ammonia pressure is built up that prevents one to reach the desired temperature by steam injection. An intermediate venting of the autoclave is necessary, resulting in loss of time and in increased steam demand. The advantage is the proper separation of all impurities.

A second possibility is to add the mother liquor immediately after the digestion but prior to the purification steps. The disadvantage here is the presence of ammonium ions. As already pointed out, ammonium ions favor the condensation of tungsten ions, and this changes the properties and behavior of the isopolytungstate solution in the solvent extraction process. Therefore, this type of mother liquor recycling affords a separate pretreatment to expel the ammonia. This can be done by addition of NaOH and subsequent heating. The ammonia evolved can be fed into the ammonia recovery system of crystallization. The treatment produces a precipitate which can be separated by filtration.

A third possibility is to further crystallize the mother liquor yielding a more impure APT [5.55]. Also, treatment of mother liquor with hexamethylene tetramine to precipitate tungsten was proposed [5.56].

5.3. THE HIGHLY PURE INTERMEDIATES

5.3.1. Ammonium Paratungstate (APT) [5.51–5.53, 5.59]

Today, APT is the most important and almost exclusively used precursor for tungsten products. Only in tungsten melting metallurgy (Chapter 8; production of steel, stellites, etc.) and for producing WC directly from ore concentrates (Chapter 9; Menstruum WC) are other starting materials used (ferrotungsten, scrap, melting base, tungsten ores). Intermediates, such as tungsten trioxide, tungsten blue oxide, tungstic acid, and ammonium metatungstate can be derived from APT as shown in Fig. 5.15, either by partial or complete thermal decomposition or by chemical attack.

Although a hexahydrate and a decahydrate exist, only the tetrahydrate $(NH_4)_{10}[H_2W_{12}O_{42}] \cdot 4H_2O$ forms under industrial conditions, since the hexahydrate is unstable at temperatures exceeding 96 °C, while the decahydrate crystallyzes only from solutions at room temperature.

Characterization. The important physical and chemical properties have been already treated in subsection (b) of Section 4.2.7.3. Here, only parameters relevant for the further processing are discussed.

APT is a white, crystallized powder. The average crystal size of commercial products ranges between 30 and 100 μm. The SEM image in Fig. 5.16 reveals mainly faceted

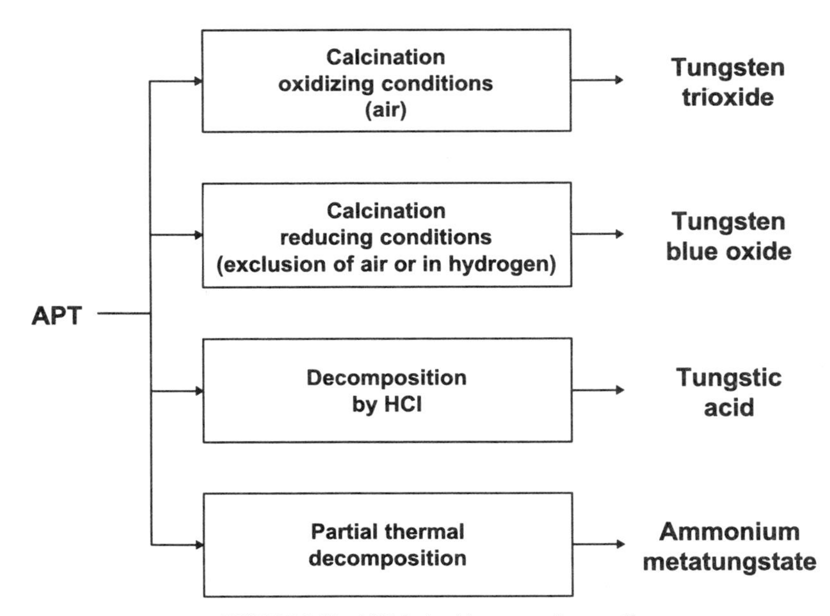

FIGURE 5.15. APT-derived important intermediates.

FIGURE 5.16. SEM image of APT crystals.

crystals and only few intergrown or agglomerates. A typical grain size distribution of a crystallized APT is shown in Fig. 5.17. For special purposes, classified APT is also available. Specification of some physical properties and impurity concentrations as common today are given in Table 5.7. They reflect the high standard of the technical APT production.

Specially purified APT (by multiple-step liquid extraction of selected, very pure batches under clean-room conditions) for the production of 4N and 6N tungsten sputter targets shows a much lower impurity content (Table 5.8).

Solubility. The solubility in water is low: 20 g WO_3/l at 20 °C; 60 g WO_3/l at 90 °C.

Thermal Decomposition. On heating APT to 300–800 °C, ammonia and water are evolved. The decomposition becomes more complete as the temperature and duration of heating are increased. The final decomposition product is determined by the

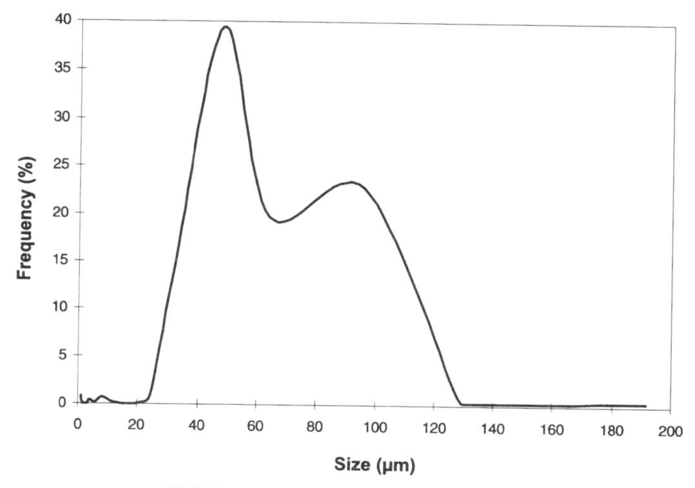

FIGURE 5.17. APT grain size distribution.

TABLE 5.7. APT Trace Impurities (upper limits in $\mu g \cdot g^{-1}$)[a]

Element	1	2	3	4	5	6
Al	5	1	5	1	5	7
As	10	4	10	7	7	5
Bi	1	0.1	—	—	1	0.5
Ca	10	1	5	1	5	5
Cd	—	—	5	—	1	—
Co	10	1	10	—	5	—
Cr	10	1	10	1	5	—
Cu	3	0.1	5	1	1	1
Fe	10	3	10	1	10	6
K	10	10	10	2	9	7
Mg	7	1	5	1	5	7
Mn	10	1	5	1	5	1
Mo	20	7	30	6	20	14
Na	10	5	10	5	9	8
Ni	7	1	5	1	5	—
P	7	5	7	7	5	7
Pb	1	0.5	5	—	1	0.5
S	7	7	7	—	5	—
Sb	8	1	—	—	2	—
Si	10	3	10	1	8	6
Sn	1	0.3	10	1	1	0.5
Ti	10	3	10	—	—	—
V	10	3	10	—	—	—
F	—	—	25	—	—	—
Li	—	—	5	—	—	—
Nb	—	—	10	—	—	—

[a] 1, Chinese National Standard GB 10116-88. 2, Excellent Chinese Grade (Hengdu Tungsten). 3, Specification Wolfram Bergba- und Hüttenges.m.b.H. 4, Osram Sylvania. 5, Fujian Xiamen Tungsten Products Plant. 6, Chinese Production, Jiangxi Province.

decomposition temperature, the retention time in the furnace, and the reduction potential of the decomposition atmosphere.

If heating takes place with exclusion of air or in an inert gas atmosphere (e.g., nitrogen), part of the ammonia evolved is cracked and the hydrogen formed can cause a slight reduction of the hexavalent tungsten matrix ("autoreduction capability of APT"; partial formation of pentavalent tungsten). The degree of reduction and the formation of compounds is determined by the decomposition conditions. The product of this type of calcination is greenish blue to flashing dark blue in color and is called Tungsten Blue Oxide (TBO).

If the decomposition is carried out under oxidizing conditions, a slight reduction can occur intermediately at low decomposition temperatures, but the final product is always tungsten trioxide (WO_3).

Besides temperature, time, and decomposition atmosphere, the amount of APT plays an important role since the mass of APT itself is responsible for producing a certain amount of ammonia and water as well as hydrogen and nitrogen. The powder layer, depending on its thickness and porosity (which increases during decomposition due to an

TABLE 5.8. APT Specification for UHP Tungsten According to the Upper Permissible Limits in UHP W Powder

Element	Concentration ($ng \cdot g^{-1}$)
Al	140
B	35
Na	35
K	35
Ca	70
Mg	70
Fe	70
Co	15
Cr	70
Ni	15
Cu	15
U	0.1
Th	0.1

increase in density), retains the gases released for some time. This fact explains the occasional contradictory literature on the decomposition of APT. For example, it is easy to understand that in a boat of comparable size 10 g or 1000 g will produce different atmospheres, especially inside the powder layers whose heights are also quite different.

Under oxidizing conditions, the APT decomposition path is as follows [5.55]:

Between 20 and 100 °C only dehydration occurs and the product is crystallized, dehydrated APT:

$$(NH_4)_{10}[H_2W_{12}O_{42}] \cdot 4H_2O \rightarrow (NH_4)_{10}[H_2W_{12}O_{42}] + 4H_2O$$

In the temperature range 180–225 °C, ammonia is released and the APT converts to amorphous ammonium meta tungstate (AMT):

$$(NH_4)_{10}[H_2W_{12}O_{42}] \rightarrow (NH_4)_6[H_2W_{12}O_{40}] \cdot 2H_2O + 4NH_3$$

Between 230 and 325 °C, ammonia as well as water vapor are evolved. The product is also amorphous:

$$(NH_4)_6[H_2W_{12}O_{40}] \cdot 2H_2O \rightarrow (NH_4)_2[W_{12}O_{37}] + 4NH_3 + 5H_2O$$

By increasing the temperature to 400–500 °C, all residual ammonia and water is released, and the reaction product is tungsten trioxide:

$$(NH_4)_2[W_{12}O_{37}] \rightarrow 12WO_3 + 2NH_3 + H_2O$$

Under a slightly reducing atmosphere between 220 and 325 °C amorphous, and above 325 °C, crystallized, ammonium tungsten bronzes form: $(NH_4)_xWO_3$. Under stronger reducing conditions, conversion to lower tungsten oxides takes place.

5.3.2. Tungsten Trioxide [5.60]

In a technical scale, WO_3 is almost exclusively produced by calcination of APT under oxidizing conditions (in air). Usual equipment consists of rotary furnaces operating at 500–700 °C. Sufficient air supply must be provided to suppress any reducing reaction by the partly cracked ammonia. The ammonia evolved can be recovered by absorption in cold water and concentrated by subsequent distillation.

The WO_3 particles are pseudomorphous to APT. This means that the shape and size of the particles are the same as the APT crystals, but they are built of very small WO_3 grains (Fig. 5.18) forming a large oxide sponge with a high degree of microporosity (specific surface area). Their grain size and agglomerate structure depend on the calcination conditions (heating rate, temperature, and time). Higher temperature and low heating rates result in coarser grains. Above 700 °C, coarse, faceted WO_3 single crystals form due to enhanced chemical vapor transport of the oxide.

Low-temperature calcined WO_3 (approximately 500–550 °C) [5.55] is highly reactive and dissolves easily in water, which is not the case for higher-temperature calcined WO_3.

For special purposes, especially in the case where a high specific surface area is necessary and APT pseudomorphology is undesirable, WO_3 can be produced also by calcination of tungstic acid.

As precursor for the W and WC powder production, WO_3 lost its importance mainly to tungsten blue oxide. WO_3 is also used as a yellow pigment.

5.3.3. Tungsten Blue Oxide (TBO) [5.59–5.65]

Industrial Production. TBO is formed by calcination of APT under slightly reducing conditions. The conversion can be performed either in multitube push-type furnaces or in

FIGURE 5.18. SEM image of WO_3. The size and shape of the APT crystals are maintained, but the pseudomorphs consist of agglomerates of very small oxide crystals with sizes between 0.1 and 0.5 μm.

rotary kilns. Various atmospheres are used. Generally, in push-type furnaces a flow of hydrogen or hydrogen–nitrogen mixtures is applied. In rotary furnaces one usually takes advantage of the reducing capacity of the gases evolved during the decomposition (H, H_2, NH_3), leading to the desired formation of reduced tungsten species.

Temperature may vary between 400 and 900 °C. Literature values can be misleading, because some are related to the real temperature of the powder layer while others are furnace temperatures measured at the wall of the heating compartment or tube. These temperatures can differ considerably, due to the overall endothermic behavior of the APT → TBO decomposition reaction. The exposure time in a rotary kiln is usually much shorter than in the pusher, and the decomposition temperature is therefore higher for obtaining a similar degree of thermal decomposition.

Chemical Composition. TBO is not a defined chemical compound but is a mixture of different constituents, such as ammonium, hydrogen and hydronium tungsten bronze phases, tungsten trioxide, tungsten-β-oxide ($WO_{2.9}$ or $W_{20}O_{58}$), and tungsten-γ-oxide ($WO_{2.72}$ or $W_{18}O_{49}$). Under more reductive conditions, even traces of WO_2 and β-tungsten can be present. The relative amounts of the various compounds in the TBO depend on the calcination parameters:

- temperature,
- heating time,
- composition and pressure of atmosphere,
- mass of APT flow with time,
- gas flow,
- layer height in the boat (pusher furnace),
- slope and rotation rate (rotary furnace).

The oxygen index (molar ratio O/W) is commonly used to characterize the degree of reduction of TBO. However, since most TBOs also contain ammonia and water in addition to W and O, a more complete description is given by $x(NH_3) \cdot y(H_2O) \cdot WO_n$.

A series of analyzed industrial samples gave the following ranges for the coefficients x and y and the index n: $x = 0.02-0.09, y = 0.02-0.14$, and $n = 2.82-2.99$.

Qualitative and quantitative X-ray analyses of the same samples revealed quite a large scatter in composition: tungsten bronzes, 0–45%; WO_3, 0–45%; $WO_{2.9}$, 5–20%; $WO_{2.72}$, 0–25%, and amorphous, 30–55%.

Amorphous species form by dehydration which, on further heating, convert into crystalline binary tungsten oxides as well as tungsten bronzes. The conversion from amorphous to crystalline is a slow process. Therefore, if the heating period is short, as in rotary furnaces, the time available for overall crystallization is insufficient. This is why rotary-furnace-derived TBO can be high in amorphous oxide when compared to TBO from pushers.

Homogeneity. The product of a rotary furnace is more homogeneous in regard to differences between individual particles, since the powder is constantly being mixed by the turning motion. The TBO particles are composed of different compounds but are similar in composition, while pusher-furnace-derived TBO particles can also differ in composition, depending on their position in the powder layer.

Physical Properties. The physical properties of the TBO, such as particle size, size distribution, and internal porosity, are of importance for the subsequent hydrogen reduction process. Size, size distribution, and shape of the particles are related to the starting APT as a consequence of the crystallization process and do not alter significantly during the decomposition process, while the internal porosity is a result of the blueing process. The degree of porosity is influenced by the calcination parameters (time, temperature). The specific surface area in commercial TBOs can vary between 1 and $15\,m^2/g$.

Both macromorphology and microporosity of the TBO particles determine the permeability and thus the diffusion resistance of a TBO powder layer, which is a determining factor for the material exchange ($H_2 \rightarrow H_2O$) during reduction.

Characterization. Industrially produced TBO is characterized chemically by the residual ammonia content and the oxygen index, leaving open the composition in regard to the different compounds and the amount of amorphous species. These determinations would be much too tedious and expensive for routine purposes. As long as APT quality and calcination conditions are kept constant, the composition will be reproducible. Consequently, in many companies it is preferred not to buy TBO but APT, and to perform the blueing in-house.

Physical characterization of TBO includes particle size and size distribution measurement by laser diffraction (macroporosity) as well as specific surface area measurement (microporosity). Particle size measurement by FSSS (Fisher Sub-Sieve Sizer), as also sometimes used, is misleading, because of the porosity of the TBO particles. An empirical relationship between FSSS and particle size measured by laser scattering can, however, be detected if the microporosity of the samples is uniform (constant blueing conditions).

5.3.4. Tungstic Acid

In order to make use of the high APT purity in modern processing, tungstic acid is produced today by treating an aqueous APT crystal slurry with hydrochloric acid. In this way APT is decomposed and H_2WO_4 is precipitated. After filtration, it must be thoroughly washed to remove ammonium chloride. The earlier process of precipitation from sodium tungstate solutions by addition of acids no longer has industrial importance.

Tungstic acid has a high active surface. The former, most important intermediate today is only used in smaller quantities for special purposes:

- production of ultrafine tungsten and tungsten carbide powders in order to circumvent the sometimes disturbing pseudomorphology of APT-derived products, and
- for tungsten chemicals.

5.3.5 Ammonium Metatungstate (AMT) [5.66–5.71]

AMT corresponds to the formula $(NH_4)_6H_2W_{12}O_{40} \cdot xH_2O$. The amount of combined water is variable. Usual technical products contain 3 to 4 molecules. Further dehydration may lead to insoluble compounds.

The importance of AMT lies in its extremely high water solubility. This is why AMT has gained increasing usage as an intermediate in a variety of applications like chemicals and catalysts.

Several methods of preparation exist:

- Acidification of an ammonium tungstate solution and subsequent evaporation.
- Partial thermal decomposition (see Section 5.3.1).
- Addition of tungstic acid to an ammonium tungstate solution to reach a 1:2 NH_3/WO_3 ratio, followed by evaporation.
- Partial replacement of ammonium ions by hydrogen ions using selective ion exchange and subsequent evaporation.

The latter three methods are preferred today for industrial production, since acidification and thermal decomposition result in lower-quality products (less purity and amorphous structures respectively).

Acidification by Electrodialysis. Ammonium polytungstate solution from the solvent extraction process, possesses a pH of approximately 9 and is circulated through an electrodialyzer as anolyte. In the electrical field, ammonium ions pass through the cationic exchange membrane forming ammonium sulfate in the circulating catholyte. The pH drops due to the decrease in the ammonium ion concentration (generation of H^+), and the reaction is stopped at pH 3. The following overall equation represents the process:

$$4(NH_4)_2WO_4 + 8H_2O \rightarrow (NH_4)O \cdot 4WO_3 \cdot 8H_2O + 6NH_4^+ + \tfrac{3}{2}O_2 + 6e^-$$

The resulting solution is evaporated to a syrup from which white crystals are recovered. The yield is better than 99%.

Properties. AMT is a white crystallized powder having a density of $4 g/cm^3$. Decomposition starts at 100 °C and, between 200 and 300 °C, it converts to the anhydrous form. Further decomposition leads to WO_3.

The solubility in water at 22 °C is 1635 g/l corresponding to 1500 g WO_3/l and, at 80 °C 2200 g/l are dissolved. The pH of the aqueous solution depends on the AMT concentration and varies at 25 °C between 2.5 and 5, indicating that different species are present simultaneously.

High solubility is the prerequisite for the production of highly concentrated, alkali-free tungsten solutions. They are necessary to impregnate suitable carrier materials like γ-Al_2O_3 in catalyst production.

5.4. TUNGSTEN METAL POWDER PRODUCTION

5.4.1. General

The manufacture of tungsten metal powder is a crucial step in tungsten metal and alloy production, since the powder properties significantly affect the properties in subsequent operations, such as pressing, sintering, and metalworking. Between 70 to 80% of tungsten worldwide is produced via powder metallurgy and thus passes through this important stage. In the past, advances in powder technology have greatly contributed to the development of tungsten and its alloys, as well as today's high standard of product quality. Powder grades are tailormade for the subsequent applications, and the powder industry is facing a competitive market where the stringent fulfillment of exacting demands is an important part of business success.

The powder is characterized by chemical (purity), physical (grain size, size distribution, shape, agglomeration, etc.), and technological properties (fluidity, compaction density, etc.), which are influenced by the production process and which can be controlled—to a certain extent—by the process parameters.

Today, production of tungsten metal powder is accomplished almost exclusively by the hydrogen reduction of high-purity tungsten oxides. Reduction of the oxides by carbon, common in the early years of metal production, is presently used only for the production of tungsten carbide (direct carburization). The hydrogen reduction of tungsten halides (Axel Johnson process) has not become established on a large scale.

The common starting materials are tungsten trioxide (WO_3) and tungsten blue oxide (WO_{3-x}), the latter being the most widely used material. Tungstic acid (H_2WO_4) is used only for selected metal grades.

In principle, APT can also be directly reduced without any prior calcination step. The disadvantage of direct reduction is the formation of ammonia which has to be scrubbed, but a certain amount of ammonia cracks and dilutes the hydrogen by nitrogen. Consequently, from time to time, part of the contaminated, circulating hydrogen must be vented, thus increasing costs.

Reduction is carried out in pusher furnaces in which the powder passes through the furnace in boats or in rotary furnaces (see below). Walking beam furnaces or furnaces with internal band conveyors are less often used. Fluidized-bed reactors are still not in commercial use, except for the production of nanophase W or WC/Co powder precursors (see Chapter 9). Furnaces are provided with several temperature zones controlled between 600 and 1100 °C. A large excess of hydrogen is used, which is recycled to the furnace after purification. The flow of hydrogen is usually in a countercurrent direction, more rarely cocurrent. The hydrogen acts not only as a reducing agent but also carries away the water formed.

The reduction of tungsten oxides by hydrogen to tungsten metal is, in some respect, a unique process. It offers the possibility to produce tungsten powder of any desired average grain size between 0.1 and 10 μm (and, in the case of doped oxides, even up to 100 μm), starting from the same oxide precursor. Individual tungsten particles form during reduction as a result of chemical vapor transport of tungsten (vaporization/deposition process), which is responsible for the final powder characteristics. For principles of the reactions occurring during reduction as well as thermodynamic, kinetic, and morphological aspects, see Section 3.3.

By changing the reduction parameters, powder characteristics like average grain size, grain size distribution, etc. can be regulated. Temperature and humidity (i.e., the water vapor partial pressure prevalent during reduction) are the two main parameters in steering the average grain size of the W powder, the latter being related to a number of oxide and process-related variables as indicated in Fig. 5.19 and discussed briefly below. The reason for the strong influence of the humidity on powder grain size originates in the strong dependence of humidity on the nucleation rate of the metal phase and the high mobility of tungsten due to the presence of a volatile tungsten compound ([$(WO_2(OH)_2$]). The lower the humidity, the higher the nucleation rate (under isothermal conditions) and the smaller the grain size.

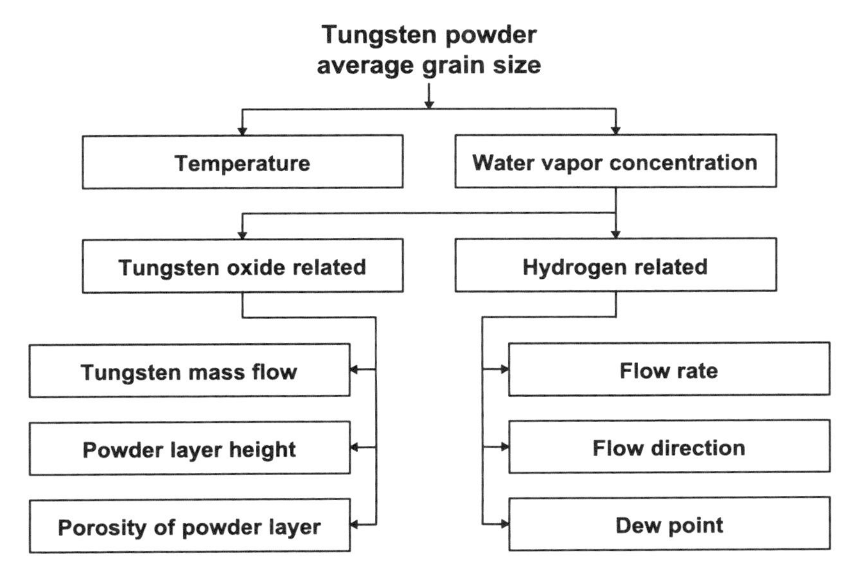

FIGURE 5.19. Average grain influencing reduction parameters.

Temperature. Temperature influences the rate of all reactions occurring during reduction, hence the dynamic humidity and partial pressure of the volatile $[WO_2(OH)_2]$ which forms during reduction and which is responsible for the chemical vapor transport (CVT) of tungsten. Temperature and tungsten particle size are directly proportional while temperature and time required for final reduction are inversely proportional.

Oxide Feed. The tungsten mass flow determines the amount of H_2O liberated during the entire reduction process. The higher the flow, the larger the grain size.

Tungsten Powder Layer Height. During reduction and accompanying water formation, the powder layer exerts a considerable diffusion resistance against water removal from the layer. The higher the layer, the greater the diffusion resistance and the more slowly the reaction water will be removed. The local humidity is higher at the bottom of the layer than at the top. It is this local humidity which determines the nucleation and growth conditions of the metal particles formed at a particular temperature. The layer height is directly proportional to powder grain size.

Porosity of the Powder Layer. The porosity of the powder layer, and thus its permeability, is determined by the macroporosity (intermediate space between the oxide particles) and by the microporosity (porosity of the individual oxide particles). For parameters influencing porosity, see Section 5.3. The higher the porosity of the powder layer, the better the material exchange $H_2O \rightarrow H_2$ during reduction and the less the grains of the tungsten particles will grow, resulting in a smaller particle size.

Hydrogen Flow Rate. A higher hydrogen flow enhances the material exchange due to the more rapid removal of water vapor. Therefore, the flow is inversely proportional to the average grain size.

Hydrogen Flow Direction. Cocurrent hydrogen flow with respect to the tungsten flow generates a higher dynamic humidity at the later part of the reduction, while countercurrent flow (which is the standard condition) provides higher humidity during the early reduction stages.

Hydrogen Dew Point. The dew point of the incoming hydrogen influences the overall humidity during reduction. More "wet" hydrogen enhances the tungsten particle growth.

It is important to understand that the final average grain size of the tungsten powder is a consequence of combining all the aforementioned parameters. The general rules are:

Small grain size: low temperature, dry hydrogen, high hydrogen flow rate, low dew point, small powder layer height, high porosity, low oxide feed.

Large grain size: high temperature, wet hydrogen, low hydrogen flow rate, high dew point, large powder layer height, low porosity, high oxide feed.

Empirically based equations for calculating the reduction time and average particle size have been derived for rotary furnaces [5.73] but have never been applied industrially. The main difficulty is that the properties of the raw material play a crucial role in the reduction process, and these characteristics are not well enough represented by corresponding equations. This is particularly true for smaller grain sizes. For larger grain sizes, the influence of the oxide precursor is less pronounced.

In industrial practice, the choice of proper reduction parameters is based exclusively on empirical experience. Besides the average grain size, the reduction parameters also influence the grain size distribution, agglomeration, apparent density, and grain morphology.

Grain Size Distribution. Grain size distribution is to a great extent the consequence of powder layer height. The growth conditions for the individual particles are different and depend on their position within the powder layer. The humidity is higher in the interior and decreases as one approaches the surface. This gradient results in larger grain-sized particles inside and smaller grain-sized particles at the surface or surface-neighboring areas. It is easy to understand that the distribution is broader for high powder layers and closer for lower layers. In any case, the distribution can be improved (made closer) by using "wet" hydrogen, since the water vapor gradient from inside to outside the layer will be decreased.

Agglomeration. One may distinguish between three different types of agglomeration:

- Reduction performed at low temperature and dry conditions results in the formation of metal sponges, which are pseudomorphous to the oxide precursor. The very fine crystals (some tenths of a micron or less) stick together in a more or less loose manner. Usually, this type of agglomeration is called pseudomorphology (Fig. 5.20). The finer the crystal size, the higher the strength of these agglomerates.
- Bigger tungsten crystals (from approximately 1 to several μm), grown individually grown by CVT, are welded together. This type of agglomeration is looser than the above-described pseudomorphology and occurs mainly under dry reduction conditions (Fig. 5.21). It can be influenced by the dew point of the incoming hydrogen. Higher dew point results in less agglomeration. The strength of this agglomeration type increases with temperature.
- Reduction under wet conditions and high temperature (high concentration of $[WO_2(OH)_2]$ in the vapor phase) results in strongly intergrown crystals (Fig. 5.22).

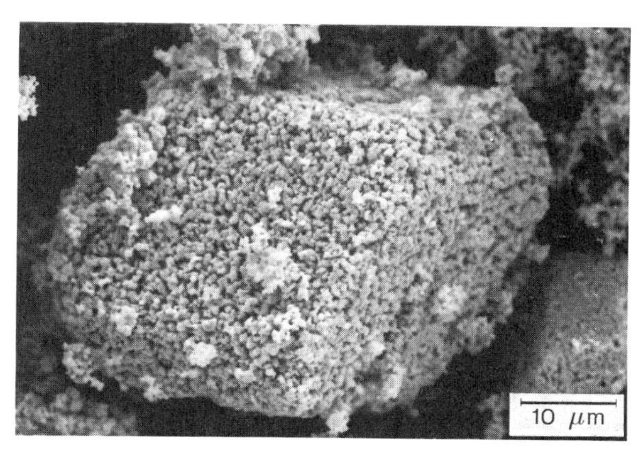

FIGURE 5.20. Fine tungsten powder agglomerated in APT pseudomorphs; average grain size of the powder is below 0.5 μm.

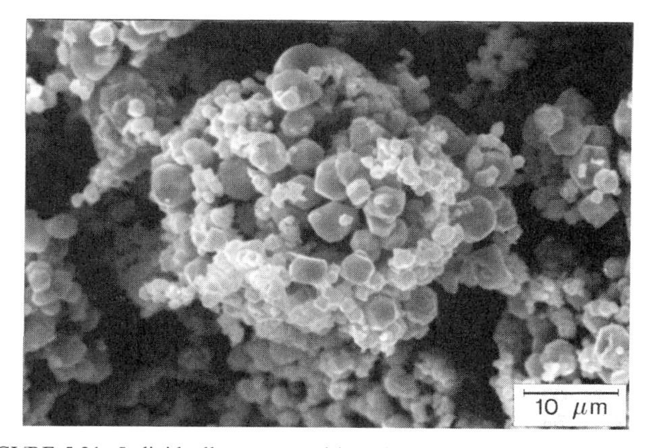

FIGURE 5.21. Individually grown and loosely agglomerated tungsten crystals.

FIGURE 5.22. Coarse intergrown tungsten crystals.

Agglomeration is closely related (inversely proportional) to the apparent density of the tungsten powder. Correspondingly, the apparent density can be influenced within certain limits by the hydrogen dew point. Agglomeration is a prerequisite for good compactability of the tungsten powder.

Morphology. As noted earlier, low temperature and dry conditions largely suppress any CVT of tungsten and lead to the formation of metal sponges, which are pseudomorphous to the oxide precursor (APT, H_2WO_4). They consist of very fine, polygonal and polycrystalline metal particles. With increasing temperature and humidity, individual tungsten grains form by CVT over comparably large distances. The particles are faceted and commonly exhibit the characteristic shape of the cubic metal (Fig. 5.23). Well-faceted crystals, showing growth steps and being partly intergrown, are characteristic for very humid conditions (high temperature, large powder layer height; Fig. 5.24).

FIGURE 5.23. Small cubic tungsten particles obtained at a reduction temperature of 800 °C (powder layer height, 2 mm).

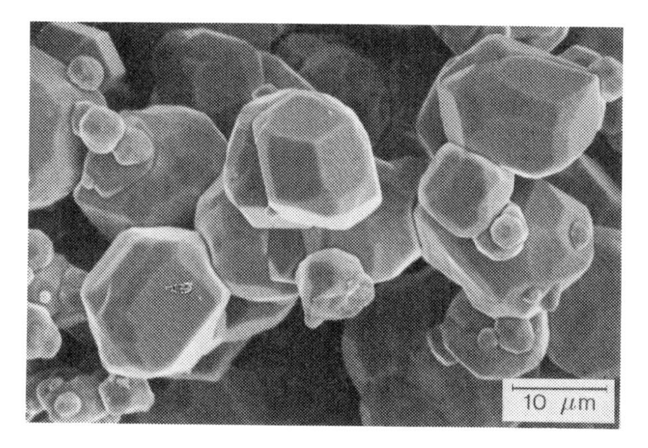

FIGURE 5.24. Well-faceted tungsten crystals obtained at a reduction temperature of 1000 °C (powder layer height, 20 mm).

5.4.2. Push-Type Furnace

Metal boats are charged with oxide to a height ranging from a few mm up to several cm and are pushed in stages through the furnace in corrosion-resistant steel tubes at specific time intervals. By introducing a new boat into the tube, the row in front is pushed forward by the length of a boat. Hydrogen in excess flows either co- or countercurrent to the tungsten flow direction. The hydrogen is not only responsible for the reduction process itself but serves also to remove the water vapor formed and also acts as protecting atmosphere in the cooling zone. The "wetted" hydrogen leaving the furnace is dried to a desired dew point and recycled to the furnace (Fig. 5.25). As indicated, hydrogen having higher dew points can also be fed into the furnace.

Hydrogen has to be applied in large excess, which guarantees a fast flow over the powder layer. The excess depends on the desired grain size (smaller for coarse and higher for fine powder). The range is somewhere between 2.5 and 40 times stoichiometric.

Multitube furnaces (14 to 18 tubes arranged in two rows) are frequently in use today. A photograph of an industrial aggregate is shown in Fig. 5.26. The boat material, in most cases, is an iron alloy high in Ni and Cr (Inconel). More seldom, because of the high price, boats are made of TZM (molybdenum alloy with Ti, Zr, and C) or pure tungsten.

The big disadvantage of the iron alloys is that diffusion of the elements occurs into the contacting tungsten powder layer. In this respect, Ni is the most dangerous element although widely used. Ni rapidly diffuses over the tungsten grains, thereby weakening the surface of the bottom and wall of the boats. With time, a Ni, Fe, Cr, and W containing scale is formed. This scale sticks more or less firmly to the boat. After several travels through the furnace, it gets thicker and partly breaks off, contaminating heterogeneously the tungsten powder. Bigger-scale particles can be separated by the always applied screening process following the reduction, but the smaller particles remain in the tungsten powder. The higher the temperature and humidity, the more pronounced the scale formation. Cast alloy material (coarse microstructure) shows enhanced scale formation compared to boats made of rolled sheet. Alloys containing Co instead of Ni are more resistant, but the high price of Co makes them unacceptable for boats. Co containing alloys are only in use as tubes in rotary furnaces.

The furnaces are either gas fired or electrically heated in three or four separate zones. Furnace temperatures range between 600 and 1100 °C. For smaller and medium W grain

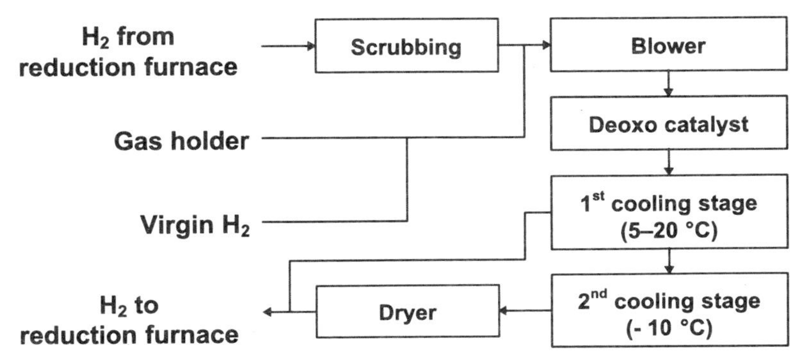

FIGURE 5.25. Hydrogen regeneration.

FIGURE 5.26. Push-type reduction furnaces. By courtesy of Wolfram Bergbau- und Hüttenges. m.b.H., Austria.

sizes, a temperature profile is preferred in order to decrease the time necessary for the last reduction step from WO_2 to W (slow reduction rate). For larger grain sizes (>6 μm), isothermal reduction conditions are applied.

The reduction is commonly carried out in one stage. Alternatively, a two-stage reduction sequence can be applied instead. In this case, the first reduction stage takes place at lower temperature (500–700 °C; formation of brown oxide, WO_2) and the second stage at 600–1100 °C (formation of tungsten metal).

In industrial practice, the boats are loaded with a certain oxide weight (layer height) and pushed through the furnace with a given temperature profile and hydrogen throughput. After dynamic equilibrium is reached, the particle size of the metal powder is measured. If the powder does not meet the requirements, parameter adjustments such as change in temperature, boat load, hydrogen throughput, or push time are introduced.

Typical reduction conditions and furnace capacities for various grain sizes are summarized in Table 5.9.

Subsequent to reduction, the powders are screened on 60 mesh (sometimes also on 200 mesh) to eliminate contaminants stemming from furnace or boat materials and are blended to form a homogeneous powder batch. No special atmosphere is necessary for handling, since the powder surfaces are rapidly saturated with oxygen and water vapor. However, below 1 μm, the powders may be pyrophoric and precautions are necessary, in particular below 0.5 μm. Reduction under cocurrent hydrogen flow is the most effective method to avoid burning of the fine powders. Already, during the cooling stage in the furnace, the powder is contacted to “wet” hydrogen, and the surface is saturated when the

TABLE 5.9. Push-type Furnace Reduction Conditions and Capacities for Different Grain Sizes[a]

FSSS grain size (μm)	Temperature (°C)	Capacity ($kg \cdot W \cdot h^{-1}$)
0.7	600–800	40–50
1.5	700–900	90–110
2.5	800–950	100–120
5.0	900–1000	130–150
12.0	900–1000	140–160

[a] By courtesy of Wolfram Bergbau- und Hüttenges.m.b.H., St. Martin i.S. A-85452.

powder leaves the furnace. Under countercurrent flow conditions, the powder has to be slowly saturated with oxygen. This can be achieved either by an inert gas storage (nitrogen or argon containing small amounts of oxygen) or by exposing the powder to the atmosphere in small portions in order to omit local overheating. This can be done by leaving the powder in the boat for approximately 30 min.

It is obvious that the furnace capacity for smaller grain sizes, especially for submicron tungsten powder, is low. Only very thin powder layers can be applied to retain the grains from growing. In order to improve the capacity, a double or triple boat technique was invented. The reduction boat is topped with one or two upper boats in a way permitting hydrogen flow between the boats, so that the capacity also for smaller grain sizes could be increased considerably.

Modern furnaces are fully automated, which means that all variables can be set and controlled. Loading, pushing, and discharge of the boats is done by machine.

The advantage of the push-type furnace in comparison to rotary kilns is its flexibility in switching from one condition (grain size) to the next and in its high capacity, especially for finer powder qualities. Disadvantages are higher energy consumption, broader grain size distribution, more contamination by the scale from the boats, and higher maintenance costs.

5.4.3. *Rotary Furnace* [5.72–5.74]

In contrast to the pusher furnace, where a static powder layer passes the furnace, a dynamic powder flow exists in the rotary kiln. Furnace rotation and the incline continuously move the powder through the hot zone. Feed rate, rotational speed, incline, and lifters inside the tube determine the depths of the dynamic powder layer. High oxide feed, longer retention time (low rotation rate and low inclination), and higher temperature result in coarser tungsten grains.

The powder layer is constantly disturbed by the rotating motion and powder from inside moves to the surface, and vice versa. Therefore, the water-vapor retaining capability of the layer is less compared to a static layer, and the material exchange rate $H_2O \rightarrow H_2$ is enhanced. The humidity in a rotary furnace powder layer is thus lower and reduction proceeds under drier conditions as compared to the pusher. Lifters inside the tube combined with stepwise turning motion make sure that a powder bed is maintained for a certain time giving the chance to build up higher humidity both temporarily and locally.

In order to maintain the humidity for a certain grain growth, only a small excess of hydrogen is necessary (between 2 and 3 times stoichiometric). This corresponds to the lowest values in pusher furnaces, applied only for very coarse powders. Moreover, the influence of the H_2/WO_3 molar ratio on grain growth is less pronounced.

The hydrogen is cleaned and dried and recycled to the furnace after preheating. Due to the constant motion of the powder bed, fine particles are swept away by the hydrogen flow and form a dust, which must be removed by filtering devices.

Very coarse blue oxide (derived from classified APT) has improved flow characteristics and helps to prevent powder from sticking to the tube wall.

Due to the "dry" conditions, rotary-furnace-produced tungsten powder shows a high degree of agglomeration, lower apparent density, and closer grain size distribution. The upper size limit is approx. 6 μm.

A photograph of a modern rotary furnace is shown in Fig. 5.27.

Furnaces in use possess different sizes: inner diameter between 300 and 750 mm and heated drum length from 3000 to 8000 mm. The most common equipment has an inner

FIGURE 5.27. Rotary reduction furnace. By courtesy of Wolfram Bergbau- und Hüttenges. m.b.H., Austria.

TABLE 5.10. Rotary Furnace Conditions and Capacities for Different Grain Sizes[a]

FSSS grain size (μm)	Temperature (°C)	Capacity ($kg \cdot W \cdot h^{-1}$)
0.7	600–800	18
1.3	600–850	27
4.0	800–920	58

[a] By courtesy of Elino Industrie-Ofenbau, Carl Hanf GMBH + CO, Düren D-52355.

diameter of 600 mm and a heated length of 6500 mm. Typical conditions and furnace capacities for various tungsten powder grain sizes are presented in Table 5.10.

Advantages of the rotary furnace compared to a pusher are less specific energy consumption (valid only for median grain sizes), while automation is less complicated and consequently maintenance is cheaper. Finally, the product is more homogeneous and less contaminated, but the particle morphology may differ significantly from powders produced in pusher furnaces.

5.4.4. Tungsten Powder

Technical tungsten powder grades prepared by hydrogen reduction are commercially available in average grain sizes from 0.1 μm (100 nm) up to 100 μm. The whole palette of grain sizes find their application in cemented carbides. The main portion of tungsten powder produced (80%) is directed into that production.

The starting powder for the powder metallurgical production of pure tungsten (ductile tungsten) and sintered tungsten alloys commonly covers grain sizes between 2 and 6 μm. Finer or coarser powder would be unsuitable. Fine powder has too high a sintering activity, not permitting the evaporation of trace impurities during the sintering process, while coarser powder leads to incomplete sintering under usual conditions.

Extremely coarse powder gained by classification (to separate any finer particles) exhibits excellent flow characteristics and is used in plasma spraying.

Chemical Properties: Purity. The purity of the tungsten powder is of particular importance in PM manufacturing of tungsten metal, since during subsequent sintering further purification through evaporation is only possible to a certain extent. The demand for purity of tungsten powder has increased steadily during the last three decades. Considerable improvements in hydrometallurgy have led to concentrations fairly below 10 μg/g for most of the elements. This trend with time can be demonstrated by comparing today's usual specifications with those given in the last book on "Tungsten" by Yih and Wang, published in 1979 (Table 5.11).

The reason for this enhanced demand for powder purity originates in the fact that remaining impurities after sintering greatly affect the workability (grain boundary strength, recrystallization temperature, etc.) and properties of the final product.

By considering the impurity-to-tungsten ratio (disregarding oxygen and ammonia) on the long path from ore concentrates to compact tungsten metal, one observes a constant increase in purity up to the stage of APT crystallization. At this stage, the maximum purity

TABLE 5.11. Tungsten Powder Specifications 1975–1995 (upper limits in $\mu g \cdot g^{-1}$)[a,b]

Elements	1	2	3
Al	10	3	10
Fe	50	8	10
Ni	30	5	10
Mo	100	16	20
Si	10	5	10
Ca	20	3	10
Co	10	5	2
Cr	10	8	10
Cu	10	2	2
Mg	10	5	2
Mn	10	3	2
Pb	10	1	2
Sn	10	1	10

[a] 1, Yih and Wang, 1975. 2, Chinese Specification, 1995. 3, Data Sheet, 1993, Wolfram Bergbau- und Hüttenges.m.b.H.
[b] Lower impurity levels in recent years are partly a consequence of improved production techniques and partly due to the higher sensitivity of advanced analytical methods.

is more or less reached. Consequently, the purity of the tungsten powder depends mainly on the APT cleanliness. During APT processing to tungsten powder, the purity already decreases again. Sources of contamination are contacts with metallic tubes or boats in the respective furnaces. Slightly enhanced, overall concentrations of elements, such as Fe, Ni, Cr, and Co, are the consequence. This type of impurity occurs heterogeneously and represents small areas of locally high concentrations of foreign elements. If they are big enough, they might act as the origin of sintering defects.

Moreover, volatile elements or compounds present in the hydrogen atmosphere can be adsorbed by the tungsten powder during reduction. A typical example are alkali metals.

Physical Properties. The relevant physical properties are average particle size, particle size distribution, apparent, tap, and compact or green density, specific surface area, degree of agglomeration, and morphology. All of them are closely related to each other and can be influenced by the oxide properties and reduction conditions. They represent the important criteria for further processing and are responsible for the compactability, sintering behavior, dissolution reactions during liquid-phase sintering, and carburization reactions. Physical properties, methods of determination, and standardized procedures are outlined in Table 5.12.

Average particle size and particle size distribution: "As supplied" tungsten powder is always more or less agglomerated, depending on reduction conditions. Therefore, the measured value for the average particle size in this stage does not always correspond to the real particle size and can even be misleading. For example, certain submicron powders show "as supplied" grain sizes of 1–2 μm, but after deagglomeration these values drop to 0.4–0.5 μm.

TABLE 5.12. Important Physical Properties of Tungsten Powder

Parameter	Method of determination	Remarks
Average grain size		
as supplied	Fisher Subsieve Sizer	ASTM No. B 330–88 (reapproved 1993)
lab milled	Fisher Subsieve Sizer	after standardized milling according to ASTM No. B 430-90
	SEM	for average grain size $\leq 0.5\,\mu m$
Grain size distribution	Gravity sedimentation turbidimentry	ASTM No. B 430–90
	Gravity sedimentation X-ray absorption	ASTM No. B 761–90
	Laser scattering	ASTM No. B 822–92
Apparent density	Scott volumeter	ASTM No. B 329–76
Tap density		ASTM No. B 527–93
Press density	By measuring the height of a cylindrical compact prepared under standardized conditions (powder weight, compacting pressure and time)	ASTM No. B 331–85 (reapproved 1990)
Specific surface	BET	ASTM No. C 1096–86 (reapproved 1992)
Strength of compacts		ASTM No. B 312–82 (reapproved 1988)
Morphology	SEM	

For production control in the range of $1-10\,\mu m$, the "as supplied" value mostly suffices, because production conditions are kept constant within close limits. However, for submicron-sized powders as well as for a correct characterization of coarser powders, only the "lab milled" powder should be used.

Normally, the particle size distribution is a function of particle size. The bigger the particle size, the broader the distribution. For a given particle size, the distribution can be made narrower during production by application of wet hydrogen and/or doping the oxide with alkali compounds. Typical particle size distributions are shown in Fig. 5.28.

Particle size distribution measurements should also be made from "lab milled" samples only. The particle size distribution influences the compacting behavior and green density as well as the sintering properties.

Agglomeration. The difference between "as supplied" and "lab milled" (deagglomerated) particle sizes is a measure of the degree of agglomeration. Agglomeration is very important for the strength of the green compacts and is therefore a necessary property for powder going into ductile tungsten production.

Much effort was expended in the past to elucidate the dependence of the compactability of tungsten powders on grain size and grain size distribution. Low grain sizes result in too low green densities due to high friction between the particles. The closer the grain size distribution, the poorer the particle packing. So it is evident that broader grain size distributions, or even blends of powders having different average grain sizes, result in better packing and higher strength of the green compacts.

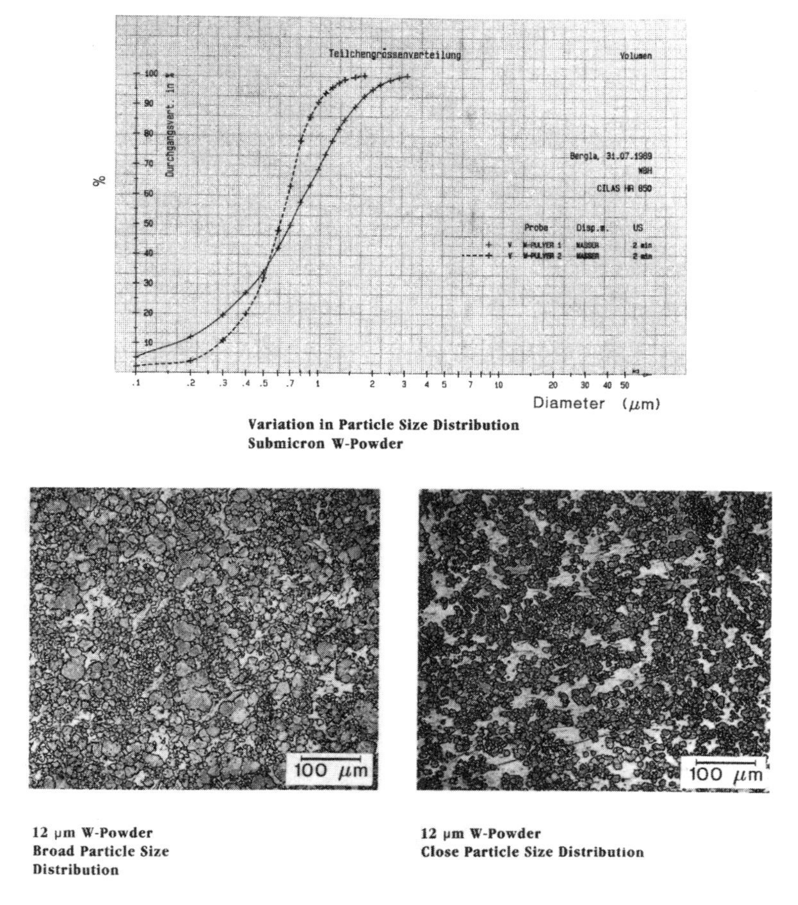

Variation in Particle Size Distribution
Submicron W-Powder

12 μm W-Powder
Broad Particle Size Distribution

12 μm W-Powder
Close Particle Size Distribution

FIGURE 5.28. Particle size distribution of fine and coarse tungsten powders.

Compaction and green strength is also influenced by the particle morphology. More irregular shapes, which cause interlocking, improve the strength.

Specific surface area. The specific surface area is an important criterion for the sintering activity (solid state sintering), dissolution processes (liquid phase sintering), and reaction with gaseous or solid substances during carburization. Commonly, it is in the range of 0.01 m^2/g (coarse powders) up to 12 m^2/g (very fine powders).

Based on the specific surface area, an average particle size can be calculated. However, attention must be paid to the surface roughness of the particles. In particular, very fine powder particles do not exhibit "smooth" but "rough" surfaces, or even exhibit a certain degree of microporosity, hence increasing the specific surface area by a factor of 2–4.

Apparent and tap density. Both densities increase with increasing average grain size, but are additionally influenced by the grain size distribution, particle shape, and degree of agglomeration.

Compressibility. This is the ratio of green density to apparent density. The ratio increases with increasing pressure until a limit is reached. Compressibility of a powder is an important criterion for press and die design.

5.4.5. *Reduction of Doped Tungsten Oxides*

Powder properties can be significantly influenced by the addition of a third element (see also Section 3.3). This can be used industrially for the production of special W grades, as mentioned below.

Tungsten blue oxide is treated with aqueous solutions of the respective element compounds.

Sodium Doping. Addition of sodium compounds (concentrations equal to 50–200 μg/g Na) is a usual practice in producing coarse tungsten powder. At 1000 °C and wet conditions, average grain sizes of 10–25 μm (depending on the layer height) can be attained. The high temperature guarantees the extensive evaporation of sodium at the end of the reduction process. Only for special purposes must a subsequent acid or water leach be applied to decrease the residual sodium concentration.

Lithium Doping. Lithium compounds are preferred to sodium for very coarse tungsten powder grades (50–100 μm); Fig. 5.29. Reduction conditions are the same as above. Due to the high lithium boiling point, it is not volatilized during reduction like

FIGURE 5.29. Coarse W powder obtained by Lithium doping (average particle size 50 μm). By courtesy of Wolfram Bergbau- und Hüttenges. m.b.H., Austria.

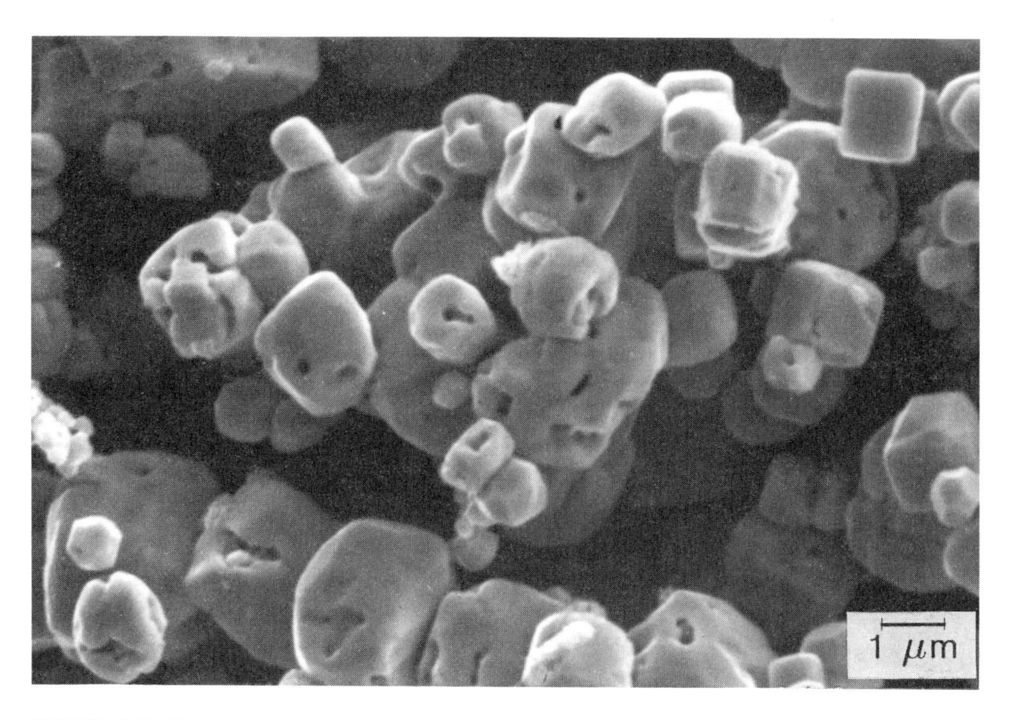

FIGURE 5.30. HF-washed NS-doped tungsten powder. Large dopant phases which were not completely incorporated during W crystal growth are removed by the acid washing treatment, leaving behind large holes in the crystal surfaces.

sodium. The powder has to be leached by hydrochloric acid and water subsequent to reduction. Finer fractions are separated by screening.

Chromium Doping. Additions of chromium compounds can be used to produce very fine metal powder grades ($<0.5\ \mu m$). Chromium oxide is formed during reduction and acts as a grain refiner. Such powders are subsequently carburized to ultrafine WC. Special precautions have to be taken because of the high pyrophoricity of the metal powder.

NS or AKS Doping [5.75, 5.76]. For the production of lamp filaments (incandescent or halide lamps), a special doping procedure is carried out which is commonly called NS or AKS doping. NS stands for the non-sag properties of the coiled lamp filament, while AKS represents the doping elements used in this process: Al, K, and Si.

TBO is doped with about 3000–3500 μg/g K, 2000–2500 μg/g Si, and 400–500 μg/g Al in the form of potassium silicate and aluminum chloride or nitrate solutions.

In reduction of the doped oxide, part of the dopants is incorporated during the CVT growth of the metal particles in the form of silicates. Excess dopant, which remains on the tungsten crystal surfaces, is removed by subsequent leaching of the powder in HCl and HF acids, while the dopants which are internally trapped are retained. The amount of incorporated dopants, the size of the inclusions, and their chemical composition can be influenced by the reduction conditions within certain limits. About 100–150 ppm K, 60–100 ppm Al, and 200–300 ppm Si commonly remain in the acid-washed powder.

During sintering, the silicates dissociate thermally. The smaller Al, Si, and O atoms escape via diffusion through the tungsten crystal lattice while the large K atoms remain in the form of potassium bubbles. It is this potassium that enables the formation of rows of smaller potassium bubbles during wire fabrication, and consequently the formation of a long-grained interlocking microstructure during subsequent filament operation, which is the key to the non-sag characteristics (see Section 6.2.1).

The grain-growth-enhancing effect of K makes it possible to conduct the reduction at a lower temperature than normally applied for W powder of 3–5 μm average grain size (i.e., the optimal grain size for powder-metallurgical processing). The hydrogen gas has to be kept separated from hydrogen for reduction of undoped powder, because it is contaminated by K.

NS–W powders exhibit a characteristic morphology, exhibiting holes in the crystal surfaces (Fig. 5.30), which form during HF-washing of the powders due to the dissolution of partly intergrown silicate phases.

5.5. POWDER METALLURGY [5.1, 5.2, 5.77–5.84]

Although the Wollaston process of producing platinum metal from platinum powder is regarded as the birth of modern powder metallurgy (1808–1815), it was the pioneering work of W. D. Coolidge on the production of ductile tungsten wires in 1909–1913 which led to its first commercial application [5.82]. Over the years, only few changes were made in the industrial production of tungsten metal. Powder metallurgy is still today the main route in tungsten and tungsten alloy manufacture. Unlike other refractory metals, such as Zr, Hf, Nb, or Ta, melting technology has assumed no industrial importance for metal production, since the very high temperature, necessary for melting, and the resulting coarse microstructure of the "as cast" material make further processing both difficult and costly.

Powder metallurgy comprises two steps: *compaction* and *sintering*.

5.5.1. Compacting

Tungsten powder is consolidated into a compact by two main routes: pressing in rigid dies (uniaxial pressing) and isostatic pressing in flexible molds (compaction under hydrostatic pressure). Other techniques, such as powder rolling, cold extrusion, explosive compaction, slip casting, vibratory compaction, or metal injection molding, have gained no industrial importance.

Tungsten powder is not easy to compact due to its relatively high hardness and difficult deformation. Nevertheless, in most cases compaction is performed without lubricant to avoid any contamination by the additive. The resulting compacts are generally sufficiently strong so that they can be handled without breaking. For machining the part, it must be pre-sintered beforehand.

5.5.1.1. Die Pressing. Pressing of powders in rigid dies is carried out either in mechanical or hydraulic presses. The pressure is applied from the top, or from the top and bottom (double action presses). Die and punches are made of high-speed tool steel or (more rarely) hardmetal. Mechanical presses (pressing forces up to 1 MN) are used for small parts and high production rates. They allow a higher degree of dimensional precision and are well suited for process automation. Hydraulic presses are mainly used for simple preforms. Large presses with up to 30 MN (3000 t) pressing force are used for pressing of plates which are to be rolled to sheet metal. The size and shape of the compact are limited by the capacity of the press and also by the geometry of the part.

Due to the friction between the powder and the die wall and the nature of the load distribution inside the die, the pressing density is not the same all over the compact. This is more pronounced for large parts and large part heights and can lead to crack formation and/or distortion of the pressed compact during sintering. Large and critical parts are therefore commonly pressed isostatically.

Typical compaction pressures are in the range of 200–400 MPa (2–4 t/cm^2) but can reach 1000 MPa (using hardmetal dies and punches). The green density (compact density) is in the range of 55–65% of the theoretical density (75% at most) and it depends upon the applied pressure, particle size, size distribution, particle shape, and size of the compact. There are several theoretical equations relating green density and applied pressure, but in practice empirical relations are used.

The relationship between the average particle size and the green density as well as the compressive strength of compacts is shown in Fig. 5.31 for a constant pressure. Although the green density increases as average particle size varies from 1 to 9 μm, the compressive strength exhibits a maximum between 3 and 6 μm. This maximum corresponds to the preferred particle size range for most tungsten compacts.

5.5.1.2. Cold Isostatic Pressing. In isostatic pressing, the powder is filled into flexible molds made of rubber or elastomers and subjected to hydrostatic pressure. The pressure is commonly in the range of 200 to 400 MPa. As a result of the uniform pressure, a much higher uniformity in density is achieved. Isostatic pressing has gained much

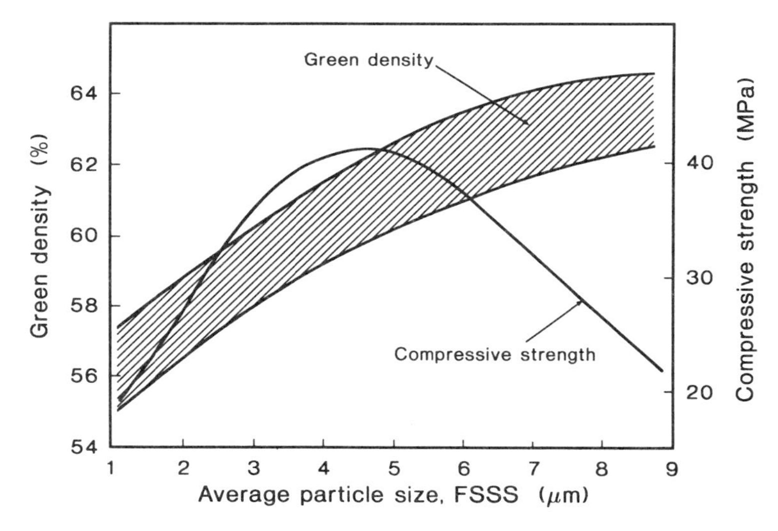

FIGURE 5.31. Relationship between average grain size and green density as well as compressive strength for a constant compacting pressure (241 MPa) [5.1].

importance during the last 30 years, because it offers several further important advantages compared to rigid die pressing:

- Lower pressure required for a certain green density.
- Higher strength of the compacts.
- More free choice in dimension (ratio of diameter to lengths).
- Parts with undercuts and reentrant angles can be pressed.
- Thin-walled tubes can be produced.
- Very large parts can be compacted.

Less precise dimensional control and a much lower rate of production are the two main disadvantages.

Isostatic pressing is carried out by two different techniques: *wetbag* pressing or *drybag* pressing. In wetbag pressing, the powder is filled into flexible molds which are sealed outside the pressure vessel. Several molds (either of the same shape or of different shape) are then immersed in a fluid, most commonly water, and pressure is applied isostatically. Wetbag pressing is the most common technique for producing forging or rolling preforms, where an even and high green density is more important than dimensional control. Nevertheless, wetbag pressing is also used for more complex geometries and even near-net shape parts. The size of the part is limited by the size of the pressure vessel. Tungsten ingots of up to 1000 kg are produced via wetbag pressing.

In drybag pressing, the elastomeric mold is fixed into the pressure vessel. The mold is filled with powder and sealed with a cover plate. Then pressure is applied between the mold and the vessel wall. After pressure release, the cover is removed and the part removed. Then the procedure starts again. Pressure vessels with both top and bottom plate are also in use, and allow more rapid removal of the compact.

Drybag pressing is used for simple shapes, such as plugs, and high production rates (mass production). However, only one compact (with one specific shape) can be pressed at a time.

5.5.2. *Sintering* [5.78–5.84]

5.5.2.1. General. In order to increase the strength of the green compacts, they are subjected to heat treatment, which is called sintering. The main aim of sintering is densification in order to provide the metal with the necessary physical and mechanical properties and a density which is suitable for subsequent thermomechanical processing. Sintering of tungsten is commonly carried out in a temperature range of 2000 up to 3050 °C under flowing hydrogen, either by direct sintering (self-resistance heating) or indirect sintering (resistance element heating systems). The density thereby obtained should be a minimum of 90% of the theoretical density, but is commonly in the range between 92 to 98%.

The main driving force for sintering is the lowering of free energy, which takes place when individual particles grow together, pores shrink, and the high surface area of the compact (i.e., its high excess surface energy) decreases. The decrease in surface area is accomplished by diffusional flow of matter into the pore volume under the action of capillary forces (surface tension forces). Besides shrinkage, recovery (change of subgrain structures and strain relief), recrystallization (formation of strain free crystals low in dislocation density), and grain growth occur during sintering, also contributing to the minimization of free energy.

Sintering is commonly regarded as taking place in three stages [5.79, 5.83]:

- During the *early stage*, necks are formed between individual particles and grow by diffusion, increasing the interparticle contact area. The powder aggregate shrinks, involving center to center approach of the particles. In this stage, the degree of densification is still low and the pore structure is open and fully connected.
- With increasing neck formation (*intermediate stage*), the necks become blunted and lose their identity. The pores are assumed to be cylindrical. Their radii vary along their lengths and, with increasing shrinkage, the pore channels break up into small, still partly interconnected segments. During this stage (channel closure stage), pronounced densification occurs and significant grain growth occurs concurrent with shrinkage.
- Finally, in the last stage (*isolated pore stage*), the pore segments further break up into chains of discrete, isolated pores of more or less spherical symmetry. This stage occurs when about 90% of the theoretical density is achieved. The sintering density then approaches asymptotically the practical limit of 92–98%.

Investigations have shown that the densification is controlled by grain boundary diffusion over most of the densification range, unless at very high densities it becomes controlled by lattice diffusion [5.83].

Since the motion of grain boundaries, necessary for grain growth, is impeded by the presence of pores, grain coarsening proceeds at a higher rate above 97% density. Grain sizes of the as-sintered ingots are commonly in the range of 10 to 30 μm.

Besides temperature and time, several other parameters influence densification, such as powder particle size, green density, sintering atmosphere, powder purity, compact size/weight, heating rate, thermal gradients, and the presence of insoluble phases such as oxides (ThO_2, La_2O_3, CeO_2, ZrO_2) or metallic potassium (NS-tungsten).

Temperature is the most important parameter affecting densification. Higher sintering densities can be obtained much more rapidly by increasing the sintering temperature than by prolonging the sintering time. Below 1900 °C little densification occurs, unless very long sintering times are applied. For example, more than 50 h are required to obtain 92% density for a tungsten powder of 4 μm particle size at 1800 °C [5.83]. At 2400 °C, the time necessary to obtain high densities decreases to 1–2 hours and, at 3000 °C, 20–30 minutes. The finer the starting tungsten powder, the more rapid the densification at a given temperature [5.1].

The influence of temperature and time on densification can be estimated by using so-called density diagrams (sintering maps) [5.85], which are based on approximate sintering models. Nevertheless, empirical rate equations are used for industrial purposes to calculate the necessary sintering times at different temperatures.

Tungsten sintering, in practice, is always performed in reducing atmosphere which removes the oxygen coating of the powder particle surfaces. High-purity dry hydrogen is commonly used. Under vacuum or in inert gas, sintering is retained by residual oxygen, and the desired density will not be achieved.

The high temperatures used for sintering also lead to a significant purification of the metal by volatilization of impurities. Interstitials, such as oxygen and carbon, cannot be removed completely and have to be held at a low level; otherwise, they significantly affect the workability of the parts. Metallic elements, such as Fe, Ni, Cr, Nb, etc., are evaporized but remain partly in the ingot, forming solid solutions. If Fe or Ni is present in the form of large, heterogeneous contaminations (such as iron particles stemming from reduction boats), these can lead to local melt formation and subsequent formation of large voids, which do not close during sintering. Elements which have practically no solubility in tungsten, such as alkali or earth-alkali metals, are volatilized. Grain boundary diffusion and carrier distillation effects most likely play an important role in "cleaning" the grain boundaries.

Since the ductility of tungsten is very sensitive to most of the impurities, purification is important. Therefore, special care must be taken so that, during sintering, evaporation can take place to the desired extent (i.e., as long as there is an open porosity). If the ingot densifies too quickly, impurities can be trapped. Due to the higher sintering temperature, direct sintering is more effective in cleaning than indirect sintering.

The sintering of tungsten can be enhanced by the addition of small amounts of alloying elements (0.5–1%), such as Ni or Pd. This phenomenon is called "activated sintering." It is explained by the enhancement of grain-boundary diffusion in tungsten due to the presence of the respective element in the grain boundary. High densities (up to 99%) can be obtained even at 1100 °C (at this temperature, tungsten compacts are commonly presintered). However, since the alloys so produced are rather brittle, activated sintering has never been used in industry.

The sintering of doped tungsten is a peculiar case in sintering of tungsten. This includes dispersion-strengthened materials such as thoriated tungsten or tungsten with additions of CeO_2, La_2O_3, and ZrO_2 as well as NS (non-sag) tungsten used for lamp filaments.

The oxide dispersoids are stable at the high sintering temperatures and do not dissolve in the tungsten matrix. They pin the grain boundaries during the later stages of sintering, and in that way they significantly restrict grain coarsening. So, for example, a rod containing 0.75% of thoria has a grain size of 5000 to 10,000 grains per square millimeter, as compared to 1500 grains per square millimeter for a similar rod of pure tungsten [5.1].

NS-doped tungsten powder contains small inclusions of potassium aluminosilicates, which were incorporated during the reduction process. During sintering, the silicates dissociate thermally and submicron potassium bubbles form in the tungsten ingot. Similar to the oxides, these bubbles pin the grain boundaries and inhibit grain coarsening during sintering. Since potassium is gaseous during sintering, the bubbles are under high pressure, which is balanced by the surface tension of the pore. They can be seen as small pores in the fracture surfaces of NS-doped tungsten besides the significantly coarser residual sintering pores, as characteristic for undoped tungsten (see also Section 6.2.1) [5.86]. They constitute the starting point for subsequent formation of rows of bubbles during thermo-mechanical processing.

Up to the sixties, the only sintering method used in practice was direct sintering. Although still in use for the production of doped tungsten, it has lost its importance. From then on, mainly because of the increasing demand for larger parts and the higher capacity of the aggregates, indirect sintering furnaces were developed. This technique is used nowadays as the main route for producing pure tungsten [5.80].

5.5.2.2. Direct Sintering [5.81, 5.87]. A schematic drawing of a direct sintering furnace is given in Fig. 5.32. The tungsten bars are commonly mounted in pairs or several

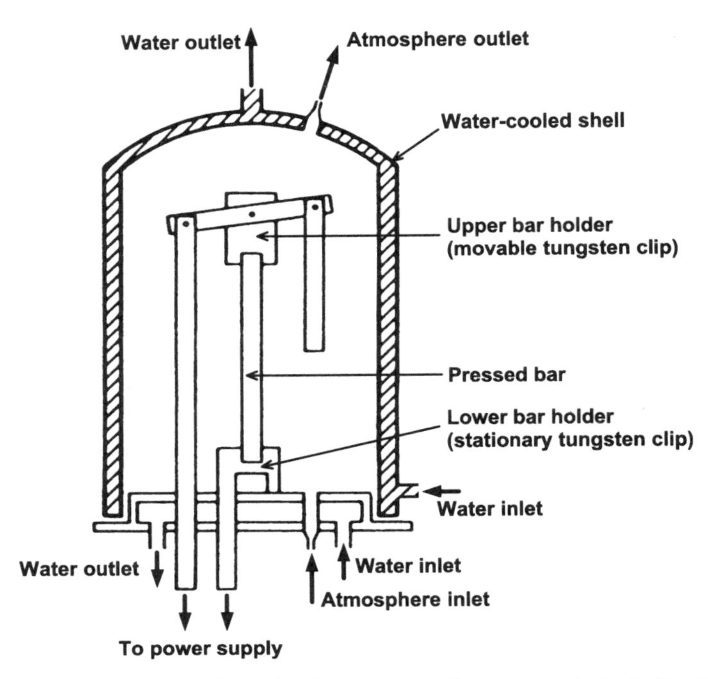

FIGURE 5.32. Schematic drawing of a direct sintering aggregate. By courtesy of Metals Handbook, ASM International, Ohio.

units in a vertical, free-hanging position in a water-cooled chamber. They are clamped at the top using water cooled copper contacts with tungsten inserts. The bottom clip is made of tungsten and must be able to move as the bar shrinks. An electric current of several thousand ampere is passed through it in order to heat it up to about 3000 °C by Joule's energy. Sintering is conducted under dry hydrogen. Several such sintering aggregates are commonly connected in series so that several bars can be sintered simultaneously (Fig. 5.33).

In order to achieve an appreciable strength of the green compact, which is necessary for handling, it must be presintered. Presintering is done by heating to 1100–1300 °C in muffle furnaces under hydrogen atmosphere. By reducing the oxide film covering the powder particles, the cohesive strength between them is increased.

The electric current is used to control the sintering progress and as an indicator of the prevailing sintering temperature. The current is kept low at the beginning due to the higher resistance of the bar (less contact between the grains). It is then gradually raised up to 2700–3000 °C. Proper hold points are important to allow the outgassing of impurities. This is particularly important for the sintering of NS-tungsten, because both Al and Si evaporate rapidly above 2150 °C [5.82]. If densification occurs too rapidly, excess dopants will be trapped and may cause difficulties during subsequent working or even bursting of the bar during sintering. Commonly, less than 10 ppm Si and Al but 60 to 100 ppm K remain in the ingot, the latter in the form of desired potassium bubbles.

Depending on the bar size, the heating up time is between 20 and 60 minutes and the holding time between 30 and 60 minutes. Typical bar dimensions are 15-25 mm ×

FIGURE 5.33. Modern industrial aggregate for direct sintering of tungsten. By courtesy of OSRAM GmbH, Germany.

15–25 mm × 600–920 mm. Typical weights are between 1.5 and 6 kg. The density of sintered bars is between 88 and 96% of the theoretical value. It is always lower for NS-tungsten (88 to 92%) than for pure tungsten, due to the presence of trapped potassium.

Advantages compared to indirect sintering are short sintering times; temperature inside the bar is higher than on the surface, which favors the diffusion and evaporation of impurities; higher temperature results in a better cleaning effect; a simpler aggregate with comparatively low maintenance costs.

Disadvantages are the dimensional restrictions (only simple shapes can be sintered); low capacity; the loss of the "heads" (clipped parts) which did not obtain the temperature due to the heat sink effect of the contacts.

5.5.2.3. Indirect Sintering. A schematic view of a furnace with resistance-heating elements is presented in Fig. 5.34. A modern industrial aggregate is shown in Fig. 5.35. The green compacts (no presintering necessary) are placed inside a cylindrical or basket-like heating element of the furnace (constructed of Mo or preferably W). In a radial direction to the outside, the furnace is adapted with radiation shields (inner shields made of W, outer sheets made of Mo), which protect the furnace wall and concentrate the heat to

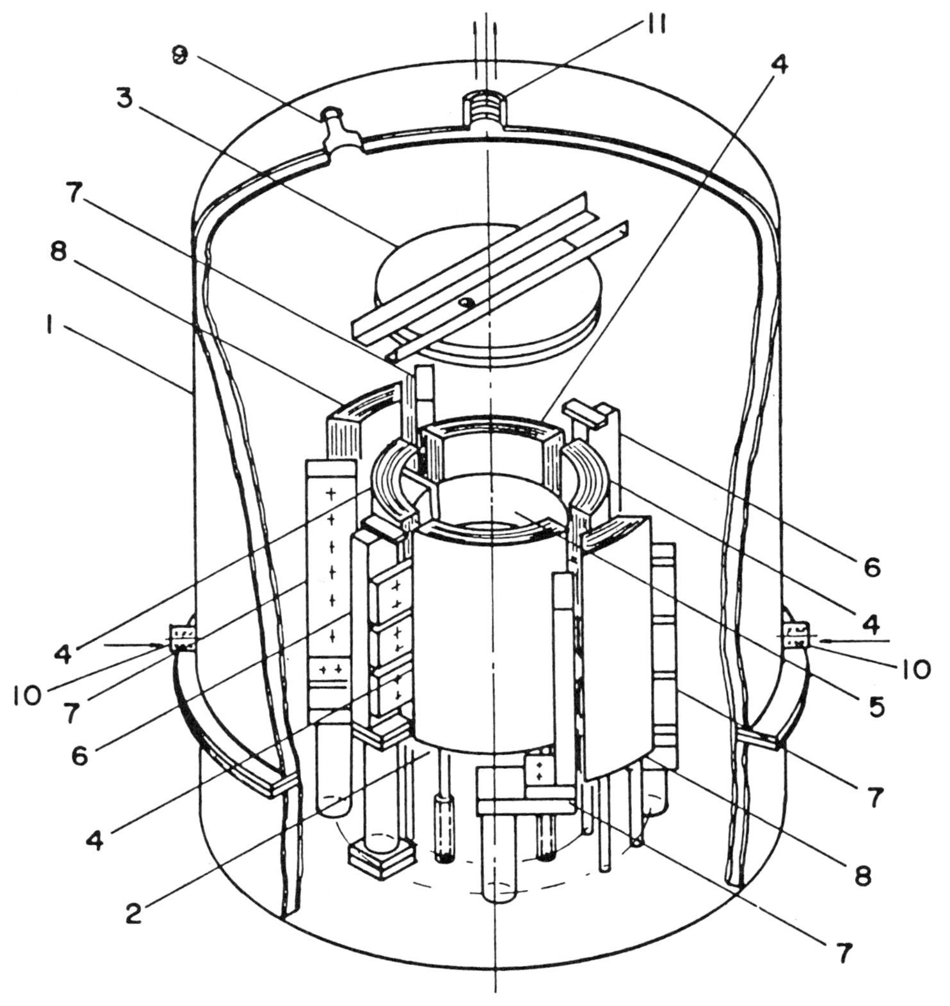

FIGURE 5.34. Schematic view of an indirect resistance-heated sintering furnace [5.1]: (1) furnace assembly with dome, (2) bottom heat shield, (3) top heat shield, (5) heating elements, (6) element support posts, (7) element clamping posts, (8) heat shields, (9) pyrometer window, (10) cooling inlet, (11) cooling outlet.

FIGURE 5.35. Modern industrial aggregate for indirect sintering. By courtesy of Plansee AG, Reutte, Austria.

the center. A vacuum system is necessary to empty the furnace prior to hydrogen flooding. Maximal dimension depends on furnace size. Compacts of any desired shape can be sintered. In order to achieve even shrinkage, the compacts have to be placed on green tungsten shims [5.1]. A slow heating rate is essential; otherwise, surface densification will occur too early, not allowing the outgassing of the interior. Internal stress will be built up, resulting in cracks. This is particularly important for large parts, such as tungsten billets for forging or rolling. Holding times are essential. In practice, the sintering schedule is adjusted to the specific requirements. Proper furnace loading is also important, since the shielding of parts by others can lead to different densification rates and less uniformity [5.84].

Common sintering temperatures are between 2000 and 2700 °C (max). Sintering times of 8 to 24 hours, or even more, are common. Furnaces having uniform hot zones of up to 1.2 m and 0.14 m^2 are available. Weights in the range of a thousand kg can be sintered in one batch. For very high loads, deformation of the part can take place due to gravity.

Advantages are: no loss of material, no dimensional restrictions, and high capacity. Disadvantages are: longer heating times, less purification, lower efficiency of heating, and higher maintenance costs.

5.5.3. Fabrication of Tungsten

5.5.3.1. Fabrication of Wrought P/M Tungsten [5.88–5.90]. With only some exceptions, tungsten is used in the form of pore-free preforms ("wrought" P/M tungsten). To obtain a completely dense material, as well as the desired shape and mechnical properties, a complex, multistage, hot and cold forming process is required.* The most important forming techniques for tungsten are *rolling* (for rods and sheet products), *round forging* (for large diameter parts), *swaging* (for rods), *forging* (for large parts), and *drawing* (for wires and tubes). Secondary forming processes include *flat rolling of wires*, *flow turning*, *spinning*, *deep drawing* and *wire coiling*. For a detailed description of the most important forming processes, we refer the reader to the book, *Tungsten*, by Yih and Wang [5.1].

In general, plastic forming of tungsten is difficult and needs experience. In the as-sintered condition, tungsten is brittle except at quite high temperatures, because it is recrystallized (coarse grained) and not fully dense. Unlike most metals, the low-temperature ductility of tungsten increases with progressive deformation, because embrittlement is due to grain boundary segregations of interstitial soluble elements, such as oxygen, carbon, and nitrogen. With the breakdown of the coarse microstructure during deformation, these impurities are distributed over a larger intergranular area, which makes the material more ductile and less sensitive to cracking during forming at lower temperatures.

The aim of primary working (ingot breakdown) is threefold: (1) to eliminate the residual porosity, (2) to convert the original massive form (sinter ingot) to the desired preshape (sheet, rod, tube), and (3) to refine the grain size in order to improve the formability during subsequent cold forming.

The first forming step is usually carried out at 1500–1700 °C. As a result of the low specific heat and high thermal conductivity of tungsten, several reheating stages are necessary in the first stages of shaping, because the heat is lost rapidly at these temperatures and the ingot cools down rapidly. If the metal is worked at too low a temperature, cracks and splits will easily develop. As the forming process continues, the forming temperature is reduced progressively since the recrystallization temperature decreases as deformation proceeds. Tungsten is generally worked below its recrystallization temperature, because recrystallization is combined with grain boundary embrittlement. With increasing work hardening during deformation, both the hardness and strength of the worked part increase significantly, and intermediate stress relief annealings are necessary to minimize the hazards of cracking (laminating) and to avoid overstraining the working tool.

With a high degree of deformation, the grains become more fibrous due to the unidirectional nature of the processing (sheet rolling, rod extrusion, or swaging) and mechanical anisotropy develops. Strong crystallographic orientations and planar distributions of dispersed, insoluble phases (in the case of doped tungsten) are characteristic of the stage of heavily worked tungsten. Figure 5.36 shows the wrought structure of a cold-rolled tungsten sheet. Table 5.13 gives an example of a rolling schedule for tungsten sheet.

The grain size and grain structure of the final product have a great influence on the mechanical properties. They can be controlled by the type of deformation, degree of deformation, and annealing processes both during and after machining.

Forming of tungsten is commonly carried out without protecting atmosphere. In air, tungsten is readily oxidized. Tungsten trioxide forms on the surfaces of the worked piece

* In a metallurgical sense, hot-working is considered as working above the recrystallization temperature. This is not the common case for tungsten, where working is usually performed below the recrystallization temperature. Thus, from this point of view, tungsten is commonly cold-worked, although very high temperatures are applied for deformation.

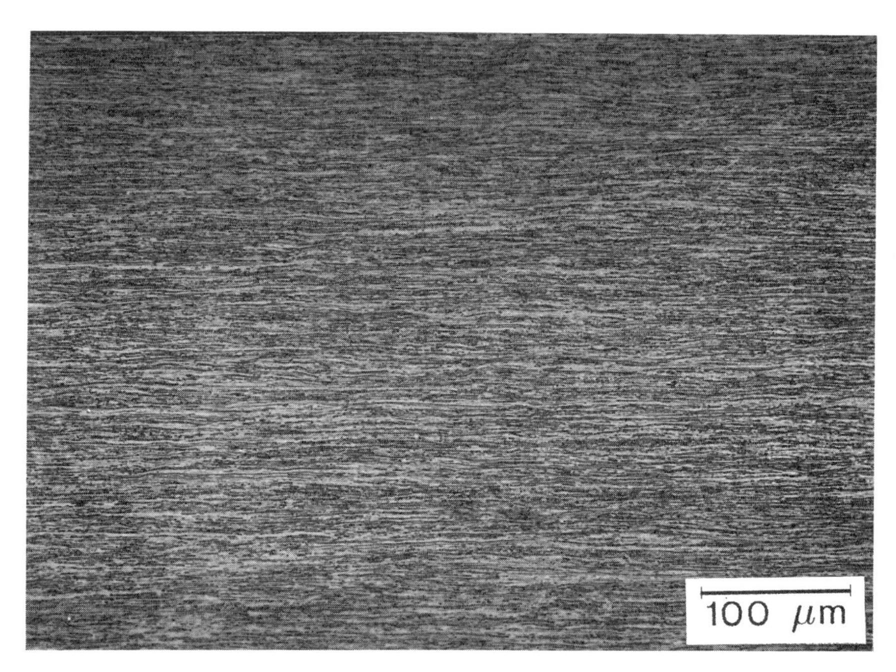

FIGURE 5.36. Typical wrought microstructure of rolled tungsten sheet. By courtesy of Plansee AG, Reutte, Austria.

TABLE 5.13. Example of Rolling Schedules for Tungsten Sheet [5.1][a]

Pass number	Thickness after pass (mm)	Preheating temperature (°C)
	Starting thickness 24.4 mm	
1	20.3	1450
2	15.9	1400
3 (Cross roll)	12.7	1400
4	10.2	1400
	Anneal 5 min at 1300–1350 °C	
5	8.26	1350
6	6.86	1300
7 (Cross roll)	5.46	1300
8	4.45	1300
	Anneal 5 min at 1250–1300 °C; caustic pickling and condition	
9	3.51	1250
10	2.82	1200
11	2.26	1200
12	1.80	1150
13	1.63	1150
Anneal 5 min at 1150 °C; caustic pickling and acid etching		

[a] Source: G. C. Bodine jr. (1963), Tungsten Sheet Rolling Program, Final Report, Fansteel Metallurgical Corp.,North Chicago, Illinois.

and, above 800 °C, it volatilizes. The oxide layer acts as a protective layer against contamination from the working tools and is removed at certain stages of deformation by pickling and/or machining. Intermediate annealing and stress-relieving annealing is performed under hydrogen to avoid enhanced oxidation of the metal and sublimation of WO_3.

5.5.3.2. Shaping [5.88–5.90]. Compared to ductile metals and alloys, the fabricability of tungsten is rather poor. Tungsten should always be heated before shaping. The temperature range for forming has a lower limit, set by the brittle-to-ductile transformation temperature, and an upper limit, set by the recrystallization temperature. This temperature is mainly dependent on the purity, the history of deformation, and heat treatment of the material. As an example, Fig. 5.37 shows the recommended temperature ranges for bending and punching of tungsten sheet as a function of sheet thickness [5.89]. Highly deformed products, such as thin tungsten wires, ribbons, or foils, are ductile at room temperature.

Thin, strongly deformed sheet and foil have a pronounced structure in the longitudinal direction due to elongation of the grains during rolling. The bending properties along the direction of rolling are therefore different from those across it. Therefore, tungsten sheet should always be bent in a way such that the bending edge is perpendicular to the rolling direction. If bending in the longitudinal direction cannot be avoided, owing to

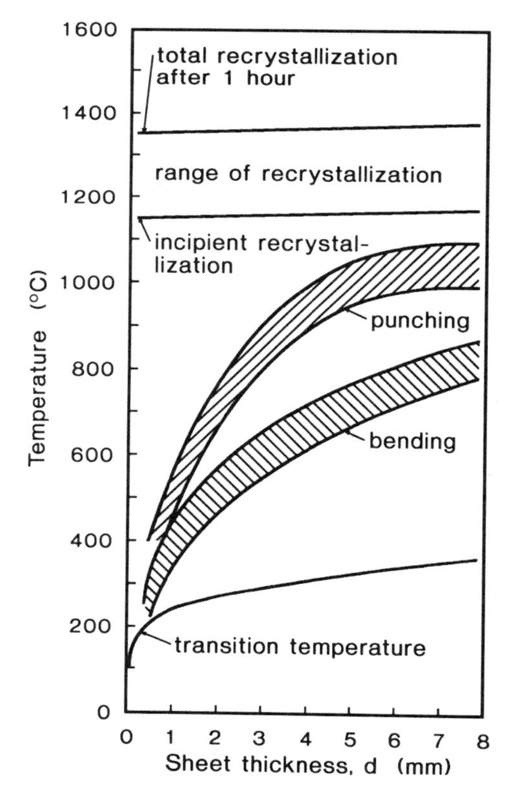

FIGURE 5.37. Transition temperature, recrystallization temperature, and recommended temperature range for bending and stamping of tungsten sheet as a function of thickness [5.89].

the design, much higher bending temperatures are required. At high temperatures, tungsten sheet can be *stamped*, *punched*, and *sheared*. Sharp tools are essential to clean cutting action without sheet cracking or delamination. Tungsten cylinders and cones can be formed by *spinning*, *flow turning*, or *forging*. The use of stress-relieved tungsten is suggested for optimum fabrication results.

Tungsten can be *drilled*, *turned*, *milled*, *planed* and *ground*. However, machining operations require experience and close adherence to optimum conditions. Preheating to about 200 °C and the use of cemented carbide tools are recommended. Complex shapes and holes can be formed by spark erosion; the tungsten workpiece forms the anode and the working electrode the cathode.

Tungsten (pure W, doped W, W–Re) is commercially available as wrought products (sintered bars, billets, forgings, rods, pins, disks, cylinders, tubes, sheet, strips, ribbons, wires) as well as in the form of a broad variety of shaped parts (crucibles, springs, heating elements, rivets, etc.).

Cold-rolled sheet is supplied in thicknesses from 0.025 to 0.4 mm (up to 500 mm × 1000 mm). Hot-rolled tungsten is available with thicknesses from 0.4 to 8.0 mm (maximum lengths, 1500 mm). Tungsten wires are manufactured from about 1 mm down to 5 μm.

5.5.3.3. Bonding of Tungsten [5.89].

Mechanical Joining. Mechanical joining, such as riveting, tacking, or lacing, comprises the simplest methods of joining tungsten, provided the joint need not be impermeable to liquids and/or gases. Tungsten and molybdenum parts can be used. Mechanical fastening is used for constructional parts, such as heating elements, containers, large shields, etc.

Brazing. The parts to be joined must be free from grease, oils, oxides, or other impurities. As tungsten is very sensitive to oxidation, brazing must be carried out under a protective gas, hydrogen, or in vacuum. If maximum strength of the joint is required, tungsten has to be brazed below its recrystallization temperature. Typical brazing metals and temperatures are: Rh (1970 °C), Ni (1430 °C), or AgCu20Pd15 (700–900 °C). More detailed information is available [5.78, 5.89]. Brazing of tungsten to ceramics, graphite, and silicon has gained importance for the fabrication of refractory-metal composites.

Welding. Tungsten possesses only moderate welding properties. Welding must be carried out under controlled weld atmospheres, preferably in a dry box, since any contamination by oxygen will reduce the ductility of the joint. Before welding, the metal has to be degreased and pickled, commonly with a mixture of nitric acid and hydrofluoric acid (90/10 vol%). The weld seams have a coarse-grained structure in the hot-fusion zone, owing to recrystallization, and can therefore withstand only low mechanical stresses. Tungsten can be welded by *tungsten inert gas* (TIG) *welding*, *laser beam welding* (for thin parts), *plasma welding*, *friction welding* and *electron beam welding*, the latter being the preferred fusion-welding method. Tungsten can be welded to W–Re and Mo–Re alloys. W–Re alloys (in particular W–26Re) are recommended as filler metal for welding tungsten. Tungsten can be *diffusion welded* to tungsten and other metals. Despite the high melting point of tungsten, temperatures of 1300–2000 °C and pressures of 2–20 $N \cdot mm^{-2}$ give satisfactory joints. Intermediate layers of Ni, Pt, Rh, Ru, and Pd accelerate the diffusion process. Tungsten wires can be *spot-welded* to other metals under a protective gas blanket.

5.6. ALTERNATIVE PROCESSES [5.1, 5.2]

The following methods have been applied to tungsten production: arc melting, electron beam melting, hot pressing, isostatic hot pressing, explosion compacting, slip casting, plasma spraying, chemical vapor deposition, powder rolling, and injection molding. Below, we restrict ourselves to the methods of technical importance.

5.6.1. Electron-Beam Zone Melting

Electron-beam zone melting is the preferred method to produce larger, single crystals of tungsten. A polycrystalline tungsten rod is mounted in a vertical position in a stainless steel bell jar and connected as anode. An electron gun (cathode) traverses along the rod by means of a screw drive mechanism. On melting in a high vacuum, a narrow, liquid zone is formed and held between the two solid rod parts by surface tension forces. As the zone slowly moves through the rod several times (zone floating), impurities are concentrated on both the starting and terminal end of the rod. Since the molten zone is quite narrow and the temperature gradient in its vicinity rather steep, nucleation starts essentially at one point and a single crystal is the result. The concentration of many impurity elements can be decreased considerably by zone melting. Al, Ca, Co, Cu, Fe, Mg, Mn, Ni, Si, Na, Ti, Ta, Re, Mo, and Nb concentrations can be reduced below 1 μg/g. Interstitials remain at the following μg/g levels: O_2 2, N_2 5, C 10, and H 1.

Single crystals produced by this technique represent the most ductile tungsten currently available. They are free of grain boundaries, which cause brittleness.

5.6.2. Plasma Spraying

Tungsten layers can be sprayed with suitable guns on different base materials. Easily flowing tungsten powder is used as spraying material. Ar or Ar–N_2 mixtures are preferred as protective atmosphere in closed chambers. Besides coatings, tungsten parts can be produced by spraying tungsten layers on an adequately formed nucleus until the desired thickness is reached.

One may distinguish between three basic types of substrates:

- *expendable* substrates are removed after spraying by selective chemical etching, melting, etc;
- *reusable* substrates are used for mass production;
- *permanent* substrates become an integral part of the article.

Density ranges from 50–95% of the theoretical value. Higher density is attained by uniform substrate temperature and low deposition rate. Density and strength might be increased by post-sintering in hydrogen. Low-pressure plasma spraying is mainly used for producing intricate shapes, which otherwise cannot be manufactured.

5.6.3. Chemical Vapor Deposition

For the deposition of tungsten layers from the vapor phase, an easily evaporable tungsten compound is mixed with a carrier gas and/or a reducing gas and decomposed or reduced at a higher temperature on a suitable substrate. Possible tungsten compounds are halides (preferably WF_6 and WCl_6), tungsten hexacarbonyl [$W(CO)_6$], and organometallics. Until now, only the halides are technically important. Extensive investigations in the direction of organometallics have been undertaken in order to overcome the disadvantages of the halides: the high-temperature necessary for reduction and the difficulties caused by corrosion. For more details see Sections 3.5.2 and 3.5.3.

Sheets, produced via CVD, exhibit a quite different microstructure as compared to wrought P/M products in terms of grain size and grain orientation. Both parameters can be influenced by the deposition conditions. Very fine grained CVD tungsten can be obtained by changing the growth conditions during deposition, for example, by brushing, which increases the nucleation rate and disrupts the columnar grain growth. The method also offers the possibility to deposit layers of desired crystal orientation or even single crystal layers.

CVD of tungsten and tungsten–rhenium (co-deposition of WF_6 and ReF_6) is of commercial importance for the production of X-ray targets. W–Re layers of up to 1 mm are deposited onto graphite disks with diameters between 100 and 150 mm [5.78]. Furthermore, CVD can be used to produce cylindrical or conical shells, cones (such as shaped charge liners), crucibles, and thin-walled tubes.

Low-pressure CVD of tungsten is tested for via plugs, low-resistance contact barriers, and interconnects for high density silicon integrated circuits. Depending on the substrate and the deposition conditions, either α- or β-tungsten, or a mixture of both, forms. This is of interest, because the resistivity of β-W is about 10 times higher than that of α-W.

5.6.4. Physical Vapor Deposition (Sputtering)

Difficulties combined with the elevated temperature in case of CVD can be eliminated by PVD (sputtering). The substrate in this process remains fairly cold.

An ion beam (preferably noble gas) is generated in a high vacuum chamber by discharge. It is focused on the water-cooled tungsten sputter target where, by its impact energy, it frees tungsten particles which are deposited on the substrate.

Improved design of sputtering equipment somewhat allows the substitution of CVD tungsten. Sputter targets used for thin layers in microelectronics manufacture are made of high or ultrahigh purity W, W–10%Ti, and WSi_x (see Section 5.7.7).

5.7. SPECIAL TUNGSTEN FORMS AND QUALITIES [5.1, 5.2, 5.90, 5.91–5.95]

5.7.1. Globular Tungsten Powder

This substance can be produced from fine powder in a plasma jet (150–200 μm). Monodisperse single crystal powder (containing small amounts of Ni) can be obtained by acid leaching of sintered W–Ni–Cu parts.

5.7.2. Coarse Tungsten Powder

Good flowing tungsten powder can be obtained by:

- reduction of WO_3 or blue oxide at high temperature, high humidity, and alkali metal doping, and subsequent classification;
- disintegration and classification of presintered parts;
- fine tungsten powder +0.2% Ni is compacted and sintered at 1840 °C for 3 hours. Afterward, the Ni is dissolved by nitric acid.

5.7.3. Single Crystals

These crystals can be produced by:

- Annealing of deformed wires or rods by secondary recrystallization and extreme crystal growth; thoriated tungsten wires (cross-section, e.g., 100 μm) are pulled through a narrow zone at a temperature of above 2000 °C at the same time the crystal grows, forming a single crystal up to several meters long (Pintsch procedure [5.96].
- Zone melting.

Single crystals, due to their high cost are used only in special cases like basic investigations. The outstanding properties are a lack of grain boundaries and a very small number of lattice defects. They are therefore easy to deform, show great ductility, and possess no embrittlement up to the highest temperature.

5.7.4. Tungsten Thin Films

The preparation of tungsten thin films is possible by different techniques, such as electrolysis from salt melts, CVD, PVD and plasma spraying, diffusion, plating, electrophoresis, and metallizing. The technically important techniques are only CVD, PVD, and plasma spraying.

5.7.5. Porous Tungsten

The production of structural parts from porous tungsten affords special technological procedures. Starting materials are mainly coarse or globular powders, which are compressed by special methods, sometimes with the addition of pore-forming materials which evaporate during sintering. Also in sintering, special procedures such as infiltration, evaporation, surface treatment, etc. are necessary. Powder grain size, compacting pressure, sintering temperature, and time determine pore size and permeability. At densities between 50 and 78% all pores are open; at >90% density, all are closed.

Parts with an open coherent porosity of >20% are used in vacuum tubes and aeronautics because of refractoriness, shape stability, and chemical resistance. Examples are porous cathodes impregnated with alkaline earth oxides as electron-emitting sources in special tubes, thermoionic converters heated by nuclear energy, and ionic propulsion units with porous tungsten plates as ionic sources for Cs vapor of high temperature. Those emitter plates are made of globular tungsten powder (7 μm) density 75–85%, pore size 2–30 μm, and pore number 1.4×10^6–$8 \times 10^6 \cdot cm^{-3}$.

Porous tungsten can be infiltrated either to machine it to rigid dimensional tolerances, or to fabricate composite materials. Copper, which is used as infiltrant for machining, can be completely removed afterward by heating in vacuum by evaporation.

5.7.6. High-Purity Tungsten

High-purity or ultrahigh-purity tungsten (5N or 6N) is used in the form of thin films in the electronic industry as the preferred material for gate electrodes, interconnects, and barrier metal because of its high resistance to electron migration, high temperature stability, and tendency to form stable silicides. In principle, tungsten thin films can be produced either by CVD or by PVD. In regard to CVD tungsten thin films, we also refer the reader to Sections 3.5.2 and 3.5.3 as well as to Section 5.6.4. For PVD-produced tungsten thin films, sputter targets of high purity (6N) are commercially available today.

It can be seen in Fig. 5.38 that the concentration of trace impurities in sputter targets has shown a downward trend since 1988. This is partly due to improvements in production methods, which in fact caused higher purity. But it is also somewhat a consequence of increased sensitivity of the analytical methods and the equipment used nowadays. For example, some years ago it was only possible to say that the concentration is less than 1 μg/g or less than 0.1 μg/g. Meanwhile, application of more advanced techniques has changed the situation drastically. In high-purity tungsten, metal levels of trace elements are in the ng/g level or even below, which affords special analytical requirements such as working in clean room laboratories, prevention of any other contamination, and correct calibration. The most important analytical methods applied today for that purpose are glow discharge mass spectrometry, secondary ion mass spectrometry, isotope dilution, and trace matrix separation combined with inductively coupled plasma atomic emission spectrometry.

As an example, it was possible by isotope dilution analysis to determine 15 ppq of uranium in an ultrapure tungsten sample. This corresponds to 15×10^{-15}%. These

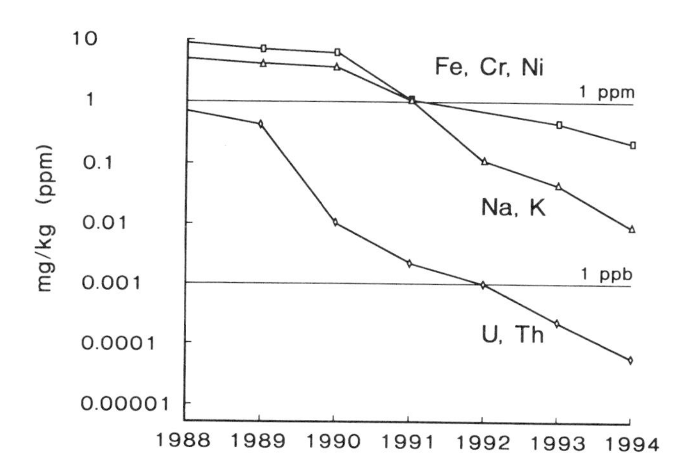

FIGURE 5.38. Ultrahigh-purity tungsten: improvement in purity from 1988 to 1994 (target data); production takes place under clean room conditions [5.96].

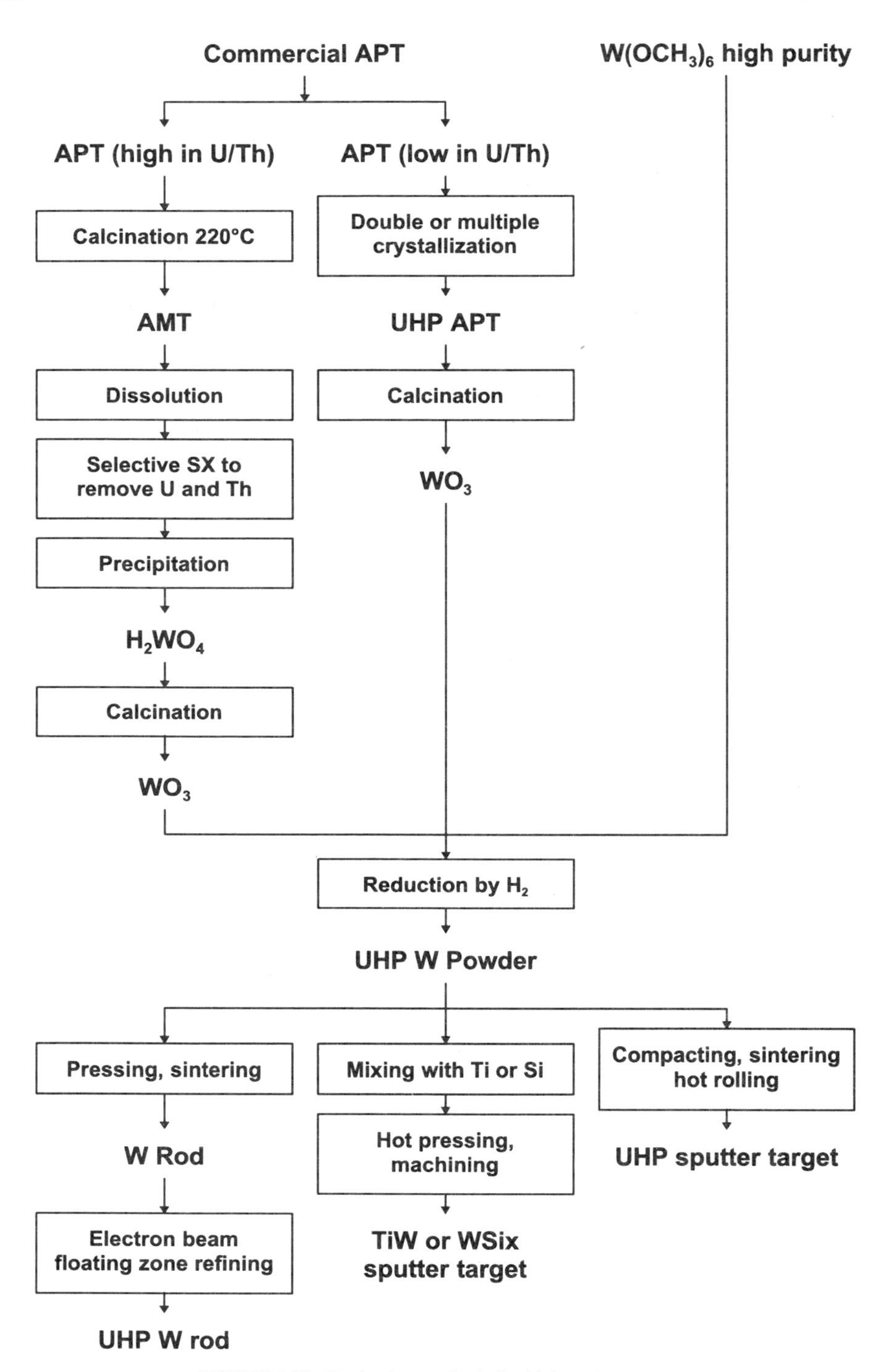

FIGURE 5.39. Production methods for high-purity tungsten.

methods are generally very complex and expensive. Comparative analyses often show significant scatter.

Of all impurity elements, in particular uranium and thorium contents must be extremely low. These naturally radioactive elements cause "soft errors" in memory circuits due to α-particle emission.

The diverse production methods for high-purity tungsten are summarized in Fig. 5.39. The raw material is always APT with different purity depending on the uranium and thorium content. It depends on the ore deposit whether these two elements are present or not. If the concentrations are unacceptably high after dissolution, a selective solvent extraction process removes the main portion of these two elements.

The APT is calcined to AMT at 220 °C and dissolved in water while adding hydrogen peroxide and sulfuric acid. The traces of uranium and thorium are extracted selectively by 0.5 M di-2-ethylhexyl phosphoric acid and 0.125 M tri-n-octyl-phosphine oxide dissolved in kerosene.

APT, having acceptably low uranium/thorium concentrations, can be further purified by double or multiple crystallization to remove other impurities.

The WO_3 gained after calcination in both cases is reduced by hydrogen to tungsten powder of high purity. Typical analyses of W and WSi_x powder are shown in Table 5.14. This can be used for the production of tungsten, tungsten silicide, or TiW sputtering targets.

If further purification is desired, rods are produced by compacting and sintering and are subjected to electron beam floating zone melting. This procedure does not further reduce the concentrations of uranium and thorium, but does reduce the concentrations of many other elements as shown in Table 5.15. Moreover, residual impurity elements are distributed homogeneously.

Powders having sufficient purity can be processed to sputter targets directly. Tungsten sputter targets have usually been manufactured by hot pressing. However, it was not possible to maintain the high-purity level of the powder. A much better way to achieve

TABLE 5.14. Purity of Ultrahigh-Purity Grade Tungsten and Tungsten Slicide Powder [5.78]

Trace impurity	Tungsten powder (<μg/g)	Tungsten silicide powder (<μg/g)
Al	0.20	0.50
B	0.05	0.50
Na	0.05	0.10
K	0.05	0.50
Ca	0.10	0.50
Mg	0.10	0.10
Fe	0.25	1.0
Co	0.02	0.05
Cr	0.25	1.0
Ni	0.15	0.30
Cu	0.02	0.20
U	0.0003	0.0003
Th	0.0003	0.0003

TABLE 5.15. Impurity change from APT via Tungsten Powder and a Tungsten Rod after Electron Beam Zone Refining [5.77]

Element	APT	W powder concentrations ($\mu g \cdot g^{-1}$)	Refined rod
Fe	6.0	8.0	2.9
Ni	1.3	2.0	<1.0
Al	3.7	9.0	<1.0
Na	7.7	7.0	0.013
Mo	15.8	9.0	<1.0
K	3.5	3.2	0.014
Cr	2.8	3.9	0.34
Si	7.1	11.0	<1.0
Mn	3.2	4.1	<1.0
U	0.062	0.0011	0.0011
Th	0.140	0.01	0.01

high purity in the sputter target is to employ a special sintering process which reduces the contents of gaseous impurities and provides a homogeneous distribution of all remaining impurities.

In general, it is important to maintain clean room conditions throughout the production steps.

REFERENCES FOR CHAPTER 5

5.1. St. W. H. Yih and C. T. Wang, *Tungsten*, Plenum Press, New York (1979).

5.2. *Gmelin Handbuch der anorganischen Chemie*, 8th ed., Syst. No. 54, *Tungsten*, Suppl. Vol. A1 (1979).

5.3. R. G. Woolery, in: *Proceedings of the 2nd Tungsten Symposium, San Francisco*, pp. 52–63, Mining Journals Books Ltd., London (1982).

5.4. Wu Weisun, in: *Proc. 2nd Tungsten Symposium, San Francisco*, pp. 64–70, Mining Journals Books Ltd., London (1982).

5.5. G. J. Willey, in: *Proc. 2nd Tungsten Symposium, San Francisco*, pp. 21–28, Mining Journals Books Ltd., London, (1982).

5.6. E. Lassner, "From Tungsten Concentrates and Scrap to Highly Pure APT," in: *The Chemistry of Non Sag Tungsten* (L. Bartha, E. Lassner, W. D. Schubert, and B. (Lux, eds.), pp. 35–44, Pergamon Press, Oxford (1995).

5.7. E. Lassner and B. Kieffer, *Conference Proceedings on Advances in Hard Materials Production, London*, MPR Publishing Services Ltd., Shrewsbury (1986).

5.8. R. F. Hogsett, D. K. Huggins, and L. B. Beckstead, US Patent 4.338,287 (1982).

5.9. E. Lassner, in: *Extractive Metallurgy of Refractory Metals* (H. Y. Sohn, O. N. Carlson, and J. T. Smith, Eds.), *Proc. TMS-AIME 110th Annual Meeting*, pp. 269–272, Chicago (1981).

5.10. L. R. Quatrini, M. B. Terlizzi, and B. E. Martin, US Patent 4.353,878 (1982).

5.11. L. R. Quatrini and B. E. Martin, US Patent 4.353,789 (1982).

5.12. L. R. Quatrini, M. C. Vogt, and B. E. Martin, US Patent 4.353,881 (1982).

5.13. L. R. Quatrini, US Patent 4.353,880 (1982).

5.14. J. Zhou and J. Xue, in: *Proc. 5th Int. Tungsten Symp. Budapest*, pp. 73–86, MRP Publishing Services Ltd., Shrewsbury (1990).

5.15. V. Zbranek, Z. Zbranek, and D. A. Burnham, US Patent 4.092,400 (1978).

5.16. N. N. Maslenitskii, *Tsvetn. Met.* **4–5** (1939), 140–143.

5.17. B. Burwell, US Patent 3.256,057 (1966).
5.18. N. N. Maslenitskii and P. M. Perlov, in: *Proc. 5th Int. Mineral Processing Congr., Group VII*, p. 839, Pergamon Press, New York (1960).
5.19. P. Queneau, Thesis, University of Minnesota, MN 687369 (1967).
5.20. A. N. Zelikmann and G. E. Meerson, *Metallurgiya, Redkikh Metallov*, Ch. 1, Metallurgiya, Moscow (1973).
5.21. P. B. Queneau, L. W. Beckstead, and D. K. Huggins, US Patent 4.325,919 (1982).
5.22. P. B. Queneau, L. W. Beckstead, and R. Hogsett, Int. Patent C01G41/00; Int. Publ. No. WO82/02540 (1982).
5.23. P. B. Queneau, D. K. Huggins, and L. W. Beckstead, US Patent 4.320,096 (1982).
5.24. H. Li, M. Liu, and Y. Li, in: *Proc. of the 1st Int. Conf. on the Metallurgy and Materials Science of W, Ti, Re and Sb, Changsha,* Vol. 1, pp. 192–197, International Academic Publishers, Pergamon Press, Oxford (1988).
5.25. H. Li, M. Liu, P. Sun, and Y. Li, *J. Cent. South Univ. Technol.* **2** (1995), 16–20.
5.26. H. Li, M. Liu, and Z. Si, Chinese Patent 85100350.8, 1985-04-01.
5.27. C. J. Smithells, *Tungsten*, Chemical Publishing Co., New York (1953).
5.28. Société Electrométallurgique de Saint Etienne, U.K. Patent 695.843 (1953).
5.29. M. Karczynski, Polish Patent 48.426 (1964).
5.30. K. Vadasdi, "Effluent-Free Manufacture of Ammonium Paratungstate by Recycling the Byproducts," in: *The Chemistry of Non-Sag Tungsten* (L. Bartha, E. Lassner, W. D. Schubert, and B. Lux, eds.), pp. 45–60, Elsevier Science Ltd., Oxford (1995).
5.31. E. Lassner, *Oesterr. Chem. Z.* **80** (1979), 111–115.
5.32. P. B. Queneau, L. W. Beckstead, and D. K. Huggins, US Patent 4.311,679 (1982).
5.33. N. N. Maslenitskii and P. M. Perlov, in: *Proc. 5th Int. Mineral Processing Congr., Group VII*, p. 839, Pergamon Press, New York (1960).
5.34. Gong Bofan, Huang Weizhuang, and Zhang Qixiu, *Int. J. Refract. Met. Hard Mater.* **14** (1996), 319–324.
5.35. A. I. Bellingham, US Patent 3.939,245 (1976).
5.36. C. R. Kurtak, US Patent 3.158,438 (1964).
5.37. T. K. Kim, R. W. Mooney, and V. Chiola, *Sep. Sci.* **3** (1968), 467–478.
5.38. T. K. Kim, J. E. Ritsko, M. B. Macinnes, and M. C. Vogt, US Patent 4.379,126 (1983).
5.39. M. B. Macinnes, R. P. Macclintic, and T. K. Kim, US Patent 4.360,502 (1982).
5.40. C. Burwell, US Patent 3.256,057 (1966).
5.41. T. K. Kim, US Patent 4.175,109 (1970).
5.42. L. W. Beckstead, H. Dale, and D. K. Huggins, US Patent 4.328,190 (1982).
5.43. T. K. Kim, M. B. Macinnes, R. P. Macclintic, and M. C. Vogt, US Patent 4.374.099 (1983).
5.44. M. B. Maccines and T. K. Kim, *Int. J. Refract. Met. Hard Mater.* **5** (1986), 78–81.
5.45. T. Zuo, Q. Zhao, S. Li, and J. Wang, in: *Proc. of the 1st Int. Conf. on the Metallurgy and Materials Science of W. Ti, RE and Sb, Changsha* Vol. 1, pp. 11–20, International Academic Publishers, Pergamon Press, (1988).
5.46. J. Zhou and J. Xue, in: *Proc. 5th Int. Tungsten Symp. Budapest*, pp. 73–86 MPR Publishing Services Ltd., Shrewsbury (1990).
5.47. Huang Weizhung, Zhang Qixu, Gong Bofang, Huang Zhaoying, and Luo Aiping, *Int. J. Refract. Met. Hard Mater.* **13** (1995) 217–220.
5.48. Hu Zhaorui and Yun Lui, in: *Application of Ion Exchange in Tungsten Hydrometallurgy in Ion Exchange for Industry* (M. Streat, ed.), p. 385, Ellis Horwood, Chichester (1988).
5.49. Zhao Heng, in: *Proc. of the 7th Int. Tungsten Symposium, Goslar*, pp. 142–149, ITIA, London (1996).
5.50. Chen Zhouxi *et al.*, Chinese Patent 88105712.6.
5.51. J. W. van Put, T. W. Zegers, A. van Sandvijk, and P. J. M. van der Straten, in: *Proc. 2nd Int. Conf. on Separation Science and Technology* (E. B. Baird and C. Vijayan, eds.), pp. 387–394, C.S.Ch.E.M.H.I., Hamilton, Canada (1989).
5.52. J. W. van Put, P. M. de Konig, A. van Sandvijk, and G. J. Witkamp, in: *Proc. 11th Symp. on Industrial Crystallization* (A. Mersman, ed.), pp. 647–652, Garmisch-Partenkirchen (1990).
5.53. J. W. van Put, W. P. C. Duyvesteyn, and J. van Vliet-Jahnberg, in: *Proc. 12th Plansee Seminar* (H. Bildstein and H. M. Ortner, eds.), pp. 433–435, Metallwerk Plansee, Reutte, Austria, (1989).
5.54. A. W. Bryson, D. Glassner, and W. F. Lutz, *Hydrometallurgy* **2** (1976), 1985.

5.55. J. W. van Put, Thesis, Delft University Press (1991).
5.56. J. R. Scheithauer, C. D. Vanderpool, M. B. Macinnes, and J. H. Miller, US Patent 4.624,844 (1986); Europ. Patent 0194.346A1.
5.57. J. B. Goddard, US Patent 4.326,061 (1982).
5.58. J. B. Goddard, *SME-AIME Annual Meeting*, Los Angeles, Reprint
5.59. H. J. Lunk, B. Ziemer, M. Salmen, and D. Heidemann, in: *Proc. 13th Int. Plansee Seminar* (H. Bildstein and R. Eck, eds.), Vol. 1, pp. 38–56, Metallwerk Plansee, Reutte, Austria (1993).
5.60. AMAX Metals Group, *Tungsten Chemicals*, W-17 (1984).
5.61. E. Lassner and W. D. Schubert, "Tungsten Blue Oxide," in: *The Chemistry of Non-Sag Tungsten* (L. Bartha, E. Lassner, W. D. Schubert, and B. Lux, eds.), pp. 111–118, Elsevier Science Ltd., Oxford (1995).
5.62. H. J. Lunk, B. Ziemer, B. Salmen, and D. Heidemann, *Int. J. Refract. Met. Hard Mater.* **12** (1993/94), 17–26.
5.63. W. D. Schubert and E. Lassner, *Int. J. Refract. Met. Hard Mater.* **10** (1991), 171–183.
5.64. J. W. van Put and T. W. Zegers, *Int. J. Refract. Met. Hard Mater.* **10** (1991), 115–122.
5.65. Z. Zou, E. T. Wu, and C. Qian, in: *Proc. 11th Plansee Seminar* (H. Bilstein and H. M. Ortner, eds.), pp. 337–348, Metallwerk Plansee, Reutte, Austria (1985).
5.66. V. Chiola, J. M. Laferty, and C. D. Vanderpool, US Patent 3.175,881 (1965).
5.67. V. Chiola and P. R. Dodds, US Patent 3.591,331 (1971).
5.68. J. O. Hay and R. J. Grodek, British Patent 1.267,585 (1972).
5.69. T. K. Kim, J. M. Laferty, M. B. McInnis, and J. C. Patton, US Patent 3.857,928 (1974).
5.70. C. D. Vanderpool, M. B. Macinnes, and J. C. Patton, US Patent 3.936,362.
5.71. D. E. Collier, C. J. Couch, and D. N. Hingle, *Hydrometallurgy*, pp. 1–9, 81/G5.
5.72. G. Schwier, in: *Proc. 10th Plansee Seminar* (H. M. Ortner, ed.), Vol. 2, pp. 369–383, Metallwerk Plansee, Reutte, Austria (1981).
5.73. H. Palmour III, PM 94, *Refractory Metals*, pp. 2025–2028.
5.74. V. Reich, PM 94, *Refractory Metals*, pp. 111–113.
5.75. L. Bartha, E. Lassner, W. D. Schubert, and B. Lux, eds., *The Chemistry of Non-Sag Tungsten*, Elsevier Science Ltd., Oxford (1995).
5.76. E. Pink and L. Bartha, eds. *The Metallurgy of Doped/Non-Sag Tungsten*, Elsevier, New York (1989).
5.77. W. D. Coolidge, *J. Am. Inst. Elec. Eng.* **29** (1909), 953.
5.78. E. Pink and R. Eck, "Refractory Metals and their Alloys," in: *Materials Science and Technology* (R. W. Cahn, P. Hassen, and E. J. Kramer, eds.), Vol. 8, pp. 589–641, VCH, Weinheim (1991).
5.79. F. V. Lenel, *Powder Metallurgy*, Metal Powder Industries Federation, Princeton (1980).
5.80. Metallwerk Plansee GmbH, *Refractory Metals*, Verlag Moderne Industrie AG & Co., Landsberg/Lech, Germany.
5.81. J. A. Mullendore, in: *The Metallurgy of Doped/Non-Sag Tungsten* (E. Pink and L. Bartha, eds.), pp. 61–81, Elsevier, New York (1989).
5.82. S. Yamazaki, in: *The Metallurgy of Doped/Non-Sag Tungsten* (E. Pink and L. Bartha, eds.), pp. 47–59, Elsevier, New York, (1989).
5.83. B. P. Bewlay and C. L. Briant, in: *The Chemistry of Non-Sag Tungsten* (L. Bartha, E. Lassner, W. D. Schubert, and B. Lux, eds.), p. 137–159, Elsevier Science Ltd., Oxford (1995).
5.84. R. F. Cheney, "Sintering of Refractory Metals," in: *Metals Handbook*, 9th ed., Vol. 7, pp. 389–393, ASM International, Metals Park, Ohio (1984).
5.85. F. B. Swinkels and M. F. Ashby, *Acta Metall.* **29** (1981), 259–281.
5.86. O. Horacsek, in: *The Metallurgy of Doped/Non-Sag Tungsten* (E. Pink and L. Bartha, eds.), pp. 175–187, Elsevier, New York (1989).
5.87. D. J. Jones, *Journal of Less-Common Metals* **2** (1960), 76–85.
5.88. W. Rostoker, "Conversion of Refractory Metals," in: *Refractory Metals and Alloys II* (M. Semchyshen and I. Perlmutter, eds.) pp. 379–394, Interscience Publishers, Wiley, New York (1963).
5.89. *Wolfram*, Company Brochure, Metallwerk Plansee, Reutte, Austria (1980).
5.90. J. P. Wittenauer, T. G. Nieh, and J. Wadsworth, *Adv. Mater. Process.* **9** (1992), 29–37.
5.91. H. S. Yui, J. S. Kim, K. I. Rhee, J. C. Lee, and H. Y. Sohn, *Int. J. Refract. Met. Hard Mater.* **11** (1992), 317–332.
5.92. R. Huenert, G. Winter, W. Kiliani, and D. Greifendorf, *Int. J. Refract. Met. Hard Mater.* **11** (1992), 331–335.

5.93. V. Glebovsky, in: *Proc. 12th Plansee Seminar* (H. Bildstein and H. M. Ortner, eds.), Vol. 3, pp. 379–389, Metallwerk Plansee, Reutte, Austria (1989).
5.94. Yoshiki Doi, *Characteristics and Application of High Purity Tungsten and Chemical Vapor Deposited Tungstein*, paper presented at the Annual ITIA Meeting, Huntsville, USA; transcript of papers: ITIA, London (1994).
5.95. G. Winter, *Tungsten and Special Tungsten Compounds—Their Application in Modern Technology*, paper presented at the Annual ITIA Meeting, Huntsville, USA; transcript of papers: ITIA, London (1994).
5.96. G. D. Rieck, *Tungsten and its Compounds*, Pergamon Press, Oxford (1967).

6

Tungsten Alloys

A large number of tungsten alloys and composites were investigated in the past, but only some of them achieved technical importance. However, these are of significant commercial interest and will be discussed in the following sections. Alloys derived from melting metallurgy, such as ferrotungsten, melting base, steels, superalloys, and stellites, as well as the large group of cemented carbides, are not discussed here but will be treated in Chapters 8 and 9.

The aim of alloying tungsten is to improve its chemical, physical, and mechanical properties at both ambient conditions and elevated temperatures. Beyond that, it is possible to combine the useful properties of tungsten with those of the alloying additives.

Low-temperature brittleness is the most crucial aspect in the manufacture of pure tungsten metal. Therefore, in the past, much effort has been directed at lowering the ductile-to-brittle transition temperature (DBTT) and hence improving the fabricability of the metal. In this regard, tungsten–rhenium alloys have gained outstanding importance. They exhibit a significantly lower DBTT and are even stronger than unalloyed tungsten at high temperatures.

To fully realize the potential of tungsten in high-temperature applications, dispersion strengthening and precipitation hardening turned out to be the most effective way to increase the high-temperature strength and creep resistance. Non-sag tungsten alloys or thoriated tungsten, used for lamp filaments, are examples of this group of materials. Moreover, alloys based on W–Re–ThO_2 and W–Re–HfC are among the strongest alloys used today for construction parts in high-vacuum and high-temperature technology.

These tungsten alloys are commonly produced by powder metallurgical techniques. Solid-state sintering, liquid-phase sintering, and infiltration techniques are employed. Fine particle size and high purity of the powder components are a prerequisite in order to achieve an even distribution during sintering and a high degree of material purity. Powder metallurgy offers a simple route to the desired ratio of components, simply by mixing them properly. Concentration(s) of the alloying element(s) in tungsten base alloys may vary from traces in the case of microalloys (non-sag tungsten contains less than 100 $\mu g/g$ K) to more than 20 wt% (W–Re). Depending on the additives, single-phase or two-phase alloys are produced.

Applications of the alloys described here are presented in Chapter 7. More information on the metal-physical principles can be found elsewhere [6.1–6.4].

6.1. SUBSTITUTIONAL ALLOYS (SOLID-SOLUTION ALLOYS)

In the case of solid-solution alloys, the size and valence of the solute atoms determine the degree of solid-solution strengthening. Differences in atomic size of about 5% between the solvent and the solute are necessary to produce a significant strain-hardening effect [6.1].

Molybdenum, niobium, and tantalum (bcc crystal structure) form a continuous series of solid solutions with tungsten, but only Nb and Ta additions lead to a strong straining effect. Higher additions of Nb and Ta raise the recrystallization temperature but also increase the DBTT and thus decrease the workability. Therefore, none of these binary or ternary tungsten base alloys developed in the past [6.2] has attained commercial importance. The only important tungsten-base solid-solution alloy today is tungsten–rhenium.

Tungsten additions to Mo, Nb, and Ta, on the other hand, improve the high-temperature strength and creep properties of these metals. Niobium-based alloys, such as, Nb–10W–2.5Zr, Nb–28Ta–10W–1Zr, or Nb–10W–10Hf–0.1Y, are potential candidates for nuclear engineering and space applications, due to their high strength and good strength-to-specific weight ratio [6.5]. Tantalum-based alloys are used owing to their high corrosion resistance (see below). Molybdenum–tungsten alloys (Mo–10W and Mo–30W), which were used earlier in the zinc-producing industry because of their long-term corrosion resistance to molten zinc, lost their importance to Sialon (SiN/Al_2O_3). Mo–5W, formerly applied in rotating X-ray anodes as support for the W–Re layer, was substituted by TZM (Mo–0.5Ti–0.08Zr–0.05C). Recently, molybdenum–tungsten alloys have been used as sputter targets for the manufacture of flat panel displays (thin-film transistor technique).

6.1.1. Tungsten-Rhenium Alloys [6.1–6.4]

Rhenium is the most important alloying element for tungsten. Rhenium additions increase the ductility of wrought products at low temperatures and also improve their high-temperature strength and plasticity. Furthermore, Re additions stabilize the grain structure, increase the recrystallization temperature, reduce the degree of recrystallization embrittlement, and significantly improve the weldability. Finally, W–Re alloys also exhibit superior corrosion behavior than unalloyed tungsten. The only drawback of rhenium is its high price (1000 US$/kg as compared to 13.5 US$/kg for tungsten).

The improved fabricability and low-temperature ductility of W–Re alloys (the so-called rhenium ductilizing effect) is attributed to the increased solubility of interstitials (N, C, O, S, P) in the substitutional alloys. Interstitials are scavenged from low-energy positions (dislocations, grain boundaries) and distributed in the form of clusters within the lattice, so increasing the grain-boundary rupture strength. Rhenium also promotes excessive twinning, which might act as a supplementary deformation mechanism to slipping for stress relaxation, particularly at lower deformation temperatures [6.3].

The rhenium ductilizing effect is more pronounced at higher Re additions (Fig. 6.1) and reaches a maximum at the solid solubility limit (approximately 27 wt% Re). Beyond that, embrittlement occurs due to the formation of a hard and brittle σ-phase. The hardness of W–26 Re alloy is 20% larger than that of unalloyed tungsten and the hot tensile strength

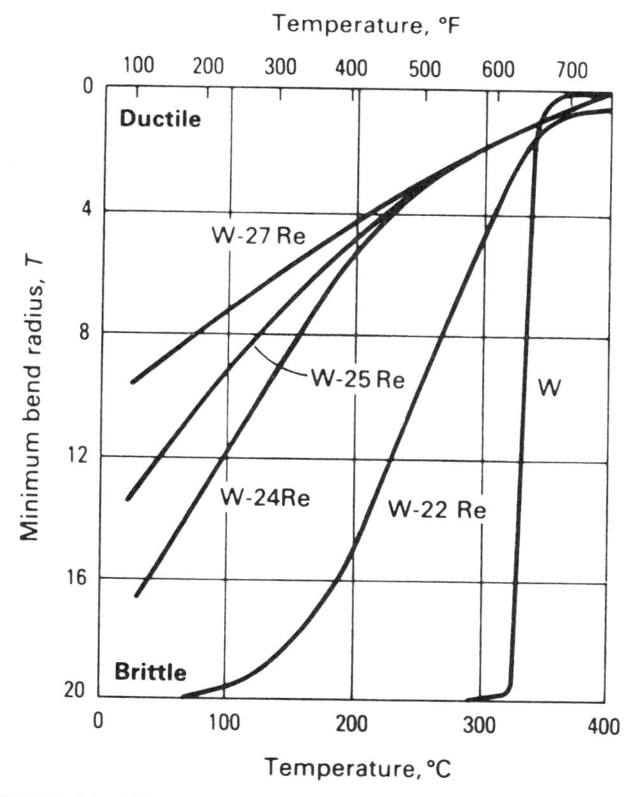

FIGURE 6.1. Effect of alloying additions of rhenium on ductility [6.3, 6.5].

is more than twice as high (Fig. 6.2). Recrystallized W–26Re exhibits a fine and even microstructure and behaves in a ductile manner at room temperature.

Compositions. The most important alloy compositions are W–(3–5)Re, W–10Re, and W–(25–26)Re.

Alloys with 5% Re exhibit the hardness minimum and highest creep strength; alloys with >8% Re show good workability (minimum forging temperature 1560 °C, annealing temp. 1600–1800 °C) and have significantly better welding properties. The 3–10% Re is extremely resistant against alternating thermal stresses which occur in X-ray anodes with high-energy electron beams.

Thermocouples based on W–Re alloys can be used to measure temperatures up to 2300 °C and, for short times, up to 2700 °C (W–3Re/W–25Re; W–5Re/W–26Re).

Production. The classical method is to mix Re and W powders in the desired ratio prior to compaction and sintering. In order to achieve a more even W–Re distribution, restricting the danger of local σ-phase formation, mixtures of tungstic acid or tungsten trioxide or tungsten blue oxide with ammonium perrhenate can be used as raw materials. These mixtures are co-reduced by hydrogen to metal powder.

Tungsten–rhenium alloys are hot-swaged and hot-rolled in air and are well suited for shaping, including tube fabrication.

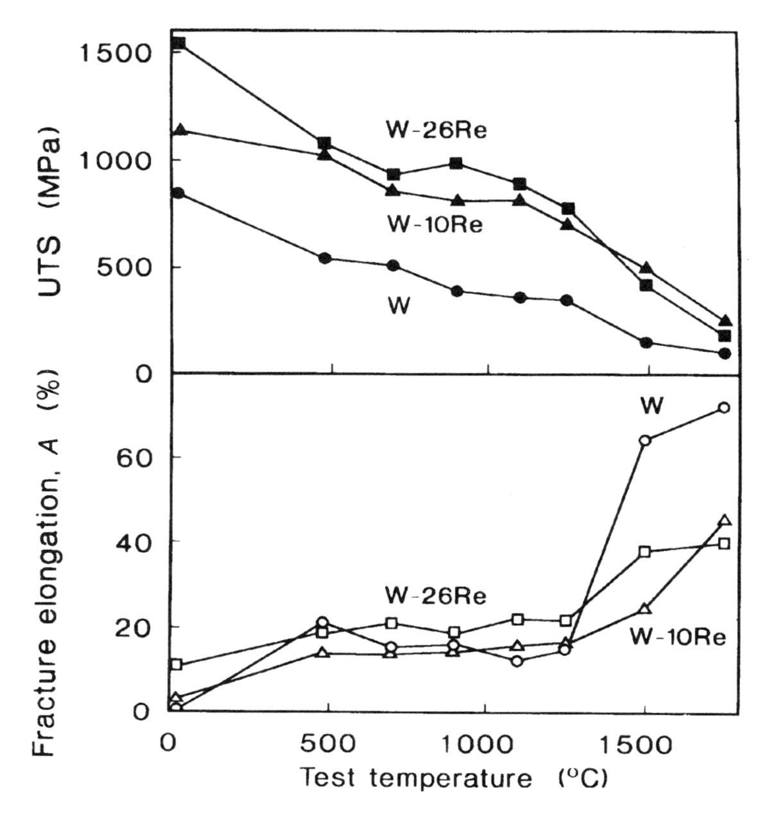

FIGURE 6.2. Tensile properties as a function of test temperature of W and W–Re disks forged by 75%, measured in the radial direction [6.1].

CVD of W–Re alloys has gained commercial importance for the production of X-ray rotating anodes, cones, tubes, and other shapes. Layer thickness for X-ray targets is 1 mm. As in CVD of pure tungsten, fluorides or chlorides of W and Re are also used as precursors. Densities over 99% of the theoretical can be obtained [6.1].

6.1.2. Tungsten–Tantalum Alloys [6.1, 6.2]

In the binary system, tungsten–tantalum unlimited solubility exists. However, only the Ta-rich side is of technical importance. Tantalum–tungsten alloys combine the good corrosion resistance and high elasticity of tantalum with the better hot strength of tungsten. Typical alloy compositions are Ta–2.5W, Ta–7.5W, and Ta–10W. They are used for construction materials in strongly corrosive media and in cases where the hot strength of pure tantalum is inadequate.

Due to the high affinity of tantalum to oxygen, nitrogen, and hydrogen, sintering can only be performed in high vacuum as is usual for pure tantalum.

Tantalum can be further strengthened by additions of Hf (solution-strengthening) or hafnium nitrides, carbides, or oxides (precipitation/dispersion-strengthening). Typical alloy compositions of this type are Ta–8W–2Hf and Ta–10W–2.5Hf–0.01C.

6.1.3. *Tungsten–Titanium Alloys* [6.6]

The alloy W–10Ti is used as a sputtering target in the manufacture of microelectronics devices, such as VLSI, ULSI (very large resp. ultralarge-scale integration), and DRAM (dynamic random access memory) chips. Thin W–Ti layers are sputtered onto silicon substrates and act as a diffusion barrier against aluminum (interconnect).

W–10Ti alloy targets are produced by blending ultrapure tungsten (see also Section 5.7.6) and titanium powders (obtained by vacuum melting and subsequent pulverizing), followed by pressure sintering. The impurity level has to be very low, in particular, with respect to radioactive elements (U, Th) and "mobile" alkaline metals (Na, K).

6.2 DISPERSION-STRENGTHENED AND PRECIPITATION-HARDENING ALLOYS

At half of the solidus temperature on the absolute temperature scale (i.e., ~1850 °C for tungsten), solid-solution alloys lose much of their strength, and dispersion-strengthened or precipitation-hardened alloys are significantly stronger and creep resistant. This is caused by the interaction between the dispersoids and dislocations, as well as subgrains and grain boundaries.

6.2.1. *Non-Sag Tungsten*

Tungsten wires are commonly used as filaments in incandescent lamps. However, due to the high working temperatures (up to 3000 °C in halogen lamps), pure tungsten wire would fail rapidly as a result of grain boundary sliding ("offsetting"). In order to increase the lifetime of such filaments even under severe conditions (e.g., in automotive applications), so-called non-sag tungsten wires are used. The term non-sag refers to the resistance of the wire against deformation ("sagging") under its own weight at incandescent temperatures (Fig. 6.3).

Non-sag tungsten, or, as it is frequently called, doped tungsten, is a remarkable kind of dispersion-strengthened material. It owes its outstanding creep resistance to rows of potassium-filled bubbles (gaseous at working temperatures) which are aligned in rows parallel to the wire axis. These bubbles ("soft" dispersoids) act as barriers for grain boundary migration in the radial direction and effect the recrystallization process in such a way as to lead to an interlocking long-grained microstructure, which gives the wire its non-sag properties (see Chapter 1; Fig. 1.9).

The discovery of NS–W was a fluke. In the early days of wire fabrication (1910–1925), it was realized that the source of the tungsten oxide used for metal powder fabrication played an important role in determining the wire quality [6.7]. Tungsten filaments, originating from material which was treated in clay crucibles, preferably those manufactured by the Battersea Company near London, were shown to be much more stable in shape at incandescent temperatures. Subsequent chemical analysis indicated that a

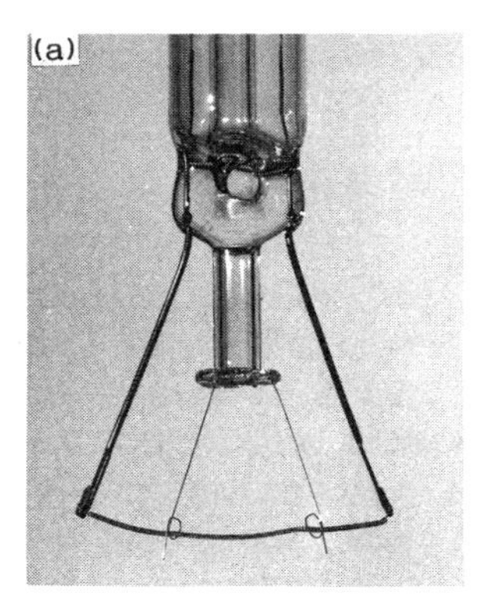

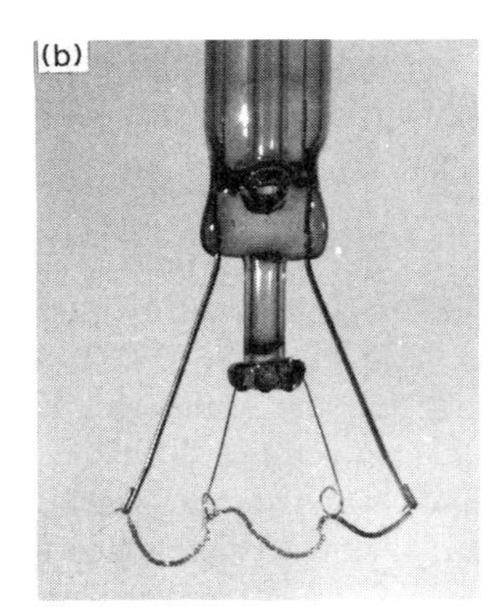

FIGURE 6.3. Coil of a 230 V/60 W lamp before use (a); sagging and deformation of the coil after several hundred hours of operation due to bad NS quality (b).

certain pick-up of potassium, aluminum, and silicon occurred during powder processing, but only potassium remained in the sintered ingot in significant amounts.

It is not clear today whether this contamination occurred during calcination of tungstic acid prior to reduction [6.8] or during the reduction process [6.7], and when the dopants were finally added intentionally in the various tungsten plants. Most of the doping and reduction technology was and still is based on empirical knowledge. Even today, different producers have their own ideas about the action of the dopants and their behavior during subsequent processing.

Although it has been known for a long time that the long-grained and interlocking microstructure was the reason for the high-temperature creep resistance of the doped tungsten wires, no explanation could at first be given for the action of the dopants. The formation of thin "tubes" of dope aligned parallel to the wire axis and the occurrence of rows of pores, which hinder the movement of dislocations and thus increase the resistance to off-setting, was first stated in the late fifties [6.9]. Later, the formation of potassium-filled bubbles was demonstrated by TEM investigations, and metallic potassium was identified as the bubble-forming material [6.10, 6.11].

Today, the desired NS properties of tungsten wires are achieved by doping the starting tungsten blue oxide with aqueous solutions of potassium silicate and aluminum chloride or nitrate (AKS doping). Only when all three elements are present during hydrogen reduction, a sufficient amount of dopant is incorporated into the metal grains as potassium aluminosilicates and cannot be removed by a subsequent acid washing. During sintering of the powder compacts by direct electrical resistance heating under hydrogen, the potassium aluminosilicates dissociate and both silicon and aluminum are removed by diffusion and volatilization. Potassium, which is insoluble in the tungsten matrix, remains in the sintered ingot forming bubbles. The potassium-filled bubbles are elongated during swaging and wire drawing and finally form the rows of tiny bubbles (typically in the range of 20–40 nm), which are essential for generating the NS microstructure.

A schematic representation of the non-sag tungsten wire fabrication process is shown in Fig. 6.4.

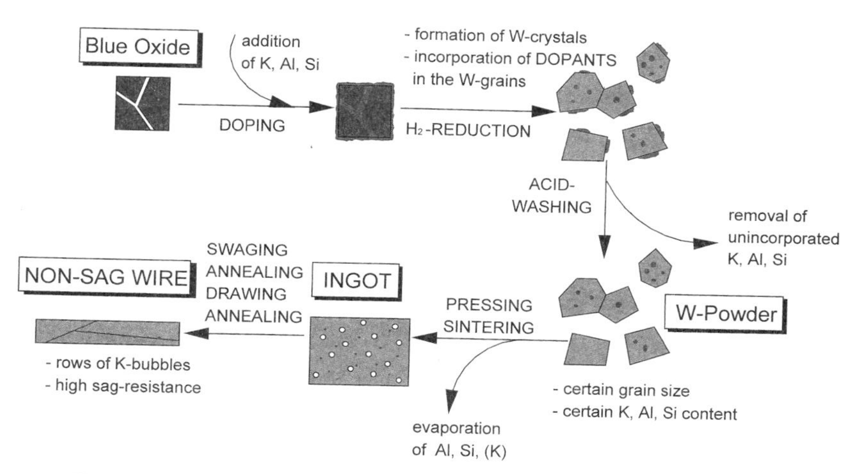

FIGURE 6.4. From tungsten blue oxide to non-sag wire (schematic presentation).

Non-Sag Tungsten Wire Fabrication

Metal Powder Production [6.12]. The production of non sag tungsten powder starts with the careful selection of an appropriate ammonium paratungstate (APT) in terms of crystal size, size distribution, particle agglomeration, chemical purity, and the subsequent "blueing" of the APT under slightly reducing conditions at 400–900 °C, in rotary furnaces or in push-type kilns. The resulting tungsten blue oxide (characterized by its residual ammonia and water content, its specific surface area, as well as its W:O ratio) is then doped in batches with aqueous solutions of potassium silicate and aluminum chloride or nitrate. Typical dopant compositions are between 2000–3000 μg/g K, 200–500 μm/g Al, and 2000–3000 μg/g Si. After drying the oxide, several batches are blended into one homogeneous lot of some 1000 kg.

Reduction is performed in multitube push-type furnaces as described in Section 5.4.2. Typical furnace temperature profiles start with low reduction temperatures of 500–700 °C and end with maximum temperatures of 800–900 °C. Alternatively, a two-stage reduction is carried out instead. In the first stage, the reduction takes place at 400–700 °C (formation of WO_2) and the second stage at 600–900 °C (formation of tungsten metal). Between the two stages, the powder is blended.

After reduction, nearly all the dopants added to the TBO appear to remain in the tungsten powder. However, only 10% are actually incorporated within the tungsten particles, and the rest is only on the particle surfaces. During subsequent washing of the powders with distilled or deionized water and concentrated HCl and HF acids, only the dopants which are entrapped internally are retained. Leaching is commonly performed in small batches, around 100 kg. The powder is then dried at 100–120 °C to remove water, screened to eliminate boat scale, and blended to form a larger, homogeneous powder lot.

The specific goal of the NS powder production is to produce a tungsten powder with an average particle size in the range 3–5 μm and typical dopant levels of 100–160 μm/g K, 20–60 μg/g Al, and 200–250 μg/g Si (analyzed after the acid-wash).

Sintering [6.2, 6.13]. The powder is pressed into compacts, usually of size (12–28) × (12–28) × (600–800) mm, and then presintered under hydrogen at 1100–1300 °C for 30–60 minutes to increase the strength of the compacts so they can be handled for the final direct sintering process. Sintering is carried out under hydrogen in bell-jar-type furnaces until the density reaches about 90% of the theoretical value (see also Section 5.5.2.2). A typical sintering schedule uses two intermediate soak points prior to the final hold at 2700–3000 °C, with carefully controlled heating rates between the holds to render the outgassing of impurities and of parts of the dopants as long as there is open porosity [6.13]. Between 1800 and 2150 °C all of the silicon, and above 2150 °C most of the aluminum, is evaporated as a result of the progressive dissociation and reduction of the incorporated potassium aluminosilicates [6.14]. About 60 to 100 µg/g K, but less than 10 µg/g Si and Al, remain in the sintered bar. It is commonly agreed today that a minimum of 50 µg/g K is necessary to produce a moderate non-sagging wire quality.

In the sintered bar, potassium bubbles appear as tiny pores of about 0.1 µm diameter in the fracture surface, besides some larger, residual sintering pores (Fig. 6.5).

Metal Working. Today, the first step in metal working is commonly hot rolling (1600–1650 °C; Kocks rolling mills) of the sintered bar down to rods of about 7–10 mm diameter. Rolling is then followed by a multiple swaging process, in which the temperature is gradually reduced as the working proceeds (1600 °C → 1200 °C). Reduction in area pass is 10% at the start of swaging and increases to 35–40% for the latter stages. Work hardening of the material requires intermediate recrystallization anneals, which are carried out between 2000–2300 °C [6.13]. Prior to annealing, the oxide scale must be removed by pickling to avoid embrittlement.

Wire drawing usually starts at a diameter of about 0.1 inch (2.54 mm) [6.2]. During the first few drawings, a 40% reduction per pass can be applied at about 1000 °C. The drawing temperature then decreases with decreasing diameter, and the reduction per pass is

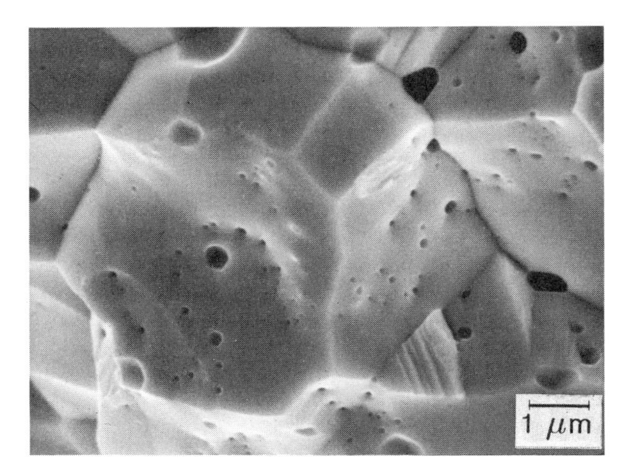

FIGURE 6.5. Fracture surface of a NS-doped tungsten ingot showing tiny potassium-filled pores besides some larger residual sintering pores. The potassium-filled pores are in the order of 0.1 µm and were formed during sintering by dissociation and reduction of aluminosilicates. By courtesy of O. Horacsek, Hungarian Academy of Science.

lowered to 10%. Intermediate in-line anneals serve only for stress relief. A graphite suspension in water is commonly used as lubricant. Cemented carbide dies are used for large-size drawing, but below 0.25 mm diamond dies are used. A typical non-sag tungsten wire processing schedule is shown in Fig. 6.6 [6.2].

Theoretically, a sintered ingot of 3 kg will yield 365,000 m of a 24-μm filament, corresponding to 500,000 coils for a 40-W incandescent lamp [6.15]. Wires for common household lamps are in the range of 17 to 45 μm, but wires can be drawn down to 10 μm. Even smaller diameters can be produced by subsequent etching.

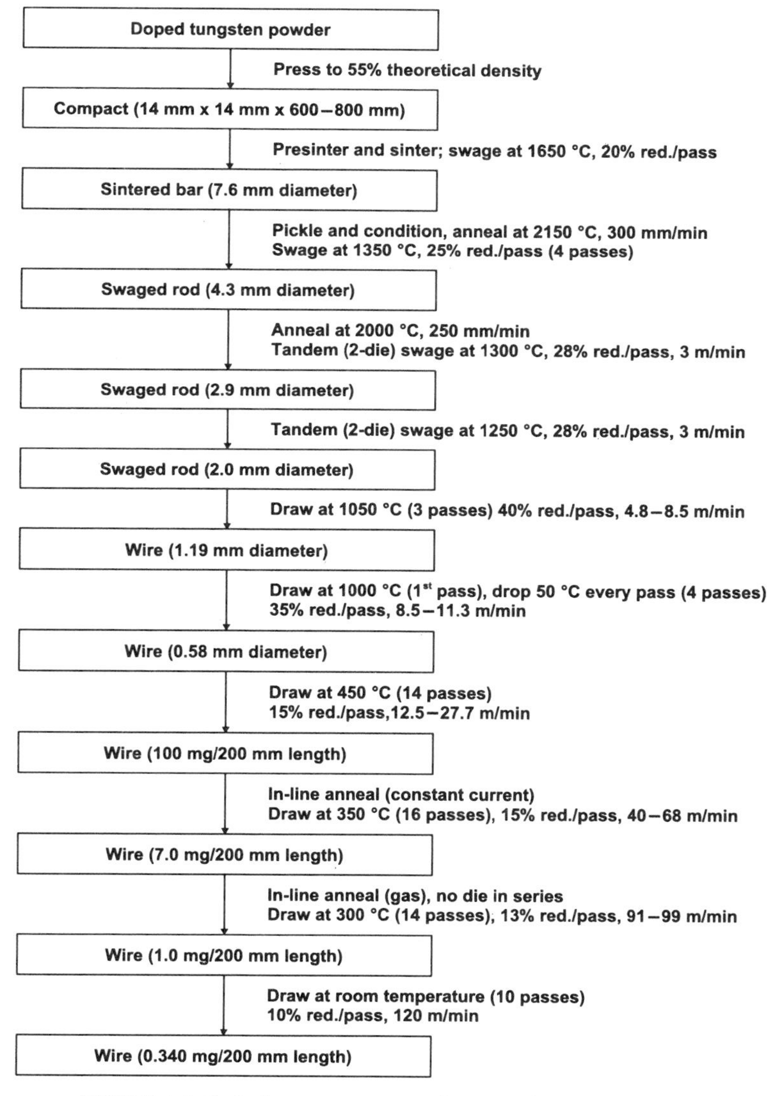

FIGURE 6.6. Typical non-sag tungsten wire processing schedule [6.2].

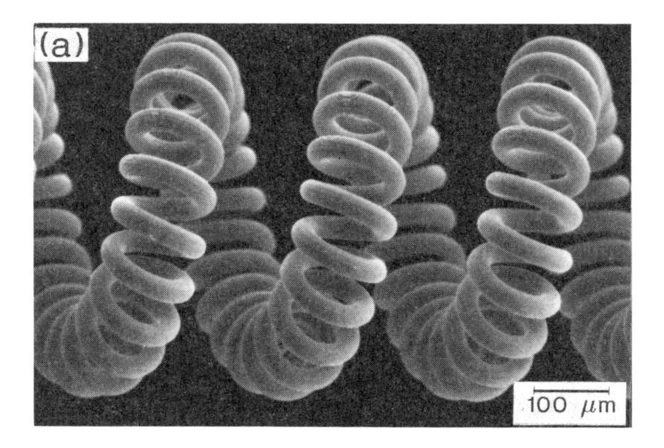

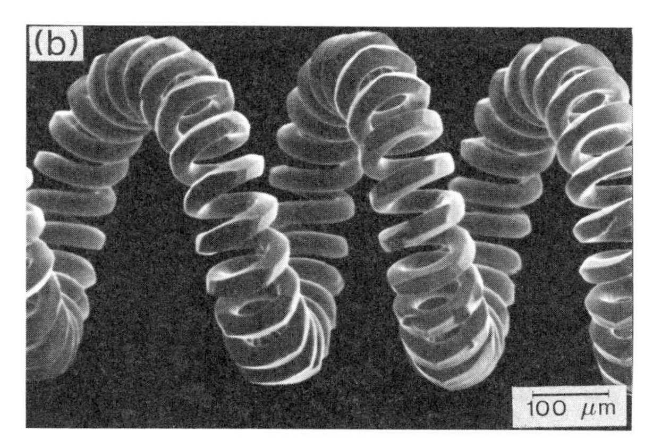

FIGURE 6.7. SEM image of a non-sag tungsten filament: (a) before use, (b) after several hundred hours of operation in a 60-W incandescent lamp. Facetting of the originally round filament occurred due to evaporation of tungsten at the high operation temperature (2400–2500 °C).

Most incandescent lamps today exhibit a coiled-coil geometry of the filament (Fig. 6.7). Coiled-coils are made by winding the filaments on mandrels of molybdenum and subsequent winding of the coil on a larger mandrel to form a secondary coil [6.2]. The coil is then heated to about 1700 °C to relieve the stresses formed during coiling and to stabilize the geometry [6.7]. The mandrels are then selectively dissolved in nitric acid or mixtures of nitric and sulfuric acid.

The last stage in filament fabrication is a heat treatment to recrystallize the wire at temperatures greater than 2200 °C* to form the characteristic NS structure. This procedure is either performed after mounting of the filament (in situ recrystallization) or in a separate

* Compared to unalloyed tungsten the recrystallization temperature of doped tungsten is shifted from 1000–1200 °C up to 2000 °C or even higher [6.1].

furnace unit prior to assembly, where the filaments are positioned in tungsten boats. The latter procedure is only used for high-performance halogen lamps [6.7].

Microstructural Evolution. As the sintered ingot is worked down into rods and wires, the residual sinter porosity is eliminated. However, the potassium-filled bubbles remain and will be elongated with increasing deformation, forming tube-like structures with an aspect ratio of 20 or even higher (Fig. 6.8). On annealing, these tubes break-up into strings of bubbles, driven by surface-energy minimization (Fig. 6.9) [6.16]. The number, size, and dispersion of the bubbles finally present in a non-sag tungsten wire is determined by the thermomechanical history of the wire. Usually, a large number of small-diameter bubbles is desired for obtaining a high NS–W wire quality.

Drawn tungsten wire has a fibrous microstructure. Under incandescent temperatures, the highly cold worked structure recrystallizes, and the fibrous microstructure is transformed into an interlocking grain structure (swallow-tail type), with grain lengths many times the diameter of the wire, which give the wire its non-sag properties. A comparison of a pure and NS-doped tungsten wire in the as-deformed condition and after recrystallization is shown in Fig. 1.9 (Chapter 1).

For vibration-resistant applications, such as traffic signal lamps, non-sag wires with (1–5 wt%) Re are used. Rhenium improves the ductility of the recrystallized wire and also increases its electrical resistance, thus allowing a thicker and mechanically more stable wire [6.17].

More than 90% of non-sag tungsten is used for incandescent lamps. Small amounts are used as defroster heating wires in automobile windshields and as heating wire coils for aluminum evaporation in metallizing applications. For the latter application, thick wires are used (0.5–1 mm). The service life of these wires is not only determined by their mechanical properties but also by their corrosion resistance, which is higher for coarser-grained doped materials with a minimum of grain-boundary volume [6.18].

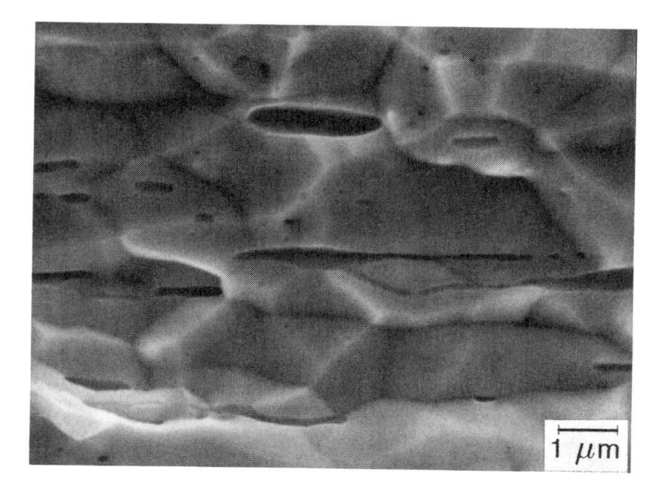

FIGURE 6.8. SEM micrograph showing cylindrical, potassium-containing tubes in the as-deformed tungsten. By courtesy of O. Horacsek, Hungarian Academy of Science.

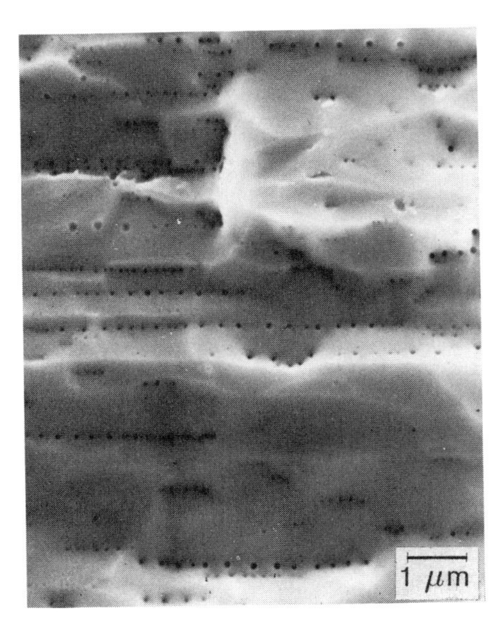

FIGURE 6.9. On annealing, the tubes break-up into strings of potassium-filled bubbles. By courtesy of O. Horacsek, Hungarian Academy of Science.

6.2.2. Alloys with Oxide Dispersoids

The addition of small amounts of finely dispersed oxides to metals, for increasing their high-temperature strength, is a common practice in physical metallurgy (ODS alloys). In the case of tungsten, this technology can be traced back to its early roots. In 1913 (i.e., long before the mechanisms of dispersion-strengthening were known), impact and vibration-resistant filaments were produced based on thoriated tungsten [6.8], basically as is still done today for rough service uses.

Thoria is insoluble in tungsten and has a very high thermal stability, which makes it an ideal dispersoid. Additions of 0.5–4 wt% significantly restrict grain coarsening during sintering, which leads to a fine-grained sintered microstructure. The thoria particles are located at the grain boundaries, thus impeding their mobility (Fig. 6.10). Thoria additions increase the recrystallization temperature and improve the hot strength and creep resistance of tungsten. Strength and fracture elongation of W–2ThO_2 are shown in Fig. 6.11 as dependent on the degree of working [6.1, 6.19]. Thoria hardening in W–Re is effective up to 2200 K (~1930 °C) [6.20].

During swaging and wire drawing, the particles are aligned in strings parallel to the drawing direction of the wire. On recrystallization, an elongated grain structure forms, but the resistance against sagging at incandescent temperatures is comparatively low. Therefore, thoriated tungsten filaments are used only where the hot strength is more important than the sagging characteristics [6.2].

Thoria additions lower the electronic work function (which is the electrical energy that an electron needs to escape from a tungsten surface) and enhances the thermionic

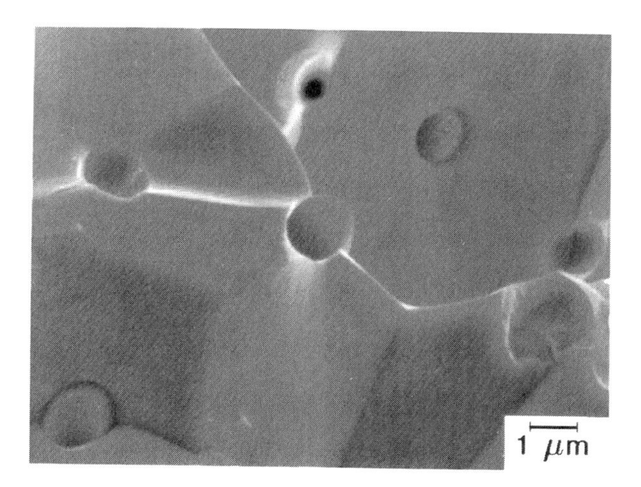

FIGURE 6.10. Fracture surface of a sintered W–2% ThO_2 ingot; SEM micrograph showing rounded thoria particles located at the grain boundaries. By courtesy of Plansee AG, Reutte, Austria.

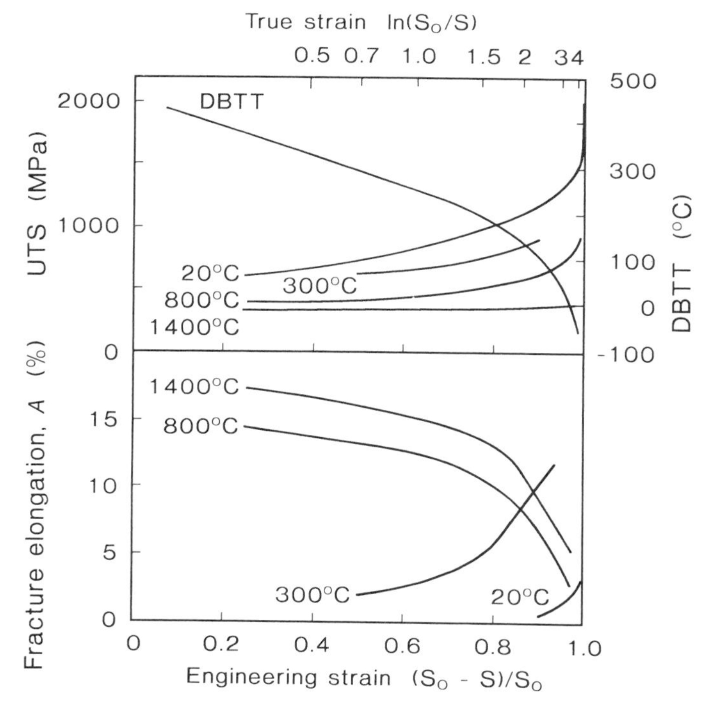

FIGURE 6.11. Ultimate tensile strength and fracture elongation at different temperature and ductile-to-brittle transition temperature as dependent on the degree of working for W–2%ThO_2 [6.1, 6.19].

emission. This is important for all types of tubes, such as radio valves and X-ray tubes, as well as for discharge lamps and welding electrodes (TIG welding*), where it improves the ignition behavior of the electrodes and gives a constant and stable arc.

Under working conditions, thoria is reduced and the released thorium atoms act as an emitter [6.21]. The reason for the formation of a monoatomic layer of metallic thorium is still not yet fully understood. It is assumed that the reduction of thoria by tungsten according to

$$2W + xThO_2 \rightarrow 2WO_x + xTh$$

is favored by the simultaneous evaporation of the tungsten oxide, because of its high vapor pressure [6.20].

Electrodes for high electron emission must be activated before use. This is done by rapid heating above 2000 °C and subsequent activation annealing at 1500–1800 °C [6.2, 6.22]. Alternatively, the activation can be performed in a carburizing atmosphere, which leads to a slight carburization of the surface. Both procedures enhance the formation of metallic thorium. A steady transport of thorium by grain boundary diffusion to the surface and an appropriate microstructure on the electrode tip are of particular importance for a good welding performance [6.23].

Production. Thorium is either added to the oxide as a solution of thorium nitrate or chloride, or as colloidal hydroxide, or it is mixed in the form of very fine thoria particles with the tungsten powder. Sintering is commonly carried out between 2600–2800 °C [6.2]. Thoriated tungsten is stronger than unalloyed tungsten and also has a lower DBTT. Therefore, working has to be carried out at comparatively higher temperatures and in-process annealing is required more often; otherwise, cracking and particle/matrix decohesion can occur [6.2].

Thorium is a radioactive element and a natural α-emitter (specific activity 4250 Bq/g and half-life 1.4×10^{10} a). Health and environmental considerations (evaporation and inhalation of thorium during welding; disposal of grinding dust) have therefore in the past led to a partial replacement of thoriated tungsten for welding operations and in the lighting industry by thoria-free ODS tungsten alloys [6.24].

Fine oxide dispersions of yttrium, lanthanum, zirconium, hafnium, cerium, and erbium behave similarly to thoria, although they possess different emission characteristics, depending on temperature. This affords an adaptation of the operating conditions compared to W–ThO_2. Electrodes, based on W–(1–2wt%)CeO_2 and W–(1–2 wt%)La_2O_3 have demonstrated good arc striking, excellent arc stability, and low electrode consumption and have even outperformed thoriated tungsten in welding applications [6.23, 6.24]. They can be used for both AC- and DC-TIG welding. Electrodes with 0.8 wt% ZrO_2 are used for special welding operations [6.22].

6.2.3. *Alloys with Carbide Dispersoids*

Hafnium carbide (HfC) is the most potent strengthener for tungsten at elevated temperatures [6.4]. Particle size and interparticle spacing must be small in order to lead to

* Tungsten-Inert Gas-Welding.

an effective strengthening. This occurs as a result of dislocation pinning through the fine (50–100 nm) HfC particles, which inhibits the slip of atom planes and the migration of grain boundaries [6.25]. A uniform distribution of fine HfC particles can be obtained from supersaturated solid solutions ("precipitation hardening"). Other carbides, such as TaC, NbC, and ZrC, were also tested but proved to be less effective, especially at temperatures above 1900 °C [6.2]. The outstanding action of HfC is attributed to its high thermodynamic stability and its comparatively low solubility and diffusivity in tungsten at high temperatures.

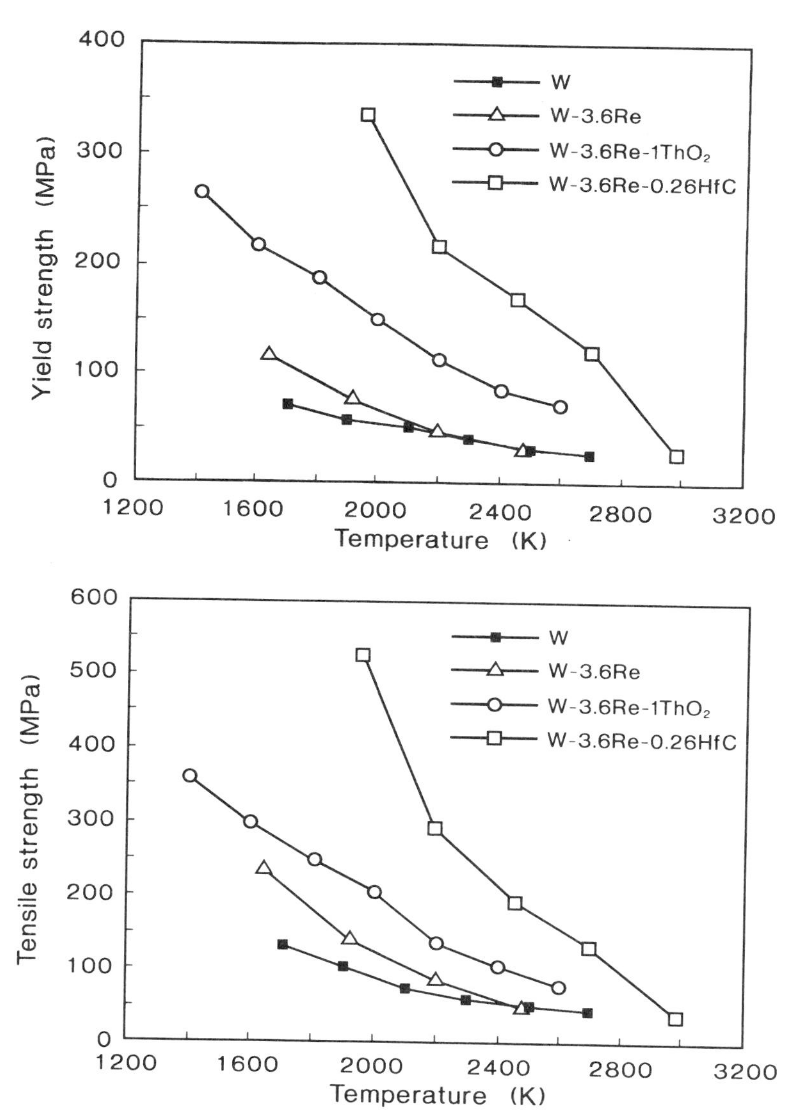

FIGURE 6.12. Yield strength and ultimate tensile strength of a W–3.6Re–0.26HfC alloy as a function of temperature, in comparison to unalloyed W, W–3.6Re, and W–3.6Re–1ThO_2 [6.20].

Significant strengthening is observed up to 2700 K (2427 °C). Above this, rapid particle coarsening occurs, and the strengthening effect fades. Additions of rhenium provide ductility at low temperatures by decreasing the DBTT.

W–Re–HfC alloys cannot be produced by powder metallurgical techniques owing to the high oxygen affinity of hafnium. They are produced by arc melting.

Yield strength and ultimate tensile strength of a W–3.6Re–0.26HfC alloy as a function of temperature are shown in Fig. 6.12, in comparison to pure W, W–3.6Re, and W–3.6Re–1ThO$_2$.

W–Re–HfC alloys are the strongest man-made metallic materials at temperatures above 2000 K [6.25].

6.3. TUNGSTEN COMPOSITES

Tungsten metal exhibits outstanding thermal properties, which makes it attractive for a broad range of applications. However, for certain applications, its electrical and thermal conductivity, sensitivity toward oxidation, and poor workability are unsatisfactory. These limitations have led to the development of two-phase alloys, in which the useful properties of tungsten are combined with those of the additive.

Two important alloy systems belong to this group and are described below:

- The so-called heavy metal alloys, based on W–Ni–Fe and W–Ni–Cu–(Fe). They are used wherever high density, excellent mechanical properties, and good workability are required.
- W–Cu and W–Ag alloys, in which the high electrical and thermal conductivity of copper or silver is combined with the high hardness and wear resistance of tungsten.

6.3.1. Tungsten Heavy Metal Alloys [6.1, 6.2, 6.26, 6.27, 6.29]

The term *tungsten heavy metal* or *heavy metal* is used for a group of two-phase composites, based on W–Ni–Fe and W–Ni–Cu. They are characterized by a high density and a novel (unique) combination of strength and ductility.

Tungsten is the main component of the alloys (typically present in the range of 90 to 98 wt%) and the reason for their high density (between 17 and 18.5 g/cm^3). Nickel, iron, and copper serve as a binder matrix, which holds the brittle tungsten grains together and which makes the alloys ductile and easy to machine. A typical microstructure is shown in Fig. 6.13. It consists of spherical tungsten grains (20 to 60 μm in diameter), which are embedded in a tough, metallic matrix. While the grains are nearly pure bcc tungsten, the fcc binder matrix contains about 20 wt% W in solid solution.

Heavy metals are used for applications, where the high specific weight of the material plays an important role. They are used as counterweights, rotating inertia members, X-ray and γ-radiation shields, as rigid tools for machining, as well as for defense purposes (kinetic energy penetrators, fragmentation devices, etc.).

Fabrication [6.26, 6.27]. Heavy metals are produced by conventional P/M techniques. A flow chart of the fabrication process is shown in Fig. 6.14. Elemental powders (W,

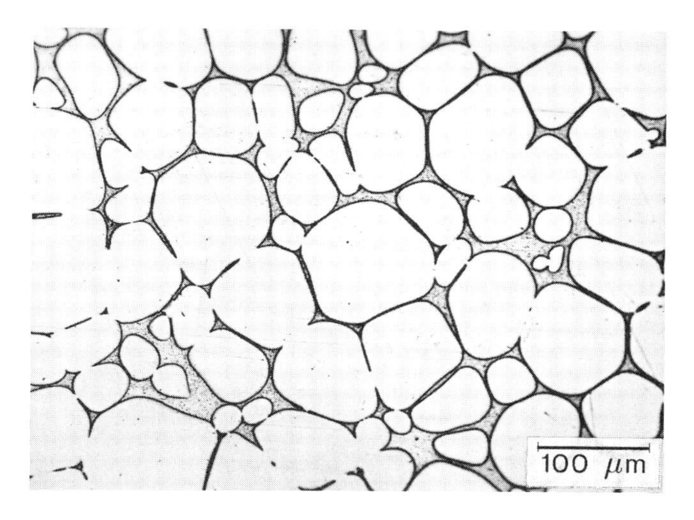

FIGURE 6.13. Optical micrograph of a 95 wt% tungsten heavy metal alloy with 3.2 wt% Ni and 1.8 wt% Cu. By courtesy of Plansee AG, Reutte, Austria.

Fe, Ni, Cu) are blended in mixers or ball mills to the desired ratio, compacted to form a green body, and subsequently liquid-phase sintered. Assuming proper manufacturing conditions, they exhibit full or near-theoretical density in the as-sintered condition.

Powder particle sizes are in the range of 2 to 6 μm. Both die pressing and isostatic pressing (dry- and wet-bag pressing) are in use. No lubricant is commonly added, since the green strength is high enough to handle the compacts. Powder injection molding (PIM) is used for applications where net shaping is desired and large quantities of complex parts are produced [6.28].

Sintering is commonly carried out in molybdenum-wire resistance-heated furnaces under hydrogen or nitrogen mixtures (dissociated ammonia) but can also be performed in vacuum units. The use of wet hydrogen has become industrial practice to suppress hydrogen embrittlement (water vapor porosity) [6.29]. The temperature/time program of the sintering cycle must be adjusted to the composition and size of the sintered parts. A

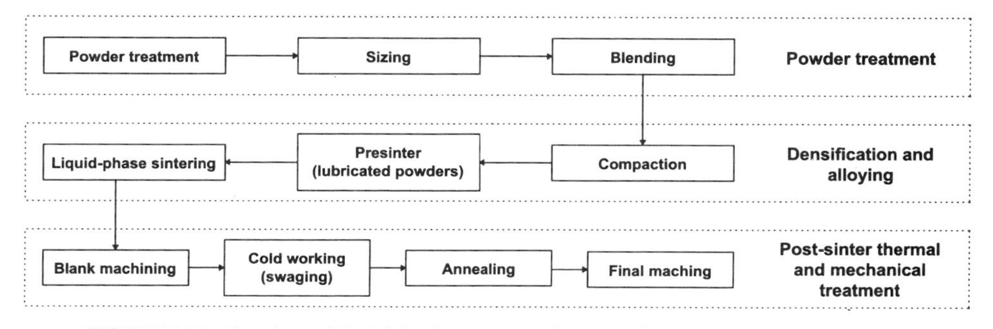

FIGURE 6.14. Flowchart of the fabrication process of tungsten heavy metal penetrators [6.26].

cleaning step in hydrogen at 1000 °C is commonly performed to render the outgassing of volatile compounds. High-purity powder grades must be used for sintering. Otherwise, blistering will occur on liquid-phase sintering, and interface precipitations will occur on cooling.

Isothermal sintering is carried out above the eutectic temperature, typically between 1440 and 1500 °C, but can be as high as 1600 °C. The higher the tungsten content of the alloy, the higher the temperature. Tungsten–nickel–copper alloys are sintered at somewhat lower temperatures than tungsten–nickel–iron alloys. Sintering times are between 30 minutes and two hours. Prolonged sintering under dry hydrogen leads to swelling of the parts and significant embrittlement due to pore coarsening by Ostwald ripening and coalescence. Residual porosities larger than 0.5% drastically reduce the ductility [6.30, 6.31].

On cooling, a significant amount of tungsten remains in solid solution, depending on the binder composition and the cooling rate. The solubility is highest in binary W–Ni alloys (up to 40 wt%; resulting in low ductility), but additions of Fe and Cu depress it to lower values (typically 20 to 25 wt% W), providing a tough and ductile binder matrix.

Shrinkage porosity can form on cooling, in particular in W–Ni–Cu alloys and at high cooling rates; furthermore, impurities (P, S, C) segregate at the tungsten matrix interface. Both effects must be controlled by proper manufacturing conditions, since they significantly lower the ductility.

Optimal mechanical properties require subsequent heat treatment of the alloys. Solution annealing at 900 to 1300 °C and subsequent quenching to avoid impurity segregation and formation of intermetallic phases significantly improves ductility [6.26]. Heavy metals can be cold-worked (swaged, rolled) to increase hardness and strengths at the expense of elongation and toughness. Aging at 500 °C after 25% deformation (cold work) is a compromise to achieve both high ultimate tensile strengths and elongation. Conventional procedure is a double swaging–heat treatment cycle (deformation 25–30%) [6.32]. The structure shows that the original spheroids formed during liquid-phase sintering have been transformed to ellipsoids on deformation (Fig. 6.15). Recently, the advantage of large deformation levels (up to 95%) yielding in a fibrous microstructure (Fig. 6.16) has been demonstrated [6.32]. Heavy metals exhibit a ductile-to-brittle transition temperature. In comparison to pure tungsten it is, however, not a sharp transition but spreads over several hundred degrees [6.1].

Although a considerable amount of densification already occurs during solid-state heating, isothermal liquid-phase sintering is a prerequisite to obtain near-theoretical density and a high degree of microstructural homogeneity. Rapid final densification occurs on formation of the liquid phase under the action of capillary forces. Particle rearrangements, solution/reprecipitation processes (Ostwald ripening), and coalescence contribute to a higher packing density and a significant grain coarsening during sintering. The original tungsten powder grain size of 3–5 μm is transformed during sintering to rounded tungsten particles (spheroids) of at least 10 times the original grain diameter. Shape accommodation (formation of polyhedras with rounded corners) plays an important role in high tungsten alloys, where only a small quantity of liquid is available for achieving full density.

Depending on the binder composition, heavy metals can be classified into two main groups [6.2, 6.27]:

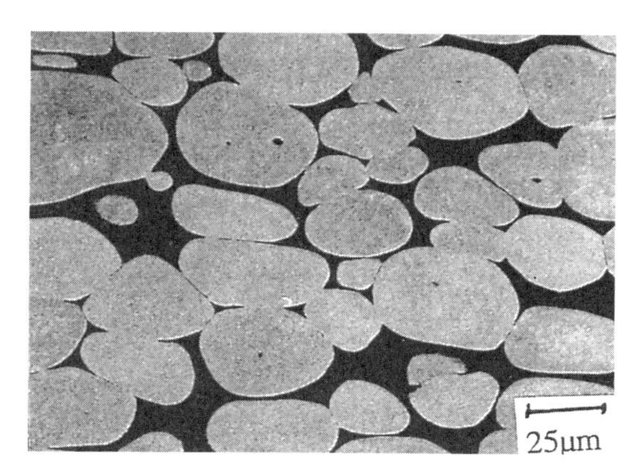

FIGURE 6.15. Microstructure of a W–Ni–Fe–Co alloy after swaging by 30% [6.32]. By courtesy of A. R. Bentley, M. C. Hogwood, and M. Power, Defence Research Agency, Kent, England.

(a) W–Ni–Fe (Ni: 1–7 wt%, Fe: 0.8–3 wt%, Mo: 0–4 wt%)

This group is ferromagnetic. Typical nickel-to-iron ratios range between 1:1 and 4:1. The preferred ratio is 7:3, since this composition avoids the formation of intermetallic phases [6.26]. W–Ni–Fe alloys exhibit excellent strength/ductility combinations and can be cold-worked to a reduction of 60% without intermediate annealing. Molybdenum acts as grain refiner. Higher additions of iron and/or additions of molybdenum cause a significant matrix-strengthening effect and improve high-temperature strength (Gyromet grades) [6.2]. Other additives, such as Co, Ta, and Re, act as grain refiners and increase hardness and strength, but lower the ductility [6.29]. Rhenium additions are of interest in net-shape production, since their high strength in the as-sintered condition (1180 MPa) does not require any post-sintering treatment [6.33].

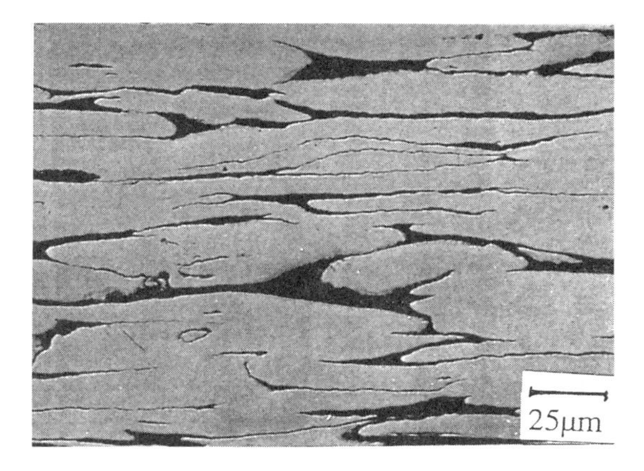

FIGURE 6.16. Microstructure of a W–Ni–Fe–Co alloy after defromation by 80% [6.32]. By courtesy of A. R. Bentley, M. C. Hogwood, and M. Power, Defence Research Agency, Kent, England.

(b) W–Ni–Cu (Ni: 1–7wt%, Cu: 0.5–3 wt%, Fe: 0–7 wt%)

Members of this group are nonmagnetic and exhibit a higher electrical conductivity. The nickel-to-copper ratio ranges from 3:2 to 4:1. W–Ni–Cu alloys exhibit lower strength and ductility than comparable W–Ni–Fe alloys. Due to the low melting point of copper, low heating rates are required to obtain full density .

Properties: [6.1, 6.34, 6.35]. Some properties of tungsten heavy metal alloys are summarized in Table 6.1. Ultimate tensile strengths (660–1350 MPa), yield strengths (565–300 MPa), and fracture elongations (5–30%) can vary widely. In general, mechanical properties are very sensitive to processing conditions, impurities, and microstructure. Problems with controlling porosity, impurities, and microstructural homogeneity are therefore common in heavy alloy fabrication [6.31]. Residual sinter porosity and the formation of interface precipitates are the main reason for inferior material properties.

With increasing tungsten content, the contiguity of the tungsten grains increases (i.e., the W–W interfacial area as a fraction of the total interface area) and both strength and ductility decrease [6.36]. This "microstructural limitation" of the properties is demonstrated in Fig. 6.17 for alloys of different tungsten contents [6.1, 6.31, 6.37]. While the

TABLE 6.1. Properties of Tungsten Heavy Metal Alloys[a]

Properties	1	2	3
Density ($g \cdot cm^{-3}$)	17.0–18.5	17.0–18.5	17.1–18.6
Young's modulus (GPa)	350–400	320–380	—
Shear modulus (GPa)	—	125–160	—
Poisson's ratio	0.28–0.29	0.28–0.29	—
Hardness			
HV10	270–360	270–470	—
HB30	—	250–450	—
RC	—	—	28–40
Tensile strength (MPa)			
20 °C	870–1000	660–1350	880–1320
100 °C	—	560–730	—
500 °C	340–650	340–610	—
1000 °C	220–260	90–260	—
Yield strength (0.2%) (MPa)	600–700	565–1300	750–1240
Fracture elongation (%)	10–30	—	5–30
Compressive strength (MPa)	3500–4500	3500–5500	—
K_{ic} ($MPa \cdot mm^{-1}$)	800–6300	—	—
Compressibility (% of length)	—	—	45–60
Charpy value (notched)	0.9–2.8	—	—
Specific electrical resistivity ($\mu\Omega \cdot m$)	0.10–0.18	0.10–0.18	—
Mean linear coefficient of thermal expansion at 20–800 °C ($mm^{-1} \cdot K^{-1}$)	$5.2–6.5 \times 10^{-6}$	$5.2–6.5 \times 10^{-6}$	—
Half-value thickness			
against ^{60}Co	—	8.5–9.7	—
against ^{137}Cs	—	4.5–5.2	—

[a] 1. Pink and R. Eck, "Refractory Metals and their Alloys," in: *Materials Science and Technology* (R. W. Cahn, P. Haasen, and E. J. Kramer, eds.), Vol. 8, pp. 591–638, VCH, Weinheim, (1996).
2. Metallwerk Plansee, Densimet Schweremetall Legierungen, Company Brochure, Reutte, Austria (1982).
3. Ashot Ashkelon, Tungsten-Based Products, Company Brochure.

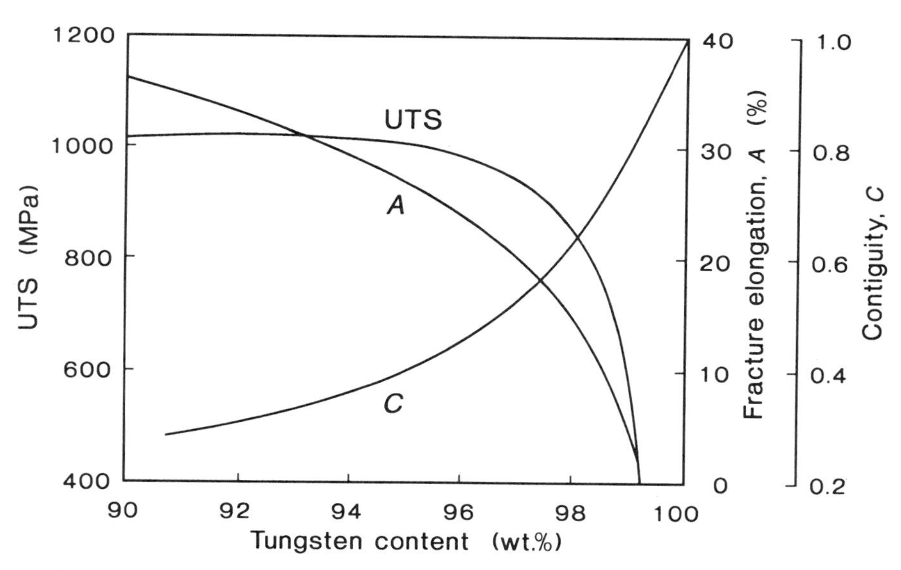

FIGURE 6.17. Optimum values of room-temperature strength and fracture elongation and their correlation with the contiguity factor for tungsten heavy alloys [6.1, 6.31].

trend is most pronounced for the fracture elongation, the tensile strength exhibits a maximum at about 85 vol% W (93 wt%) before it significantly decreases. At 99 wt% W, the elongation is zero and the tensile strength is about 400 MPa [6.38].

Recently, it was demonstrated that large deformation levels yield exceptionally high tensile strength levels (up to 1700 MPa), combined with fracture elongations above 15% [6.32].

Besides the high density and the unique combination of high strength and ductility, there are other attributes which make heavy metals a versatile product:

- the high modulus of elasticity (much higher than steel),
- excellent vibration damping characteristics (for chatter-free heavy machining),
- its good machinability,
- the high absorption ability for X-rays and γ-rays,
- good thermal and electrical conductivity,
- low electrical erosion and welding tendency,
- good corrosion resistance.

Research in tungsten heavy alloys was previously boosted by the ballistic application as anti-amor kinetic energy penetrator. Numerous papers have appeared in the last years and reviews have recently been published [6.29, 6.39, 6.40]. Although the basic requirements are the same for civil as for defense applications (i.e., high density, strength, and elongation), there are two more important factors which must be considered for this specific application: the behavior of the material under high strain rate conditions and their ballistic performance (i.e., their penetration ability). In particular, the latter aspect is of critical importance and, to a certain extent, still a weak point of heavy metals. Their ballistic performance is inferior compared to depleted uranium, which is still used as a standard penetrator material [6.29]. Nevertheless, recent environmental considerations have put a strong emphasis on substituting depleted uranium by heavy metals because of its radioactivity. Efforts to improve the ballistic performance through proper processing and compositional modifications have failed [6.29, 6.39]. Recent research has therefore focused on alternative matrix alloys, such as tungsten–hafnium, tungsten–uranium composites [6.29], or heavy metal alloys with a spiculating core of WC [6.41].

6.3.2. *Tungsten Copper and Tungsten Silver* [6.1, 6.2, 6.42].

Tungsten–copper and tungsten–silver composites are widely used in mechanical and electrical engineering. Typical applications include high-, medium-, and low-voltage circuit breakers (W–Cu, W–Ag), resistance welding electrodes, electrode materials for electrical discharge machining, and heat sink materials for microelectronic packaging (W–Cu). More recently, W–Cu composites have been tested as heat flux components in experimental fusion reactors [6.43] and as materials in MHD (magnetohydrodynamics) power generation systems [6.44].

They combine the high hardness, hot strength, and wear resistance of tungsten with the outstanding electrical and thermal conductivity of the two high-conductivity metals. Furthermore tungsten increases the resistance of the materials against spark and arc erosion (burn-off) and lowers the sticking and welding tendency, which are both important criteria for heavy-duty electrical contacts, where switching currents can be up to 100 kA and arc temperatures of a few 10,000 K can occur within a few milliseconds [6.45]. Under arcing conditions, the contacts are cooled through the melting and evaporation of silver or copper ("transpiration" cooling), an effect which was earlier used for rocket nozzle throat liners (for example for underwater-launched ballistic missiles) made of W–Ag [6.2].

W–Cu and W–Ag composites are not real alloys in the strict meaning of the word, because the mutual solubility of the components is practically zero. Therefore, they are also sometimes called "pseudoalloys."

Production. Powder metallurgy is the only viable way to produce composites of high quality. The method of production depends on the composition ratio. Materials with 10 to 40 wt% copper (20 to 50 wt% silver) are commonly produced by infiltration, while at higher copper or silver contents the powders are blended, pressed, and subsequently solid-state sintered [6.42, 6.46].

Infiltration of porous structures. A porous tungsten skeleton is first produced by pressing either pure tungsten powder or a mixture of tungsten powder and small amounts of the additive element with subsequent sintering in a reducing atmosphere at comparably low temperatures (1200–1800 °C). The porosity of the compact can be varied by the pressing and sintering parameters, as well as by the grain size distribution of the starting tungsten powder. It determines the percentage of infiltrate. For higher copper and silver contents, loose tungsten powder can also be infiltrated.

Infiltration is accomplished by immersion, dipping, or flooding, either in vacuum, inert gas, or, preferably, in reducing atmosphere at 1150–1250 °C. Oxide layers disturb the infiltration process and must be removed prior to infiltration by reduction with hydrogen. Trace impurities can impair the wetting between tungsten and the melt (for example, silicon) but can also improve it (nickel, cobalt, iron) [6.47]. Densities of 96% up to near-theoretical can be obtained. The presence of some closed pores in the sintered skeleton prevents full density.

Machining is required to finish the infiltrated parts.

Sintering. Liquid-phase sintering of low-copper (silver) composites results in inferior product qualities, due to a higher degree of residual porosity. Shrinkage during sintering is difficult to control, and large distortions of the parts commonly occur. However, very fine powder grades, which render a better particle rearrangement during sintering, can be sintered to full density [6.48]. Additions of nickel, cobalt, or iron improve the sintering behavior, but lower the conductivity. Powder injection molding of complex shapes for large volume fabrication of W–Cu parts used for microelectronic packaging and subsequent liquid-phase sintering might be an interesting alternative to infiltration [6.28].

Sintering of high-copper and silver alloys is carried out below the melting point of the two metals. In such alloys, tungsten does not form a contiguous skeleton, as in the case of infiltration. About 92% of the theoretical density can be obtained. Parts are repressed after sintering or can be hot-worked (rolled, swaged, extruded) to achieve higher densities.

A typical microstructure of an infiltrated W–20Cu alloy is shown in Fig. 6.18. It consists of tungsten particles which are embedded in a soft copper matrix. The particles form a ridged skeleton, which is an important prerequisite for obtaining optimal mechanical properties. As the components W and Cu have no mutual solubility, no appreciable change in the particle size and morphology of tungsten takes place during infiltration, i.e., the particle size and morphology of the tungsten powder approximately matches the particle size and morphology of tungsten in the W–Cu alloy, though in densified form.

Composition and Properties. The tungsten content in contact materials varies between 10 and 90 wt%, but composites between 70 and 90 wt% W, and W particle

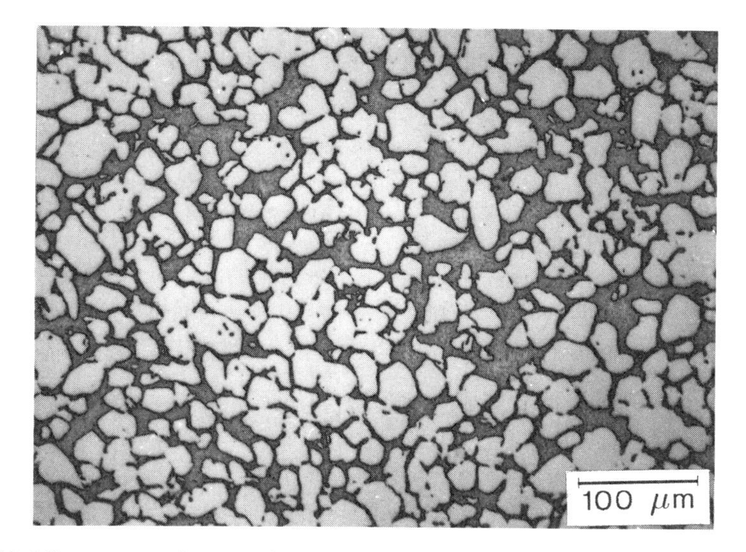

FIGURE 6.18. Microstructure of tungsten infiltrated with 20 wt% copper. By courtesy of Plansee AG, Reutte, Austria].

sizes between 5 and 50 μm are the standard materials for arcing contacts [6.42, 6.45]. Compositions between 25 and 40 wt% Cu are preferably used for electrical discharge machining [6.49]. For special switches, WC–(20–65) Ag and WC–(30–50)Cu composites are used.

W–Ag exhibits better oxidation and corrosion resistance than W–Cu and has higher current-carrying capacity. They are used in switching devices operated in air. However, the costs are significantly higher. W–Cu composites are preferably used in vacuum, inert gas, and oil-switching devices.

Some important electrical and mechanical properties of W–Cu and W–Ag alloys are summarized in Table 6.2 (see page 279).

At room temperature the composites behave in a brittle manner. However, they are readily machinable using cemented carbide tools (ISO K10). The arcing contacts can be clamped, brazed, or directly infiltrated onto the circuit-breaker components (commonly pure copper or copper alloys).

6.3.3. Tungsten-Fiber-Reinforced Composites

Tungsten metal and tungsten alloy wires have attracted interest for fiber reinforcement of high-temperature materials, in particular of superalloys. However, until now, they have remained "exotic" materials with only a few specific applications or for model studies of mechanical behavior and are not produced on a large scale. This is at least partially due to the difficulties encountered during composite fabrication and the resulting high cost of the material.

When a ductile and oxidation-resistant matrix is used, it protects the tungsten from oxidation. The specific strength of the composite is much higher than that of the matrix alloy, especially at elevated temperature. Strength and ductilities of typical tungsten-wire reinforced superalloys, measured in the wire direction, are shown in Fig. 6.19 for the temperature range of 20 to 1000 °C, together with a micrograph of a composite sectioned perpendicular to the fiber axis. Other than in the wire direction, the properties of the materials are dominated by the respective matrix [6.50]. Typical fiber diameters are in the range of 100 to 500 μm.

Most studies on wire reinforcement were performed on superalloy matrices (Incoloy[a], Waspaloy, FeCrAlY), in particular at temperatures between 800 and 1200 °C, where the usual strengthening mechanisms in superalloys begin to fail [6.50]. The most promising research today is on the fabrication of high-pressure turbopump blades for the space shuttle main engine [6.51, 6.52]. Reinforcement by NS tungsten fibers was shown to be superior in regard to stress-rupture properties as compared to undoped tungsten, but even higher strength was obtained using a W–Re–HfC wire [6.52]. This wire has a tensile strength of more than 6 times the strength of the strongest nickel– and cobalt–base superalloys at ~1100 °C [6.51]. A Waspaloy/W–24Re–HfC wire reinforced composite was shown to be most promising [6.52].

Other interesting matrices are niobium alloys (investigated for long-term, high-temperature applications in space power systems) and copper and silver for electrical applications and heat exchangers [6.50, 6.51].

Tungsten fiber composites can be produced either by infiltration of the fiber bundle by the molten matrix, or by bringing together the fibers and matrix in the solid state and

TABLE 6.2. Important Properties of W–Cu and W–Ag Pseudoalloys

Properties[a]	W20Cu	W25Cu	W30Cu	W35Cu	W40Cu	W50Cu	W20Ag	W35Ag
Density ($g \cdot cm^{-3}$)	15.4 15.4	14.5	14.0	13.5 13.5	12.7	11.6	15.6 15.6	14.0 14.0
Young's modulus (GPa)	225 230	225	230	200 185	180	150	205 210	132 135
Tensile strength at 20 °C (MPa)	440 400–460	390	390	340 340–380	290	250	340 240–470	290 300
Hardness HB 10	180 180	160	150	150 130	140	120	180 180	110 100
Mean linear coefficient of thermal expansion ($mm^{-1} \cdot K^{-1}$)	— 8.5	—	—	—	—	—	— 8.7	—
Specific electrical resistivity ($\mu\Omega \cdot m$)	0.050 0.050	0.042	0.040	0.037 0.038	0.033	0.026	0.050 0.050	0.037 0.040
Electrical conductivity (% IACS)[b]	35	36	43	47	51	58	35	47
Thermal conductivity ($W \cdot m^{-1} \cdot K^{-1}$)	134 134	138	147	155 155	176	195	138 138	159 154

[a] First Line: Metallwerk Plansee Elmet Kontaktwerkstoffe, Company Brochure, Reutte, Austria (1982).
Second Line: E. Pink and R. Eck, Refractory Metals and their Alloys, in: *Materials Science and Technology* (R. W. Cahn, P. Haasen, and E. J. Kramer, eds.), Vol. 8, pp. 591–638, VCH, Weinheim (1996).

[b] % IACS = % International Annealed Copper Standard.
100% IACS = 58.00 $\mu\Omega^{-1} \cdot m^{-1}$; 1/% IACS = 1.7241 $\mu\Omega \cdot m$.

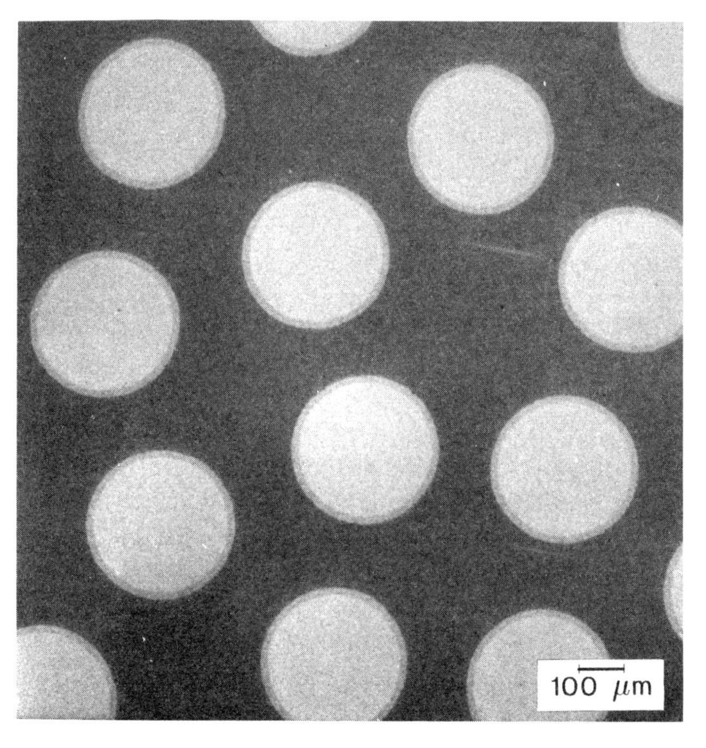

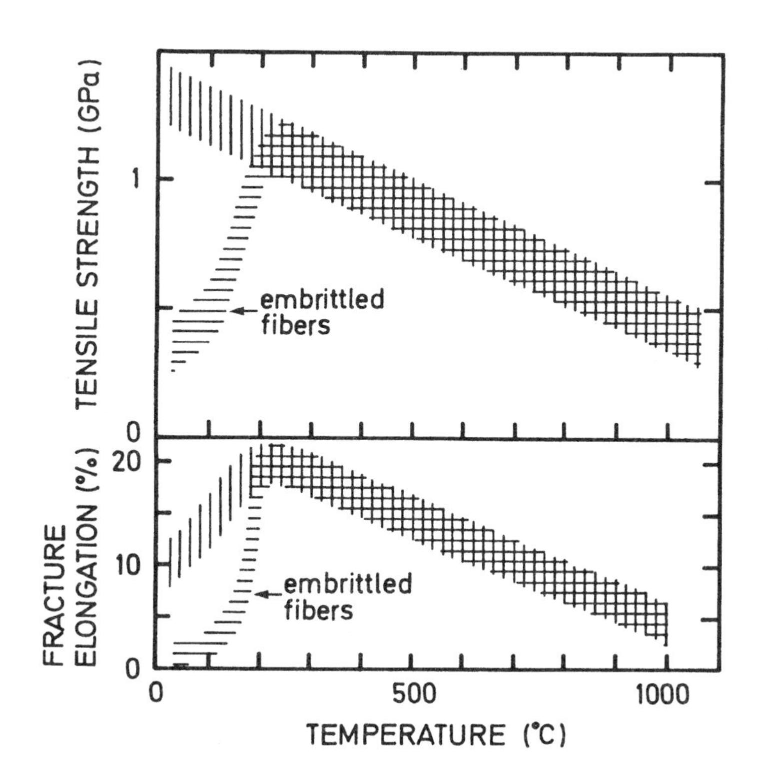

FIGURE 6.19. Micrographs of a tungsten-wire reinforced superalloy sectioned perpendicular to the wire direction (left); strength and ductilities of typical tungsten-wire reinforced superalloys with 40 vol% fibers (right) [6.50]. By courtesy of R. Warren, Lulea University of Technology, Sweden.

subsequent solid-state consolidation at comparatively low temperatures (approximately 1100 °C). Excessive fiber/matrix interactions are the main disadvantages of the liquid-phase route, leading to the partial dissolution of tungsten in the matrix, the formation of brittle intermetallic phases, and a partial recrystallization of the tungsten wire, making it weak and brittle. Diffusion barrier coatings (refractory oxides, carbides, or nitrides) have therefore been developed, but further increase the cost of the material. For solid state processing, the danger of reaction between matrix and fibers is less, due to the lower temperatures. It is essential to have the matrix either in wire or in sheet, foil, or powder form. Consolidation is achieved by mechanical pressing and subsequent solid state sintering or by hot pressing.

REFERENCES FOR CHAPTER 6

6.1. E. Pink and R. Eck, "Refractory Metals and Their Alloys," in: *Materials Science and Technology* (R. W. Cahn and E. J. Kramer, eds.), Vol. 8, pp. 591–638, VCH, Weinheim (1996).
6.2. S. W. H. Yih and C. T. Wang, *Tungsten*, Plenum Press, New York (1979).
6.3. G. T. Hahn, A. Gilbert, and R. I. Jaffee, in: *Refractory Metals and Alloys II* (M. Semchyshen and I. Perlmutter, eds.), Vol. 17, pp. 23–63, Wiley, New York (1962).
6.4. W. D. Klopp, *J. Less-Common Met.* **42** (1975), 261–278.
6.5. S. Geraldi, *Metals Handbook*, 10th edition, Vol. 2, pp. 565–571, ASMI, Materials Park, Ohio (1990).
6.6. O. Kanou, in: *Proc. 7th Int. Tungsten Symposium*, pp. 410–418, ITIA, London SW90QT, UK (1996).
6.7. C. L. Briant and B. P. Bewlay, *MRS Bulletin*, pp. 67–73 (August 1995).
6.8. T. Millner and J. Neugebauer, in: *The Metallurgy of Doped/Non-Sag Tungsten* (E. Pink and L. Bartha, eds.), pp. 1–13, Elsevier, London (1989).
6.9. G. D. Rieck, *Tungsten and its Compounds*, Pergamon Press, Oxford (1967).
6.10. D. M. Moon, R. Stickler, and A. L. Wolfe, in: *Proc. 6th Int. Plansee Seminar* (F. Benesovsky, ed.), p. 67, Springer-Verlag, Vienna–New York (1969).
6.11. D. B. Snow, *Metall. Trans.* **3** (1972), 2553.
6.12. W. D. Schubert, B. Lux, and B. Zeiler, *Int. J. Refract. Met. Hard Mater.* **13** (1995), 119–135.
6.13. J. A. Mullendore, in: *The Metallurgy of Doped/Non-Sag Tungsten* (E. Pink and B. Bartha, eds.), pp. 61–80, Elsevier, London (1989).
6.14. Sh. Yamazaki, in: *The Metallurgy of Doped/Non-Sag Tungsten* (E. Pink and B. Bartha, eds.), pp. 47–59, Elsevier, London (1989).
6.15. O. Schob, in: *The Metallurgy of Doped/Non-Sag Tungsten* (E. Pink and L. Bartha, eds.), pp. 83–111, Elsevier, London (1989).
6.16. O. Horacsek, in: *The Metallurgy of Doped/Non-Sag Tungsten* (E. Pink and L. Bartha, eds.), pp. 175–187, Elsevier, London (1989).
6.17. S. Leber, R. J. Arena, and D. L. Bly, *Metals Handbook*, 9th edition, Vol. 7, pp. 629–631, ASMI, Materials Park, Ohio (1984).
6.18. R. Eck, in: *The Metallurgy of Doped/Non-Sag Tungsten* (E. Pink and L. Bartha, eds.), pp. 285–291, Elsevier, London (1989).
6.19. R. Eck and E. Pink, in: *Proc. 13th Plansee Seminar* (H. Bildstein and R. Eck, eds.), Vol. 1, pp. 16–25, Metallwerk Plansee, Reutte, Austria (1993).
6.20. A. Luo and D. L. Jacobson, in: *Proc. 13th Plansee Seminar* (H. Bildstein and R. Eck, eds.), Vol. 1, pp. 263–277, Metallwerk Plansee, Reutte, Austria (1993).
6.21. W. J. van den Hoek and A. G. Jack, "Lamps," in: *Ullmann's Encyclopedia of Industrial Chemistry*, Vol. A15, pp. 115–150, VCH, Weinhein (1996).
6.22. *Electrodes for TIG Welding*, company brochure, Metallwerk Plansee, Reutte, Austria (1986).
6.23. O. Prause, in: *Proc. 7th Int. Tungsten Symposium*, pp. 289–293, ITIA, London SW90QT, UK (1996).
6.24. J. Resch and G. Leichfried, in: *Proc. 7th Int. Tungsten Symposium*, pp. 294–311, ITIA, London SW90QT, UK (1996).

6.25. K. S. Shin, A. Luo, and D. L. Jacobson, *JOM*, pp. 12–15 (August 1990).
6.26. T. W. Penrice, *Metals Handbook*, 9th edition, Vol. 7, pp. 688–691, ASMI, Materials Park, Ohio (1984).
6.27. R. F. Cheney, *Metals Handbook*, 10th edition, Vol. 2, pp. 392–393, ASMI, Metals Park, Ohio (1990).
6.28. R. M. German, in: *Proc. 14th Plansee Seminar* (G. Kneringer, P. Rödhammer, and P. Willhartitz, eds.), Vol. 1, pp. 194–206, Plansee AG, Reutte, Austria (1997).
6.29. W. D. Cai, Y. Li, R. J. Dowding, F. A. Mohamed, and E. J. Lavernia, *Rev. Particulate Mater.* **3** (1995), 71–132.
6.30. R. M. German and K. S. Churn, *Metall. Trans.* **15A** (1984), 747–752.
6.31. R. M. German, L. L. Bourguignon, and B. H. Rabin, *J. Met.* **37**, No. 8 (1985), 36–39.
6.32. M. C. Hogwood and A. R. Bentey, in: *Tungsten and Refractory Metals 2* (A. Bose and R. J. Dowding, eds.), pp. 37–45, MPIF, Princeton, N.J. 1995).
6.33. A. Bose, G. Jerman, and R. M. German, *Powder Metall. Int.* **21**, No. 3 (1989), 9–13.
6.34. *Densimet Heavy Metal Alloys*, company brochure, Metallwerk Plansee, Reutte, Austria (1982).
6.35. *Tungsten-Based Products*, company brochure, Ashot Ashkelon, Israel.
6.36. H. Danninger and B. Lux, in: *Proc. 14th Plansee Seminar* (G. Kneringer, P. Rödhammer, and P. Wilhartitz, eds.), Vol. 1, pp. 315–328, Metallwerk Plansee, Reutte (1997).
6.37. B. H. Rabin, A. Bose, and R. M. German, *Int. J. Powder Metall.* **25**, No. 1 (1989), 22–27.
6.38. R. M. Germain, in: *Tungsten and Tungsten Alloys—1992* (A. Bose and R. J. Dowding, eds.), pp. 37–45, MPIF, Princeton, N.J. (1992).
6.39. R. J. Dowding, M. C. Hogwood, L. Wong, and R. L. Woodward, in: *Tungsten and Refractory Metals 2* (A. Bose and R. J. Dowding, eds.), pp. 3–9, MPIF, Princeton, N.J. (1995).
6.40. L. S. Magness and D. Kapoor, in: *Tungsten and Refractory Metals 2* (A. Bose and R. J. Dowding, eds.), pp. 11–20, MPIF, Princeton, N.J. (1995).
6.41. L. Ekbom, L. Holmberg, and A. Persson, in: *Tungsten and Tungsten Alloys—1992* (A. Bose and R. J. Dowding, eds.), pp. 551–558, MPIF, Princeton, N.J. (1992).
6.42. Y. Shen, P. Lattari, J. Gardner, and H. Weigard, *Metals Handbook*, 10th edition, Vol. 2, pp. 840–868, ASMI, Materials Park Ohio (1990).
6.43. Y. Itoh, M. Takahasi, and H. Takano, *Fusion Eng. Design* **31** (1996), 279–289.
6.44. L. C. Farrar and J. A. Shields, Jr., *JOM*, pp. 30–35 (August 1992).
6.45. L. Zehnder, in: *Proc. 7th Int. Tungsten Symposium*, pp. 317–335, ITIA, London SW90QT, UK (1996).
6.46. H. Mayer, in: *Pulvermetallurgie und Sinterwerkstoffe* (F. Benesovsky, ed.), pp. 179–189, Metallwerk Plansee, Reutte (1982).
6.47. H. Danninger and B. Lux, in: *Proc. 13th Plansee Seminar* (H. Bildstein and R. Eck, eds.) Vol. 1, pp. 486–500, Metallwerk Plansee, Reutte (1993).
6.48. J. L. Johnson, K. F. Hens, and R. M. German, in: *Tungsten and Refractory Metals 2* (A. Bose and R. Dowding, eds.), pp. 245–252, MPIF, Princeton, N.J. (1994).
6.49. *Electrode Materials for Electrical Discharge Machining*, company brochure, Metallwerk Plansee, Reutte, Austrialia (1985).
6.50. R. Warren, in: *The Metallurgy of Doped/Non-Sag Tungsten* (E. Pink and L. Bartha, eds.), pp. 293–301, Elsevier, London (1989).
6.51. T. Grobstein and D. W. Petrasek, *Metals Handbook*, 10th edition, Vol. 2, pp. 582–584, ASMI, Materials Park, Ohio (1990).
6.52. F. J. Ritzert and R. L. Dreshfield, in: *Tungsten and Tungsten Alloys—1992* (A. Bose and R. J. Dowding, eds.), pp. 455–462, MPIF, Princeton, N.J. (1992).

7

Tungsten and Tungsten Alloy Products

Besides tungsten metal, only those alloys which were described in Chapter 6 are treated in this chapter. Alloys produced by melting metallurgy as well as the cemented carbides (hardmetals) and their application are treated in Chapters 8 and 9, respectively.

The application of tungsten metal which, at the beginning of the 20th century, was used only as a filament in incandescent bulbs and as alloying element in steel, as well as of its alloys, is nowadays very widespread and covers quite different fields such as lighting, electronics, high-temperature technology, medicine, aviation, military uses, sports, and so on.

In order to facilitate use of this survey, the application of tungsten and of its alloys is divided into 13 different fields. The material used for the application(s) is given in parentheses. (Note: HM is used for **H**eavy **M**etal).

7.1. LIGHTING APPLICATION [7.1–7.7]

About 4–5% of annual worldwide tungsten production is consumed by the lighting industry. The largest market is still for incandescent lamps (worldwide production in 1995: $\sim$10.5 billion lamps) [7.8]. However, more than 70% of artificial lighting is today generated by fluorescent lamps [7.1] and this portion is steadily increasing. Modern production units can produce more than 6000 lamps per hour [7.1]. Tungsten is used in the form of wires, coils, and coiled coils in incandescent lamps, and as electrode in low- and high-pressure discharge lamps.

The most important artificial light sources are summarized in Table. 7.1 in terms of their main lighting characteristics and applications. From this table it is clear that the efficiency of the light sources (efficacy in $lm \cdot W^{-1}$) has significantly improved over the years, in particular through the development of discharge lamps. Modern fluorescent lamps consume 80% less energy than an incandescent lamp of similar light output. Their cost effectiveness can save more in energy and replacement than their initial purchase price. Calculations have indicated that more than 40% of the energy used worldwide for lighting

TABLE 7.1. Standard Lamp Types, Lamp Efficacy, and Typical Applications[a]

Type	Maximum lamp efficacy ($lm \cdot W^{-1}$)	Burning life (hours)	Main characteristics	Typical applications
Edison lamp	1.4	40	Carbon filament/1879	
Incandescent lamps	22	1,000	Easy to install, wide variety; low cost price	General lighting in the home; decorative lighting
Halogen	27	2,000	Compact; high light output; longer life	Accent lighting; floodlighting; background lighting
Fluorescent	104	7,500	Wide choice of colors; economical in use	Lighting in public buildings; streetlighting; home lighting
Gas discharge				
High-pressure mercury	63	12,000	High efficacy; reasonable color quality	Sports grounds; industry; residential areas
Metal-halide	94	6,000	Excellent color rendering	Floodlighting; plant irradiation
High-pressure sodium	125	12,000	Very high efficacy; good color rendering	Public lighting; floodlighting; industrial lighting
Low-pressure sodium	200	10,000	Extremely high efficacy; poor color rendering; monochromatic light	Wherever energy/cost effectiveness is important, i.e., highway lighting

[a] By courtesy of Philips Lighting B.V., The Netherlands [7.5].

could be saved simply by changing from standard to more advanced light sources, without losing light quality [7.5].

Filaments in incandescent lamps (NS–W). Incandescent bulbs of various shapes and colors dominate the market worldwide in regard to general lighting. They are used as household lamps, automotive lamps, traffic signal lamps, reflector lamps, etc., but also for many thousands of specialty lamps. Wire diameters are commonly in the range of 17 to 50 μm, but can be as low as 6 μm or as high as 1 mm. Power ratings are from a few watts (watch lighting) to several thousand watts (lighthouse lamps; up to 10 kW).

The overwhelming majority of filaments are made of NS–W (sag-resistant tungsten). Only for special shock and vibration resistant lamps are $W–ThO_2$ or W–Re wires used. Every year, about 20 billion meters of lamp wire are drawn, a length which corresponds to about 50 times the earth–moon distance [7.8].

Large bulbs are used for conventional incandescent lamps to spread the evaporating tungsten over a large area in order to minimize wall blackening. Operating temperatures are between 1700 and 2500 °C. Household lamps have inert gas fillings ($Ar–N_2$) and operating pressures of about 1 bar.

Filaments in halogen lamps (NS–W of highest non-sagging quality). Halogen lamps are used more and more for indoor and outdoor applications due to their higher luminous efficiency as compared to conventional lamps. Bulb wall blackening is avoided by the addition of halogen compounds (commonly mono-, di-, or tribromomethane) and a higher inert gas pressure. Therefore, halogen lamps can be operated at higher temperatures (up to 3000 °C) and have about twice the lifetime (Table 7.1) of conventional incandescent lamps.

The principle of the *halogen cycle* is shown schematically in Fig. 7.1 [7.9]. In the vicinity of the filament, where the temperature is above 1200 °C (area C), only atoms of tungsten and halogens are present. Near the quartz bulb wall (area A in Fig. 7.1) a tungsten–halogen compound forms, which diffuses into the hot zone, where it dissociates. The "brought back" tungsten is deposited on the filament. A minimum bulb temperature of 250 °C and a maximum bulb temperature of about 900 °C is necessary to maintain the tungsten regenerative cycle.

In commercial bulbs tungsten–oxygen–halogen compounds will form rather than tungsten–halogen compounds, because traces of oxygen are always present. Nevertheless, the principle of regeneration is the same.

In case of iodine or bromine additions, the deposition of tungsten does not occur at the place where it was evaporated, so hot spots will form by thinning in time, finally resulting in failure of the lamp. Computer calculations have demonstrated that fluorine additions would result in a selective deposition of tungsten directly on the hot spot. However, such additions are not practicable, because they would destroy the glass bulb [7.1].

The variety of coils used in incandescent and halogen lamps is shown in Fig. 7.2.

Electrodes in fluorescent lamps (coated W or NS–W). Fluorescent lamps are low-pressure mercury discharge lamps which radiate in the UV region. The ultraviolet radiation is converted into light by means of a fluorescent layer (phosphor). Tungsten coils, coated with a mixture of Ca, Ba, and Sr compounds, are used as electrodes (emitter).

Electrodes in high-intensity discharge lamps (coated NS–W, W–ThO_2, NS–W–ThO_2, W–CeO_2, W–La_2O_3, and porous W). These lamps are light sources with high output, high luminous efficiency, and long operating time (Table 7.1). Their main applications are lighting of roads, outdoor areas, halls, shopping windows, agriculture, horticulture, and in photography and medical technology. They were only introduced recently for automotive lighting (Xe microdischarge lamps).

The following lamp types are in use:

- High-pressure mercury lamps (mercury vapor pressure during operation: $\sim 4 \times 10^5$ Pa).
- Low-pressure sodium lamps (sodium vapor pressure ~ 0.5 Pa).
- High-pressure sodium lamps (sodium vapor pressure 10^4 Pa).
- Metal halide gas discharge lamps.

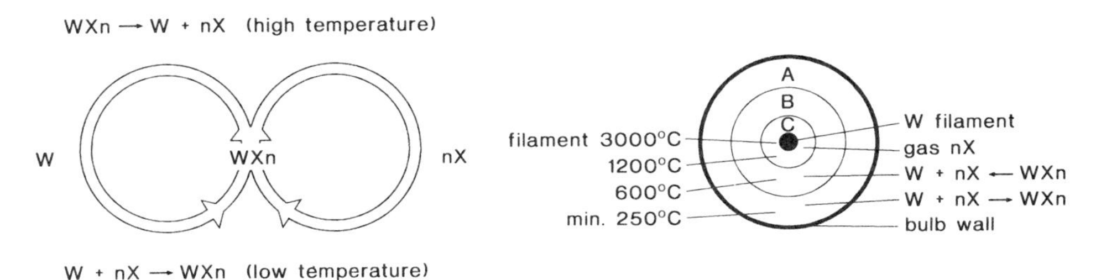

FIGURE 7.1. Schematic presentation of the "halogen cycle" in a halogen lamp [7.9]. (W...tungsten; *n*X...halogen; WX*n*...tungsten–halogen compound). The cycle is described in the text.

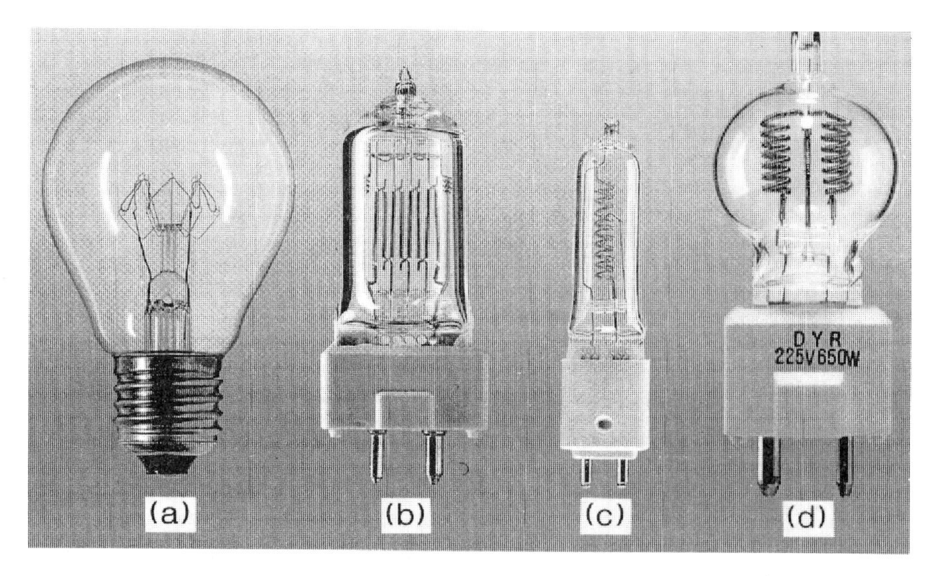

FIGURE 7.2. Tungsten filaments of different shape used in (a) traffic signal lamps, (b) projection/overhead lamps, (c) entertainment lighting/studio/film lamps, (d) video/photo lamps. By courtesy of Philips Lighting B.V., The Netherlands.

Coiled tungsten electrodes (Fig. 7.3) [7.10] or tungsten rod electrodes, coiled with a thinner tungsten wire or coil, all coated with emissive material, are used as electrodes. Exceptions are metal–halide lamps where thoriated tungsten or W–CeO_2 are used. Emitters usually contain Ba, or mixtures of Ba, Ca, and Sr compounds (tungstates, aluminates, tantalates) to lower the work function. Electrode temperatures are below 2000 °C. All lamps have noble gas fillings (either neon/argon or xenon) to facilitate ignition.

Figure 7.4 shows a section through a metal–halide lamp [7.11]. A short description of the principle is given below.

Electrodes for Xe arc lamps and mercury Xe arc lamps (NS–W, W–ThO_2, W–La_2O_3, porous W). Xe arc lamps are used as white light source for film and video projection as well as in scientific instruments (Fig. 7.5). Mercury–xenon arc lamps are used as an UV light source in photolithography.

Lamp support wire or rod (W, NS–W).

7.2. ELECTRICAL ENGINEERING [7.7, 7.12, 7.13]

Rods as electrical contacts (W).

Vacuum-tight conducting connections (W).

Contact rods to Hg switches (W).

Electrical contacts for high-frequency switching (W). Tungsten is an ideal contact material for switches with high operation frequency, such as car horn contacts, ignition

FIGURE 7.3. Triple-coil tungsten electrodes coated with emissive material, used in low-pressure sodium lamps [7.10]. By courtesy of Philips Lighting B.V., The Netherlands.

contact breakers, voltage regulators for automotive generators, etc. Optimal properties may be achieved by a fine grained microstructure.

Electrical contacts in high-performance switches (W–Cu, W–Ag, WC–Ag). Tungsten-containing contact materials are used for high-voltage (72.5–800 kV), medium-voltage (5–38 kV), and low-voltage ($\leq$1 kV) switches. The magnitude of switching currents can attain 100 kA, and arc temperatures in excess of 10,000 K can prevail for milliseconds during switching [7.13]. Evaporated copper (W–Cu) or silver (W–Ag, WC–Ag) cool the tungsten matrix ("sweat" cooling).

Typical shapes used in high-voltage switches are shown in Fig. 7.6.

Heating filaments for defrosting automobile shields (NS–W).

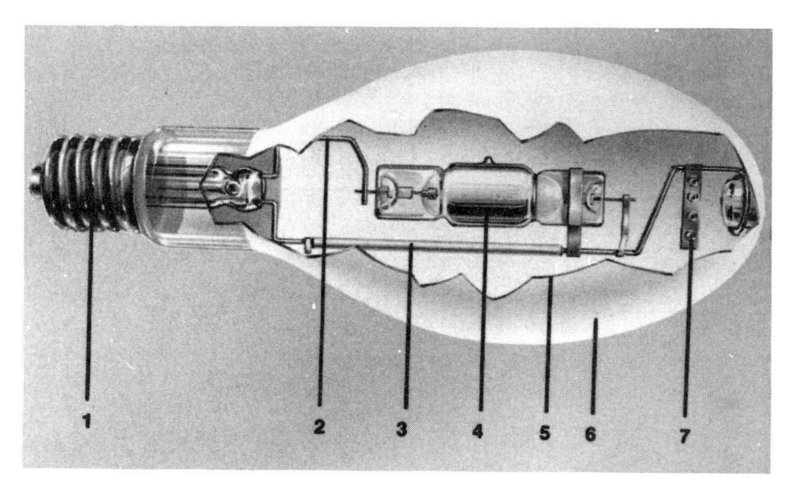

FIGURE 7.4. Section through a metal–halide lamp: (1) mechanically fitted E40 screw base, (2) lead-in wire/support, (3) sleeve protecting the support, (4) quartz discharge tube, (5) internal phosphor coating, (6) elliptical hard glass outer bulb, (7) getter ring for maintaining a clean gas atmosphere. By courtesy of Philips Lighting B.V., The Netherlands.

Light is obtained from an electrical discharge occurring between two spiral-like tungsten electrodes inside a quartz tube. Within the tube is a mixture of two gases (neon and argon) to facilitate ignition. Also contained within the tube are metallic mercury and three metals in iodide form (indium, thallium, sodium). After ignition, the temperature increases and the three iodides as well as the mercury evaporate. After dissociation of the iodides, the metals alone are excited and produce specific radiation mainly in the visible part of the spectrum. The internal fluorescent coating converts UV radiation into visible light which further improves the color impression [7.11].

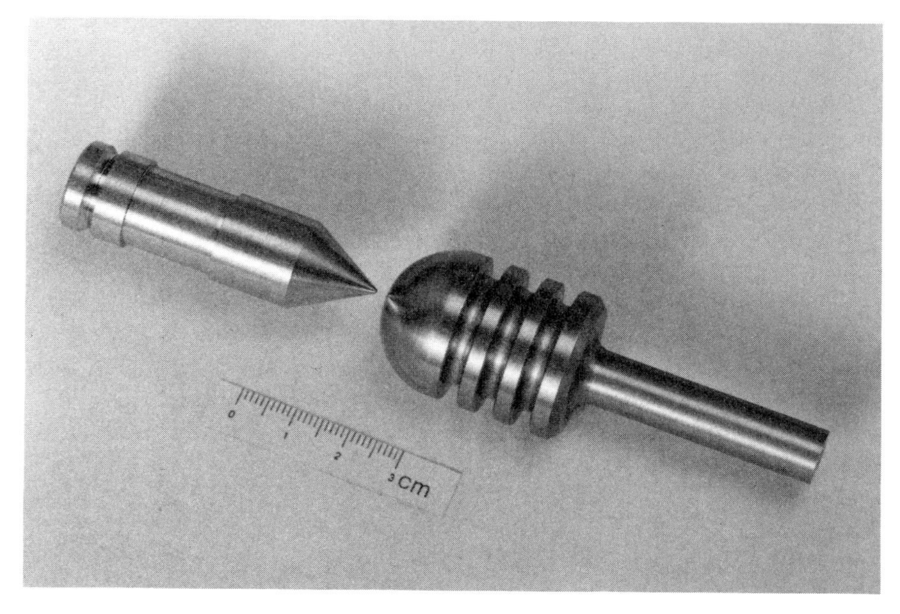

FIGURE 7.5. Thoriated tungsten cathode (left) and thoriated tungsten anode (right) for xenon short-arc lamps used for wide-angle cinema projection. By courtesy of Plansee AG, Austria.

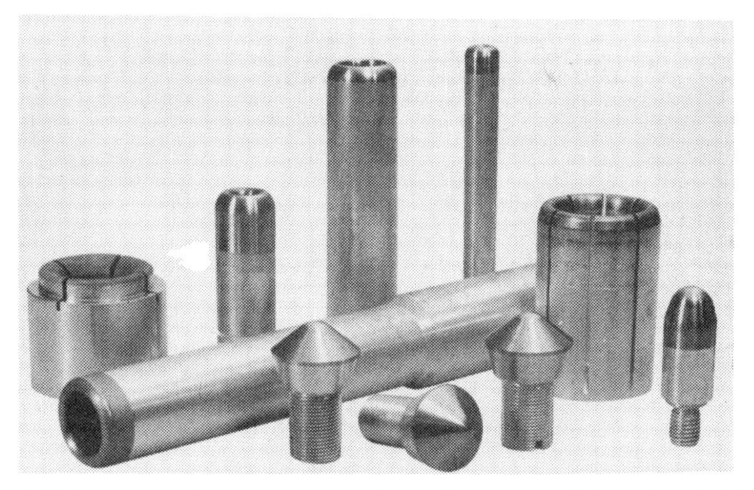

FIGURE 7.6. Contacts for high-voltage switch gear. By courtesy of Plansee AG, Austria.

7.3. ELECTRONICS [7.7]

Electron emitter (W, W–ThO_2, W–La_2O_3, porous W). Other elements are better emitters, but W has an extremely low vapor pressure and can be used at higher temperatures (it keeps the filament free of gas contamination, which would decrease emission). Therefore, W is practically the only material used in this field. Tungsten can be used as cold (field emission) and as hot emitter (thermionic emission). The principles are explained in Section 1.2.7. Field emission cathodes are used in field emission electron microscopy (FESEM). Examples of thermionic emission are manifold:

- Heater wires of oxide-coated metal cathodes in electronic tubes (W, W–ThO_2, porous W infiltrated with barium–aluminum compounds). BaO coating reduces the work function and thus increases the emission.
- Directly heated cathodes for transmitter tubes, rectifier and electrometer tubes (W–ThO_2, infiltrated porous W).
- Cathodes of electron guns (W, W–ThO_2); for example, for electron beam evaporation.
- Filaments for magnetrons (NS–W–ThO_2) for microwave ovens.
- Heavy-duty cathodes of high-power pulsed microwave tubes (impregnated porous W); cathodes of this type are shown in Fig. 7.7.
- Filaments of transmission and scanning electron microscopes (W, W–ThO_2).

Grid wires in vacuum tubes (NS–W–ThO_2, W–ThO_2).

Vacuum tight conducting connections (W). Ground pins for glass seals in high vacuum tubes.

Tungsten charger wires (Corona wires) for laser printers, photocopiers, FAX machines, etc. (W, gold-plated, Pt–clad W).

Ni or Au coated rounds (W). They serve as supports for silicon single-crystal plates used in transistors, diodes, thyristors, and rectifiers.

FIGURE 7.7. Porous tungsten cathodes with large surfaces for high-power microwave tubes. By courtesy of Plansee AG, Austria.

Heat sinks (W–Cu) [7.14]. Downscaling and higher integration of electronic devices (LSI via VLSI to ULSI) generates very high and localized heat release. An effective and fast heat transfer to heat sinks of W–Cu composite materials of different W/Cu ratios is necessary. The following properties are of importance:

- High thermal conductivity and high thermal stability enabling a co-firing with the ceramic substrates.
- Low thermal expansion coefficient matching that of the co-fired ceramic capsule.
- Good machinability and smooth surface finishing.

Different heat-sink geometries are shown in Fig. 7.8. They are used for submounting chips, as substrates for IC, as base metal for H/F power transistors, as motherboard for multichip LSI, and housing for H/F power amplifiers.

Thin films in microelectronic and optoelectronic devices (W, WSi_x, W–Ti). The important properties are listed below.

Examples are source, drain, and gate metallization in transistors and diffusion barriers, ohmic contacts, and interconnections (lines, vias, plugs) in integrated circuits. The favorable electrical, mechanical, and chemical properties are:

- high activation barrier for self diffusion,
- high resistance to electron migration under high current densities,
- low thermal expansion coefficient only marginally changing with temperature, matching very well that of Si,
- formation of ohmic contacts with Si and III/V semiconductors (GaAs or InP),
- the only metal not reacting with III/V semiconductors up to 600–700 °C,
- does not form silicides <600 °C in contact with pure or n^+/p^+ doped Si layers.

Thin films are deposited by CVD or sputtering.

FIGURE 7.8. W–Cu heat sinks for electronic devices. By courtesy of Sumitomo Electric Industries, Ltd., Japan.

7.4. HIGH TEMPERATURE TECHNIQUE [7.7, 7.15]

Heating element in high-temperature furnaces (W, W–La_2O_3). The elements are fabricated from wire, mesh, sheet, strip, or open seam tube for temperatures up to 2800 °C (for typical examples, see Figs. 7.9 and 7.10). Due to the high electrical conductivity of tungsten, large currents and low voltages must be applied. Furnaces operate either in a high vacuum or in a protective atmosphere.

Radiation shields in high-temperature furnaces (W). Shields closest to the tungsten heating element are made of tungsten sheets.

FIGURE 7.9. Spacer parts, heating elements, and heat radiation shields of tungsten and molybdenum for a high-temperature furnace; height of the heating zone, approximately 440 mm. By courtesy of Plansee AG, Austria.

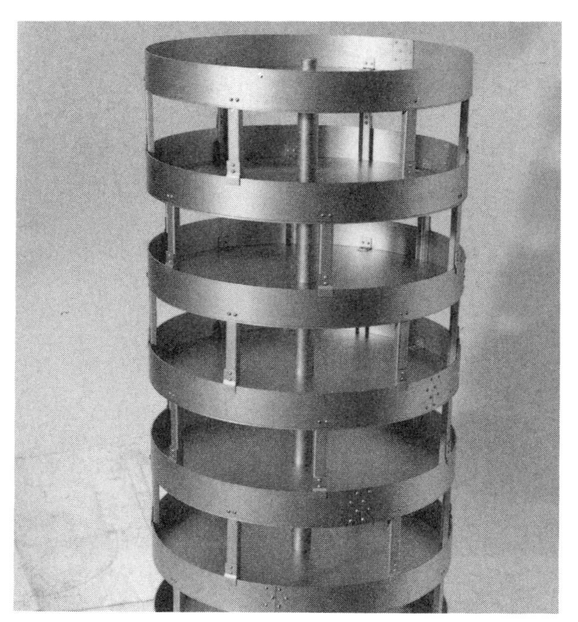

FIGURE 7.10. Tungsten/molybdenum furnace for heat treatment of lamp components; charge carrier and heating elements are made of tungsten. By courtesy of Plansee AG, Austria.

Charge carriers for high-temperature furnaces (W). Other furnace components are rivets, pins, washers, spacers, and annealing boats (W).

Susceptors in high-frequency furnaces (W, W–La_2O_3).

Thermocouples (W/W–Re). These are employed for temperature measurement in high-temperature furnaces in hydrogen or inert atmosphere or in vacuum, and in combustion chambers of turbine engines or reactors (up to 2600 °C). Couples mostly used are W–5Re/W–26Re and W–3Re/W–25Re.

Ionization chambers (W) for implantation of ions in the doping of silicon wafers.

Applications which might gain more importance in the future are *Construction material for the plasma chamber in magnetohydrodynamic power generation* (W and W–Cu) and *target plates in fusion reactors* (W, W–La_2O_3) [7.16, 7.17].

Recent plasma experiments and theoretical and numerical studies show that tungsten may be the best, if not the only, material to withstand the extraordinary operating conditions in a nuclear fusion reactor divertor. The divertor, being that part of the vacuum vessel where the plasma particles interact with the first wall, and where a large fraction of the fusion heat is removed, consists of water-cooled copper heat-exchanger elements covered with a plasma facing armor. The plasma particles (electrons, protons, and α-particles) are directed by the magnetic field toward the divertor target plates, where they are neutralized and pumped. The convective heat flux reaches 20 $MW \cdot m^{-2}$ and the attendant surface temperature more than 3000 °C. Therefore, a suitable armor material must have a high thermal conductivity (in order to transfer high heat fluxes), low thermal expansion coefficient and low Young's modulus (in order to keep thermal stresses low), and a high melting point and low sputtering yield (in order to keep erosion low). Although tungsten does not have as high a thermal conductivity and as low a Young's modulus as carbon–carbon composite materials, which are foreseen for the sections of the divertor with the highest heat flux, many experts believe that, in the long run, reasonable lifetimes will only be achieved by tungsten divertor plates, which have the lowest erosion rates of all materials in sections of the divertor with relatively low plasma temperature but high particle density.

In the current ITER (International Thermonuclear Experimental Reactor) divertor design (November 1997; Fig. 7.11), almost the whole divertor surface is covered by tungsten, representing more than 100 tons in the case of the ITER construction. During ITER operation, the tungsten target plates will have to be replaced between 5 to 8 times due to sputtering erosion [7.16].

7.5. WELDING, CUTTING, PLASMA SPRAYING, SPARK EROSION, AND VAPOR DEPOSITION [7.7, 7.18]

Welding electrodes (W, W–ThO_2, W–La_2O_3, W–CeO_2, and W–ZrO_2). Their application is in inert gas welding (TIG) and plasma welding. W–ThO_2 or W–CeO_2 carry 50% more current than W, start more easily, and produce a more stable arc.

Hot cathode emitter for electron beam welding and cutting (W–ThO_2).

Electrodes (W–ThO_2, W–Cu) *and nozzles in plasma guns* (W–ThO_2). The guns are used in plasma melting, arc cutting, and plasma spraying.

Electrodes for resistance welding (W–Cu, W–CeO_2, W–La_2O_3).

Electrodes for electrical discharge machining, arc erosion, and wire eroding (W,

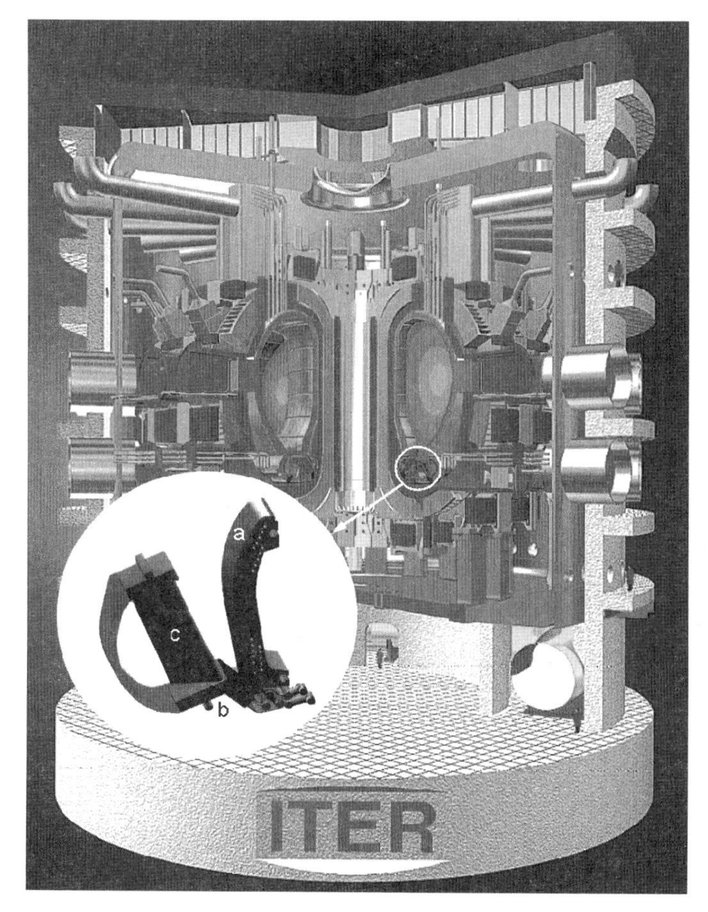

FIGURE 7.11. Schematic view of the International Thermonuclear Experimental Reactor (ITER); 100 t of tungsten are used for the ITER construction. By courtesy of Plansee AG, Austria.

W–Cu, W–ThO_2). Electrical discharge machining allows close dimensional tolerances, accuracy of form, and a high-quality surface finish to be achieved. Both hardened steels and hardmetals can be machined (Fig. 7.12). Tungsten in wire eroding is not only used as a guide and contact for the wire electrodes, but also as erosion wire.

Heating coils or boats for vacuum metallizing (W and NS–W). Tungsten is an excellent material for evaporation sources. Coils (Fig. 7.13) and boats (Fig. 7.14) of different size and shape are used. The main application is to evaporate aluminum, but a variety of other metals and metal compounds can be evaporated too, as listed in Table 7.2. Evaporization is performed for metallizing different products such as reflector lamps, head lamps, television picture tubes, electronic components, plastic, etc.

Disks as sputtering cathodes (high purity W, W–Ti, W–Si_x, MoW). Sputter targets are used to produce thin layers in integrated circuits (VLSI) by PVD. Recently, W–Mo sputter targets were used for the manufacture of flat panel displays [7.19].

FIGURE 7.12. Electrical discharge machining of a compacting die used in the production of gear parts. By courtesy of Plansee AG, Austria.

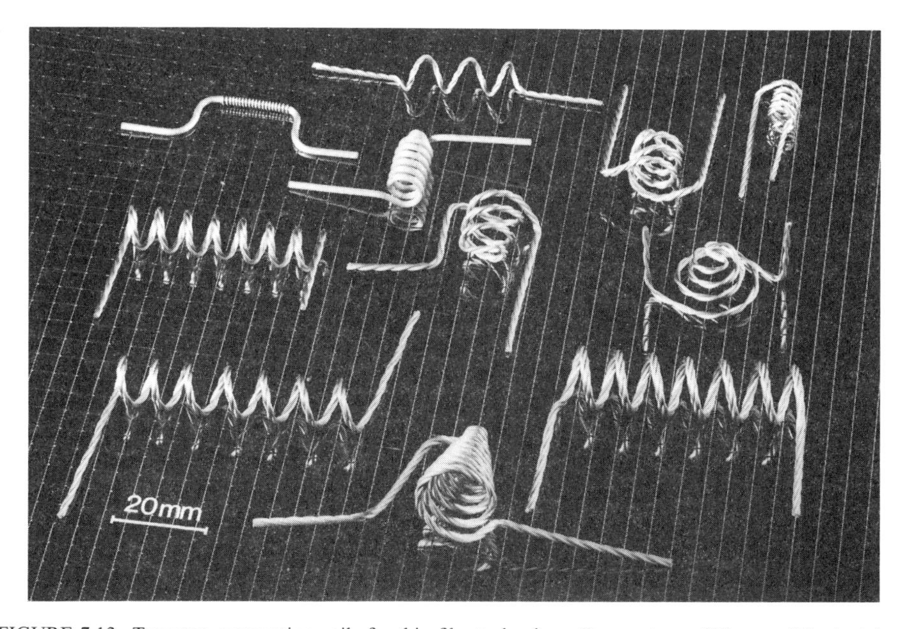

FIGURE 7.13. Tungsten evaporation coils for thin-film technology. By courtesy of Plansee AG, Austria.

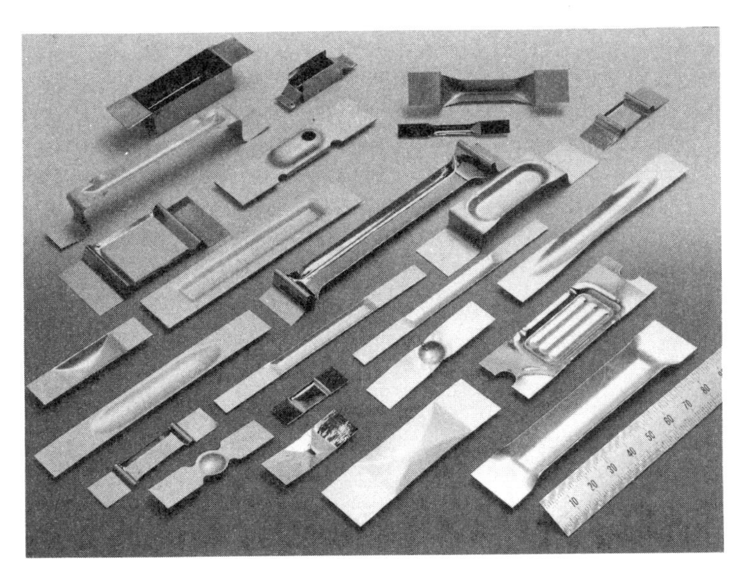

FIGURE 7.14. Tungsten, molybdenum, and tantalum evaporation boats. By courtesy of Plansee AG, Austria.

TABLE 7.2. Chemical Compounds which can be Evaporated with Tungsten Evaporation Sources[a]

Ag	Cr_2O_3	MgO	Se
Al	CsF	Mn	Si
Al/0.1–2% Si	Cu	MnS	Sm_2O_3
Al/1–4% Cu	Er_2O_3	NaF	Sn
Al/4% Cu/1% Si	Eu_2O_3	Nb	Ta_2O_5
Au	Fe	Nb_2O_5	Tb_2O_3
BaF_2	Fe_2O_3	Nd_2O_3	Te
$BaTiO_3$	Ga	Ni	Ti
BeO	GaAs	Ni/Cr	Ti_2O_3
Bi	GaP	Pb	TiO
Bi_2O_3	Gd_2O_3	PbF_2	TiO_2
CaF_2	Ge	PbO	V
Cd	GeO_2	PbS	WO_3
CdS	In	PbSe	Y
CeO_2	In/Sn	Pd	Yb_2O_3
Co	In_2O_3	Pr_2O_3	Zn
Cr	La_2O_3	Pt	Zn/Se
Cr-SiO	Lu_2O_3	Sb	Zr
Cr_2C_3	LuF_3	Sb_2O_3	ZrO
	Mg	Sc_2O_3	ZrO_2

[a] By courtesy of Plansee AG, Austria.

7.6. X-RAYS, RADIATION, MEDICAL ENGINEERING [7.7]

Static and rotating anodes in X-ray tubes (W and W–Re). For high performance tubes W–Re is preferred. These anodes are used in modern medical technology (e.g., computerized tomographic scanning equipment). Only the surface layer consists of W/3–10Re while the support body can be a Mo–Ti–Zr–C microalloy, called TZM, or graphite. The W–Re layer can be produced either by powder metallurgical techniques or by CVD. Compared to pure tungsten, which was employed in former times, the W–Re alloy allows higher electron currents yielding in shorter irradiation times. The important property is the increased stability to extreme alternating thermal stresses. Roughening of the heating zone is much retarded in comparison to W, hence increasing the lifetime of the tube.

Rotating anodes and a complete X-ray tube are presented in Fig. 7.15.

Directly heated cathodes in X-ray tubes (W).

Components for X-ray and radiation shielding (W, HM, HM–polymer) [7.20–7.23]. Typical examples are containers or flasks for radioactive materials and liquids, shielded syringes (Fig. 7.16), shielding construction parts such as collimators in computer tomographic scanners (Fig. 7.17), radiation therapy instruments, and containers and shielding for oil prospecting using radioactive sources (Fig. 7.18).

Infrared reflector lamps (NS–W) used in medicine and health promotion (for example, treatment of hyperbilirubinaemia in a baby).

FIGURE 7.15. Rotating X-ray target with graphite body and tungsten–rhenuim focal track. By courtesy of Plansee AG, Austria.

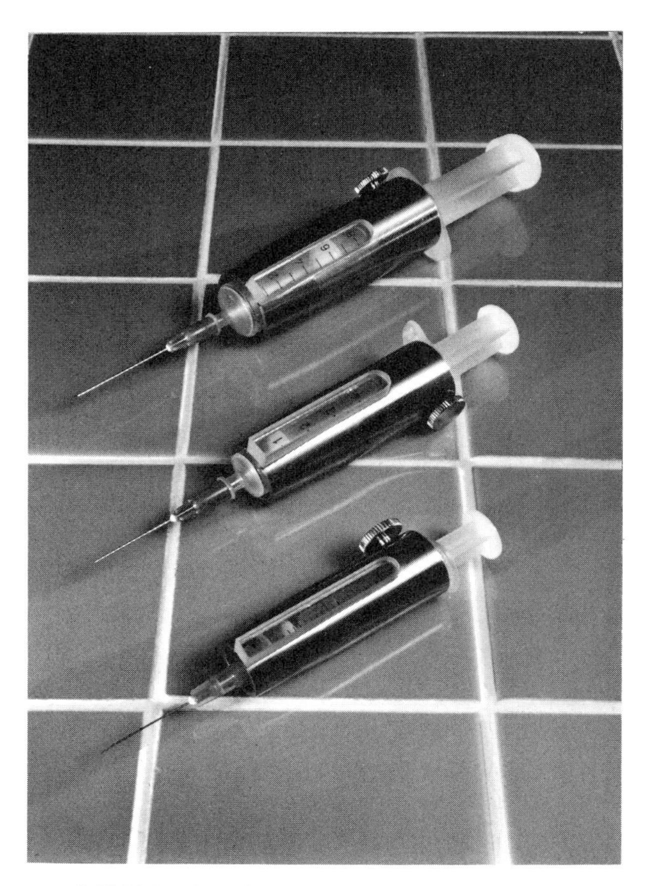

FIGURE 7.16. Heavy metal shielded syringes for applications in nuclear medicine. By courtesy of Plansee AG, Austria.

7.7. MECHANICAL AND ENGINE ENGINEERING [7.20, 7.21]

Construction parts (W) for high-temperature nuclear reactors, fusion reactors, plasma technique, and ion propulsion.

Combustion chamber of turbo engines (WRe).

Gyroscope rotors (HM) (Fig. 7.19).

Actuator in self-winding watches (HM).

Floats in flow meters (HM).

Governor balance weights (HM).

Anti-vibration tool holders (HM). Heavy metal is an excellent material for damping torsional vibrations. Tools with longer shafts are typical applications. Examples are turning shanks, boring bars, endmills, grinding quills, pinion type cutter supports, etc. (Fig. 7.20).

Counter balance weights (HM).

Flywheel weights (HM).

FIGURE 7.17. Collimator platelets for computerized tomography. By courtesy of Plansee AG, Austria.

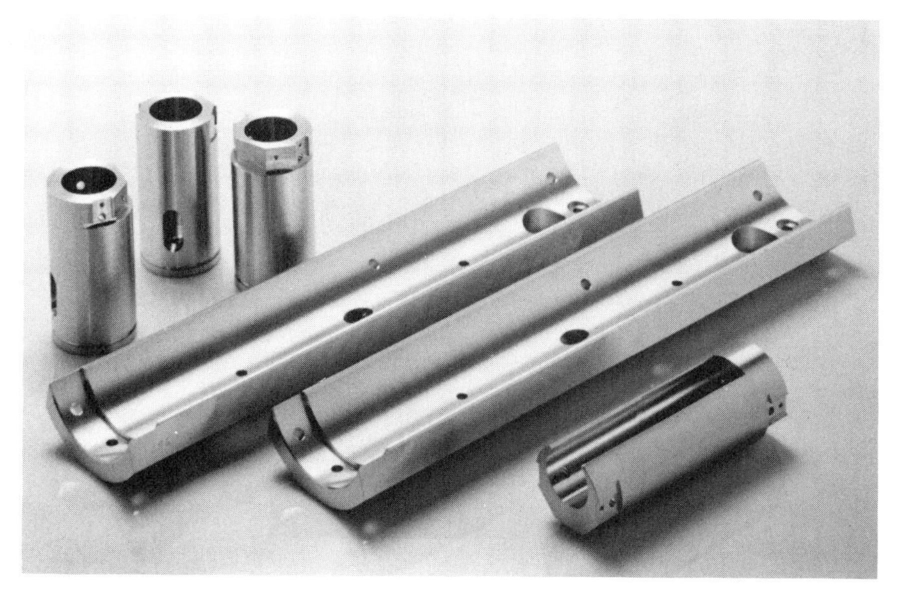

FIGURE 7.18. Heavy metal containers and shieldings for oil prospection using radioactive sources. By courtesy of Plansee AG, Austria.

FIGURE 7.19. Heavy metal rings for gyroscope systems. By courtesy of Plansee AG, Austria.

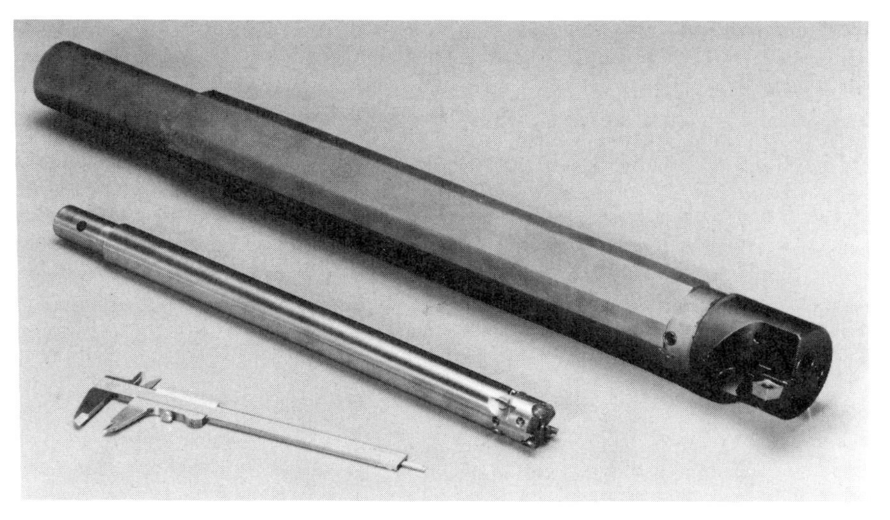

FIGURE 7.20. Heavy metal boring bars with simple and multiple cutter heads. By courtesy of Plansee AG, Austria.

7.8. LEISURE TIME AND SPORTS EQUIPMENT (FIG. 7.21)

Professional darts (HM).
Hunting cartridge shots (HM).
Diving weights (HM).
Golf club components (HM, W–Cu).
Hand weights for joggers (HM).

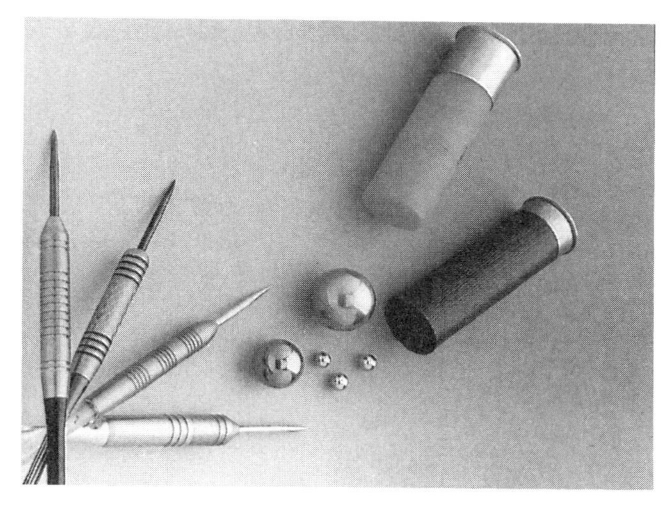

FIGURE 7.21. Professional darts and hunting cartridge shots. By courtesy of Ashot Ashkelon, Israel.

FIGURE 7.22. "Safeweight" nontoxic angler weights produced from tungsten-loaded polymers using injection molding or extrusion. By courtesy of Royal Ordnance Speciality Metals Ltd., England.

Tennis racket balance weights (HM).
Ski balance weights (HM).
Saddle weights (HM).
Fishing weights (Fig. 7.22).
Counter balance weights in Formula 1 racing cars (HM).

7.9. CHEMICAL INDUSTRY AND METALLURGY [7.23, 7.24]

Components for the production of quartz glass, transparent alumina tubes, and ceramic single crystals (W, W–La_2O_3). Continuous production of quartz glass tubes is conducted at temperatures higher than 2000 °C in order to keep the viscosity of SiO_2 sufficiently low. Tungsten crucibles, drawing dies, mandrels, and heating elements are

FIGURE 7.23. Tungsten–thoria guide nozzles. By courtesy of Plansee AG, Austria.

applied. Melting crucibles in seamless version are among the largest parts of tungsten produced in one part (weight about 800 kg) [7.25].

Electrodes for melting of caolinit (W).

Spinning nozzles for ceramic wool production (W).

Crucibles for processing of used nuclear fuels and for handling of metal melts (W, NS–W).

Tooling material for brass and aluminum die casting.

Spring material in corrosive media (Ta–2.5W, Ta–10W). Preferably for use in mineral acids and halide vapor.

7.10. SPACE AVIATION [7.7]

Ignition tubes for the main rocket engine (W–Re).

Guide nozzles of satellites for reliable positioning and stabilizing (W–ThO_2, Ta–2.5W; Fig. 7.23).

Thermionic emitter in spacecrafts (W, CVD–W, W–Re). The important properties are high-temperature resistance and stability against cesium vapor.

Ion propulsion jet (porous W). Emitters of electrostatic ion propulsion systems.

Construction parts in space nuclear reactors (W and tungsten fiber reinforced Nb matrix alloys).

7.11. MILITARY APPLICATIONS [7.22, 7.26]

High kinetic energy penetrators in tank ammunition (HM, W–Hf). A schematic drawing of a whole cartridge is shown in Fig. 7.24. Upon firing, when the projectile exits the muzzle, gas and aerodynamic forces cause the sabot segments to separate from the subprojectile. Muzzle velocity is $\sim$1700 m/s. The fin-stabilized subprojectile continues on its extremely flat trajectory to the target. Penetration is achieved by the kinetic energy of the high-density penetrator. Differently shaped and sized penetrators can be seen in Fig. 7.25.

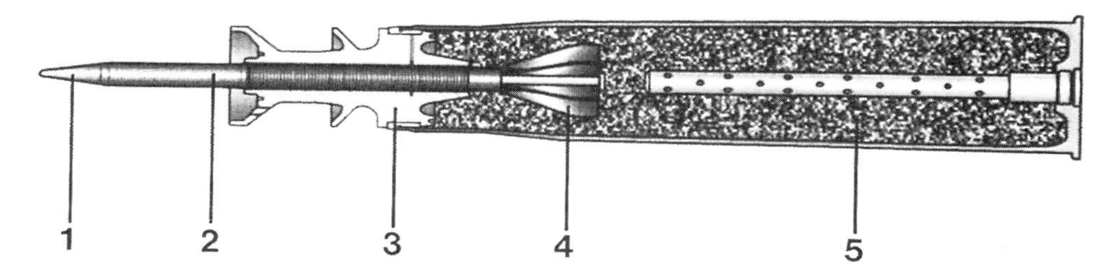

FIGURE 7.24. Schematic drawing of a heavy metal tank ammunition cartridge: (1) windshield, (2) tungsten heavy-metal penetrator (subprojectile), (3) three-section sabot, (4) stabilizing fin, (5) propellant. By courtesy of TAAS—Israel Industries Ltd., Israel.

FIGURE 7.25. Selection of large-caliber kinetic energy penetrators; medium-caliber kinetic energy penetrators and ammunition, as well as tungsten alloy cones for shaped charges. By courtesy of Royal Ordnance Speciality Metals, England.

Armor plating (HM).
Scatter grenades (HM) (fragmentation devices).
Shaped charge liners (W); Fig. 7.25.
Counterweights in tanks (HM).
Nozzles in air to air rockets and jet vanes (W, W–Cu, W–ThO_2, Ta–2.5W).

7.12. AVIATION [7.24]

Rim components of flywheels or rotating members in governors (HM).

Counterweights in static and dynamic balancing application in airplanes and helicopers (aircraft balance weights) (HM). For typical shapes, see Fig. 7.26.

Instrument balance weights (HM).

Airborne antenna bases (HM).

7.13. LASER TECHNIQUE [7.7]

Construction parts in gas lasers (W-Cu).

Cathodes (porous W impregnated with Ba aluminate). Coiled cathodes for laser resonators are shown in Fig. 7.27.

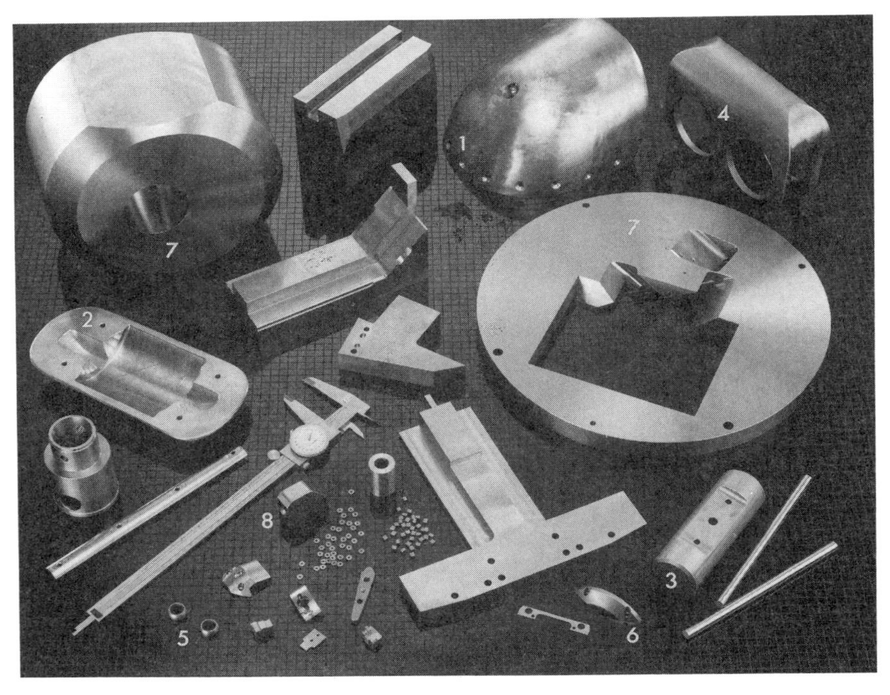

FIGURE 7.26. Heavy metal parts used in aviation and radiation shielding: (1) passenger aircraft elevator weight, (2) passenger aircraft rudder weight, (3) helicopter vibration dampener weight, (4) helicopter rotor vibration dampening weight, (5) rotor weight, (6) missile component, (7) nuclear medicine shielding, (8) radiation shielding; rest: miscellaneous parts. By courtesy of Teledyne Advanced Materials, USA.

FIGURE 7.27. Coiled cathodes of porous impregnated tungsten for laser resonators. By courtesy of Plansee AG, Austria.

REFERENCES FOR CHAPTER 7

7.1. W. J. van den Hoek and A. G. Jack, "Lamps," in: *Ullmann's Encyclopedia of Industrial Chemistry*, Vol. A15, pp. 115–150, VCH, Weinheim (1990).
7.2. C. L. Briant and B. P. Bewlay, *MRS Bulletin*, pp. 67–73 (August 1995).
7.3. F. Mertens, in: *Proc. 7th Int. Tungsten Symposium, Goslar*, pp. 277–293, ITIA, London (1997).
7.4. L. Bartha, in: *Mater. Res. Soc. Symp. Proc.* **322** (1994), 531–536.
7.5. *Philips Light... More Than Meets The Eye*, company brochure, Philips Lighting, The Netherlands (3/87).
7.6. *Lighting Technology*, company brochure, Plansee AG, Austria (9/95).
7.7. *Tungsten*, company brochure, Plansee AG, Austria (3/95).
7.8. F. Mertens, *Philips Lighting*, private communication.
7.9. *Philips Lighting*, product information, The Netherlands (9/88).
7.10. *SOX-E Low Pressure Sodium Lamps*, company brochure, Philips Lighting, The Netherlands (4/86).
7.11. *Metal Halide Gas-Discharge Lamps*, company brochure, Philips Lighting, The Netherlands (4/86).
7.12. H. Kippenberg, in: *Proc. 7th Int. Tungsten Symposium, Goslar*, pp. 312–316, ITIA, London (1997).
7.13. L. Zehnder, in: *Proc. 7th Int. Tungsten Symposium, Goslar*, pp. 317–335, ITIA, London (1997).
7.14. *CMSH*, Sumitomo Electric Industries, Ltd., Japan (12/91).
7.15. *High-Temperature Furnace Construction*, company brochure, Plansee AG, Austria (11/96).
7.16. L. Plöchl, Plansee AG, private communication.
7.17. D. Naujoks, K. Asmussen, M. Bessenrodt-Weberpals, S. Deschka, R. Dux, W. Engelhardt, A. R. Field, G. Fussman, J. C. Fuchs, C. Garcia-Rosales, S. Hirsch, P. Ignacz, G. Lieder, K. F. Mast, R. Neu, R. Radtke, J. Roth, and U. Wenzel, ASDEX Upgrade Team, *Nucl. Fusion* **36** (1996), 671.
7.18. *Electrode Materials for Electrical Discharge Machining*, company brochure, Plansee AG, Austria (9/85).
7.19. *Thin Film Technology*, company brochure, Plansee AG, Austria (7/97).
7.20. *Densimet W + WS*, company brochure, Plansee AG, Austria (6/95).
7.21. *Tungsten-Based Products*, company brochure, Ashot Ashkelon, Israel (11/92).
7.22. *Kinetic Energy Technology*, company brochure, Royal Ordnance Speciality Metals Ltd., England.
7.23. *Chemical Apparatus Engineering*, company brochure, Plansee AG, Austria (2/97).
7.24. E. Okorn and G. Leichtfried, in: *Proc. 7th Int. Tungsten Symposium, Goslar*, pp. 361–370, ITIA, London (1997).
7.25. G. Leichtfried, Plansee AG, private communication.
7.26. *Tank Ammunition*, company brochure, TAAS–Israel Industries Ltd. (8/94).

8

Tungsten in Melting Metallurgy

8.1. TUNGSTEN IN STEEL

8.1.1. Introduction

At the beginning of the 20th century steel was the biggest tungsten consumer. Since the development of cemented carbides in 1927, the proportion of total tungsten consumed in steelmaking has fallen progressively to the current value of 22%. Nevertheless, steel is today still the second largest tungsten consumer.

The early tungsten-bearing steels were produced by adding tungsten powder to the steel melt—a rather expensive procedure. In 1914 the master alloy *ferrotungsten* was introduced to the market. Ferrotungsten could be produced in a much cheaper way directly from ore concentrates, and provided a lower melting point for faster dissolution and higher tungsten yield compared with the use of tungsten powder. In 1940 a second master alloy produced from secondary raw materials—called *tungsten melting base*—came into use. Tungsten can also be added to the steel melt in the form of tungsten scrap or direct scheelite concentrates.

Tungsten alloying is important mainly for the following steel grades:

- cold work tool steels (0.5 to 3 wt% W),
- hot work tool steels (1.5 to 9 wt% W),
- high speed tool steels (2 to 18 wt% W),
  and to a lesser extent
- stainless steels and constructional steels (<1 wt% resp. $\ll 1$ wt%).

High speed steels (HSS) consume by far the most tungsten in steelmaking, although they account for less than 0.02% of the total steel production (1997: about 700 million metric tons). In fact the tungsten story in special steel making is the story of high speed steel [8.8].

By shifting the eutectoid point to lower carbon concentrations, tungsten increases the amount of undissolved and excess carbide in the hardened steel. Tungsten is a carbide-forming element and binds the carbon to form straight tungsten carbide or combines with other alloying elements to form complex carbides (M_6C, $M_{23}C_6$). The chemistry of these carbides depends on the concentration of carbon and the alloying elements present as well as on the heat treatment applied. For the same level of carbon in the steel tungsten will produce a larger carbide volume than other alloying elements [8.3].

In addition to improving hardness, tungsten and complex carbides also subscribe to retaining a fine-grained structure in steels as the excess carbides retard the grain growth during austenizing. The resulting structure can be defined as fine grained with evenly distributed, very fine carbide precipitations. The influence of tungsten on carbide formation is shown in Fig. 8.1.

The effects of tungsten on steel properties in terms of carbide volume, shape and distribution, and fine-grained structure are:

- more carbides and harder carbides increase hardness and wear resistance;
- fine carbides and a small grained structure result in higher toughness for a certain hardness level.

Good wear resistance is the most important property of tool steels, and tungsten, next to vanadium, is the most effective carbide-forming element for increasing wear resistance. Furthermore, tungsten significantly improves the hot hardness and hot strength of steel, and leads to increased yield strength and tensile strength without detriment to ductility (elongation and area reduction on breaking) and notched bar fracture toughness. The high hot and red hardness (which is the maintenance of hot hardness over long-term exposure to high temperatures) and the high hot strength are the result of a microstructure containing numerous high-alloy carbides, and a composition heavily alloyed with diffusion-slowing

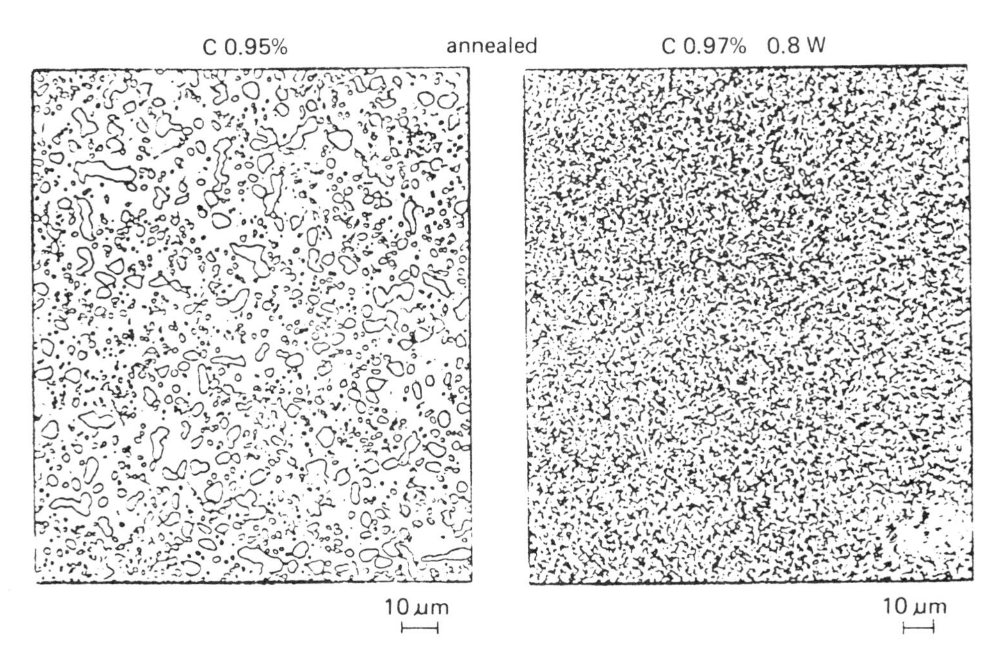

FIGURE 8.1. Influence of tungsten on carbide formation in steel [8.3].

elements and carbon. Hardening is achieved by high-temperature austenitization to dissolve the alloying elements, followed by transformation during quenching and tempering to martensite with precipitated fine carbides. This produces a material which will retain most of its hardness up to about 550 °C, and regain it again on cooling to room temperature.

Tungsten supply in the past was sometimes uncertain, and prices were dictated by the supply/demand balance. In times of shortage, as in World War II, prices increased considerably. This led to the introduction of molybdenum as a substitute for tungsten in high speed steels (most of the early grades contained up to 18 wt% tungsten), with the advantage that only half the amount is needed compared to tungsten. However, molybdenum-bearing steels generally require more sophisticated and expensive heat-treatment facilities. The most common grades of high speed steels (HSS) currently used in the western world contain both tungsten and molybdenum together with vanadium. The US market, and also Japan, mainly asks for the low-tungsten–high molybdenum grades (typically 2 wt% W; 9 wt% Mo, and 1 wt% V; *M1-type*); while in Europe the 6 wt% W, 5 wt% Mo, and 2 wt% V grade (*M2-type*) is predominant [8.8]. In the eastern countries, especially in the former Soviet Union and the Peoples Republic of China—countries which have traditionally been self-sufficient in tungsten—little tungsten substitution has occurred.

While tungsten will continue to play an important role in high speed steels, these materials are under constant attack from the extending range of application of cemented carbides, particularly in the area of cutting tools. This trend will accelerate with the modernization of machining plants in developing countries, and with the improvements in performance and tool life provided by the coating of high speed steels with hard layers such as TiN [8.8, 8.9]. Already a considerable percentage of drills sold today in the western world are coated.

From an environmental standpoint, cemented carbide tools are preferable to high speed steel because any tungsten in steel is lost by dilution (see Section 11.1) in contrast to cemented carbide scrap, which is a valuable source for recycling [8.10].

8.1.2. Master Alloys

Tungsten in most cases is added to the steel melt as master alloy. The two usual types are ferrotungsten and melting base. Due to their lower melting point, master alloys dissolve and distribute more readily in the steel melt as pure tungsten. Moreover, they are cheaper because they are produced either directly from ore concentrates in case of ferrotungsten or from scrap materials in case of melting base alloys.

8.1.2.1. Ferrotungsten. Properties. Commercial ferrotungsten contains between 75 and 80 wt% W. It has a steel gray appearance and a fine grained structure consisting of FeW (δ-phase) and Fe_2W (λ-phase). It is supplied in 80–100 mm lumps. Standard compositions are given in Table 8.1.

At 80 wt% W, the liquidus temperature is >2500 °C and the solidus temperature about 1640 °C. The density is 15.4 $g \cdot cm^{-3}$.

Production. The raw materials for ferrotungsten production are either rich ores or ore concentrates of the minerals wolframite, huebnerite, ferberite, and scheelite. Also, artificial scheelite or soft scrap can be used.

TABLE 8.1. Composition of Ferrotungsten According to ISO 5450 [8.2]

	FeW 80	FeW 80LC
W (wt%)	70–85	70–85
Ranges (wt%)		
Si	0.6–1.0	0.6–1.0
Mn	0.6–1.0	0.6–1.0
Cu	0.20–0.25	0.20–0.25
S	0.05–0.06	0.05–0.06
P	0.05–0.06	0.05–0.06
Tolerable maximum (wt%)		
C	1.0	0.1
Al	0.10	1.0
Sn	0.10	0.10
As	0.10	0.10
Mo	1.0	0.5

Tungsten is reduced in preference to iron so that ferrotungsten with high W content can be produced even if the ore is high in iron. The WO_3 contained in these compounds can be reduced either carbothermically in electric arc furnaces or metallothermically by silicon and/or aluminum. The carbothermic or silico-carbothermic method is preferred for cost reasons and, moreover, the tolerance level for impurities like As and Sn in the raw materials is higher.

Although tungsten forms carbides with carbon, the formation of WC can be largely prevented by controlling the CO/CO_2 ratio. Thus a low-carbon ferrotungsten with maximum carbon content of 1% is produced.

Carbothermic Production. Because of the high melting temperature tapping off is not possible. Therefore, the so-called *solid block process* is normally used. The process operates in two stages. Both stages are performed in electric arc furnaces. In the refining stage an ingot of high-grade ferrotungsten and a WO_3-rich slag is produced. The ingot is removed and cleaned. In the reduction stage the WO_3-rich slag is processed to ferrotungsten of 50% W and a slag having $<1\%$ WO_3. The 50% ferrotungsten together with the outer parts of the cleaned, high-grade ingot are recycled to the refining stage. A flow sheet of the process is given in Fig. 8.2. The tungsten yield is between 97 and 98%.

Typical consumption figures per metric ton of 80% FeW are:

Tungsten concentrate (65% WO_3)	1600 kg
Charcoal	150 kg
Coke	20 kg
Fluorspar	80 kg
Quarzite/calcite	50 kg
Steel scrap	75 kg
Bauxite	80 kg
Electrodes	120 kg
Electric energy	7500–8500 kWh

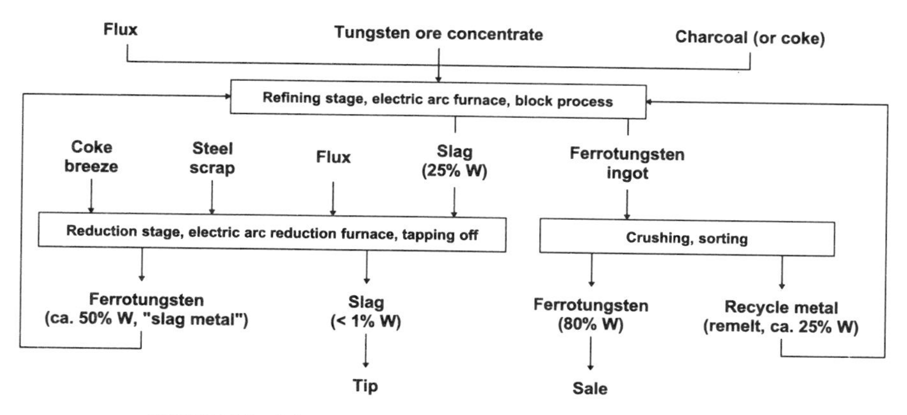

FIGURE 8.2. Scheme of carbothermic ferrotungsten production [8.2].

Carbothermic–silicothermic production. The process is carried out in three successive stages in an electric arc furnace. Sixty percent of the oxygen in WO_3 are reduced by carbon and 40% by silicon added as 75% FeSi.

Stage 1: Reduction of tungsten ore concentrates by carbon under a WO_3-rich slag (10–16% WO_3), and scooping out ferrotungsten (75% W).

Stage 2: Reduction of the WO_3-rich slag from Stage 1 by FeSi75 and addition of scrap iron. Tapping off the low-WO_3 slag (<1% WO_3). The metal phase (50–70% W) remains in the furnace.

Stage 3: Refining of the low-W metal phase by addition of ore concentrates. A WO_3-rich slag is formed (18–25% WO_3), and the W content of the metal phase increases.

Advantages are:

- continuous operation responsible for prolonged lifetime of the furnace lining;
- lower electric energy consumption.

Disadvantages are:

- heavy labor of the scooping operation;
- accumulation of some ferrotungsten in a "furnace sow" or "salamander" which can be recovered only after furnace relining.

Typical consumption figures per metric ton of 80% FeW are:

Ore concentrate (60% WO_3)	1527 kg
FeSi75	130 kg
Scrap iron	80 kg
Coke	120 kg
Electrodes	85 kg
Electric energy	4000–4200 kWh

Metallothermic Production. Metallothermic reduction requires purer raw materials, because it proceeds very rapidly and some of the impurity elements are reduced as well as tungsten. For example, about 50% of the Sn and As contained in the raw materials end up in the ferrotungsten, most of which would be evaporized during the longer-lasting carbothermic process.

Concentrates are mixed with coarsely powdered Al and Si (ratio Al/Si 70:30), and are charged into a refractory-lined furnace vessel. The vessel is preheated to 400–500 °C or, alternatively, an Al booster is added. The reaction is started by igniting initiators (e.g., BaO_2 and Al), and is completed in terms of several minutes. After cooling, the vessel is removed and the metal block (700 to 2000 kg) is separated. The yield is about 96%.

The metallothermic reduction process has lost much of its importance in recent years owing to the increased costs of Al and Si powders and the necessity for using pure and expensive raw materials.

8.1.2.2. Melting Base. Properties. This type of master alloy can have various compositions, depending on the scrap material used. Typical examples are listed in Table 8.2.

Production. Different types of tungsten-containing scrap are mixed to meet the desired composition. Reductive melting is done in electric arc furnaces as usual in steel melting. The product is finally granulated (<10 mm) by casting on a rotating disc and quenching in water.

8.1.3. Tungsten Alloyed Steels. Tungsten addition to the steel melt.

Tungsten sources which can be added to the steel melt are:

Tungsten metal scrap	90 to 98 wt% W
Ferrotungsten	75 to 80 wt% W
Scheelite ore concentrates	55 to 70 wt% W
Melting base	up to 36 wt% W

Tungsten metal scrap and ferrotungsten, due to their high percentage in tungsten, show poor solubility and this may cause bath inhomogeneities. The tungsten recovery using this type of tungsten addition lies at approximately 95% in case of low carbon and slag levels.

Tungsten addition as scheelite permits rapid attainment of alloy homogeneity at a recovery of 94%. The costs are significantly lower but the addition is limited to 5% W in steel. Higher amounts increase the slag volume and thus decrease the recovery.

Addition of melting base offers the advantages of known chemistry, easy charge, and good solubility.

TABLE 8.2. Typical Analyses of Melting Base Alloys (wt%)[a]

Designation	W	Mo	Cr	V	Co	Ni	S	P	C
TMB	28–38	0.5	0.2	0.2	max 0.6	0.4	0.04	0.04	1.2
TMB+Co	24–30	0.6	0.5	0.2	5–10	0.4	0.04	0.04	1.4
FeWMo	7	5.5	0.5	2	0.6	0.4	0.04	0.04	1.2
FeWMoCo	7	5.5	0.5	2	5–10	0.4	0.04	0.04	1.2
Darby Melt	10	0.4–1.5	3	1.2	—	—	0.03	0.02	0.4–1.3

[a] Treibacher Chemische Werke, Lieferprogramm Wolfram-Melting Base, Treibach, Austria (1991).

In general, a high melting temperature followed by a low teeming temperature results in a homogeneous ingot [8.7].

The quality of tool steels has greatly improved in recent years by new melting and processing routes. Most of the special steels are made by vacuum arc melting (ingot casting) and subsequent hot working (rolling, forging). However, electroslag refining (ESR) and, to a lesser extent, vacuum arc remelting (VAR) are increasingly important for producing high-quality steels with significantly lower nonmetallic inclusion contents.

Hardening and tempering. Tungsten-containing steel affords a certain carbon level and furthermore a careful heat-treatment practice. One disadvantage of tungsten steel is that annealing prior to hardening, if not done with utmost care, will cause formation of tungsten carbides, rich in carbon, which do not dissolve during hardening. Hence the carbon level in the austenite is reduced and the martensite does not present the desired hardness. If hardening is performed from a sufficiently high temperature the tungsten carbides fully or partly dissolve in the austenite prior to quenching. After quenching, they remain in supersaturated solution in the martensite. During tempering below 400 °C tungsten does not diffuse and prevents diffusion of other elements, too. Very fine and evenly distributed tungsten alloy carbides precipitate from the martensite matrix above 400 °C causing a precipitation hardening effect. In this way high-temperature hardness (around 500 °C) is increased. The effect of tungsten on hardness during tempering is shown schematically in Fig. 8.3. At higher temperatures the carbides coalesce and the effect gets lost.

In very hard tool steels, for example, cold work steels tempered at low temperature, the wear resistance is originated mainly by hypereutectoid carbide which is not affected by hardening. These carbides are very small and evenly distributed and very hard (heavy-duty

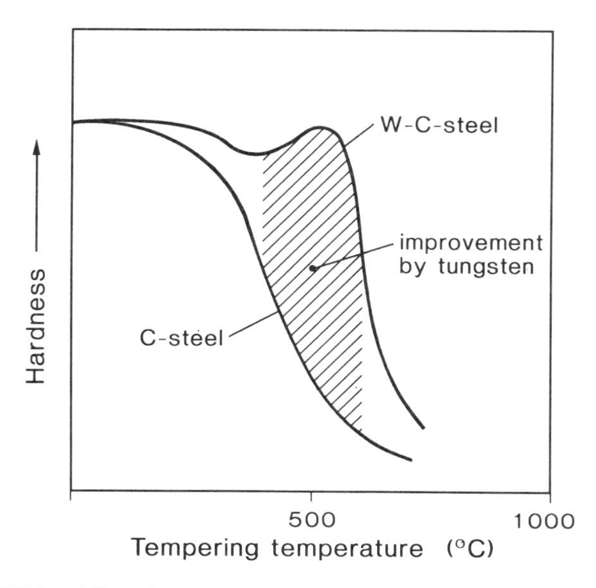

FIGURE 8.3. Effect of tungsten on hardness after tempering (schematic) [8.3].

cold work tools with high hardness, good wear resistance and toughness at ambient or slightly higher temperature). Fine grained martensite adds to this effect.

High Speed Tool Steels (HSS). Tungsten is the most prominent alloying element in HSS on a worldwide basis, although its originally large contribution has reduced in western countries. About 95% of HSS is used for machining tools, the rest for chipless forming (punching tools).

The basic analysis of the three main groups of HSS is:

wt%	C	Cr	W	Mo	V
Group I	0.7	4	18	—	1
Group II	0.8	4	—	9	1
Group III	0.8	4	6	5	2

Experience showed that all three groups behave similarly, but they differ slightly in respect of susceptibility to decarburization, carbide formation, and heat treatment. They can be hardened and are austenized at about 80 °C below their melting temperature prior to air cooling for hardening. All show appreciable secondary hardening after tempering (550–600 °C). This is their most important feature, responsible for the retention of hardness at high temperature occurring during cutting and machining.

The first group (molybdenum free) has been in service for more than 80 years, possessing all the characteristic properties of a HSS, such as red hardness, high toughness, and good wear resistance. The advantage of this group of steels as compared to the others is that the heat treatment is much more tolerable in regard to temperature control and equipment. It is still the most popular group in the eastern countries, but not in the west. Table 8.3, row 1, shows that in practice a wide variation of tungsten containing HSS without Mo exists.

The second group containing only Mo and no W is used more rarely.

Group III is most popular in western countries and has proved most satisfactory in practical service. Also, within this group a broad variation in the alloying elements W, Mo, V, and Co is in use as can be seen in Table 8.3, row 2.

Also low-carbon, precipitation-hardened steel can be produced. It owes its substantially higher hardness to the precipitation of intermetallic phases at annealing temperatures above 600 °C. Figure 8.4 compares the hardness to tempering-temperature dependence of high- and low-carbon HSS. However, costs as well as machining and forming difficulties have prevented its use on a large scale. The analysis is given in Table 8.3, row 3.

In producing HSS, the maximum hardness after tempering is obtained if the hardening temperature is sufficiently high and the holding time long enough to insure uniform heating. Higher hardening causes an increase in hardness after tempering, as will be shown in Fig. 8.5. It must be chosen in such a way that sufficient carbides dissolve, but no overheating and grain coarsening may occur.

About 8% of HSS are produced today by powder metallurgy (P/M). This amount is steadily increasing, in particular in the field of cutting tools (gear cutter, endmills, taps, and broaches) [8.10]. More carbides can be added compared to conventional HSS and a finer particle size and more uniform distribution of the carbides can be achieved. However, the costs are still higher than for conventional tool steels, which retards a more general applicability [8.10].

TABLE 8.3. Analyses of Tungsten-Containing Steels (wt%) [8.2, 8.4]

Row No.	Steel type	W (%)	C (%)	Cr (%)	V (%)	Mo (%)	Co (%)	Ni (%)	Others (%)
1	Tungsten high speed tool steels	12–18	0.7–1.5	4	1–5		5–9		
2	Molybdenum high speed tool steels	1.5–6	0.8–1.5	4	1–5	3.5–9.5	0–12		
3	Low-carbon high speed tool steels	11	0.2			7	23		
4	Chromium hot work tool steels	1.5–7	0.35–0.55	5–7	0–0.4	0–1.5			
5	Tungsten hot work tool steels	9–18	0.25–0.50	2–12	0–1				
6	Molybdenum hot work tool steels	1.5–6	0.60–0.65	4	1–2	5–8			
7	High creep resistant steels (martensitic)	0.5	0.22	12		1		0.8	
8	High creep resistant steels (austenitic)	2.5–2.8	0.10–0.12	16–21		0–3		13–20	0.5 Ti
9	Valve steels	1–2.5	0.12– 0.83	14–21		0–3	0–20	0.7–20	0–9 Mn, 0–1 Nb, 0–0.5 N_2
10	Oil hardening cold work tool steels	0.5–1.8	0.9–1.2	0.5–0.8			0.25 opt.		
11	High-C high-Cr cold work tool steels	1	2.25	12					
12	Carbon–tungsten tool steels	1.3–3.5	1.00–1.25	0–0.75					
13	Permanent magnet steels	5–7	0.70–0.75						
14	Shock-resisting tool steels	1–2.5	0.50	0.75–1.5					

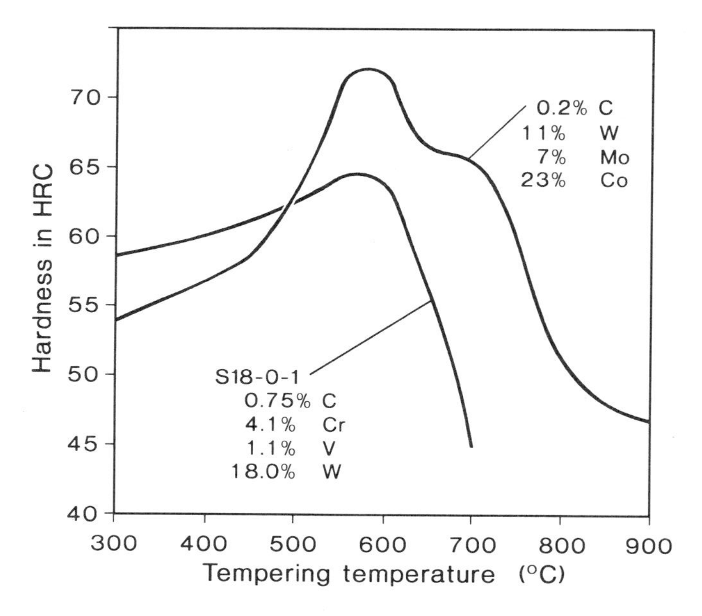

FIGURE 8.4. Effect of tempering on hardness of a low and high carbon HSS [8.3].

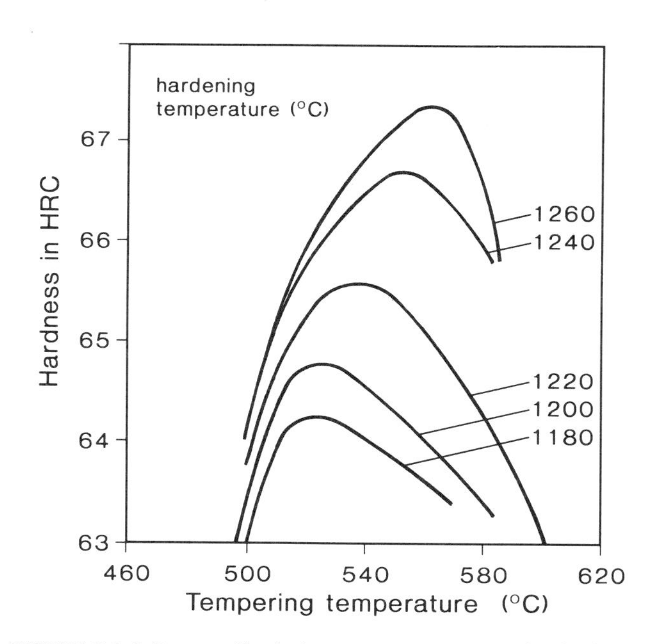

FIGURE 8.5. Influence of hardening temperature on tempering hardness [8.3].

Hot Work Tool Steels. This steel type is mainly used for producing casting and pressing tools in light and heavy machinery. It was important in the development of metal extruding and die casting. For application where very high temperature and severe wear exists, tungsten steels are favorable. Analyses of three different hot work steel groups are given Table 8.3, rows 4–6.

High-Temperature Steels. For applications affording superior high-temperature strength and high-temperature creep resistance, austenitic and Cr-alloyed tungsten steels containing also Mo and V are in use (see Table 8.3, rows 7 and 8).

Valve steels, which have to meet stringent requirements regarding hot hardness, fatigue strength, and wear resistance, also belong to this group. Cr and Ni account for the matrix (martensite or austenite), carbide and nitride for additional hardness, and tungsten for wear resistance. Analyses are given in Table 8.3, row 9.

Cold Work Tool Steels. Historically, this was the first group of tungsten-bearing steels. They are hypoeutectoid, hypereutectoid high Cr, high C steels. The tungsten function is to increase the carbon proportion and carbide hardness, to result in a fine structure and improved toughness. Here, too, a carefully controlled heat treatment is necessary. Moreover, tungsten accounts for the secondary hardening effect on tempering. Typical alloy compositions are given in Table 8.3, rows 10 and 11.

Some further steels containing tungsten are:

- carbon tungsten tool steel (Table 8.3, row 12),
- permanent magnet steels (Table 8.3, row 13),
- shock-resistant tool steels (Table 8.3, row 14).

8.2. SUPERALLOYS

Superalloys are complex alloys containing up to 16 and more elements. There exist three basic types: iron, nickel, and cobalt base alloys. Their important properties are:

- high-temperature strength;
- high creep strength at high temperature;
- high thermal fatigue resistance;
- good oxidation resistance;
- excellent hot corrosion resistance;
- air melting capability;
- air or argon remelting capability;
- good welding properties;
- ease of casting.

The group of iron-based alloys are of minor importance. The composition ranges for Ni- and Co-based alloys (cast and wrought) are summarized in Table 8.4. It can be seen that the carbon content is in the range of 0.25–1.0% for cast Co-based alloys and 0.05–0.4% for hot worked.

The hardness is a result of carbide precipitation and solid solution strengthening. The presence of carbides at the grain boundaries is important for the mechanical properties, because they retard grain boundary migration.

TABLE 8.4. Composition Ranges of Ni- and Co-Based Superalloys (concentrations in wt%) [8.4, 8.5, 8.6]

	Nickel base alloys		Cobalt base alloys	
Element	Cast	Wrought	Cast	Wrought
Ni	58–73	45–68	0–10.5	10–22
Cr	0–16	12–25.5	3–29	20–22
Co	0–20	0–15	52–68	39–52
Mo	0–6.5	3–9	—	0–4
W	2–20	1–6	4.5–25	4–15
Ta	0–9	0–3.5	0–9	—
Nb	0–2.5	0–3	0–2	0–4
Al	3–6.5	0–4.6	0–4.3	0–0.5
Ti	0–5	0–5	0–1	0–1.3
C	0.12–0.20	0.05–0.32	0.18–1.00	0.05–0.4
B	0.010–0.020	0–0.020	0–0.010	—
Zr	0–1.5	0–0.10	0–2.25	—
Fe	—	0–1.5	0–2.5	0–4
Mn	—	0–1.5	0–0.75	0–1.5
Si	—	0–1.2	0–0.75	0–0.50
Y	—	—	0–0.1	—
Re	—	—	0–2.0	—

Tungsten accounts for

- solid solution strengthening;
- strengthening by formation of intermetallic compounds (e.g., Co_3W);
- formation of carbides.

Superalloys are produced via master alloys. In case of nickel-based alloys they include NiW and more complex nickel-based alloys containing Cr, Ta, and Mo. Some examples are given in Table 8.5. All raw materials have to be of high purity. Melting (performed in induction furnaces), casting and cooling takes place in vacuum. Master alloys are supplied in small lumps or as powder.

The most important applications of superalloys are in aircraft engines, marine vehicles, and stationary powder units as turbine blades and vanes, and in sheet form for exhaust case assemblies and burner liners. Moreover, they are used in furnace combustion tubes, muffles, cracking and reformer tubes.

8.3. STELLITE ALLOYS

Stellites are hard-facing and wear-resistant alloys. They have been developed since 1920 and today more than 20 different qualities are on the market. There are two main groups:

- alloys of Co, Cr, W, and C,
- alloys of Co, Cr, W, Mo, Ni, Fe, and C.

TABLE 8.5. Examples of Master Alloys Used in Superalloy Production (concentrations in wt%) [8.6]

Element	NiW	$NiCrAl_{12}MoW$	$NiCrAl_8WTi_{4.4}Ta_{3.5}Mo$
W	55	7	7
Ni	45	36	46.6
Mo		4	3.5
Cr		40	26
Co		1	1
Al		12	8
Ti			4.4
Ta			3.5
S	0.001	0.001	
P	0.002	0.015	
C	0.002	0.1	

Special hard-facing qualities contain in addition small percentages of Si and B. Typical analyses of tough, medium, and hard alloys are given in Table 8.6 including hardness at ambient and high temperature.

The microstructure is defined by hard carbide particles in a Co-rich solid solution matrix. The wear characteristics of the matrix are influenced by a phase change from a strain-induced cubic to a hexagonal form. Toughness and abrasion resistance are due to the volume fraction of carbides and their morphology.

They are prepared from Co–Cr–W master alloys by addition of the other components in pure quality.

For hard facing, they are used either as welding rods or as powders atomized from a homogeneous melt. Hard facing is applied to valves, valve seats, bearings, elevators, rock crushers, mixing components, marine propeller shafts, automotive water or fuel pump shafts, knives, and blades.

Massive forms are in use as cast only. Finished shape can be attained by grinding, since they are not amenable to hot rolling or forging. The properties of stellite metal cutting tools are in the middle between HSS and cemented carbides. Other applications are wear and corrosion resistant parts, like pump sleeves, valves, valve seats, extrusion dies, and mill guides.

TABLE 8.6. Analyses and Hardness of Stellite Alloys (Co content is balance) [8.5]

					HV ($N \cdot mm^{-2}$)	
Alloy	Cr (wt%)	W (wt%)	C (wt%)	Others	20 °C	900 °C
Group I	30	4–17	1–3.2			
Examples						
tough	30	4	1		375	100
medium hard	30	9	1.5		470	150
hard grade	30	18	2.5		600	200
Group II	25–30	0–19	0.1–2	3.22 Ni, 3–10 Mo, 5–20 Fe		

REFERENCES FOR CHAPTER 8

8.1. E. Lassner, H. Ortner, R. M. Fichte, and H. U. Wolf, "Wolfram und Wolframlegierungen und -verbindungen," in: *Ullmanns Encyklopedie der Technischen Chemie*, Band 24, *Wolfram*, pp. 457–488, Verlag Chemie GmbH, Weinheim (1983).
8.2. E. Lassner, W. D. Schubert, E. Loderitz, and H. U. Wolf, "Tungsten, Tungsten Alloys and Tungsten Compounds," in: *Ullmann's Encyclopedia of Industrial Chemistry*, 5th ed., Vol. A27, pp. 229–266, VCH, Weinheim (1996).
8.3. M. Kroneis, "Tungsten in Steel," in: *Proc. 1st Int. Tungsten Symposium, Stockholm*, pp. 96–107, Mining Journals Books Ltd., London (1979).
8.4. S. W. H. Yih and Ch. T. Wang, *Tungsten*, Plenum Press, New York (1979).
8.5. J. D. Donaldson, "Cobalt and Cobalt Compounds," in: *Ullmann's Encyclopedia of Industrial Chemistry*, 5th ed. (W. Gerhartz, ed.), Vol. A7, VCH, Weinheim (1986).
8.6. F. W. Strassburg, "Nickel Alloys," in: *Ullmann's Encyclopedia of Industrial Chemistry*, 5th ed. (B. Evers, St. Hawkins, and G. Schulz, eds.), Vol. A17, VCH, Weinheim (1991).
8.7. F. B. Pickering, "Constitution and Properties of Steels," in: *Materials Science and Technology*, Vol. 7 (R. W. Cahn, P. Haasen, and E. J. Kramer, eds.), VCH, Weinheim (1995).
8.8. H. J. Fleischer, *Proc. 4th Int. Tungsten Symposium, Vancouver*, pp. 40–49, MPR Publishing Services Ltd., Shrewsbury, England (1988).
8.9. G. Kientopf and R. Anderson, *Proc. 7th Int. Tungsten Symposium, Goslar*, pp. 254–274, ITIA, London, UK (1996).
8.10. A. Egami, *Proc. 7th Int. Tungsten Symposium, Goslar*, pp. 13–23, ITIA, London, UK (1996).

9

Tungsten in Hardmetals

9.1. INTRODUCTION [9.1–9.6]

Cemented carbides represent a group of hard and wear-resistant refractory composites in which hard carbide particles are bound together or are "cemented" by a ductile and tough binder matrix. Although the term "cemented carbide" is still widely used, mainly in English-speaking countries, and well describes the nature of the composite, they are even better known internationally as "hardmetals." The latter term will be used in the following.

Hardmetals combine the high hardness and strength of the covalent carbide(s) (WC, TiC, TaC) with the toughness and plasticity of the metallic binder (Co, Ni, Fe). This unique combination of hardness and toughness makes them outstanding as tool materials in the manufacturing industry. Their application is extremely widespread and includes metal cutting, machining of wood, plastics, composites, and soft ceramics, chipless forming (hot and cold), mining, construction, rock drilling, structural parts, wear parts, and military components. The metal cutting group accounts for about 67% of the total use, followed by mining (13%), machining of wood and plastics (11%), and construction (9%) [9.1].

Straight grades. Tungsten carbide (WC) is the basic and most widely used hard component. Cobalt was found to be the best binder metal. WC–Co grades, consisting of two phases (WC and Co; Fig. 9.1), are often referred to as unalloyed grades, straight tungsten carbide grades, cast iron cutting grades, or edge wear-resistant grades. The cobalt content varies from 3–13 wt% in cutting applications and goes up to 30 wt% for wear resistant parts. The average WC grain size varies from submicron to approximately 30 μm [9.2]. Regarding consumption and broad applicability, this group is the most important (more than 80%). Alternative binder metals in straight grades can be nickel (corrosion resistant alloys; woodcutting alloys), steel, or combinations of the iron binder metals.

Tungsten–(titanium–tantalum–niobium)–carbides. WC–Co alloys have been modified in the past to produce a variety of hardmetals with improved high-temperature properties, such as improved oxidation resistance, hot hardness, hot strength, and diffusion resistance against iron alloys, by introduction of additional carbides of titanium, tantalum, niobium, chromium, vanadium, and molybdenum. The ideal microstructure of this second large group of hardmetals consists of three phases: hexagonal WC, cubic mixed crystal carbides, and binder (Fig. 9.2) These grades are also referred to as steel cutting grades, or

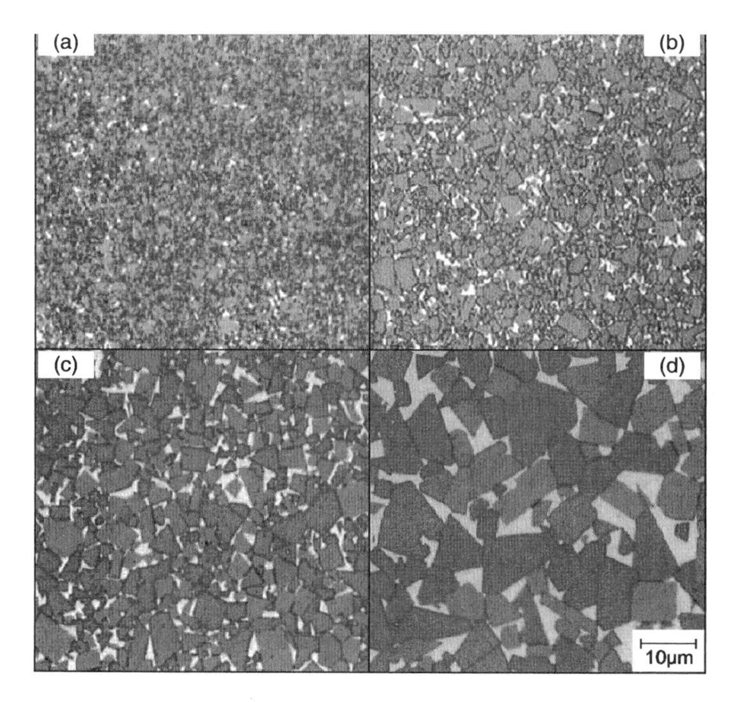

FIGURE 9.1. Microstructures of straight WC–(9–10)wt% Co grades; all samples etched with Murakami's reagent; 1000×. (a) Submicron grade for finish machining and chipless forming. (b) Fine/medium grade for wear parts and chipless forming. (c) Coarse grade for impact drills in medium/hard rock drilling. (d) Very coarse grade for chisels in coal and vein mining. By courtesy of Böhlerit Ges.m.b.H & Co. KG., Austria.

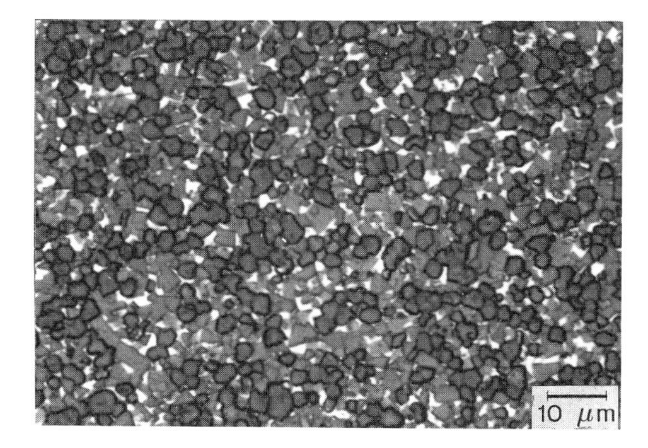

FIGURE 9.2. Microstructure of a 68.5WC–21(Ti,Ta,Nb)C–9.5Co steel cutting grade (ISO P25); etched with Murakami's reagent; 1000×. The gray faceted grains are WC; the dark gray and globular grains are the (W,Ta,Ti,Nb)–solid-solution carbides.

crater-resistant grades. Their basic composition contains 3–12 wt% Co, 60–85 wt% WC, 4–25 wt% TiC, and up to 25 wt% TaC.

Besides WC-based hardmetals, Ti(C,N)-based grades are also used, particularly in Japan. These so-called *cermets* contain no, or only minor additions of, WC. They will not be discussed in this book.

Mechanical properties. The mechanical properties of hardmetals depend primarily on composition, binder content, carbide grain size, and grain size distribution. Variations in the carbide-to-binder ratio, as well as in the average grain size of the carbide phase(s), allow adjustment of the desired properties of the composite in terms of hardness, strength, rigidity, toughness, etc. They form the basis for its widespread applications. One example may facilitate easy comprehension! Simply by changing the average WC grain size in a straight WC-6 wt% Co alloy, the Vickers hardness can be increased from 1170 HV30 (6 μm WC) to 1600 HV30 (1.7 μm WC), and even to 2300 HV30 in the case of ultrafine alloys (0.2 μm WC).

Generally speaking, a decrease in WC grain size is connected with an increase in hardness, compressive strength, and also bending strength, but gives a decrease in impact strength, rupture strength, and fracture toughness. An increase in cobalt content yields lower hardness, lower modulus of elasticity, and compressive strength, but it increases bending strength and fracture toughness.

As a result of the strong adhesion between the WC grains and the Co binder, straight WC–Co grades exhibit the best hardness-to-toughness ratio of all hard materials produced today. This explains their outstanding role in the manufacturing industry.

Hardmetals today. Changes in composition or grain size of hardmetals, in order to improve the performance of tools, are of minor importance today (with the exception of ultrafine grades). Most of the recent advancements in the manufacturing industry were made either by coatings (CVD, MT-CVD, PVD) or by optimization of the tool-design and tool-holder systems. This is easy to understand. Development of coatings which increase tool lifetime by a factor of 2–4, or tools with multiple cutting functions, based on already available ("well known") alloys, will save more time and money in manufacturing than would be possible by further (limited) alloy improvements.

The machining industry is the best example. The dominant tool material today is the coated hardmetal which, in combination with indexable insert technology, has had an enormous impact on the machining industry since its introduction in the late 1960s [9.3].

However, despite these changes in the tool industry, hardmetals have not lost their importance. On the contrary, larger amounts are produced than ever before and, in their various areas of application, their proportion is steadily increasing. The reason for this situation is manifold:

- The availability of high-quality raw materials (WC, Co, WC–TiC, TaC, etc.).
- The high standard of manufacturing (spray drying, pressure sintering, etc.) and high product reliability (important for automized manufacture).
- Their high strength and rigidity (two to three times that of steel), excellent thermal conductivity, and low thermal expansion, making them the ideal coating substrate (excellent adhesion).
- The possibility of tailormaking the substrate simply by variations in powders and/or in sintering technology (recent examples: gradiented structures and WC plate reinforcement).

- Unlimited possibilities in tool design by CAD/CAM and subsequent realization by P/M techniques (near net shape production, powder injection molding, etc.).
- Last, but not least, the possibility of material reclamation through well-developed recycling processes (zinc reclaim, bloating, etc.).

Manufacture of hardmetals [9.4]. The manufacture of hardmetals is based on powder metallurgical techniques, which include several steps. Each step must be carefully controlled to achieve a final product with the desired properties. These steps are:

- Preparation of WC powder.
- Preparation of other carbide powders.
- Production of grade powders (blending, powder milling).
- Powder consolidation.
- Liquid-phase sintering.
- Postsinter operations.

This chapter will follow the different stages of powder metallurgical manufacture. More emphasis will be put on the description of WC powder production methods and qualities, and the preparation of graded powders. Less emphasis will be put on sintering, hardmetal qualities, and applications. In this context, we refer to several excellent books and review articles dealing particularly with hardmetal technology, properties, and applications [9.1, 9.2, 9.4, 9.7–9.9].

Economic importance. About 23,000 tons/year of hardmetal are currently used in the western world, which correspond to more than 60% of tungsten demand [9.1]. The world market volume for hardmetal cutting tools in 1995 was estimated to be roughly 9 billion DM [9.10]. The price of hardmetals, on average, equals that of silver. For special tools, like coated indexable inserts, the price is about double that of silver [9.11].

9.2. RAW MATERIALS

9.2.1. Tungsten Carbide Powder

Tungsten carbide powder can be prepared from different raw materials and by various production methods, as outlined in Fig. 9.3. Only methods used today on a technical or pilot plant scale are considered. The conventional procedure covers the widest range of powder qualities in regard to particle size (0.15 → 12 μm), and consequently the biggest percentage of WC presently produced, while all other methods yield special coarse or fine-grained powders.

9.2.1.1. Conventional Production (Carburization) [9.7]. Tungsten powder is reacted with carbon at temperatures between 1300 and 1700 °C in a hydrogen atmosphere. The average particle size and size distribution of the starting W powder determine the particle size and size distribution of the WC (Fig. 9.4). Only a slight increase in size occurs due to the change in density from 19.3 $g \cdot cm^{-3}$ (W) to 15.7 $g \cdot cm^{-3}$ (WC). In addition a certain agglomeration (local sintering), always occurs, especially at higher temperatures.

As carbon source, high-purity carbon black is mostly used and, more rarely, graphite, due to its higher price and lower reactivity. Thermal carbon black is preferred to flame

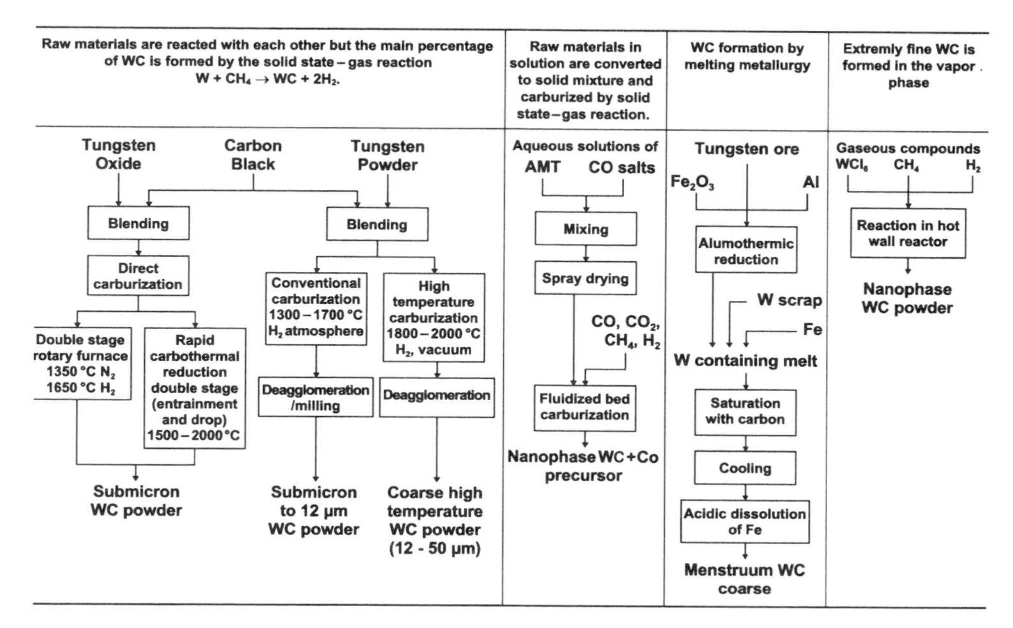

FIGURE 9.3. Routes of WC powder production; conventional manufacture and alternative processes, including WC/Co precursors.

carbon black because of its higher purity. Pelletized qualities have a smaller apparent density and are therefore easier to handle.

In any case, carbon black is always more impure than tungsten powder, particularly in regard to the alkali metals, Ca, Si, Mg, and S. Part of these trace elements are volatilized during carburization (the percentage depending on temperature). This is why finer WC powders (lower carburization temperature) are usually more impure than coarser powders.

The two components (W and C) must be blended thoroughly prior to carburization. This is done in different types of equipment, like V or double cone blenders, mixing ball mills, or high-energy mixers. An even blend is of importance because, during carburization, carbon atoms can only move via diffusion or as methane molecules over very short distances. Pelletizing or compacting enhances diffusion and increases the furnace capacity.

The carbon balance is of utmost importance and depends on several parameters and properties:

- The particle size of the starting tungsten powder; the smaller the size, the higher the oxygen content. Part of this oxygen is responsible (especially in case of submicron powder) for a certain carbon consumption during heating, although hydrogen as a reducing agent is present.
- Content of volatile substances in the carbon black.
- Temperature.
- Hydrogen flow rate and its moisture content.
- Type and duration of mixing.

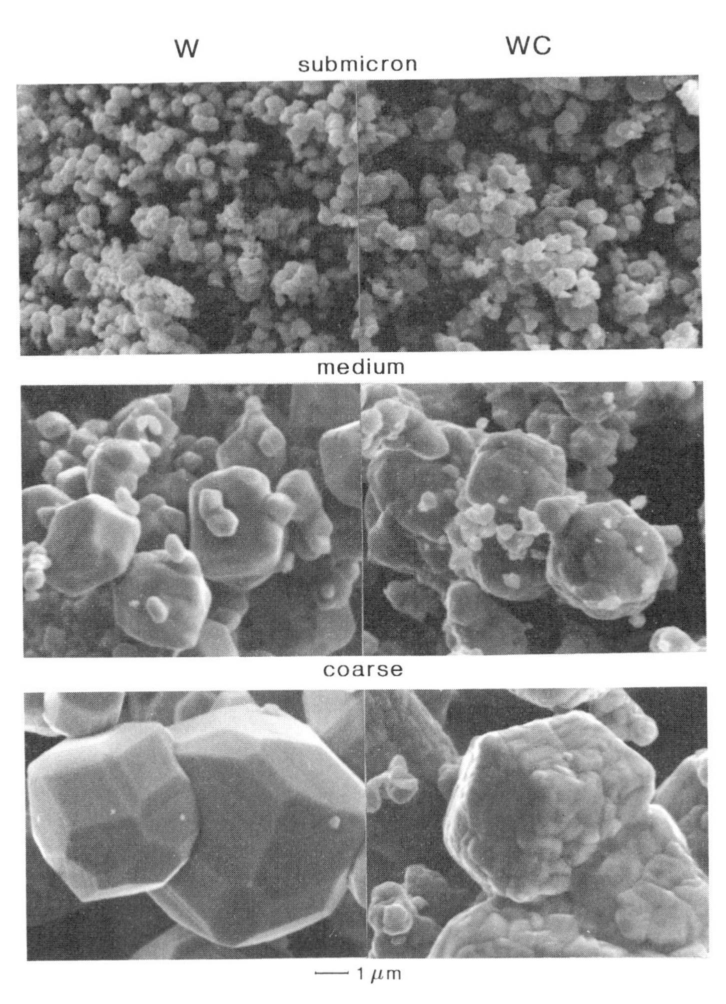

FIGURE 9.4. SEM images of three different WC powders (right), and of the respective W powders (left), from which they were made. Note that the particle size of the carbide corresponds well with the particle size of the metal.

In any case, the difference between the percentage of carbon added and the percentage in the final product can only be determined by practical experience. Nevertheless, it is possible to produce WC powders with a desired carbon content to an accuracy of +0.01 wt% C by empirical knowledge of the carbon loss or uptake. It is easy to understand that all raw material properties, as well as production parameters, must be kept strictly constant in order to meet the above accuracy.

The desired final carbon content of the WC powder depends on the further production mode of the hardmetal producer (mainly powder milling and sintering conditions), and varies between slightly understoichiometric to stoichiometric (6.13 wt%) to slightly overstoichiometric.

In producing submicron WC powder, small amounts of grain-growth-inhibiting substances (inhibiting WC grain growth during carburization and especially during hardmetal sintering) are sometimes added to the W+C charge prior to blending. The usual chromium or vanadium are added either as oxide or as carbide. The addition of oxide must be considered in calculating the carbon balance, because additional carbon will be consumed for reduction and carburization of the metal oxide.

Carbon vessels are filled with the powder blend. Depending on the furnace type, either boats or boxes made of dense graphite are used. The vessels are covered with a graphite cover to avoid any contamination, and pass through the furnace.

Push-type furnaces equipped with heated tubes or channels are mainly used (Fig. 9.5). Construction material for the tubes and channels is either alumina or graphite, and the heating elements are made of molybdenum wire or graphite. Both materials have advantages, but also significant disadvantages which shorten their lifetime. The advantage of graphite is its high chemical stability against trace elements which evaporize during carburization; its disadvantage is the slow but constant reaction with hydrogen and water vapor. In contrast, alumina is very stable against hydrogen and water vapor, but reacts with alkali metals (evaporating from the powder blend), which finally weaken the ceramic by decreasing the melting point.

Molybdenum heating elements, in combination with alumina tubing or muffles, are limited in temperature to a maximum of 1650 °C.

All this taken together illustrates that a carburization furnace will in any case, be a compromise in regard to lifetime.

Induction-heated muffle furnaces with graphite susceptors are used more seldom.

The furnace tubes and the heating elements are swept with dry hydrogen, which acts as protective atmosphere for the product as well as for the sensitive furnace parts.

FIGURE 9.5. Push-type carburization furnaces used for conventional powder carburization (maximum temperature 1650 °C). By courtesy of Wolfram Bergbau und Hüttenges.m.b.H, Austria.

Moreover, it carries away a certain amount of the impurities which evaporate from the product and leads to a purification. Finally, it favors the carburization reaction by intermediately forming methane molecules (see also Section 3.6.1). The latter is of special importance in carburizing coarser tungsten powder.

Carburization temperatures vary between 1300 and 1700 °C, mainly depending on the average particle size of the powder. The smaller the particle size, the lower can the temperature be maintained. Lower carburization temperature leads to a higher degree of lattice defects, and consequently to a higher reactivity during sintering, which is undesirable especially for submicron grades. On the other hand, very fine powders tend to grow already during carburization at higher temperature. Therefore, a compromise has to be made in carburizing submicron powder.

Usual retention times in the hot furnace zone are between 1 and 2 hours. The exothermic heat of reaction can be used in the rear part of the hot zone to maintain the temperature without any heating.

After heating is completed, the vessels pass a cooling zone, still under hydrogen, and are discharged at room temperature. More modern furnaces are equipped with locks, and charging and discharging is done automatically. By using locks, no air is allowed to enter during loading and unloading, thus avoiding reactions with oxygen or moisture. For WC powders with average particle sizes below 0.5 μm (ultrafine grades), special care must be taken due to their pyrophorictiy, and handling is commonly done under inert gas.

During the carburization, the single WC particles agglomerate. The higher the temperature and the finer the powder, the stronger the agglomeration.

Further treatment of the WC powder depends on whether or not the powder is processed in the same factory. In case of in-house processing no further treatment is necessary. Deagglomeration is carried out during the graded powder manufacture.

In case the WC powder is processed elsewhere, a rigid specification must be observed not only in regard to the carbon content as already described, but also in regard to a series of physical properties, such as:

- Average particle size.
- Particle size distribution.
- Apparent (bulk) density.
- Homogeneity.

To meet these demands, the powder can be deagglomerated or milled. A more or less loose deagglomeration can be performed by little jaw or impact crushers. A more rigid deagglomeration consists of short-time (some minutes) milling in a ball mill. Procedures of these types will suffice for powder grain sizes between 1 and 12 μm, unless higher apparent density values are unnecessary.

For submicron WC powders, an extended milling process is applied, especially in cases where the subsequent wet milling procedure for graded powder preparation is not very intense (attritor milling). The WC milling can be performed either in optimized ball mills (hardmetal balls; optimal ball milling conditions avoid any contamination from the steel walls and keep the abrasion of the hardmetal balls at a minimum) or in jet mills in combination with a sifter. The main reason for this milling is to destroy any coarser WC particle (<2 μm) which might be responsible for coarse WC crystals in the sintered structure. Furthermore, heterogeneous impurity particles (graphite from the carburizing vessel and Fe–Ni–Cr containing particles from the reduction boat scale) are finely divided and distributed. This type of milling does not effectively influence the WC average particle size.

The lower limit in average WC particle size is about 0.15 μm (150 nm). To achieve this limit, very dry reduction conditions must be applied during W powder manufacture in order to avoid particle coarsening by CVT (see Chapter 3). This decreases the product capacity and significantly increases the product cost. However, to date, there is still limited demand for such extreme grades.

The physical parameters are not only responsible for the microstructure after sintering, but also for the shrinkage behavior during the sintering period. Therefore, they have to be kept constant within very close limits.

By far, the biggest percentage of WC is produced by this method.

9.2.1.2. High Temperature Carburization [9.12]. Preparation of the W+C blend is the same for this type of carburization as for the conventional type. The main difference with respect to the foregoing method is the carburization temperature ranging between 1800 and 2200 °C. Furnace types in use are push-type graphite tube furnaces, working semicontinuously with the powder in graphite vessels, and batch-type furnaces (powder in trays) either inductively or resistance heated. A modern semicontinuous aggregate is shown in Fig. 9.6.

High-temperature carburized powders are usually coarse (10–50 μm). Sometimes, 5–10 μm powder is also treated that way.

At temperatures ≥1900 °C, extensive grain (crystallite) coarsening takes place, most likely by coalescence, yielding in WC crystallite sizes up to 50 μm. Smaller, average-sized powder (5–10 μm) can partly consist of single-crystal particles.

FIGURE 9.6. High-temperature carburization furnace with graphite resistance heating system; capable of 1500 to 2400 °C continuous operation; automated introduction and discharge of the powdered product loaded in graphite carriers. By courtesy of Harper International, USA.

The push-type furnaces work in a flowing hydrogen atmosphere, while in batch-type furnaces a reduced pressure hydrogen atmosphere is commonly used. In case of batch carburization and especially for coarse powder, it is advantageous to precarburize the material first at 1600 °C under flowing hydrogen, then to crush and blend it, and finally to carburize at high temperature. The precarburization is mainly a cleaning step, and the subsequent crushing and blending results in an improved carbon distribution.

The percentage of high-temperature carburized WC is low. Main applications are in road planning tools and soft rock drilling tools. Part of the coarse crystallized WC for these applications is also supplied by Menstruum WC.

9.2.1.3. Direct Carburization [9.13–9.16]. An alternative for producing WC powders on an industrial scale is the *direct carburization process*, which was patented in Japan. It has been exclusively used there for several years to produce submicron and ultrafine WC powders of high quality. Other than the conventional process, where steering of the WC particle size is mainly conducted via the metal powder process, this is done through the quality of the oxide and carbon source as well as the process parameters.

The necessary equipment consists of mixer, pelletizer, drier, and two rotary furnaces, as shown schematically in Fig. 9.7.

Raw materials such as WO_3 derived from tungstic acid or low-temperature calcined WO_3 from APT (via AMT, see Section 5.3) can be used. The procedure is as follows.

Eighty-four parts per weight WO_3 and 16.5 parts carbon are mixed for 1 hour in a Hensell mixer. Twenty-three parts water are added and the mixture is extruded (1.2 mm diameter) and cut every 3rd mm and dried (<0.1% H_2O). The pellets are fed into the first rotary furnace and flow automatically from there into the second by gravity.

Typical furnace conditions are: *stage 1*—nitrogen atmosphere, 1350 °C (reduction); *stage 2*—hydrogen atmosphere, 1650 °C (carburization).

The exclusion of hydrogen in the reduction step avoids any tungsten crystal growth via the volatile $WO_2(OH)_2$. The reaction path follows the sequence: $WO_3 \rightarrow (WO_{2.9}) \rightarrow WO_{2.72} \rightarrow WO_2 \rightarrow W \rightarrow W_2C \rightarrow WC$. The intermediately formed, needle-like $WO_{2.72}$ decomposes to ultrafine WO_2 nuclei, which are further transformed to WC of

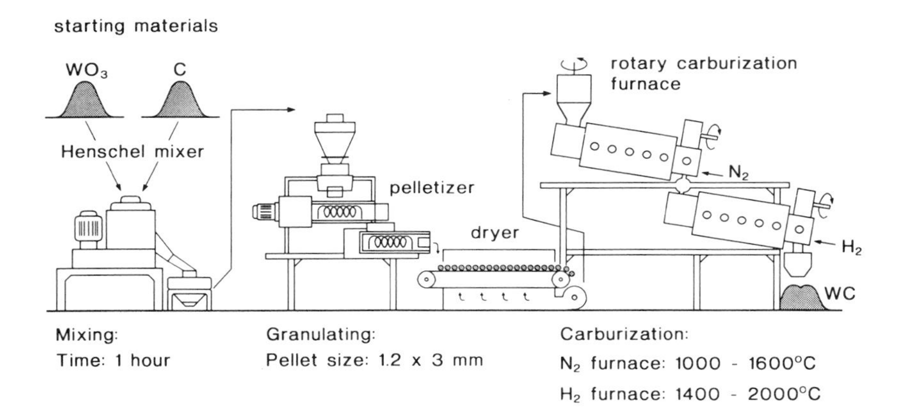

FIGURE 9.7. Schematic presentation of the direct carburization process [9.13–9.16]. By courtesy of Tokio Tungsten Co., Ltd., Japan.

approximately the same size. To meet the carbon balance (around 6.13 wt% C), precise control of temperature and atmosphere is necessary throughout the two-stage procedure.

The finest powders available (Cr_3C_2 doped or undoped) are characterized by a close particle size distribution and a specific surface area between 3.0 and 3.5 $m^2 \cdot g^{-1}$. The real particle size ranges around 0.15 μm, as it can be seen from Fig. 9.8.

FIGURE 9.8. SEM image of an ultrafine WC powder grade produced by the direct carburization process; 20,000×. By courtesy of Tokio Tungsten Co., Ltd., Japan.

9.2.1.4. Alternative Processes.

Menstruum tungsten carbide production (Kennametal Inc.) [9.17, 9.18]. The process is also called a "macro process" and is unique in that it produces WC directly from ore concentrates, either scheelite, ferberite, or wolframite. A mixture of ore concentrate with Fe_3O_4, Al, CaC_2, and/or C reacts exothermally after initiation (by using a starter) to form WC, Fe, CaO, and Al_2O_3. The metallic components gather in the bottom part of the melt and the oxidic in the less dense slag at the top. The process represents an aluminothermic reduction of iron and tungsten oxides. The tungsten metal formed intermediately in the molten iron is carburized and, due to the limited solubility of WC in Fe, it crystallizes. The following chemical reactions proceed (Me can be Fe or Mn):

$$3Fe_3O_4 + 8Al \rightarrow 9Fe + 4Al_2O_3$$

$$MeWO_4 + 2Al + CaC_2 \rightarrow Me + WC + CaO + Al_2O_3$$

$$6CaWO_4 + 3CaC_2 + 10Al \rightarrow 6WC + 9CaO + 5Al_2O_3$$

The reactants are proportioned to develop a self-sustained exothermic reaction at about 2500 °C. A typical charge of 72 t of reactants is divided into smaller portions contained in sheet aluminum bags for progressively charging and maintaining the reaction. The reaction is performed in circular kilns. The furnace has to withstand the high temperature and attack of the reactants. It is designed to minimize heat losses by using graphite slabs for the inner wall backed by insulating carbon. Complete reduction and carburization is finished within 60 minutes and yields about 22 t of WC.

Gangue elements, like Ca, Mg, Si, etc. which remain in the oxidic state, as well as the Al_2O_3 formed are contained in the slag. Carbide-forming elements, like Ti, Nb, Ta, and Mo, appear in the crystal mass and should be contained in the ore concentrates only in low concentration. Elements which are reduced but do not form carbides are dissolved in the iron.

Part of the tungsten can also be supplied by scrap. In this way the amount of Al necessary for reduction can be decreased, because scrap contains the tungsten already in the elemental state and dissolves easily in the Fe melt.

The cooled mass consists of the upper slag layer and the metallic part further down. The latter contains about 65% WC (Fig. 9.9) in addition to Fe and Mn and the excess of metallic Al. After separation of the slag, the metallic part is crushed, washed with water to remove the excess of CaC_2, and finally treated with mineral acids to dissolve the Fe phase, while WC remains unattacked.

The WC consists of coarse, well-faceted crystals (40 mesh and below) and contains approximately 0.2% Fe. The carbon content is stoichiometric and no nitrogen, free carbon, or η-carbides are contained.

The main difference with respect to conventionally prepared WC powder is not only the coarser crystal size, but that the carbides crystallize from the melt and are practically free of internal subboundaries. The latter aspect improves the fracture strength of the carbide [9.2].

A severe disadvantage of the process is the generation of heavy metal salts, mainly $FeCl_3$, the disposal of which is an environmental problem. Another disadvantage is that no grain size steering is possible, besides milling of the final WC. Further processing of milled Menstruum WC affords special skill, because a milled 5-μm powder behaves differently from a conventionally prepared grade.

Menstruum WC is used in hardfacing as well as in high-impact nonmachining applications.

FIGURE 9.9. WC crystals in a solidified thermite mass; 10×. By courtesy of Kennametal Inc., USA.

The Rapid Carbothermal Reduction (RCR) Process (The Dow Chemical Company) [9.19, 9.20]. In principle, the patented RCR process is another type of direct carburization. The tungsten carbide powder is continuously synthesized by an extremely rapid carbothermal reduction of WO_3 in a graphite transport reactor. Calculated heating rates are in the range of 100 million K per second by thermal radiation. The WO_3/C mixture during this process is converted to a tungsten carbide precursor (WC_{1-x}) within seconds.

There exist two basic methods of rapid heating:

Entrainment Method: The powder mixture is fed to a vertical tube furnace and falls down by gravity in an inert gas atmosphere. The heating rate is between 10,000 and 100,000,000 $K \cdot s^{-1}$ (providing small particle size). Residence times are on the order of 0.2 to 10 seconds (1800–2000 °C). A schematic sketch of a reactor is presented in Fig. 9.10.

Drop Method: The powder mixture is dropped into a preheated crucible (heating rates between 100 and 10,000 $K \cdot s^{-1}$). Exposure time may vary between 5 min to 2 h (1500 °C).

A combination of both methods offers the possibility to produce stoichiometric WC. The WC_{1-x} precursor is first prepared by the entrainment method, after which the

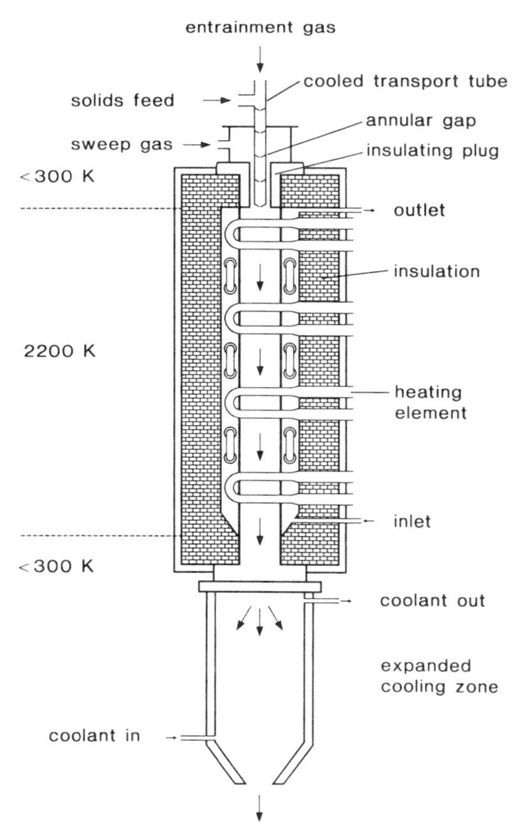

FIGURE 9.10. Continuous graphite transport reactor (entrainment process); patented by The Dow Chemical Company, USA [9.19, 9.20].

necessary amount of carbon is added by mixing, followed by the second step, generally conducted in hydrogen atmosphere, for example by the drop method.

The method allows the production of submicron WC, but is preferable for average grain sizes between 0.2 and 0.4 μm (Fig. 9.11).

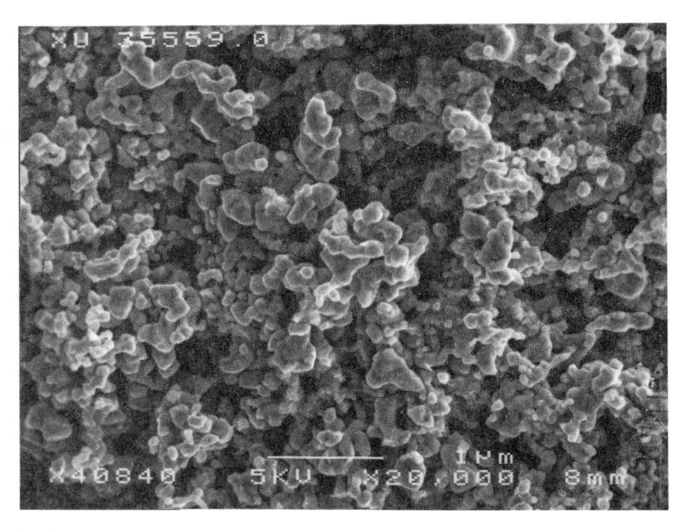

FIGURE 9.11. Ultrafine WC produced by the RCR process; 20,000×. By courtesy of The Dow Chemical Company, USA.

The application of this procedure to preparing tungsten carbide is relatively new and is presently performed on an industrial 500t/year scale.

The Chemical Vapor Reaction (CVR) Process (H. C. Starck) [9.21]. The process is based on chemical vapor reaction between a metal-containing compound (for example tungsten hexachloride) and a carbon-containing gas mixture (for example CH_4/H_2). The principle has been known for many years (see also Section 3.5.4), but has only been realized recently on a technical scale. The aim of the process is to produce a very fine (3–500 nm), homogeneous powder with a narrow particle size distribution.

The separately heated reactants are fed into a horizontal hot-wall tube reactor under defined flow conditions (Fig. 9.12). To fulfill the demand of particle homogeneity, each individual grain must have the same evolution history (i.e., exactly the same nucleation and growth condition). Therefore, nucleation and exposure time within the reactor must be well defined in terms of chemical and thermal environment. Wall reactions are avoided by shielding the wall with a narrow stream of argon, which is led into the tube through an annular slit system. This also avoids any significant decrease in flow rate of the reacting particle cloud in the wall near areas of the tube reactor.

A reverse flow filter operating at elevated temperature separates the carbide particles from the gaseous components and minimizes the amount of gases adsorbed. This is further decreased in the subsequent vacuum chamber. The powder is cooled and collected either under vacuum or in inert gas. The process still runs on a pilot plant scale and is currently

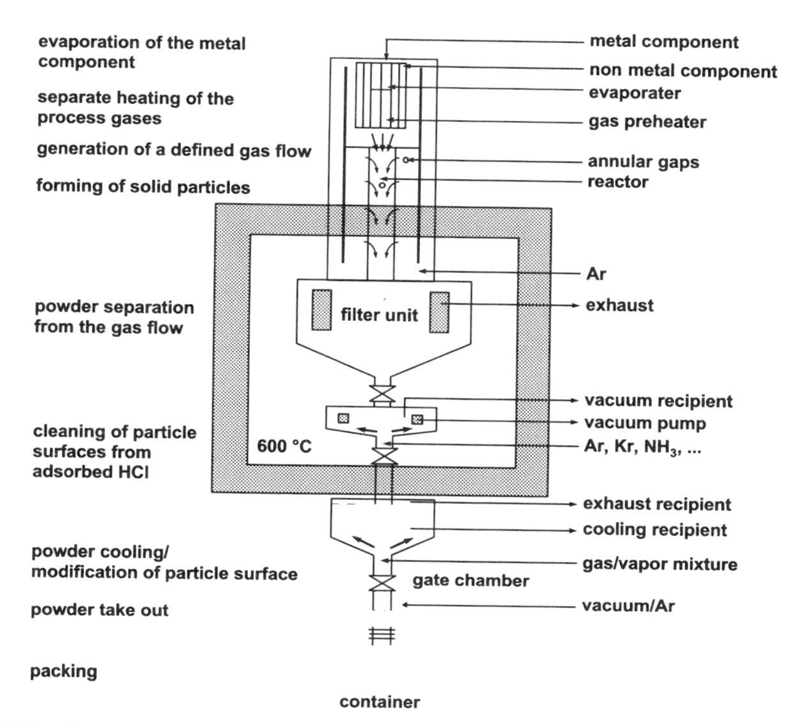

FIGURE 9.12. The chemical vapor reaction process [9.21]. By courtesy of H. C. Starck, Germany.

used to produce nanophase TiC and Ti(C,N). It is, however, also capable of synthesizing nanophase WC.

The Spray Conversion Process (Nanodyne Inc.) [9.22, 9.23]. In comparison to all the above-described methods, the process outlined here yields a final product which is already an intimate mixture of WC and Co. The process is composed of three major stages:

1. *Preparation of an aqueous solution* of W and Co in the desired W/Co ratio (with respect to the final composition of the hardmetal).
2. *Spray drying* of the solution to obtain an amorphous precursor.
3. *Thermochemical treatment* of the precursor *in a fluidized-bed reactor* and conversion into a *nanostructured* WC/Co powder.

A simplified flow sheet is given in Fig. 9.13.

Solutions of ammonium metatungstate (AMT) and soluble Co salts are mixed and spray dried, yielding hollow spheres of an amorphous oxide precursor. The conversion of the precursor to the nanostructured WC–Co powder is performed by carbothermal reduction in a fluidized-bed reactor using gas mixtures, such as CO/CO_2, CO/H_2, and CH_4/H_2. The WC particles (grain size 20–50 nm) are agglomerated in hollow, porous spheres (average size ~70 μm). Hundreds of millions of nanocrystalline WC grains embedded in a Co matrix are contained in one sphere. Any desired WC/Co ratio can be set by solution mixing.

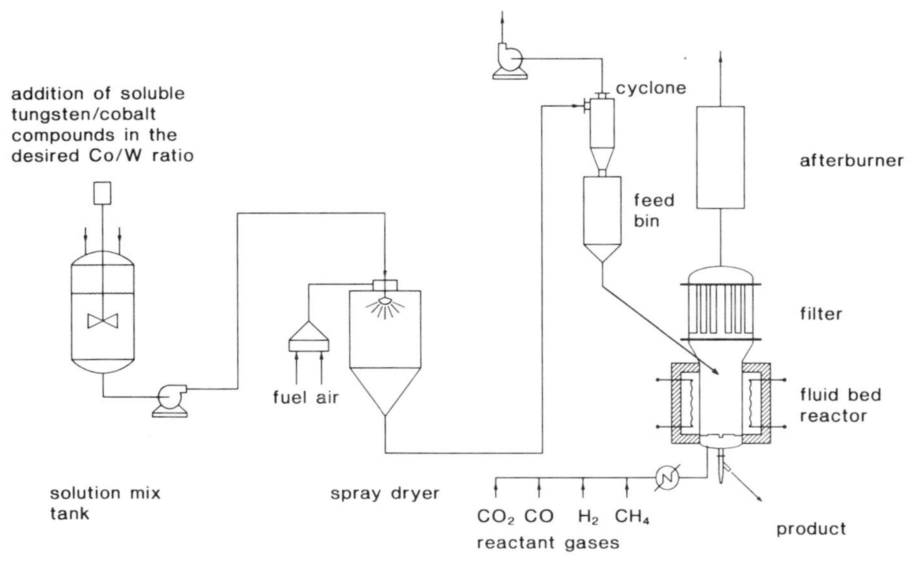

FIGURE 9.13. Schematic diagram of the spray conversion process (Nanodyne Inc., USA) [9.22, 9.23].

Powders made by spray conversion are further processed similar to conventional powders. Prior to compaction, the agglomerates are milled down in ball mills or attritors to fragments of about 0.5 μm.

9.2.1.5. WC Powder Properties and Qualities. The important properties of tungsten carbide powders today, regarded as relevant for their use in hardmetal production, are as follows.

Average particle size. The average particle size (APS) of the powder is the main parameter for WC powder classification. It is obtained by *Fisher Subsieve Sizer* measurements and is commonly given both as "as-supplied" and "lab-milled," according to ASTM standard procedures (ASTM B330, and 430; see also Chapter 5; Table 5.12).

Unfortunately, the term "particle size" is not well defined, and also depends on the corresponding measurement technique. Nevertheless, there exists an ISO standard (ISO 3252), which defines the difference between particle, grain, and agglomerate. This standard is sketched in Fig. 9.14. From this drawing, it becomes clear that the often used term "grain size" for WC powders is misleading, and should be avoided when talking about measurement values (FSSS, laser diffraction, etc.). Strictly speaking, the term "grain" is only correct in case of single crystalline particles (which is not the case for most WC grades).

For most WC powders (APS: 0.5–4 μm), there also exists a close relationship between the APS of the WC powder and the average grain size* (AGS) of the WC in the sintered hardmetal structure (as long as milling conditions in grade powder preparation and sintering conditions are kept constant). In this case, the APS of the WC powder is used as the main steering parameter for the hardmetal structure. This is shown schematically in Fig. 9.15.

* In this case the term "grain size" is correct, because it refers to single crystals.

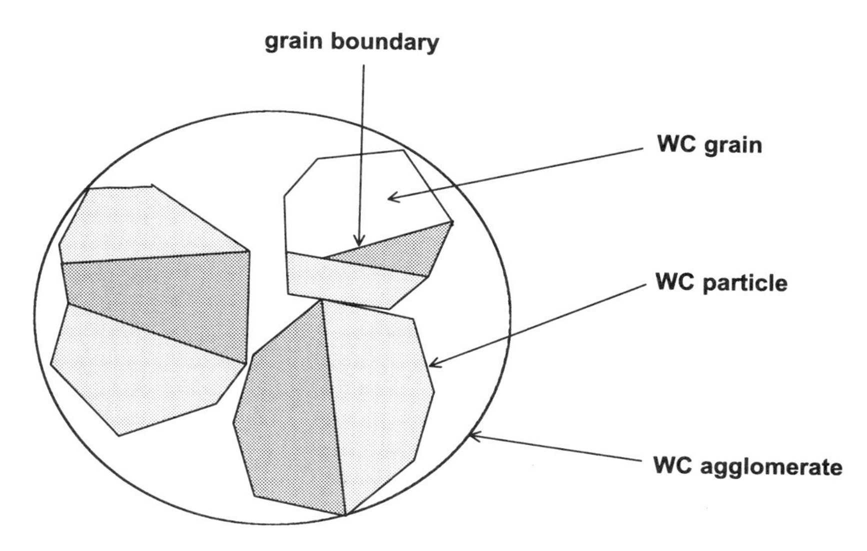

FIGURE 9.14. Physical constitution of WC powders (schematic).

Very fine WC powders (APS <0.5 μm) are commonly strongly agglomerated. These agglomerates survive any deagglomeration treatment and will finally co-determine the values as-obtained by the FSSS measurement. In this case, it seems more sensible to assess the average particle size by image analysis obtained by high-resolution scanning electron microscopy.

Particle size distribution. (For determination methods, see Table 5.12.)

The particle size distribution of the WC is very similar to that of the corresponding W powder. The latter is a consequence of the powder layer height during hydrogen reduction of tungsten oxides and can be somewhat further influenced by the reduction conditions (see Section 5.4.3).

A certain decrease in the coarse fraction can be achieved by heavy milling of the powder either after carburization or during grade powder production.

WC crystallite (grain) size. WC particles larger than 3–4 μm are commonly polycrystalline. In this case, the size of crystallites (grains; single crystal domains) will co-determine the behavior of the powder during liquid-phase sintering (Fig. 9.15). The reason for this behavior is the penetration of liquid cobalt along the grain boundaries during sintering, which leads (at least partially) to disintegration of the former larger particles. For example, if the average crystallite size of a 5-μm and 10-μm average grain size WC is about the same, the respective hardmetal microstructures will also be about the same.

The crystallite size of coarser-grained WC particles is determined by sectioning of the Cu-embedded WC powder. The size is mainly determined by the carburization temperature (compare Figs. 9.16 and 9.17), but also by the purity of the raw materials (W,C). The higher the carburization temperature, the larger the crystallites, mainly as a result of coalescence processes which are favored by the higher temperature (1650 °C → 1900 °C). Trace impurities which segregate at the grain boundaries, such as alkali metals, inhibit or retard grain boundary migration and thereby crystal growth. Therefore, a coarse WC

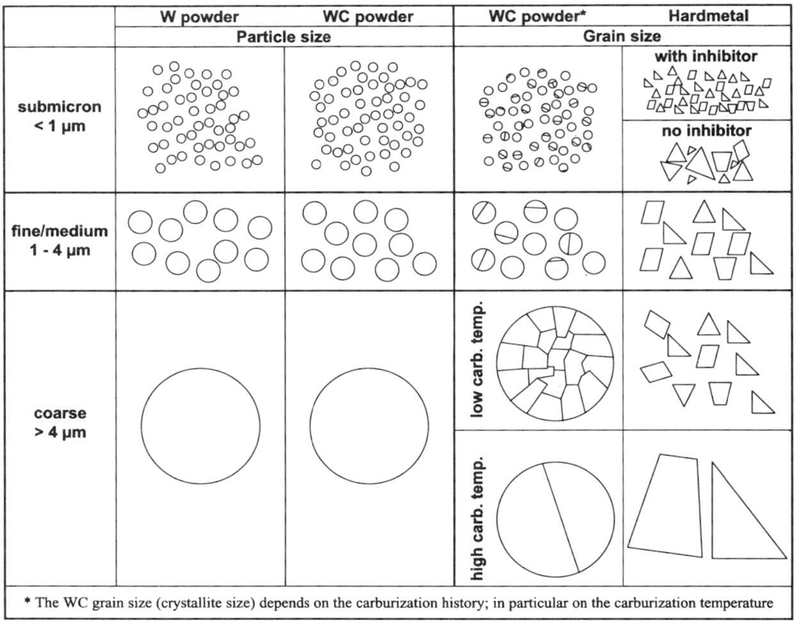

	W powder	WC powder	WC powder*	Hardmetal
	Particle size		Grain size	
submicron < 1 μm				with inhibitor no inhibitor
fine/medium 1 - 4 μm				
coarse > 4 μm			low carb. temp. high carb. temp.	

* The WC grain size (crystallite size) depends on the carburization history; in particular on the carburization temperature

FIGURE 9.15. Relationship between particle size of W powder, particle/grain (crystallite) size of the derived WC, and WC grain size in the sintered hardmetal (schematic).

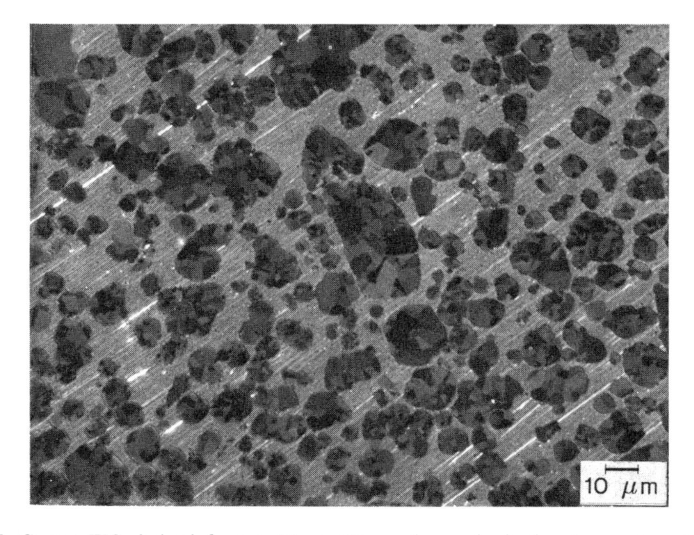

FIGURE 9.16. Coarse WC derived from a 12-μm W powder; carburization temperature 1650 °C; powder embedded in copper. Note the small crystallite size of the particles; during liquid-phase sintering the large particles will be disintegrated into the small grains (Fig. 9.15). By courtesy of Wolfram Bergbau und Hüttenges.m.b.H., Austria.

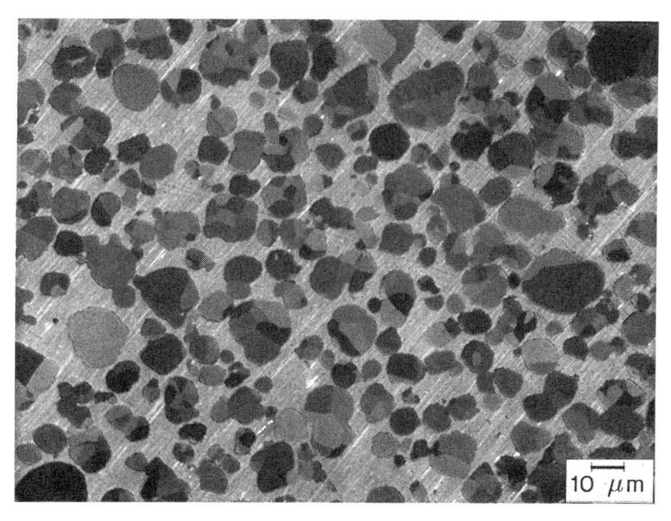

FIGURE 9.17. Coarse WC derived from a 12-μm W powder; carburization temperature 1900 °C; powder embedded in copper. The average crystallite size of the particles is much larger than that of the powder in Fig. 9.16; several particles have even become single crystals. By courtesy of Wolfram Bergbau und Hüttenges.m.b.H., Austria.

crystallite size is obtained in high-temperature carburized WC using raw materials of high purity.

Morphology and microstructure. In recent years, it has become more and more common to include microscopic investigations into routine testing of WC powders.

SEM imaging is especially useful to determine the "true" particle size and the degree of agglomeration of ultrafine powders (Figs. 9.8 and 9.11), or to reveal the polycrystalline surface structure in case of coarse powder (Fig. 9.18).

Also, light optical microscopy can be helpful to detect coarse particles in submicron WC (Fig. 9.19), or to investigate the internal structure of coarse WC in order to determine the frequency of particles which are not totally carburized (W_2C or $W_2C + W$ in the particle interior) (Fig. 9.20).

Apparent density (scott density) and press density. (For determination methods, see Table 5.12.)

Both properties are determined by the average particle size and size distribution of the powder, as well as by its agglomeration and the particle shape. Agglomerates and their strength are influenced by the carburization temperature. Powder milling, depending on its intensity, leads to a more or less complete breakdown of the agglomerates, and increases the values for both the densities, although during subsequent grade powder milling further comminution may occur. The powder density values are closely related to the shrinkage during sintering. Controlled shrinkage is a very important property in "near-net-shape" sintering of hardmetal parts.

Carbon content. The carbon content of a WC powder is commonly stated as total carbon and free carbon (uncombined carbon present as graphite or amorphous carbon).

FIGURE 9.18. Coarse WC powder obtained by high-temperature carburization of a 25-μm tungsten powder. The smooth surface of the particles is characteristic for the high-temperature treatment (1900 °C); nevertheless, grain boundaries are still visible which indicates that the particles are polycrystalline. By courtesy of Wolfram Bergbau und Hüttenges.m.b.H., Austria.

The difference between these values accounts for the combined carbon (chemically bound as carbide).

The total carbon content should be theoretically 6.13 wt% carbon, which means that all carbon is chemically combined as WC. However, in practice, a small amount of uncombined carbon is always present (some hundredths of a percent).

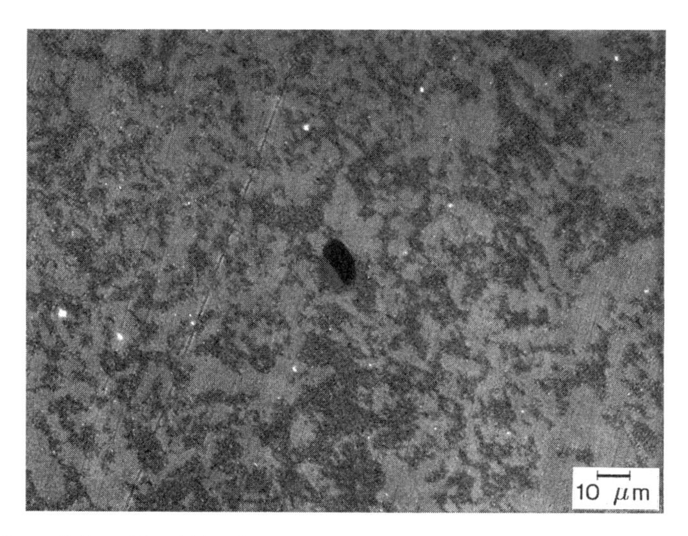

FIGURE 9.19. Coarse WC particle within a fine powder matrix; powder embedded in copper; light optical image. By courtesy of Wolfram Bergbau und Hüttenges.m.b.H., Austria.

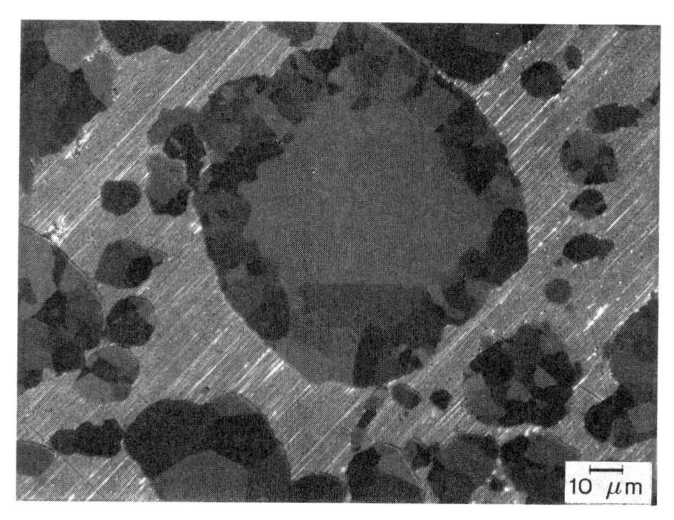

FIGURE 9.20. Very large carbide particle (>100 μm) consisting of an outer, polycrystalline ring of WC and a core of W_2C; light optical image. By courtesy of Wolfram Bergbau und Hüttenges.m.b.H., Austria.

The proper total carbon depends on further processing of the powder (compact) and is dictated by the hardmetal producer, due to its processing experience (carburizing, neutral, or decarburizing conditions).

The determination method is described in ISO 3907 and ISO 3908 and should result in a standard deviation of no more than +0.01 wt%.

Content of grain growth inhibitors. Grain growth inhibitors like VC or Cr_3C_2 are sometimes added prior to carburization to minimize particle/grain coarsening. This procedure has become industrial practice for so-called ultrafine grades (sintered structures with grain sizes ≤0.5 μm). Further additions are commomly made to the grade powder milling process. Typical additions to the powders prior to carburization are in the range of 0.1 to 1 wt% VC resp. Cr_3C_2.

Trace impurities. Trace impurities play an important role in hardmetal manufacture, since they can cause numerous types of defects (porosity, inclusions, origin of local grain growth, etc.). They should be kept as low as possible in the powder, in particular when a detrimental action has already been demonstrated (as, for example, in the case of Ca, Si, Al, and S).

The impurity content of a WC powder is the result of three main factors: (1) the purity content of the starting W powder; (2) the impurity content of the carbon black or gaphite, and (3) the carburization temperature (the higher the temperature, the more the trace impurities are evaporated during carburization). A typical analysis is shown in Table 9.1.

Although the above characterization in regard to physical and chemical properties seems to be quite rigid and complete, in some cases unexpected and unsatisfactory results may occur during the further hardmetal production. The reason for this behavior is based on still insufficient specification, because the carbon black quality, carburization temperature, as well as type and duration of milling after carburization are not clearly defined, and can be changed by the supplier or may differ between diverse suppliers.

TABLE 9.1. Trace Impurities in WC Powder (upper limits in $\mu g \cdot g^{-1}$)[a]

Element	Concentration
Al	20
As	10
Ca	20
Cd	2
Cu	2
K	10
Li	2
Mg	2
Mn	2
Mo	30
Na	20
P	20
Pb	2
Sn	10
Si	10
Ti	10

[a] Product data sheet, Wolfram Bergbau- und Hüttenges.m.b.H.

Conventional tungsten carbide powder. The particle size range of these qualities is between 1 and 10 μm (the majority between 1 and 2.5 μm). The average particle size has to be met usually within the limits of + 10% relative. Particle size distributions for a given average particle size can be adjusted within certain limits already during tungsten powder reduction (see Section 5.4.4).

In order to minimize testing operations for the hardmetal producer, it has become common in recent years to prepare homogeneously blended lots of 5 t or even 10 t. In blending, it is strictly forbidden to use smaller lots of strongly diverging properties to meet the specified values.

Submicron tungsten carbide powder. In principle, every WC powder having an average particle size below 1 μm can be regarded as "submicron." However, it has become common practice to further classify these powders into the following three groups:

Submicron WC	<1.0 to 0.5 μm,
Ultrafine WC	<0.5 to 0.10 μm,
Nanosized WC	<100 nm.

Fine-grained WC powder grades have gained significant importance during the last 10 years as demonstrated in Fig. 9.21. While submicron qualities still dominate the market, there is growing interest in using even finer grades for specific applications (machining of highly abrasive materials). Ultrafine grades are offered today by several producers, based on conventional powder processing (Fig. 9.22) or alternative routes (Figs. 9.8 and 9.11). Nanosized powder precursors are also available, but until now could not be transformed into pure nanocrystalline composites by conventional liquid-phase sintering due to their high interface energy, which causes rapid grain coarsening. Alternative consolidation techniques are necessary in order to exploit their potential.

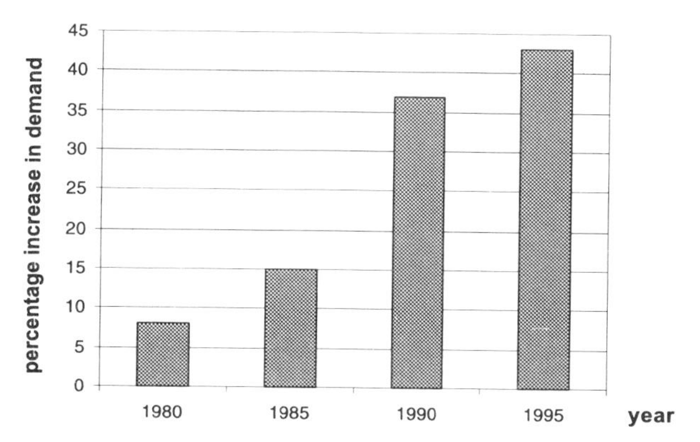

FIGURE 9.21. Demand in submicron WC powder during the period 1980–1995 based on current industrial statistics.

Coarse tungsten carbide powder. Coarse WC powders are available in the particle size range of >8 to 50 μm. Their crystallite sizes vary from <10 μm up to about 50 μm and mainly depend on the carburization temperature (strictly speaking, the maximum crystallite size is determined by the particle size).

9.2.2. *Other Carbide Powders* [9.7, 9.8]

The following carbide powders are used in hardmetal manufacture: WC–TiC, WC–TiC–TaC(NbC), TaC, TaC(NbC), as well as Mo_2C, Cr_3C_2, and VC (the latter are used in small concentrations as WC grain growth inhibitors).

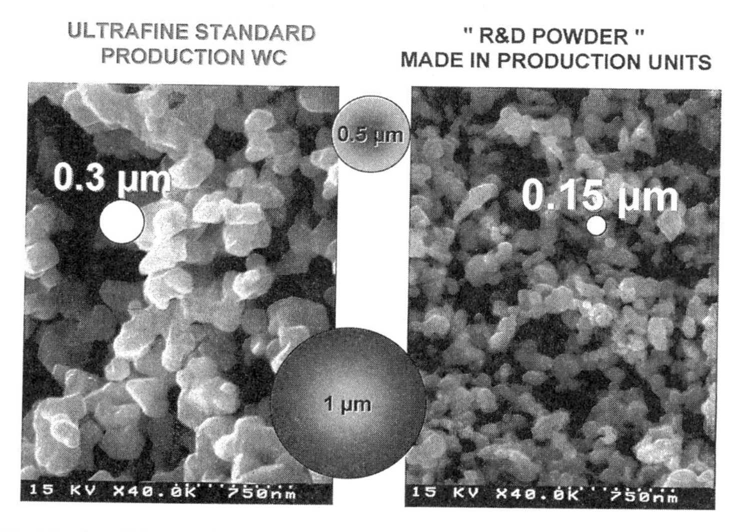

FIGURE 9.22. Ultrafine WC powders produced by conventional (reduction/carburization) manufacture; 40,000×. The size of the spheres emphasizes the degree of fineness of the development grade. By courtesy of Wolfram Bergbau und Hüttenges.m.b.H., Austria.

WC–TiC mixed crystal carbides (solid solutions). The raw materials are W powder (1–2 μm), TiO_2, and carbon black. The proper removal of oxygen is essential for obtaining a high-quality (pore-free) hardmetal. The most common compositions are WC/TiC 70:30 and 50:50 (wt%). The carburization is carried out mainly in vacuum aggregates, more rarely under hydrogen. Resistance-heated (carbon or graphite heating elements) or induction-heated furnaces are in use. Carburization temperatures are up to 2100 °C. A double carburization for proper carbon adjustment and for the total removal of residual oxygen and nitrogen is preferable. The second step should be performed at a temperature 200 K higher than the first.

The powder properties are either directed by powder milling of coarser powders (high-temperature carburized), or by steering the particle size through the oxide precursor and subsequent soft carburization (lower temperature, but longer carburization times).

WC–TiC–TaC(NbC) mixed crystal carbides (solid solutions). The raw materials are the same as for WC–TiC grades, with addition of Ta_2O_5. Any furnace equipment capable of producing a good quality of WC–TiC is suitable. Total reduction and complete diffusion can only be reached by high temperature, good vacuum, and long exposure time. Carburization temperatures are about the same as for WC–TiC.

Nb, which is a natural companion of Ta, can be tolerated up to 30%, but the exact content must be known for optimal carbon adjustment. Typical grades contain 33–50 wt% WC.

TaC and TaC(NbC). The preparation is similar to that of the above-described mixed crystal carbides. The expensive TaC (<1% NbC) as well as the cheaper mixed Ta/Nb carbide (Ta/Nb = 6:1 to 1:1) are applied.

9.2.3. *Binder Metals* [9.7, 9.8]

Cobalt powder. Powder grades with average particle size between 1 and 5 μm are available for hardmetal manufacture. They are produced by hydrogen reduction of cobalt oxide, oxalate, or carbonate. Ultrafine Co powders (particle sizes of 0.3–0.5 μm) are prepared through the reduction of cobaltous hydroxide by a mixture of ethylene–glycol and diethylene–glycol ("polyol" cobalt), or by hydrogen pressure precipitation in autoclaves out of cobaltous sulfate solutions (using nucleation catalysts).

Nickel and iron powder. Fine Ni and Fe powder is obtained by hydrogen reduction of the oxide or decomposition of carbonyl compounds.

9.2.4. *Other Materials* [9.7, 9.8]

As solvents or milling liquid in grade powder milling, several organic compounds are employed: alcohols, acetone, hexane, heptane, gasoline, and tetraline. Hexane, heptane, and acetone are the most preferred today.

The most popular lubricant is paraffin wax, other than bees wax, camphor, and polyethylene–glycol. Besides its lubricating properties, its purity is essential. It must be removable by heating without any residue.

As plastifiers for extrusion, cellulose derivatives are in use.

9.3. HARDMETAL PRODUCTION

The preparation of a hardmetal specimen consists of several steps and can be performed along different paths. Basically, one may distinguish between three or four main steps:

- Preparation of powder grades; this stage includes the blending of all necessary components, and is combined in most cases with a more or less rigid powder milling. A big percentage of the grade powder is granulated.
- Consolidation by compacting or extruding, in some cases followed by shaping.
- Sintering.
- Postsinter treatment.

A generalized flow sheet, outlining the different modes of preparation, is presented in by Fig. 9.23.

9.3.1. Preparation of Powder Grades [9.1, 9.7, 9.8, 9.24]

Weighting. The necessary amounts of all constituents (WC, other carbides, grain growth inhibitors, additional carbon black, and binder metal) and of all further necessary temporary additions (organic solvent, additional organic liquids, lubricant) are weighted or measured precisely and loaded into a suitable blending–milling device. The usual equipment for that purpose consists of ball mills, vibratory mills, rod mills, or attritors (mostly ball mills and attritors).

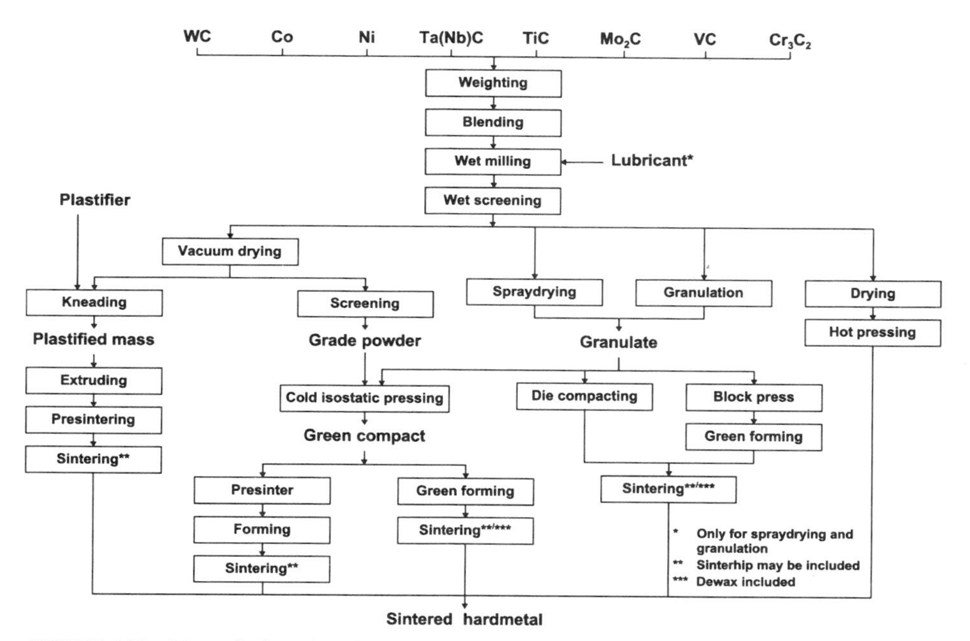

FIGURE 9.23. Schematic flow chart for hardmetal production, showing the variety of production routes.

Powder milling. The goals in powder milling can be quite different, and can vary between a soft deagglomeration of particle agglomerates combined with mixing of the components (mixing mill), and heavy milling of the powder, whereby significant particle comminution will occur. Depending on the goal, different aggregates and milling conditions are applied.

In general, a fundamental knowledge of the milling process is required. Mill dimensions, rotational speed, ball diameter, loading of milling media, and hardmetal components must exist in an optimized relationship. In this way the uptake of iron (abrasion) from the mill walls and of cemented carbide from the milling media can be kept at a minimum. The weight ratio of milling media to powder together with the milling time is responsible for the amount of disintegration. The higher the ratio, the more pronounced the milling (comminution).

Any particular mill and mill charge combination has a limiting particle size, beyond which no further comminution will occur. For most industrial aggregates, this is in the range of 0.5–1 μm [9.25]. Only in optimized mills is further comminution possible, finally reaching a size limit of about 100–150 nm [9.24]. This limit is inherent to the WC (i.e., to its mechanical properties). This kind of milling is used only by specialists.

Commonly, either ball mills or attritors are used. Ball mills that were in use at the very beginning of hardmetal production, and which had a volume of only about 10 liters, have disappeared, and modern aggregates possess volumes of up to 500 liters (charge weights up to 700 kg). This change has also significantly improved their milling efficiency (milling times of 4–6 days in a 10-liter mill was reduced to 1–3 days in a 300-liter mill). Basically, the ball mill is more effective for the coarser fraction of the powder, which is trapped between the balls or studs. This is why a wide particle size distribution can be improved by ball milling. In addition, proper ball milling provides an excellent carbide/binder distribution and minimizes the danger of cobalt "flake" formation (which results in porosities during subsequent sintering).

About 40 years ago, a large proportion of the ball mills was substituted by attritors. The advantage of the latter is a much shorter milling time (one-fifth or one-tenth of the time needed in ball mills), which is the main reason why they have become increasingly popular. This means hours of milling instead of days. However, this advantage is combined with certain disadvantages. The attritor mill produces a Gaussian particle size distribution which enhances the carbide grain growth. Furthermore, heterogeneities can be trapped in "dead areas," which can be only partly overcome by the attritor design and by pumping and recirculating the slurry from the bottom to the top of the attritor.

The choice of the correct mill and milling conditions depends on the hardmetal grade produced, mainly on the WC grain size. Attritors are used today for most powder qualities (1–12 μm). They have become increasingly popular because of the significantly shorter milling times. Optimized ball mills are still preferred for submicron qualities, which are very sensitive to all types of heterogeneities.

During milling, several processes occur simultaneously: break-up of agglomerates, blending, and comminution. In general, the surface energy of the mix is increasing, which enhances the sintering. Sometimes, the carbides are milled alone or with only a small part of the Co ("split milling") to achieve a more intense milling in a shorter time, while the rest of the binder is added only in the late part of the milling. Cobalt, by its large powder volume (low density compared to WC), decreases the probability for a WC particle to be hit by a ball, thereby prolonging the milling time. In any case, a uniform distribution of the binder is important in order to avoid increased porosities in the sintered body, which is particularly important for ultrafine grades. For these grades, longer milling times are therefore mandatory. The use of ultrafine Co powder grades has further contributed to achieving better distributions.

Grade powder milling is always conducted in an organic liquid to prevent local overheating and oxidation. The enormous milling energy, which is converted to heat, must be removed from the mill by cooling (mills equipped with water jackets).

Balls or parts of the propeller arms (in the case of attritors) can fracture during milling, leading to hardmetal fragments which are not further disintegrated. Moreover, agglomerated cobalt powders can be forged to flat platelets (flakes), which resist any further comminution. Therefore, the slurry after ball removal is screened (400 mesh) to eliminate any heterogeneities.

Lubricants. Either before milling or afterward a lubricant (wax, polyethyleneglycole, camphor, etc.) is added in amounts of 1–3 wt%. It dissolves in the solvent (acetone, hexan, alcohole) and is distributed homogeneously. It remains in the grade powder and protects it from oxidation after distillation of the solvent. The solvent is recycled to the next batch. Moreover, the lubricant improves the strength of the green compact.

Granulation. A big percentage of graded powder is granulated by either spray drying or mechanical granulation. Granulation is a necessary prerequisite for automated compacting combined with near-net shaping. Both granulation methods yield a free-flowing powder consisting of spherical particles It guarantees the highly reproductive filling of the compacting dies. Spray-dried granules are commonly hollow spheres, while granules made by mechanical granulation are full. Therefore, the filling height during compacting is higher for spray-dried powder (lower apparent density).

Spray drying is a combination of the distillative separation of the solvent and the granulation process and is the most frequently used method for granulation of powder grades on a bigger scale. In processing, the sludge is forwarded to a spray nozzle (pressure 5–10 bar). On leaving the nozzle it is finely divided into droplets (10–50 μm) forming an upward-directed jet. The jet forms a taper, roughly fitting the upper end of the chamber. The forming of granules is a rather complicated mechanism. Hot gas (100–130 °C) heats the incoming particles and the milling liquid is evaporated. Particles of different speed and moisture grade hit each other, resulting in quite spherical particles of 100–400 μm. Particles big enough settle. The particle size distribution is quite close. To obtain the desired size, the working conditions must be adjusted.

The granules collected at the bottom are evacuated into a gas-tight container. Critical properties of the granules, besides their size, are also the apparent density and hardness. The density is responsible for the filling reproducibility of the compacting die. If the hardness is too low, granules may be smashed during handling, while too high a hardness causes granules to partially survive during compacting, yielding an increased residual porosity during sintering.

Approximately 5% of the powder remains ungranulated, is separated by the cyclone, and can be used for purposes where granulation is not required. The solvent is almost totally recovered and can be recycled to the milling stage.

The capacity of a spray drier is between 120 and 180 kg $\cdot$ h^{-1}. The size of the spray drier is about 5 m in height and 2 m in diameter.

9.3.2. *Powder Consolidation* [9.1, 9.7, 9.8]

Direct compacting. Direct compacting is applied for "near-net shaping" of parts. This means that the final hardmetal specimen after sintering already has the desired dimensions and only limited surface treatment is necessary afterward. For that purpose, semiautomatic or automatic mechanical or hydraulic presses are used. The pressure application is only from one direction, resulting in a slightly anisotropic density distribution in the green compact.

Cemented carbide powders do not deform during compaction. Therefore, the maximum green density is about 65% of the theoretical density (35–50% porosity). Compacting pressures are 100–300 MPa, usually around 200 MPa.

The pressing dies (fabricated of hardmetal) are made to the shape of the desired end product, which means that the die must be greater to allow for shrinkage occurring during sintering.

The height of the green compact corresponds to 35–40% of the filling powder height in the die. Shrinkage of the compact during sintering is between 15 and 20% in each direction. Shrinkage calculations in Europe are based on the as-sintered size, while in the USA it is based on the compacted or presintered size. Thus, a value of 25% in Europe corresponds to about 20% in the USA. The percentage of shrinkage is generally smaller in the pressing direction than perpendicular to it.

Nevertheless, today's advanced granulation and pressing technology have made it possible to achieve very rigid dimensional tolerances of the sintered product. It is easy to understand that a prerequisite is to keep the properties of the raw materials, as well as the conditions during grade powder preparation and granulation, strictly reproducible.

Direct compacting is mainly used in producing mining tools, metal cutting inserts, and parts for construction applications. Around two-thirds of hardmetal production is compacted by this method.

Cold isostatic pressing. The powder is compacted in a tight rubber or plastic container inserted in a liquid under pressure. Besides wear parts and metal-forming tools, mainly bigger pieces are compacted by this method and in most cases need further indirect shaping. The subsequent forming can be done either in the "green" state or after presintering. All types of machining and grinding operations can be used. In forming, the sintering shrinkage must also be considered. The dust as well as the scrap originating from these operations is collected and recycled internally.

Extrusion. Round or profiled rods or tubes can be produced by extrusion. The amount of lubricant must be increased and additional plastifiers added. The corresponding recipes are kept secret.

Injection molding. Injection molding can be regarded as a combination of extrusion and precision die compacting. A highly plastified mixture of the hardmetal powder and lubricants is screwed under pressure into a precision die. The hardmetal-to-plastifier ratio is roughly 90:10 by weight or 50:50 by volume. The procedure is only economic in case of complex shapes and a large number of parts. Dewaxing is carried out very slowly and requires special care, as for extruded material.

9.3.3. *Sintering*

A common sintering cycle includes several stages, such as dewaxing, presintering, sintering, and cooling. For pressed-to-shape parts, it is customary to combine dewaxing, presintering, and sintering in one aggregate and heating cycle, thus minimizing energy consumption and handling. It is a prerequisite that the furnace is equipped with a wax condenser to remove the lubricant from the sintering chamber. Only for parts which must be shaped before sintering a presintering stage is performed at 700–1000 °C to give the compact sufficient strength to permit forming. This is commonly done in special furnaces.

Sintering is conducted either in hydrogen (for straight grades and low TiC-containing alloys) or, more commonly, in vacuum. For higher TiC- and TaC-alloyed grades a good vacuum is mandatory, in order to avoid residual gas porosity. In general, there is a tendency today to use vacuum sintering for all grades.

A typical sintering cycle is shown in Fig. 9.24. At first, the charge is slowly heated up to 400–500 °C to render proper dewaxing of the compacts. The temperature is then raised gradually to the optimal sintering temperature of 1350–1600 °C, which depends on the hardmetal composition and the WC grain size. Several holds at certain temperatures are common during heating to allow outgassing of carbon monoxide (which forms as a result of residual oxygen still present in the densifying compact), unless there is open porosity.

Isothermal liquid-phase sintering lasts about 1–1.5 hours.

Densification starts below the liquidus temperature during heating (nonisothermal sintering). This solid-state shrinkage is especially pronounced in submicron and ultrafine grades. It is due mainly to particle rearrangement processes, which occur as a result of the spreading of cobalt along the WC surfaces, and subsequent contraction under the action of capillary forces. Both processes are activated by the partial dissolution of WC in cobalt.

At about 1300–1340 °C (binary eutectic temperature WC–Co 1310 °C; ternary eutectic temperature W–Co–C 1275/80 °C) partial melting occurs, and more WC is dissolved until the eutectic concentration (54% Co and 46% WC) is reached [9.8]. A further increase in temperature results in additional dissolution of WC and complete melting of the binder phase. In this stage, rapid final densification occurs and the sintered body is practically pore-free.

Microstructural changes. At 1400 °C, the constitution of the molten phase corresponds to about 50% Co and 50% WC [9.8]. The liquid phase preferably dissolves the

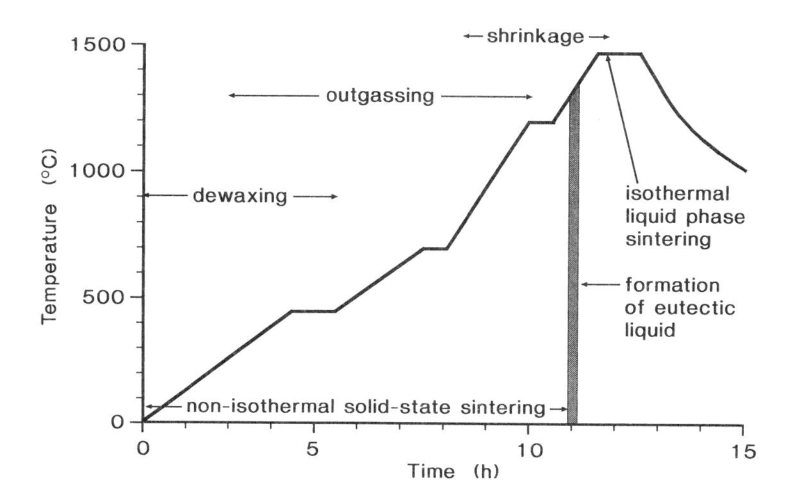

FIGURE 9.24. Representative sintering cycle of hardmetal production, indicating the different stages of sintering.

surface layers of the WC particles and grains which have been activated during powder milling, as well as bridges between agglomerated carbides. The morphology of the particles changes from rounded to prismatic (formation of low-energy interfaces). Furthermore, the melt disintegrates large polycrystalline carbide particles into individual grains by penetrating into the large-angle grain boundaries. During isothermal sintering, further adjustment of the microstructure occurs due to solution/reprecipitation and/or coalescence processes, which lead to a more uniform distribution of the cobalt binder and a decrease in carbide contiguity. This is an important aspect for obtaining optimal mechanical properties of the sintered parts. Too high a sintering temperature will promote unwanted grain coarsening.

On cooling, the dissolved WC is precipitated onto already present WC grains. Certain amounts of W and C remain in solid solution, their amount depending on the gross carbon of the respective alloy, and the cooling rate. The W content in solid solution at room temperature varies between 18–20 wt% W in substoichiometric alloys and 1–2 wt% W in stoichiometric alloys.

The solutes stabilize the cubic modification of cobalt down to room temperature. Due to the different coefficients of expansion of the carbide(s) and cobalt, the binder phase is under tensile stresses and the carbides under compressive stress. Uneven macroscopic stresses can lead to distortion of the sintered body.

Shrinkage during sintering corresponds to 17–25% on a linear scale, or 45–60% by volume. The amount of shrinkage depends on WC powder particle size, the milling, granulation, and compacting conditions, and the grade composition.

The carbon balance of a hardmetal is a very important parameter. The W:C ratio should be close to 1:1. If the carbon content is too low (strongly substoichiometric), η-phases of the M_6C type will be formed and lead to embrittlement of the alloys. If the W:C ratio is higher than stoichiometric, graphite is precipitated (C porosity), which also leads to product deterioration. Thus, only a small gap exists in terms of the carbon in which only WC and Co are present. This gap is smaller, the lower the cobalt content. Most hardmetals are slightly substoichiometric to obtain an optimal property relationship and to avoid precipitates. Tungsten in solid solution improves the strength of the binder and makes the material less prone to plastic deformation during machining (solid-solution strengthening).

The carbon balance is severely affected by any traces of oxygen present in the grade powder, but also by the sintering atmosphere. The latter can be both decarburizing as well as carburizing. Being able to maintain the right carbon balance during sintering comes from experience (empyry) and requires very strict reproducibility in all the foregoing steps. It is most difficult in very low Co grades, where the gap between graphite or η-precipitation is the smallest.

Pressure sintering. Sometimes, hardmetals exhibit residual porosity (A and B type), which can occur for several reasons. Since the 1970s pressure sintering (HIP-ing) has been applied to remove this porosity. For that purpose, the sintered specimens are loaded in a HIP-ing device and heated again under Ar or He pressure of 50–150 MPa (temperatures and commonly 25–50 °C below the vacuum sintering temperature). Hence any residual porosity can be removed. Exceptions are gas porosities, which either can occur due to incomplete outgassing of the carbon monoxide, or due to the presence of impurities.

In recent years, this rather expensive two-stage treatment (vacuum sintering and subsequent HIPping) was substituted by the sinterhip process, which was developed in the early 1980. Sinterhip represents low-pressure isostatic pressing up to approximately 7 MPa combined with normal vacuum sintering. At the end of the common sinter heating cycle, the pressure is applied as long as the binder is still liquid. Void-free specimens can be produced.

Sintering aggregates. The majority of modern hardmetal sintering furnaces are batch-type aggregates which are resistance-heated (graphite heating elements). Examples are shown in Figs. 9.25 and 9.26. More rarely, semicontinuous push-type furnaces, which operate under flowing hydrogen, or inductively heated aggregates equipped with locks are used. All of them are electronically monitored (temperature and pressure control) and equipped to work under H_2 or controlled H_2/CH_4 mixtures (sometimes pulsating). Special heat treatments are common practice today and allow for tailormade properties in different parts of the hardmetal piece (hardmetals with a gradient structure). Characteristic examples are shown in Figs. 9.27 and 9.32.

Hot pressing. Hot pressing combines the two steps of compacting and sintering into one. The graphite mold is filled with hardmetal powder. The mold itself is the resistance heating element. Pressure is applied throughout the heating period. Hot pressing includes high labor costs and is generally only applied for special parts and/or alloys, which are not easy to sinter to full density otherwise.

9.3.4. Postsinter Treatments [9.1, 9.7, 9.8]

Coating. Coating was the logical consequence in the historical development of cutting tools. Braised tools in the past were reground several times when their cutting edge was blunt, unless almost the whole tip was consumed. Later, with less specialized labor, higher costs, and pressure in time, the throwaway insert was generated. Held mechanically, indexable inserts having several cutting edges or corners came on the

FIGURE 9.25. Three-chamber vacuum sintering furnace for hardmetal production (Ipsen). By courtesy of WIDIA GmbH., Germany.

FIGURE 9.26. Sinterhip furnace for pressure sintering of hardmetals (max 60 bar). Charges of up to 1000 kg can be sintered within a 24-hour cycle. By courtesy of ALD Vacuum Technologies GmbH., Germany.

market. When they are discarded, over 99% of their original mass remained. Discarding does not mean the material is completely lost. Today, the tips are collected and sold as scrap.

Consequently, it became reasonable to apply a hard and wear-resistant surface to a less hard but much tougher substrate by CVD. It is easy to understand that for a throwaway part the lifetime is of utmost economic interest, and by the invention of coatings it could be prolonged by a factor of 2 to 5.

Coating became one of the most significant developments in the history of hardmetals which started in the early 1960s and is still underway today.

In coated parts, the carbide plays a different role than in noncoated parts, because it is no longer the active component. It is the coating which has to withstand the erosion from the machined material. The carbide part in turn has to supply the best mechanical support for the coating (rigidity, toughness, and thermal properties) and allow perfect bonding to the coating (good adhesion) in order to resist spalling. When the coating is abraded, the cutting performance decreases to the value of the noncoated carbide.

TiC layers were the first to be applied by CVD. Others, like TiN, TiCN, Al_2O_3, HfC, HfN, ZrN, AlON as well as combinations and permutations, followed. The newest generation are multilayer coatings, which are deposited on binder-phase enriched substrates (to increase toughness); see Fig. 9.27. Further examples are TiC, TiCN, TiN, sequences, or an 8-layer undercoat composition topped by AlNO. Also, 13-layer compositions can be found. The thickness of single layers was 3–7 μm and multilayer coatings range from 8–12 μm in total.

Coatings are made by CVD, MT (medium temperature)-CVD, PVD, and plasma-activated CVD. The latter technique was recently successful in producing adherent diamond layers. The keenest edges are now produced by PVD coating.

Coated hardmetal is the dominant tool material for indexable inserts and will remain so in the foreseeable future. More than 80% of all turning inserts and about 70% of milling

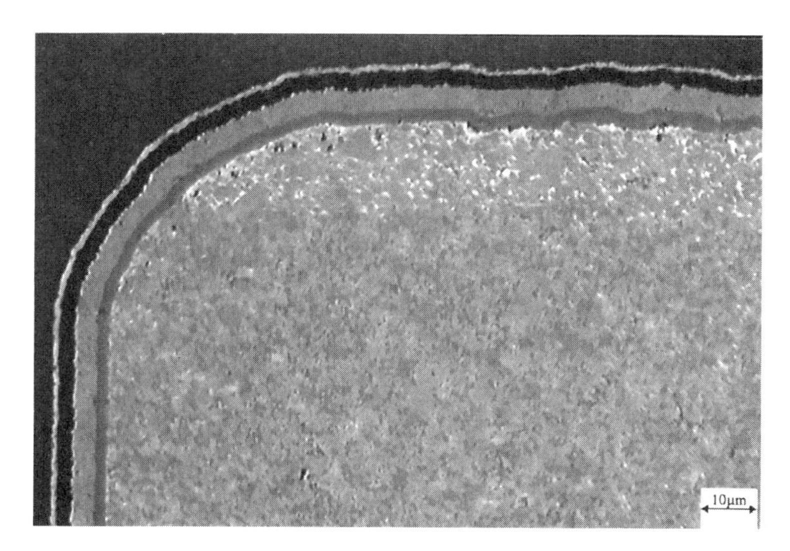

FIGURE 9.27. Cobalt-enriched hardmetal substrate, obtained by gradient sintering with CVD multilayer coating. Outer layer: TiCN/TiN; middle layer: Al_2O_3; inner layer: TiC/TiCN. By courtesy of Kennametal Inc., USA.

inserts are coated today. Also, the proportion of PVD-coated drilling tools is steadily increasing.

Other postsinter treatments. The following methods are in use: postsinter forming, spark erosion, grinding, electrochemical machining, ultrasonic machining, conventional machining, lapping, honing, and polishing.

9.4. HARDMETAL QUALITIES AND APPLICATIONS [9.1, 9.2, 9.4, 9.7, 9.8, 9.11, 9.26–9.28]

As mentioned earlier in the introduction to this chapter, the properties of WC–Co based hardmetals can be varied widely and consequently their applicability is extremely widespread. The properties are intimately connected with their microstructure (including micro- and macroporosity) and surface conditions (grinding cracks and excessive roughness). These can be influenced by several raw material properties and processing conditions:

- The average particle size, size distribution, and internal structure of the WC powder; its purity.
- The nominal composition.
- The production methods (especially powder milling and sintering).
- Excess or deficiency of carbon (stoichiometric and substoichiometric alloys).
- Postsinter treatments (HIPping, coating, grinding, polishing, etc.).

It is important to bear in mind that many of these items are interlinked.

The basic properties of WC–Co and WC–(W,Ti,Ta,Nb)C–Co grades are summarized in Table 9.2.

TABLE 9.2. Basic Data for Different WC–Co and WC–(W,Ti,Ta,Nb)C–Co HardMetal Grades [9.1]

Grade[a] (wt%)	Hardness HV30	Compressive strength ($N \cdot mm^{-2}$)	Transverse rupture strength ($N \cdot mm^{-2}$)	Young's modulus ($kN \cdot mm^{-2}$)	Fracture toughness ($MPa \cdot m^{-1/2}$)	Mean thermal expansion coefficient ($10^{-6} \cdot K^{-1}$)
WC–4Co	2000	7100	2000	665	8.5	5.0
WC–6Co/S*	1800	6000	3000	630	10.8	6.2
WC–6Co/M**	1580	5400	2000	630	9.6	5.5
WC–6Co/C***	1400	5000	2500	620	12.8	5.5
WC–25Co/M	780	3100	2900	470	14.5	7.5
WC–6Co–9.5(Ti,Ta,Nb)C	1700	5950	1750	580	9.0	6.0
WC–9Co–31(Ti,Ta,Nb)C	1560	4500	1700	520	8.1	7.2

[a] S*...submicron; M**...fine/medium; C***...coarse.

Straight Grades. In regard to quantity, this group is by far the most important. Figure 9.28 details the impressingly broad applicability of the two-component system in which, by varying the WC grain size in the sintered structure (0.4–7 μm) and the Co content (3–30%), quite remarkable property changes can be obtained.

Applications can be divided into two main fields: nonmachining and machining.

Nonmachining. Thirteen percent of tungsten consumed by the hardmetal industry, corresponding to 6% of world tungsten demand, are used in underground mining, road

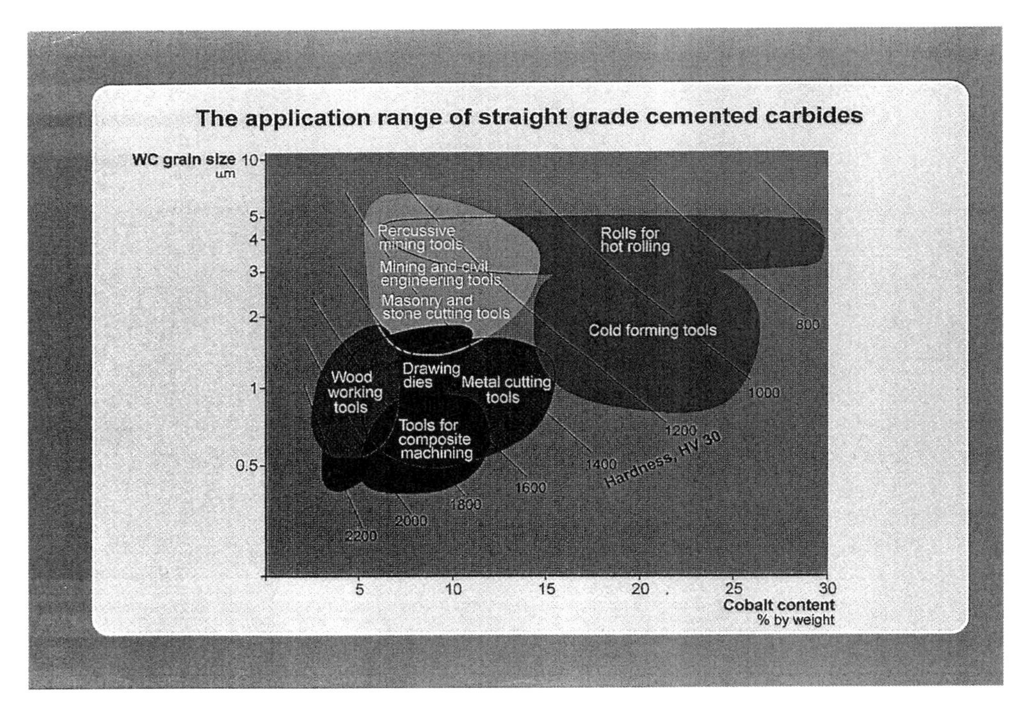

FIGURE 9.28. Application areas of straight WC–Co hardmetals. By courtesy of Sandvik Hard Materials, Sweden.

construction and maintenance, rock drilling, oil and gas drilling, as well as tunnel boring (Figs. 9.29–9.31). Consumption is dominated by coal mining.

Conventional coatings have failed to find a niche in these marketplaces, due to the aggressive nature of the applications [9.26].

Grain sizes for these applications are commonly in the medium-to-coarse size range (for certain applications up to 30 μm).

Hardmetals are also used as a support for polycrystalline diamond cutting tips, or as matrix alloy for diamond grit, used in rock drilling and oil mining [9.2].

Further nonmachining applications include metalforming (drawing dies, rod mill rolls, wire flattening rolls, slitter knives, extrusion punches, dies, etc.]; (see Figs. 9.32 and 9.33), structural components (plungers, boring bars, dies, pistons, hammers, etc.), and fluid-handling components (seal rings, valve stems, and nozzles).

Wear resistance and toughness of parts can be further improved by controlled redistribution of the binder phase ("dual property" carbide) [9.28], which is achieved during sintering. The principles of such a gradient material are demonstrated in Fig. 9.34.

Machining. The grain size of WC in the machining industry is in the range <1–2 μm. Straight grades are used for machining of cast irons, hardened steel, stainless steels, nonferrous metals, nickel-based high-strength alloys, wood, plastics, composites, etc. (see Figs. 9.35–9.37).

FIGURE 9.29. Examples of the broad variety of mining tools. By courtesy of Sandvik Rock Tools, Sweden.

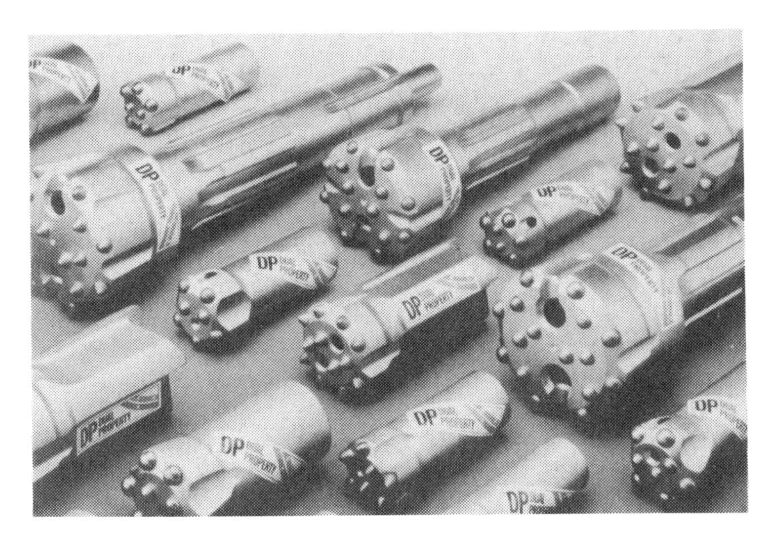

FIGURE 9.30. Modern button bit rock tools. The button bits exhibit a Co-depleted surface region, which is under compressive stresses, and an adjacent region with enhanced cobalt content (higher toughness; DP-carbides) [9.28]. By courtesy of Sandvik Rock Tools, Sweden.

FIGURE 9.31. Cutting head of a tunnel borer; WC drag bits for operation in soft material. Bore diameter, 3.5 m; weight of machine, 180 t; cutting head speed, 12.5 rpm; crowding force, 6280 kN; cutting head drives, 4×240 kW. By courtesy of Tamrock Voest-Alpine Bergtechnik Ges.m.b.H., Austria.

FIGURE 9.32. Rod mill rolls for high-speed finishing blocks; WC–(15–20)wt% Co or, alternatively, Co/Ni/Cr binder; WC grain size 3–4 μm. By courtesy of Sandvik Hard Materials, Sweden.

Submicron alloys and, more recently, so-called ulfrafine grades play an important role in this field. All these alloys have small additions of growth inhibitors (VC, Cr_3C_2, TaC) to restrict WC grain coarsening during liquid-phase sintering. VC is the most effective inhibitor. A microstructure of an ultrafine grade is shown in Fig. 9.38.

FIGURE 9.33. Integrated CIC rolls (cast-in carbide) for intermediate stands of wire rod mills or finishing stands in bar mills; up to 30 wt% binder; WC grain sizes: 3–5 μm. By courtesy of Sandvik Hard Materials, Sweden.

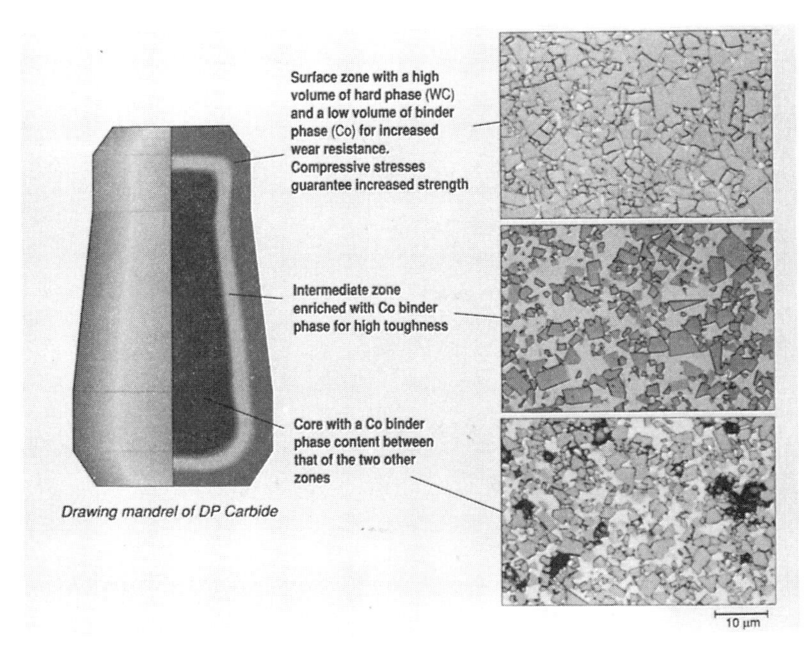

FIGURE 9.34. Schematic presentation of a Dual Property [9.28] drawing mandrel, together with microstructures of the surface/intermediate/core zone. By courtesy of Sandvik Hard Materials, Sweden.

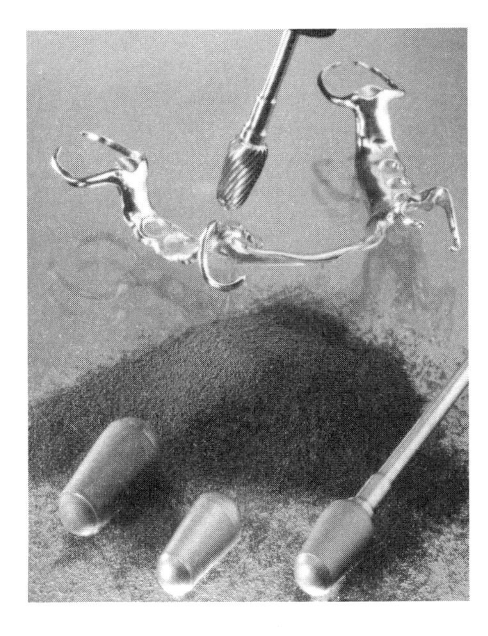

FIGURE 9.35. Dental endmill shown in different stages of manufacturing. Pressed part at the bottom left; as-sintered part in the middle; brazed onto a steel shaft (right); endmill together with the workpiece in the background. By courtesy of United Hardmetal, Germany.

FIGURE 9.36. Saw blades for wood cutting. By courtesy of Gebr. Leitz, Germany.

Plate-reinforced hardmetals have recently gained attention due to their improved hardness/toughness combinations [9.30]. Their industrial use is near (Fig. 9.39).

WC–(W,Ti,Ta,Nb)-Co grades. These grades are the working horse of the cutting materials in machining of steel, especially for long chipping alloys. They are used for all kinds of machining operations, such as turning, milling, grooving, threading, drilling, etc.

Examples are shown in Figs. 9.40 and 9.41. The large variety of indexable inserts and chip-breaker geometries can be assessed from Fig. 9.42. They demonstrate well today's high standard in the machining industry.

FIGURE 9.37. Solid tungsten carbide tools for the machining of printed circuit boards. Drill rotation speed, 10^5 rpm; feed, 2–8 m · min^{-1}; material, glass reinforced epoxy sheet. By courtesy of Kemmer Präzision GmbH., Germany.

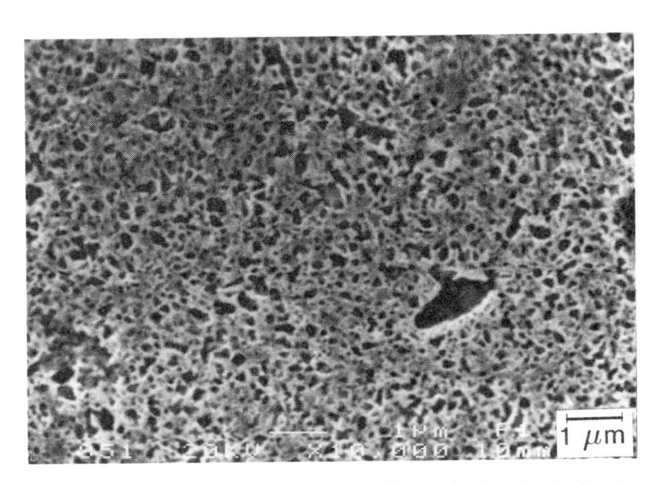

FIGURE 9.38. Microstructure of a straight WC–10 wt% Co alloy, etched with Murakami's reagent; 10,000×. Such grades are used for the machining of printed circuit boards. Note that the "giant" grain in the figure is only ~1 μm.

It would be far beyond the scope of the present book to go into more details. Larger hardmetal producers prepare several thousand differently shaped hardmetal specimens ranging in mass between less than 1 g (such as balls for ball-point pens, which are produced to the tune of 5 billion parts a year) and some hundred kg (such as dies and pistons in the synthetic diamond industry).

The quality palette of larger producers comprises up to 50–60 hardmetal grades and 8–10 different types of coatings.

Worldwide, there are around 30 big hardmetal companies with capacities between 250 and 3000 tons/year, besides a great number of smaller and very small companies (5–10 tons/year).

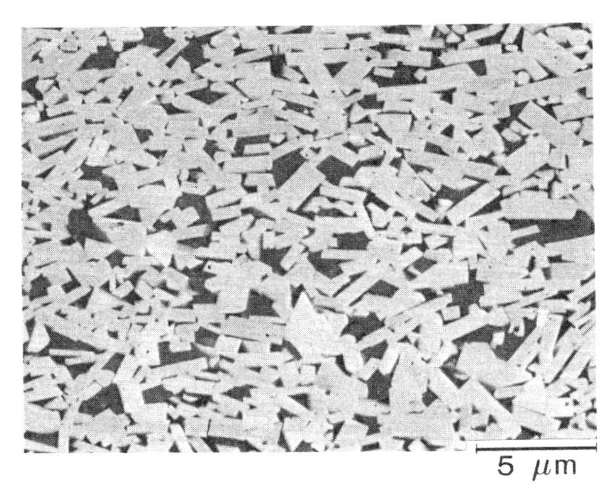

FIGURE 9.39. SEM image of a plate-reinforced hardmetal; 4,000× [9.29]. By courtesy of Toshiba Tungaloy Co, Ltd., Japan.

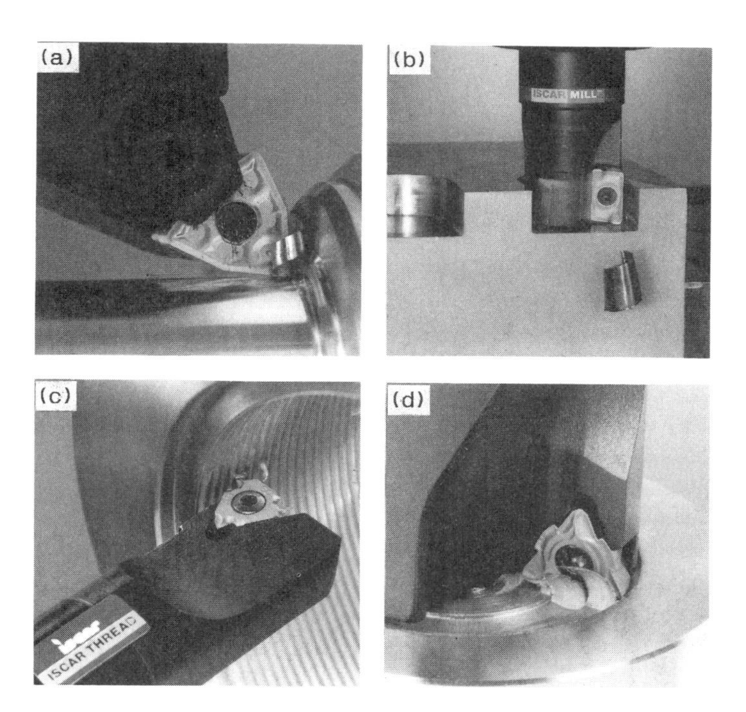

FIGURE 9.40. Hardmetal tools used in different machining operations: (a) turning; (b) slot milling; (c) threading; (d) drilling. By courtesy of Iscar Ltd., Israel.

FIGURE 9.41. Milling cutter for camshafts. The cutter is equipped with 66 inserts; milling cutters with diameters >2 m are produced today; individual tools are equipped with up to 300 inserts. By courtesy of Walter AG, Germany.

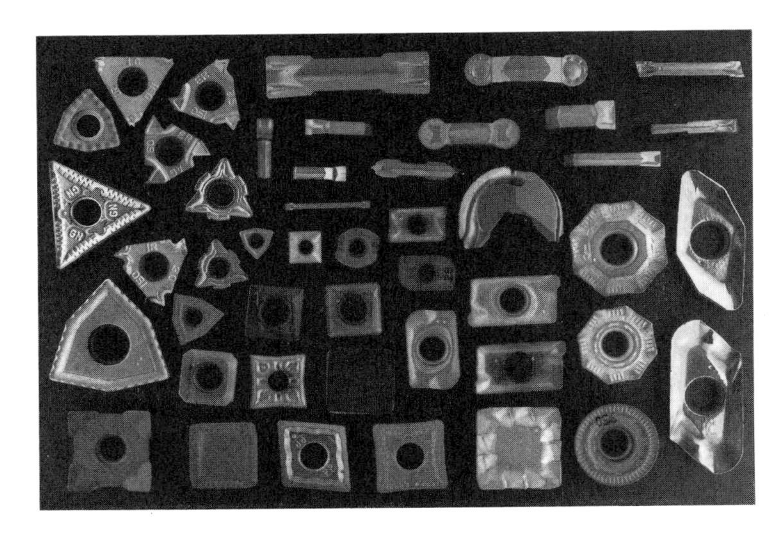

FIGURE 9.42. Indexable inserts of different shape and chip-breaker styles. Note the small inserts with edge lengths of only ~6 mm. By courtesy of Iscar Ltd., Israel.

REFERENCES FOR CHAPTER 9

9.1. H. Kolaska, *Pulvermetallurgie der Hartmetalle*, Fachverband für Pulvermetallurgie, Hagen, Germany (1992).

9.2. A. T. Santhanam, in: *The Chemistry of Transition Metal Carbides and Nitrides* (S. T. Oyama, ed.), pp. 28–52, Blackie Academic and Professional, London (1996).

9.3. S. Söderberg, in: *Proc. 7th Int. Tungsten Symposium, Goslar*, pp. 49–56, ITIA, London (1997).

9.4. A. T. Santhanam, P. Tierney, and J. L. Hunt, *Metals Handbook*, 10th edition, Vol. 2, pp. 950–977, ASMI, Materials Park, Ohio (1990).

9.5. H. Kolaska and K. Dreyer, *Metall* **45** (1991), 224–235.

9.6. R. W. Stevenson, *Metals Handbook*, 9th edition, Vol. 7, pp. 773–783, ASMI, Materials Park, Ohio (1984).

9.7. K. J. A. Brookes, *World Directory and Handbook of Hard Metals and Hard Materials*, 5th ed., International Carbide Data, Hertfordshire, UK (1992).

9.8. W. Schedler, *Hartmetall für den Praktiker* (Plansee TIZIT GmbH, ed.), VDI Verlag GmbH, Düsseldorf (1988).

9.9. P. Ettmayer, *Annu. Rev. Mater. Sci.* **19** (1989), 145–164.

9.10. F. Klocke, K. Gerschwiler, and R. Fritsch, in: *Proc. 7th Int. Tungsten Symposium, Goslar*, pp. 213–238, ITIA, London (1997).

9.11. A. Egami, in: *Proc. 7th Int. Tungsten Symposium, Goslar*, pp. 13–23, ITIA, London (1997).

9.12. B. F. Kieffer, unpublished results and personal experience of the authors.

9.13. M. Miyake and A. Hara, *J. Jpn. Soc. Powder Metall.* **26** (1979), 16–21.

9.14. M. Miyake, A. Hara, and T. Sho, *J. Jpn. Soc. Powder Metall.* **26** (1979), 90–95.

9.15. Y. Yamamoto, A. Matsumoto, S. Honkawa, S. Honkawa, and N. Shigaki, in: *Proc. of 1993 Powder Metallurgy World Congress* (Y. Bando and K. Kosuge, eds.), pp. 785–788, The Japan Society of Powder Metallurgy (1993).

9.16. Y. Yamamoto, A. Matsumoto, and Y. Doi, in: *Proc. 14th Plansee Seminar* (G. Kneringer, P. Rödhammer, and P. Wilhartitz, eds.), Vol. 2, pp. 596–608, Plansee AG, Reutte (1997).

9.17. P. M. McKenna, US Patent 3,379,503 (1968).

9.18. E. N. Smith, *Met. Powder Rep.* **35** (1980), 53–54.

9.19. A. W. Weimer, W. G. Moore, and R. P. Roach, US Patent 5,110,565 (1992).

9.20. S. D. Dunmead, W. G. Moore, A. W. Weimer, G. A. Eismann, and J. P. Henley, US Patent 5,380,688 (1995).

9.21. B. Günther, T. König, and R. L. Meisel, in: *Pulvermetallurgie in Wissenschaft und Praxis*, Band 11 (H. Kolaska, ed.), pp. 41–58, DGM Informationsges. m.b.H., Oberursel, Germany (1995).
9.22. L. E. McCandlish, B. H. Kear, and S. J. Bhatia, US Patent 5,352,269 (1994).
9.23. L. E. McCandlish and P. Seegopaul, *Proc. European Conf. on Advances in Hard Materials Production, Stockholm*, pp. 93–100, EPMA, Shrewsbury, UK (1996).
9.24. E. Rudy and B. Kieffer, private communication.
9.25. E. Lardner and S. Iggstrom, in: *Proc. 10th Int. Plansee Seminar* (H. Bildstein and H. Ortner, eds.), pp. 547–579, Metallwerk Plansee, Austria (1981).
9.26. D. Kassel, G. Schaaf, and K. Dreyer, in: *Proc. 7th Int. Tungsten Symposium, Goslar*, pp. 24–37, ITIA, London (1997).
9.27. A. D. Landsperger and T. R. Massa, in: *Proc. 7th Int. Tungsten Symposium, Goslar*, pp. 85–96, ITIA, London (1997).
9.28. G. Gille and B. Szesny, in: *Proc. 7th Int. Tungsten Symposium, Goslar*, pp. 150–170, ITIA, London (1997).
9.29. *Sandvik Hard Materials*, company brochure H-9097 ENG, Sandvik DP Carbide.
9.30. M. Kobayashi, K. Kitamura, and S. Kinoshita, European Patent Application 0759 480 A1 (1997).

10

Tungsten in Catalysis

10.1. METALLIC TUNGSTEN

10.1.1. Survey

Tungsten alloyed with nickel, cobalt, or rhodium in thin layers on alumina supports, also sulfided, is used on an industrial scale as a catalyst in crude oil processing (hydrotreating, hydrocracking, reforming, hydrodesulfurization, and hydrodenitrogenation), as well as in Fischer–Tropsch synthesis (alcohol formation from $CO + H_2$).

10.1.2. Hydrodesulfurization and Hydrodenitrogenation [10.1, 10.2]

HDS and HDN are processes for removing chemically bound sulfur and nitrogen from various petroleum feedstocks. HDS and HDN became more important in recent years since new oil drilling techniques made it possible to recover heavier crude oils. These oils are higher in organic sulfur and nitrogen containing compounds such as thiols, organic sulfides, and thiophenes, as well as amines, pyridine, and chinoline compounds. The sulfur content of crude oils varies between 0.2 and 4%, while the nitrogen content ranges from 0.1 to 0.9%. The removal of sulfur and nitrogen by hydrogen, performed at 300–400 °C and pressures up to 200 bar, can be described by the following overall equations:

$$C_aH_bS + cH_2 \rightarrow H_2S + C_aH_d$$
$$C_aH_bH + cH_2 \rightarrow NH_3 + C_aH_d$$

As catalyst, besides others, tungsten is used as nickel alloy in thin layers on alumina supports. During the reaction, tungsten is converted, at least at the surface, to WS_2, which also acts as a catalyst.

The HDS and HDN reactions occur during hydrotreating and hydroreforming of crude oil, and the catalyst also promotes hydrogenation of unsaturated hydrocarbons and hydrodeoxygenation.

There are several reasons for the reduction of sulfur and nitrogen concentrations in petroleum feedstocks:

- To minimize the amount of sulfur and nitrogen oxides, which form during combustion processes in order to reduce air pollution.
- To prevent the poisoning of catalysts used in the subsequent steps of product fuel preparation, such as reforming and cracking. Nitrogen compounds are a poison for the zeolithe cracking catalyst and large sulfur amounts are poisonous for Pt/Re reforming catalysts.
- Last but not least, new environmental regulations require lower sulfur concentrations in fuels in general.

The necessity to improve HDS and HDN has recently boosted R&D efforts. The search for better catalysts has also improved our understanding of the complicated processes which occur at the catalyst surfaces.

In order to understand the importance of these technical processes, it must be stressed, that approximately 33 million barrels per day (1 barrel $\cong$ 159 liter) are hydrotreated worldwide.

10.2. OXIDES

10.2.1. Survey

Colloidal tungsten trioxide can be used for the photocatalytic reduction of organic compounds.

$W_{20}O_{58}$ catalyzes the hydration, dehydration, hydroxylation, and epoxidation.

$DENO_x$ mixed oxide catalysts are used in power plants, waste incineration plants, hydration, aldol condensation, ROMP, and dimethyl sulfoxide synthesis.

WO_x is particularly active in hydrogen oxidation in acidic media as a cathode catalyst (x equals preferably 2.72–2.9) [10.3, 10.4].

10.2.2. DENOX SCR Catalysts [10.5–10.8]

The selective catalytic reduction (SCR) of NO_x by NH_3 using DENOX catalysts is the best developed and most widely used method for treating flue gases of thermic power plants in order to reduce the emission of toxic compounds. The efficiency, selectivity, and economics of the method are favorable. It was developed in Japan.

Commercial catalysts are composed of: TiO_2, 78%; WO_3, 9%; V_2O_5, 0–5%; SiO_2, 7.5%; Al_2O, 1.5%; CaO, 1.0%; SO_4, 0.5–2%.

TiO_2 is in the anatas modification and acts as high specific surface support; V_2O_5 and WO_3 are the catalytically active components; V_2O_5 is an oxidation catalyst; WO_3 improves the selectivity of the catalyst by suppressing any $SO_2 \rightarrow SO_3$ oxidation (by decreasing the oxidizing properties of the catalyst system); WO_3 is further responsible for the stabilization of the anatas structure and improvement of the activity of TiO_2 (specific surface, approximately 90 $m^2 \cdot g^{-1}$). The starting point in the production of the catalyst is a precipitated TiO_2 in anatas form suspended in water, which is doped with tungstate in ammoniacal solution. The WO_3 is produced by precipitation, while TiO_2 acts as crystallization nucleus and will be covered by a thin H_2WO_4 layer during which calcination converts to WO_3. In a second step, V_2O_5 and the other components are added. The remaining components act as mechanical promotors.

The ceramic catalyst is a homogeneous mixture of all components. After mixing with binders, plastifiers, and glass fiber the mass is extruded to honey-comb-structured elements, which are dried and calcined (Fig. 10.1).

SCR catalysts are full catalysts, which means that the whole mass is catalytically active.

FIGURE 10.1. DENOX catalyst. The catalyst elements exhibit a honey-comb-like structure (upper photo); view into a reactor partly filled with catalyst modules (lower photo). By courtesy of Porzellanfabrik Frauenthal GmbH and Austrian Energy and Environment SGP Wagner-Biro GmbH, Austria.

The ratio of NO/NO_2 in exhaust gases of power plants is usually about 95/5. The nitrogen oxides, during the DENOX process, are reduced by NH_3 at the catalyst surface while the following reactions occur:

$$NO_2 + NO + 2NH_3 \rightarrow 2N_2 + 2H_2O$$
$$4NO + 4NH_3 + O_2 \rightarrow 4N_2 + 6H_2O$$
$$2NO_2 + 4NH_3 + O_2 \rightarrow 3N_2 + 6H_2O$$

The temperature should be in the range between 150 and 450 °C. The side reaction $2SO_2 + O_2 \rightarrow 2SO_3$, which is 100% when using TiO_2–V_2O_5 catalysts, is less than 1% in case of TiO_2–V_2O_5–WO_3 systems.

DENOX-SCR catalysts have been used successfully for years to remove NO_x from power plant exhaust gases. New areas of application in recent times are removal of NO_x from exhaust gas of large-volume diesel engines used in ships or emergency power-generating sets.

DENOX-SCR catalysts can drastically reduce the dioxine and furane content of exhaust gases of waste incineration plants, far below the limits prescribed by law ($0.1\,ng \cdot m^{-3}$). The dioxine concentration in the emission of a DENOX-equipped waste incineration plant is 100 times lower than the concentration in nonincinerated waste.

10.3. HALIDES

WCl_6 and WCl_4O are used in combination with organometallic compounds for numerous organic syntheses (see Section 10.5).

10.4. CARBON COMPOUNDS

This section deals with carbide compounds and hexacarbonyl. Organotungsten compounds are treated separately under in Section 10.5. For information about tungsten-containing biocatalysts (enzymes), see Section 14.2.

10.4.1. Survey

WC can act as catalyst for gas-phase oxidation/reduction reactions, alcane hydrogenolysis, alcane reforming, H_2 oxidation in fuel cells, and H_2 formation from water, and also, for anodic oxidation of hydrogen and formic acid in acidic media and of hydrogen and hydrazine in alkali media. It is only active after anodic polarization in the presence of reducing agents. It is assumed that at the surface hydrogen tungsten bronzes are formed [10.4].

W_2C can be used as catalyst for the isomerization, aromatization, and hydrogenolysis of alcanes.

Tungsten hexacarbonyl $W(CO)_6$ catalyzes a wide range of organic reactions, such as the metathesis of alkenes.

10.4.2. Preparation of Carbide Catalysts

In heterogeneous catalysis, the catalytic properties of an element or a compound are always closely linked to an active and large specific surface. The production of high specific surface tungsten carbides requires special methods. Typical examples are listed below.

- Reaction of W or an oxide with a carbon-containing gas, preferably a gaseous hydrocarbon. The most important prerequisite is a fine-grained precursor [10.9, 10.10], but also the hydrocarbon exhibits its own influence. In using CH_4/H_2 mixtures, H_2 reduces the WO_3 and CH_4 acts as carburizing agent. If C_2H_6 or C_2H_4 is used, the hydrocarbon also takes part in the reduction. The temperature in this case is about 150 K lower compared to CH_4/H_2, which is of special importance during the preparation of supported catalysts (suppression of reactions with the support). Moreover, C_2H_6 or C_2H_4 originating WC exhibits 25 $m^2 \cdot g^{-1}$ specific surface compared to 10 $m^2 \cdot g^{-1}$ in case of CH_4/H_2 carburized.
- Pyrolysis of organometallic compounds containing cyclopentadiene and carbonyl groups to produce W_2C [10.11].
- Reaction of volatile tungsten chloride or oxychloride with gaseous hydrocarbons at 1400–2900 K [10.9].
- Reaction of $W(CO)_6$ with H_2 [10.8].
- Reaction of solid carbon with vaporized tungsten oxides [10.9].
- Mixing of WCl_4, suspended in tetrahydrofurane, with Li triethylborohydride and subsequent heat treatment at 723–773 K to produce W_2C [10.9].
- Reaction of W films with a gaseous hydrocarbon or by reactive sputtering [10.12].
- CO_2 laser pyrolysis of C_2H_4 mixtures with $W(CO)_6$ or WCl_6 [10.13].
- Reaction of WCl_6 and guanidine hydrochloride to W_2C free of uncombined carbon [10.11].
- Reaction of WO_3 with NH_3 at 700–1000 K to β-W_2N and subsequent carburization with CH_4/H_2 at the same temperature to β-W_2C [10.14].

During reactions with carbon-containing compounds, free (unreacted) carbon may form by cracking, which is undesirable because it changes the catalytic properties and shows an increased specific surface not due to the carbide phase. C_2 hydrocarbons produce more excessive carbon than CH_4. It can be removed in all cases by hydrogen treatment at 973 K. The carbide phase itself is not in any case pure and may consist of W_2C/WC mixtures.

10.4.3. WC [10.14, 10.15]

WC behaves similarly to platinum in many catalytic reactions (e.g., gas-phase oxidation reactions). Unlike many metal and oxide catalysts, it is capable of operating under a wider range of conditions in oxidative as well as in reductive regimes.

Kinetically, under oxidative conditions, due to an oxidic surface layer, WC acts similar to an oxide in heterogeneous oxidation reactions (the redox mechanism consisting of successive oxidation and reduction steps at the catalyst surface). Under reducing conditions, the surface of the WC becomes free of surface oxide and reveals its metal-like nature. Consequently, both reactants being simultaneously activated at the catalyst surface enable their reaction within the adsorbed layer (typical behavior of a metal catalyst).

The heat of oxygen adsorption corresponds to $362\ kJ \cdot mol^{-1}$ at 523 K and $356\ kJ \cdot mol^{-1}$ at 723 K. Compared to metallic W ($812\ kJ \cdot mol^{-1}$) it is lower, because the carbon atoms in the metal lattice reduce the adsorption affinity toward oxygen.

The oxidation of hydrogen (at excess of O_2) is generally used to characterize the total oxidation activity of a catalyst.

The specific activity is $16.7\ mol \cdot m^{-2} \cdot s^{-1} \cdot 10^7$ at 55 kPa H_2, 22 kPa O_2 at 623 K.

The specific catalytic activity in the $H_2 + O_2$ reaction at 22 kPa H_2, 2.8 kPaO_2, and 623 K declines in the order

W	$5.0 \times 10^{-4}\ mol \cdot m^{-2} \cdot s^{-1}$,
WC	$3.5 \times 10^{-6}\ mol \cdot m^{-2} \cdot s^{-1}$,
WO_3	$5.0 \times 10^{-8}\ mol \cdot m^{-2} \cdot s^{-1}$.

The reactivity of CO in the oxidation reaction to CO_2 is less compared to the H_2 oxidation.

The specific catalytic activity for the oxidation of NH_3 at 20 kPa NH_3, 10 kPa O_2 and 573 K is $0.71\ mol \cdot m^{-2} \cdot s \cdot 10^7$ (activation energy, $63\ kJ \cdot mol^{-1}$).

WC can also be applied as catalyst for the oxidative coupling of methane.

WC can therefore be regarded as a promising candidate as heterogeneous catalyst with high stability and activity for oxidation reactions, such as the conversion of natural gas to higher hydrocarbons.

High surface ($30–100\ m^2 \cdot g^{-1}$) WC, free of carbon and oxygen, is very active in neopentane hydrogenolysis. Chemisorbed oxygen inhibits the hydrogenolysis, but catalyzes isomerization reactions not only of neopentane but also, for example, of 3,3-dimethylpentane. Furthermore, dehydrogenation as well as isomerization reactions of methyl-cyclopentane occur on the WC catalyst, suggesting the presence of a bifunctional surface that catalyzes dehydrogenation and carbenium ion reactions typically occurring on reforming catalysts.

10.4.4. W_2C [10.11, 10.14]

High specific surface ($100–150\ m^2 \cdot g^{-1}$) and unreacted carbon-free W_2C can be prepared by reacting WCl_6 with guanidine hydrochloride. It shows excellent catalytic properties in unsupported form as well as supported either on SiC-coated alumina or on Zeolithe Y.

Unsupported and SiC/Al_2O_3-supported W_2C catalysts are multifunctional. n-Alcanes are transformed by aromatization, isomerization, and hydrogenolysis. Aromatization can be increased by the addition of O_2 traces.

A comparative study using the unsupported W_2C catalyst ($122\ m^2 \cdot g^{-1}$) and a commercial alumina-supported Pt catalyst (0.3%; $180\ m^2 \cdot g^{-1}$), in the isomerization of n-heptane (523 K), revealed different activities:

isomers (2-methylhexane, 3-methylhexane, dimethylpentanes, 2- and 3-methylpentanes): 81% for W_2C and 48% for Pt,

aromatics (benzene and toluene): 3% for W_2C and 32% for Pt,

hydrogenolysis products (C_1 to C_5 hydrocarbons): comparable.

Zeolithe-Y-supported W_2C shows metallic hydrogenation properties and can also be used as catalyst in the thiophene hydrogenation.

A big advantage of W_2C over the usual noble metal catalysts is the sulfur resistance. Noble metal catalysts get poisoned (inactive) by the presence of S-containing compounds in a short time. The sulfur resistance of W_2C also allows its use as catalyst for the hydrodesulfurization of thiophene.

High surface area β-W_2C, prepared from β-W_2N by subsequent carburization, shows similar behavior as described for WC in reactions with neopentane, 3,3-dimethylpentane, and methyl-cyclohexane.

10.4.5. Tungsten Carbide Catalysts in Fischer–Tropsch Synthesis [10.16]

The properties of tungsten carbide catalysts depend strongly on their surface chemistry, which can be clearly demonstrated when using them in FT synthesis.

The starting product may act as well reducing (H_2), as carburizing (CO), and as decarburizing (H_2). The carbon content of the surface, which can be changed by the above reactions, is highly responsible for the catalytic action and may vary from W to WC. Also, if supported tungsten carbide catalysts are used, the reaction between the catalyst and support during pretreatment and use may have a significant influence on the catalytic activity. Therefore, different supports lead to different catalytic properties.

By using an unsupported catalyst, a comparison of the carbides and the trioxide revealed the following results:

	WC(%)	W_2C(%)	WO_3(%)
Hydrocarbons	79.7	86.5	68.0
Alcohols	18.4	0	20.4
Carbon dioxide	2.1	13.5	11.5

W_2C favors the CO dissociation and formation of hydrocarbons with excess of CH_4 and CO_2 and no oxygenated products, due to its more pronounced metallic character as WC. W_2C is more sensitive to H_2 reduction (the surface can be transferred to W).

Alcohols are only formed on WC. The formation of alcohols is related to the surface stoichiometry of WC (presence of C vacancies; understoichiometric WC). WC also promotes higher chain length in both hydrocarbons and alcohols compared to WO_3.

When using supported catalysts, the formation of alcohols is, on the one hand, related to the surface WC stoichiometry. Only layers with carbon deficiency (vacancies) produce alcohols while high carbon concentrations produce alcanes only. On the other hand, by reaction with special supports, anionic vacancies are formed, stabilized by mixed oxides, and associated with the carbon vacancies in the mixed carbide. The optimal support for alcohol production is TiO_2. The bifunctionality is due to carbidic and oxidic phases.

10.5. ORGANOTUNGSTEN COMPOUNDS: CATALYSTS IN ROMP AND ADMET [10.17–10.20]

As already pointed out in Chapter 4, the number of synthesized metallo-organic tungsten compounds is immense and the reason why so many compounds have been prepared and investigated can only be understood if the field of application are examined more closely.

A large number of tungsten organometallic compounds plays an important role, or can be regarded as prominent candidates, as catalyst in organic syntheses. Tungsten as well as molybdenum and rhenium catalysts are closely associated with metathesis polymerization.

The term "alkene metathesis" describes a reaction mechanism shown schematically in the following equation:

R1 R3 R2 R4 catalyst R1 R3 + R2 R4

The alkylidene groups of an alkene pair are exchanged with one another in the presence of a transition metal catalyst. In recent times, not only reactions between two C=C double bonds but also between C=C and C=O bonds have been investigated. The reaction includes the cleavage of double C=C bonds and the formation of new ones. Homogeneous as well as heterogeneous catalysts are effective.

There exist two groups of catalysts:

- a transition metal compound combined with a main group metal alkyl co-catalyst,
- a well-defined transition metal carbene catalyst.

One of the most important reaction mechanisms in organic polymer chemistry that can be catalyzed by tungsten compounds is "Ring Open Metathesis Polymerization"—ROMP. It can be expressed by the overall equation:

X R catalyst R x

The chain propagation and addition-type nature of ROMP chemistry will be elucidated by the scheme:

LnM=C H R' + R → LnM C R' H R →

→ LnM= R R' C H R → R' C H R LnM R repeat → Polymer

The first step is coordination of substrate to the metal center forming the metallacyclobutane intermediate. The subsequent opening of the metallacyclobutane ring produces the first new carbene, which reacts with the next equivalent of the monomer, etc. The final polymer molecule contains an original alkylidene group at one end and the catalyst fragment at the other. The driving force for the reaction is the loss of ring monomer strain. The structure of the catalyst allows one to direct the polymerization reaction, and consequently the properties of the polymer. This is the reason for the very intensive investigations in synthesis of catalyst compounds. Today, these efforts are mainly driven by the necessity to find "economic synthesis reactions." This requires that the highest percentage of input molecules are used to produce the desired final product comprising minimization of raw material consumption and waste production.

Typical ROMP products are linear, thermoplastic polymers Polyoctenamer (from cyclooctene), Polynorbonene (from norbonene), and the network (thermoset) polymer Polydicyclopentadiene (from dicyclopentadiene) combining toughness, solvent resistance, chemical inertness, and long-time durability.

The second important reaction is the "Acidic Diene Metathesis Polymerization"—ADMET. The overall equation is as follows:

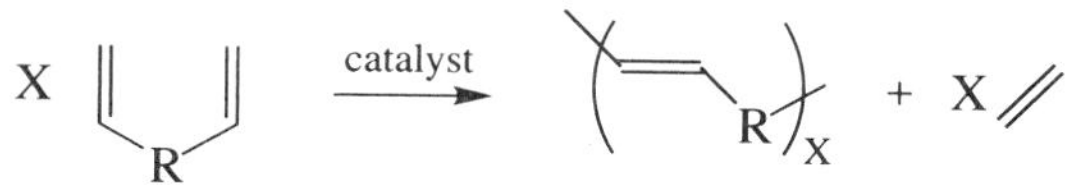

X R catalyst R x + X

The reaction is driven by a shift in equilibrium, caused by the removal of one of the reaction products.

The metathesis polymerization of acyclic alkenes is an equilibrium step propagation via, dimers, trimers, tetramers, etc. The scheme is:

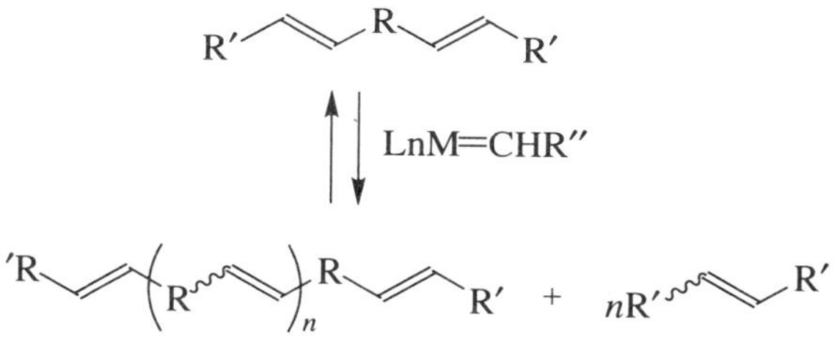

R′ R R′

LnM=CHR″

′R R R R′ + nR′ R′
n

For the equilibrium shift, it is important to remove the monomer reaction product continuously.

Typical products are Polyoctenamer (from 1,9-decadiene) and 1,4-Polybutadiene (from 1,5-hexadiene).

The main groups of metalorganic tungsten compounds in use as catalysts in ROMP and ADMET are:

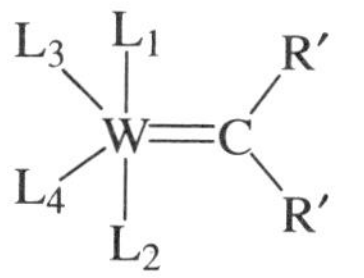

L_3 L_1 R′ W=C L_4 R′ L_2

Five-coordinate W alkylidene complexes, where L can be alkoxide or halide or SO_3CF_3, are highly effective olefine metathesis catalysts in the presence of Lewis acids. The vast number of compounds synthesized can be understood by the variation and combination of different ligands.

Tungsta Cyclobutan Complexes
They are the reaction product of

$$(RO)_2W(=NAr)(=CH{-}t\text{-}Bu) \xrightarrow[-\ CH_2{=}CH(t\text{-}Bu)]{+\ CH_2{=}CH_2\ (2\ \text{eqn.})} (RO)_2W(=NAr)(C_3H_6)$$

where

$$OR = OCMe(CF_3)_2, \qquad OC(CF_3)_2(CF_2CF_2CF_3).$$

Well-defined alkylidene metallacyclobutane catalysts exist for the polymerization of olefines (alkenes) and acetylenes(alkines).

10.6. TUNGSTATES

Tungsten bronzes M_xWO_y (Me can be Ti, V, Mn, Fe, Zn, and Ni; x can be 0–1 or 1–2, and y can be 2–4) may act as anodic catalyst in fuel cells (high electrical conductivity and corrosion resistance in acidic media) [10.4] .

Sodium tungstate bronzes with Au, Pt, or Pd additions exhibit high electrocatalytic activity (also in acidic media) in reducing oxygen [10.4], and catalyze the carbon monoxide oxidation.

Sodium salt of 12-tungstophosphoric acid acts as catalyst for isomerization, polymerization, nitrile synthesis, dehydrochlorination, dehydrogenation, ketone synthesis, ring-closure synthesis, oil bodying, aromatization, and desulfurization [10.4, 10.21].

10.7. CHALKOGENIDES [10.4, 10.8]

Tungsten disulfide WS_2 [10.8] is used for electrode activation in fuel cells. If produced by hydrogen reduction of $(NH_4)_2WS_4$, it exhibits a BET surface 70 times larger than that of commercially prepared WS_2. It acts as catalyst for the reaction

$$CO_2 + H_2 \rightarrow CO + H_2O$$

at normal pressure with a yield of more than 99.9%.

WS_2 and WSe_2 as colloidal coating are used in photocatalysis.

REFERENCES FOR CHAPTER 10

10.1. R. J. Angelici, "Hydrodesulfurization and Hydrodenitrogenation," in: *Encyclopedia of Inorganic Chemistry* (R. Bruce King, ed.), Vol. 3, pp. 1433–1443, Wiley, Chichester, New York (1994).

10.2. Shi Yahua, Kang Xiaohong, and Li Ladong, Paper presented at the ITIA Annual General Meeting Huntsville, Al, USA; transcript of papers: ITIA, London, UK (1994).

10.3. B. Broude, US Patent 3,507,701 (1970).
10.4. A. T. Vas'ko, in: *Encyclopedia of Electrochemistry of the Elements* (A. J. Bard, ed.), Vol. V, pp. 69–126, Marcel Dekker, New York (1976).
10.5. T. Nakatsuju, Paper presented at the ITIA Annual General Meeting, Huntsville, Al, USA; transcript of papers: ITIA, London, UK (1994).
10.6. Ceram Frauenthal Keramik AG, company brochure.
10.7. I. Binder-Begsteiger, E. Megla, M. Tomann-Rossos, and G. W. Herzog, *Chem. Ing. Tech.* **62** (1990), 56–64.
10.8. G. Winter, Paper presented at the ITIA Annual General Meeting, Huntsville, Al, USA; transcript of papers: ITIA, London, UK (1994).
10.9. S. T. Oyama, in: *The Chemistry of Transition Metal Carbides and Nitrides* (S. T. Oyama, ed.), pp. 1–27, Blackie Academic and Professional, (1996).
10.10. S. Decker, A. Löfberg, J. M. Basin, and A. Frennet, *Catal. Lett.* **44** (1997), 229–239.
10.11. F. Sherif and W. Vreugdenhil, in: *The Chemistry of Transition Metal Carbides and Nitrides* (S. T. Oyama, ed.), pp. 414–425, Blackie Academic and Professional, London (1996).
10.12. F. Reniers, M. Detroye, S. Kacim, P. Kons, M. Maoujoud, E. Silberberg, M. Soussi, El Begrani, T. Vandevelde, and C. Buess-Hermann, in: *The Chemistry of Transition Metal Carbides and Nitrides* (S. T. Oyama, ed.), pp. 274–289, Blackie Academic and Professional, (1996).
10.13. R. Ochoa, X. X. Bi, A. M. Rao, and P. C. Eklund, in: *The Chemistry of Transition Metal Carbides and Nitrides* (S. T. Oyama, ed.), pp. 489–499, Blackie Academic and Professional, (1996).
10.14. F. H. Ribeiro, R. A. Dalla Betta, M. Boudart, J. Baumgartner, and E. Iglesia, *J. Catal.* **130** (1991), 86–105.
10.15. N. I. Ilchenko and Yu. I. Pyatnitsky, in: *The Chemistry of Transition Metal Carbide and Nitrides* (S. T. Oyama, ed.), pp. 311–326, Blackie Academic and Professional, London (1996).
10.16. L. Leclercq, A. Almazouari, M. Dufour, and G. Leclercq, in: *The Chemistry of Transition Metal Carbides and Nitrides* (S. T. Oyama, ed.), pp. 343–361, Blackie Academic and Professional, (1996).
10.17. G. G. Hlatky, "Oligomerization and Polymerization by Homogeneous Catalysis," in: *Enclyclopedia of Inorganic Chemistry* (R. Bruce King, ed.), Vol. 5, pp. 2728–2735, Wiley, Chichester, New York (1994).
10.18. K. B. Wagener, J. M. Boncella, and D. W. Smith Jr., "Metathesis Polymerization Processes by Homogeneous Catalysis," in: *Encylopedia of Inorganic Chemistry* (R. Bruce King, ed.), Vol. 4, pp. 2242–2251, Wiley, Chichester, New York (1994).
10.19. R. H. Grubbs and S. H. Pine, "Alkene Metathesis and Related Reactions," in: *Comprehensive Organic Synthesis* (B. M. Trost, ed.), Vol. 5, pp. 1115–1127, Pergamon, New York (1991).
10.20. J. Feldman and R. R. Schrock, in: *Progress in Inorganic Chemistry* (S. J. Lippard, ed.), pp. 1–69, Wiley, Chichester (1991).
10.21. GTE Sylvania, *Technical Information Bulletin*, CM-9003 (9/80).

11

Tungsten Scrap Recycling

11.1. INTRODUCTION AND GENERAL CONSIDERATIONS [11.1 – 11.4]

The recycling of scrap containing tungsten is by no means a modern procedure. On the contrary, it has been established for many decades. Even before anyone thought about recycling of glass, paper, and aluminum—as is common today in most industrialized countries—a considerable amount of tungsten was being reclaimed from different types of scrap. The reason can be easily explained: As outlined in Chapter 2, most tungsten ores contain less than 1 wt% WO_3 and very seldom higher concentrations. In Table 11.1, different types of scrap containing tungsten and the corresponding tungsten contents are listed. It is evident that the least valuable scrap, like low-grade grinding sludge, contains about 15 times more tungsten than an average ore. Most scrap materials are even richer in tungsten than ore concentrates.

It is the high intrinsic value in anything containing tungsten, even in smaller concentrations, which makes tungsten scrap an interesting and worthy material for

TABLE 11.1. Typical Tungsten Scrap Materials

No.	Scrap type	% W	Group
1	High-purity W	≥ 99	
2	Oxide dispersed W alloys	96 – 98	
	(ThO_2, ZrO_2, CeO_2, La_2O_3)		**Hard Scrap**
3	Hard metal pieces (also containing Co, Ta)	60 – 97	(solid pieces)
4	Heavy metal W alloys	92 – 94	
5	Tungsten – copper	60 – 90	
6	Pure tungsen powder	98 – ≥ 99	
7	W grinding sludge	30 – 60	
8	W cutting sludge	70 – 80	**Soft Scrap**
9	Hard metal powder	60 – 95	(fine particles,
10	Hard metal grinding sludge	15 – 60	powder, dust,
11	Heavy metal powder	92 – 97	timings, sludges)
12	Heavy metal turnings	92 – 97	
13	W – Cu powder and green compacts	50 – 90	
14	Floor sweepings (different sources)	40 – 60	

reuse. Due to the low abundance of tungsten, which requires the mining and ore dressing of a huge ore tonnage to gain a small amount of tungsten, it was always a "valuable" metal.

Besides this intrinsic value, the tungsten world market was decisive in increasing the recycled percentage of the total tungsten production in earlier times. More recently, environmental and resource-preserving aspects became necessary.

The preservation of tungsten ore resources prolongs the availability of primary tungsten and decreases the damage to the environment caused by disposal of tailings. It also saves capital resources.

In regard to the environment, recycling requires less energy and chemical consumption, thus producing less waste. Moreover, the dumping of tungsten scrap with lower concentrations of tungsten like sludges—as exercised in the past—does not take place any more in the industrialized countries.

Because of economic considerations, the tungsten in cemented carbide scrap is not the only value of interest. The value of the other constituents, such as Co and Ta, are also important. Both metals, at least at this time, are more expensive than tungsten.

Last but not least, the pricing policy of "virgin" W products and the price stability of the recycled products influences tungsten recycling in general.

In discussing tungsten recycling, it might first be of interest to elucidate the tungsten flow in general. The questions are: how much tungsten is supplied by primary (ore) and secondary (scrap) sources and how big are the losses? Which losses are unavoidable and which can be partially avoided by increased or improved recycling in the future? An estimated balance was published in 1987 and is shown in Fig. 11.1. The figures certainly fluctuate with time but should also be valid today within a certain measure. As one can see, about 1/3 of the total tungsten demand is supplied by scrap. This proves that scrap recycling is a very important factor in the supply of tungsten raw materials.

It must be mentioned here that any in-house reuse of valuable material, for example, powder generated during green forming of cemented carbide parts and similar pure

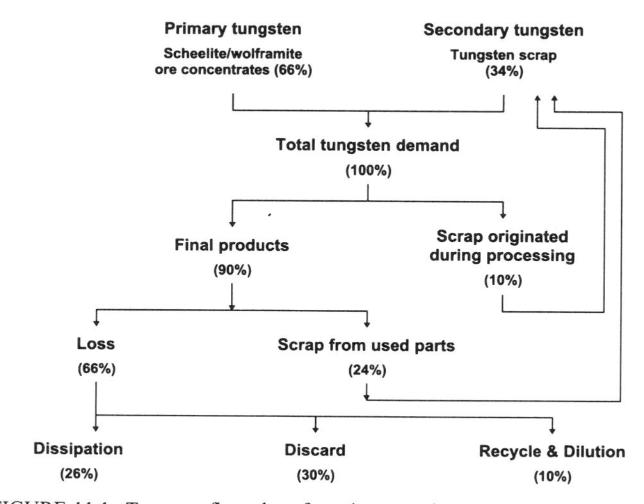

FIGURE 11.1. Tungsten flow chart for primary and secondary raw materials.

materials, will not be considered. Large amounts pass this way, but their actual volume is unknown.

The loss of over 60% is considerable. It can be divided into three different sources: *dissipation*, *discard*, and *recycling & dilution*.

Dissipate losses are wear loss of carbide products, arc erosion of electrical contacts and electrodes, oxidation losses, losses by chemical decomposition, etc. It can only be influenced to a small extent by quality improvements of the final products. However, because the quality level today is already quite high, no drastic change can be expected.

Loss by discard is mainly a matter of strict organization and today is not only a question of the price of tungsten. Typical examples for discard are noncollectable carbide products, burned out lamps and lighting fixtures, welding electrode stubs, electrical contact disks, carbide hard-facing materials, etc.

Loss by discard has decreased remarkably since 1970. A study of the U.S. Bureau of Mines revealed how drastically the discard of obsolete cemented carbide parts was reduced between 1974 and 1991 (see Fig. 11.2). It can be seen that within this period 24%, which corresponds to nearly 1/4 of the total tungsten consumed in the USA, could be recycled. Under the assumption of a steady and linear increase to the year 2010, a 54% recycle can be expected. The reasons for the drastic change are:

- change in product mix (braised tool tips versus throw away inserts),
- better awareness of the value of tungsten (improvements in communication, education, and organization),
- better understanding of the byproduct value (Ta, Co),
- improved recycling methods (direct – indirect).

Further improvements can be expected to minimize discard in the highly industrialized countries of North America, Europe, and Japan, and also in the developing countries.

Loss by recycling & dilution includes residues containing low tungsten concentrations as a result of recycling, and are not uneconomical to recover. For example, tungsten as trace constituent in recycling of tool steel, carbides, etc. is lost into the steel melt.

This last loss category is not likely to be altered. Only extremely high tungsten prices could affect the situation.

The above-mentioned USBM study, by an overall calculation, resulted in a total tungsten loss of 61% from 1955 – 1991. By subtracting the amount lost by dilution, the figure changes

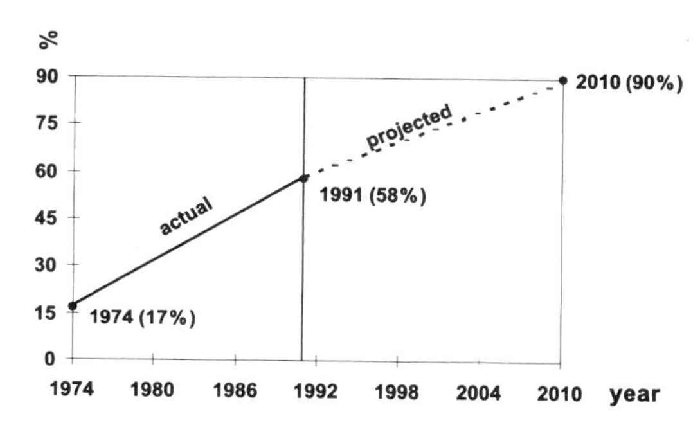

FIGURE 11.2. Percentage of tungsten recycling via cemented carbide scrap (USA).

to 52%. Regarding the USA to be an industrially advanced country, the figure of 66% loss, worldwide, in the above flow chart (Fig. 11.1), compares very favorably.

Considering recycling today, it is not sufficient to just get the material back for some useful purpose. Optimal recycling should always be done at the highest level of utilization, which includes many factors, such as conversion costs with special emphasis on energy costs, purity of the scrap material, as well as environmental reasons. For example, it is best to re-enter the same product flow, which originally generated the scrap, using a minimum of conversion steps. This is called direct recycling. Direct recycling is in any case the most economic and ecological way to reuse scrap and should always be applied, if possible. However, sometimes market conditions may dictate a less economic end use, such as when one type of scrap is in oversupply, when the optimal recycling system is unavailable, or when traders do not understand the market situation. Another case is when a fast generation of cash is more important than finding an optimal recycling possibility.

Such circumstances still direct far too much scrap to chemical processing, which is still the most important recycling system, although in most cases not the most economic and ecological one.

11.2. TUNGSTEN RECYCLING METHODS [11.1 – 11.4]

A variety of tungsten recycling methods were developed. They can be divided into four groups: hydrometallurgy, melting metallurgy, direct recycling, and semidirect recycling.

11.2.1. Hydrometallurgy

In principle, most tungsten containing scrap qualities can be processed by this method, which converts tungsten to "virgin" APT. However, it should be kept in mind that it is an expensive, high-energy-consuming and chemical-waste-producing procedure. In regard to the economy, preferably more impure scrap qualities, which are not easily converted by other methods, should pass this way. The scrap must first be oxidized by air, chemicals, or electrical energy, in order to transfer tungsten to the hexavalent state, which is soluble in alkaline solutions. From here on the entire procedure, as for ore concentrates, is required.

Typical scrap materials for hydrometallurgical treatment are grinding sludges, floor sweepings, contaminated powder qualities, tungsten – copper, and solid pieces, which can be dissolved by electrolysis.

Hydrometallurgical scrap conversion has already been discussed in detail (Chapter 5) and will not be further treated here.

In regard to tonnage of tungsten fed into this conversion type, it is the method mostly used (even for scrap) for converting in a more economic and ecologically preferable manner. One of the main reasons, besides those already mentioned in the introduction, might be that APT is the general intermediate for most tungsten products, with the exception of melting metallurgy. Furthermore, no additional contamination does occur, but in contrast purity will be increased.

TABLE 11.2. Tungsten Scrap in Melting Metallurgy

Product	Most preferable scrap type (according to No. in Table 11.1)
Superalloys	1
Stellites	1, 3, 6, 9
Cast eutectic carbide	1, 6
Menstruum WC	4, 6, 9, 11, 12
Tool steel	1, 2, 4, 7, 11, 12
Ferrotungsten	1, 2, 4, 6, 7, 11, 12
Melting base	1, 2, 3, 4, 6, 7, 8, 9, 10, 11, 12

11.2.2. Melting Metallurgy

In Table 11.2, the products are listed into which tungsten can be recycled as well as the preferable scrap materials. Superalloys, stellites, Menstruum WC, and cast eutectic carbides require very pure scrap while, for steel melting metallurgy products, impure scrap can also be used. The advantage of the latter is that all oxidic substances (such as grinding media, etc.) end up in the slag, and traces of foreign elements, which are soluble in the metallic melt, are distributed homogeneously.

Any tungsten going in this direction cannot be recycled again by other methods, with the exception of re-melting. It finally ends up in loss by recycling & dilution.

11.2.3. Direct Recycling

As direct recycling is understood, the as-supplied material is transformed to powder of the same composition by either chemical or physical treatment, or a combination of both. Prerequisites for direct recycling are:

- The composition of the scrap must be the same as for the final product.
- The scrap must be of high purity, not only in regard to foreign substances but also in quality (e.g., straight WC–Co hardmetal scrap should not contain TiC–TaC-bearing qualities, and the WC grain size should be within a small range).
- The process must offer the possibility of converting the scrap to a powder of a metallurgically acceptable form.
- No contamination by foreign materials during processing.

Typical examples, which will be discussed in more detail below, are the zinc process, or the Coldstream process for reclamation of cemented carbide scrap, and the oxidation and subsequent reduction of tungsten heavy metal turnings or thoriated tungsten (see Table 11.3).

Direct recycling is combined with a minimum of energy consumption, chemical waste, and lowest production costs. A general comparison of direct and indirect tungsten recycling is presented in Fig. 11.3.

We are still far away from full utilization of the direct recycling potential, because huge amounts of the corresponding materials are still chemically converted to APT. The reasons for that situation are:

TABLE 11.3. Direct Recycling of Tungsten Scrap

Process	Product	Scrap
Zinc Bloating/crushing} Coldstream	Hardmetal powder	Sorted hardmetal pieces
Oxidation/reduction	Heavy metal powder Thoriated tungsten powder	Heavy metal turnings Thoriated tungsten pieces, turnings, powder

- Bad organization by poor sorting, or no sorting at all, at places where scrap is generated. Not only in regard to foreign matter, but also in regard to quality (grade), purity of the scrap is one of the main demands for direct recycling.
- Fear of heterogeneous impurity content.
- Lack of knowledge of potential consumers and the different processing parameters necessary for recycling the powder in comparison to the virgin equivalent.

Any efforts to increase the percentage of directly converted material will improve the environment and the economy in general.

Zinc Process [11.1, 11.2, 11.5, 11.6]. In this process, cemented carbide scrap is treated in molten zinc or zinc vapor, which reacts with the binder to form intermetallic phases and a zinc–cobalt alloy. These reactions lead to a volume expansion of the binder and bloat the scrap. After vacuum distillation of zinc, the material is friable and can be readily disintegrated. The condensed Zn can be re-used. The reclaimed carbide/metal sponge contains less than 50 ppm Zn.

A flow sheet of the process is given in Fig. 11.4. In the first stage, the cleaned and sorted hardmetal scrap is contacted with molten Zn at 900–1050 °C in an Ar/N_2 atmosphere for several hours, making sure that all smaller pieces are completely penetrated. The second stage is the distillation of Zn in vacuum (0.06–0.13 mbar) at 1000–1050 °C, which again takes several hours. The cooled-down material is crushed, ball milled, and screened. The top screen is recycled in the next batch. The crushing and grinding involves fracture, primarily in the Co phase and not in the carbide phase.

The chemical composition of the final product is almost identical to the original material, with the following exceptions:

- pick up of approximately 0.1% Fe (during milling operations),
- depletion of carbon by 0.12–0.15%.

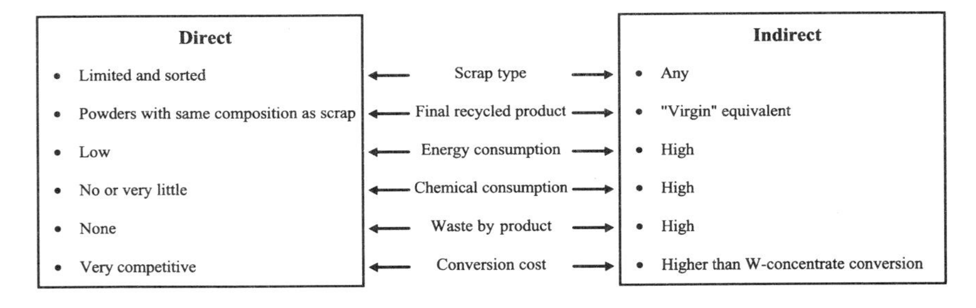

FIGURE 11.3. Comparison of direct and indirect tungsten recycling methods.

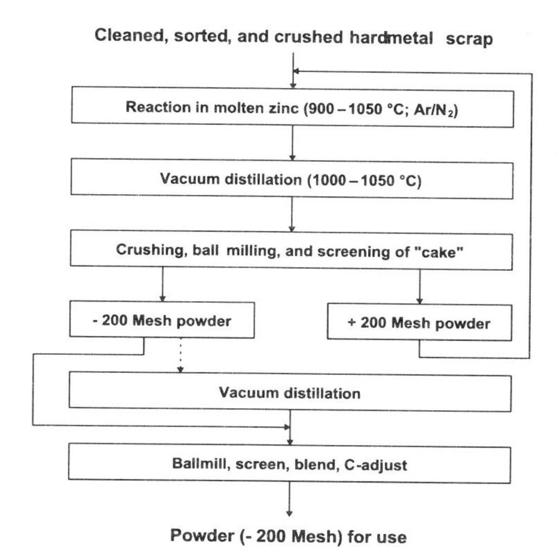

FIGURE 11.4. Flow chart of the zinc process for cemented carbide reclamation.

The basis of a successful use of the reclaimed powder involves some special skill and special equipment, such as excellent analytical capability, cemented carbide testing facilities, optimal milling equipment, and a good understanding of the shrink factor during sintering.

The energy consumption is approximately 4 kWh/kg, which compares favorably to virgin WC produced via the chemical route (12 kWh/kg). Compared to the features of indirect conversion, the cost of zinc reclamation is reduced by 20–30% for WC–Co grades and by 30–35% for WC–TiC–Ta(Nb)C–Co grades.

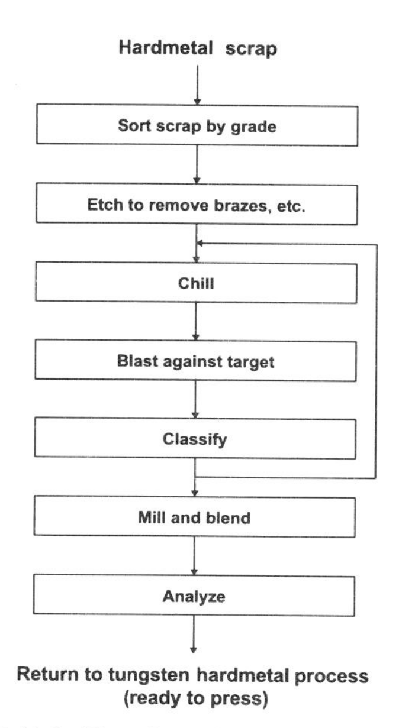

FIGURE 11.5. Flow chart of the Coldstream process.

Coldstream Process [11.7] (see the flow sheet Fig. 11.5). This process is a mechanical comminution and employs air blasts to accelerate particles against a target with sufficient energy to cause fracture (particle velocities near Mach 2). The air is cooled by expansion from the nozzles, protecting the scrap from oxidation. After screening or air classification, oversize material is recycled.

A method combining the two above-described processes is one in which zinc infiltration is followed by Coldstream crushing with a final zinc distillation step.

Bloating/Crushing. The process is based upon the embrittling effect when heating cemented carbides to about 1800 °C followed by rapid quenching. Embrittlement is the consequence of WC grain growth, formation of brittle phases, and thermally induced stresses.

Oxidation–Reduction of Heavy Metal Turnings. Heavy metal turnings can be easily oxidized by air at about 800 °C. The layer height in this step is critical, because if it exceeds a certain limit then local overheating (exothermic reaction) can cause melting. The oxidized product is subsequently milled and screened and, finally, it is reduced in hydrogen atmosphere at 900–1000 °C in the same type of push-type furnaces used for W-powder production.

11.2.4. Semidirect Recycling

Semidirect reclamation can be applied to binary- or ternary-phase alloys. One component is dissolved chemically, leaving the other phase(s) intact. By the dissolution, the integrity of the structure is lowered and attrition can take place. Typical examples are presented in Table 11.4.

In conclusion, it can be stated that the importance of tungsten scrap recycling will continue to grow considerably in the future. This will mainly be driven by economic, ecological, and natural-resource preserving considerations. The share of direct recycled scrap will increase as soon as the organization of scrap collection and separation at the places of generation has improved. Scrap should always be kept as pure as possible. This must be understood not only in regard to contamination by foreign materials, but also in regard to the grade purity. As an example, different grades of cemented carbide throw away inserts should be kept separated. The old philosophy—scrap is chunk—is not any longer valid. On the contrary, scrap is a very valuable material and has to be treated accordingly. Pure grade scrap will bring a much better price or yield and is easier to convert.

TABLE 11.4. Semidirect Tungsten Recycling

Scrap	Principle	Method
Hardmetal pieces	Selective dissolution of the Co binder	Attack by acid (H_3PO_4) Leach milling (mixed chemical and mechanical attack) Electrolysis (acidic solution)
WC or WC + mixed carbides can be recycled to hardmetal grade powder preparation.		
Heavy metal pieces	Selective dissolution of the Ni–Fe binder	Attack by acqueous $FeCl_3$ solution
Powder consists of approximately 40-μm tungsten spheres and can be used in plasma spraying.		

A further prerequisite for an increase in direct recycling is the necessity for the consumer to learn to deal with reclaimed materials, which require modified production conditions.

REFERENCES FOR CHAPTER 11

11.1. B. F. Kieffer and E. F. Baroch, in: *Extractive Metallurgy of Refractory Metals* (H. Y. Sohn, O. N. Cralson, and J. Th. Smith, eds.), pp. 273 – 294, AIME, Materials Park, Ohio (1981).
11.2. B. F. Kieffer and E. Lassner, in: *Proc. 4th Int. Tungsten Symp.*, Vancouver 1987, pp. 59 – 67, MPR Publishing Services Ltd., Shrewsbury, UK (1988).
11.3. B. F. Kieffer and E. Lassner, *Berg-und Hüttenmaemmische Monatshefte* **139** (1994), 340 – 345.
11.4. G. R. Smith, "Tungsten," *American Metal Market* (August 1993).
11.5. P. Barnard, U.S. Patent No. 3, 595484, July 27, 1971.
11.6. B. F. Kieffer, *Int. J. Refract. Met. Hard Mater.* **2** (1986), 65 – 68.
11.7. J. Walraedt, in: *Proc. 7th Int. Plansee Seminar*, Vol. IV, No. 2, pp. 1 – 12, Metallwerk Plansee, Reutte, Austria (1971).
11.8. R. Zhao, in: *Proc. 4th Int. Tungsten Symposium, Vancouver*, pp. 50 – 58, MPR Publishing Services Ltd., Shrewsbury, UK (1988).

12

Ecology

12.1. INTRODUCTION

In general, the production of a metal, its use, and the manner in which it is discarded influence the ecology more or less. These aspects grew in importance due to the steady increase in environmental awareness during the last decades. Although this awareness is more pronounced in the industrialized countries, its worldwide acceptance will be a necessity in the near future.

In regard to tungsten it can be stated that mining, ore dressing, chemical conversion, or metallurgy create more or less of the usual environmental problems. The use of the metal or its alloys in some applications leads to a dissipative loss and distribution, which is harmless and negligible. Discarding is not a big problem, because the recycling of tungsten-bearing substances has been very well organized for quite some time (see Chapter 11).

The impact on the world's ecology caused by the production of tungsten is relatively small in comparison to that of metals like iron, aluminum, copper, etc. because of the small tonnage. If we assume the following situation:

- yearly production, 40,000 t W;
- average content in the ore, 0.5% WO_3;
- average concentration in the ore, concentrate 65% WO_3;
- yield in ore dressing, 85%;
- yield in chemical conversion, 95%;

then the following quantities must be treated worldwide: 12.5×10^6 t of ore have to be mined, disintegrated, and concentrated to yield 82×10^3 t of ore concentrates and 12.4×10^6 t of tailings. The latter must be deposited.

During the chemical or melting metallurgical conversion of the concentrates, roughly 1/3 of the original weight remains as residue (slag, precipitate, and gangue). This corresponds to approximately 26×10^3 t of material, which must be dumped. Residues from chemical conversion (precipitates and gangue), which represent the majority, also contain chemicals.

Assuming that 90% of total tungsten consumption is converted chemically, it can be calculated that between 25,000 and 60,000 t of sodium hydroxide or sodium carbonate are necessary for the dissolution and about the same amount of sulfuric acid for the neutralization. This means that between 35,000 and 84,000 t of dissolved sodium sulfate are generated of which only a very small percentage is reused while the rest is disposed to the environment.

In principle, these figures can be neglected for a worldwide consideration. However, taking into account that today's supply originates from a small number of bigger mines besides a large number of smaller ones (especially in China), and that only 10–20 chemical converting plants exist, it is evident that local problems exist or may arise in the future.

Good examples of how ecological problems can be solved successfully without any dramatic influence on the economy will be presented below.

12.2. MINING AND ORE DRESSING [12.1, 12.2]

Mining activities are usually associated with the more or less severe destruction of the environment. In open pit mining they are more visible than in underground operations, because large surface areas must be removed. In both cases, the deposition of residues after separation of the ore as tailings from flotation or gravity separation represent another environmental problem. Also, ore dressing creates waste water, which in the case of flotation contains small quantities of chemicals.

Naturally, it depends on where a mine is located: whether it is far away from or close to civilian and highly populated areas. Moreover, different countries have diverging environmental regulations.

An extreme in this regard is the Mittersill Scheelite Mine, located in the Alps of central Europe, in Austria's province of Salzburg. The mine is one of the largest tungsten deposits in the Western World and is an outstanding example of how ecology and economy can be matched optimally. The open pit, the underground mine, and the ore dressing plant are situated in a National Park which mainly serves tourism (Fig. 12.1). Therefore, it is easy to understand that quite rigid and extensive environmental regulations have been imposed by the Austrian nature conservation authorities and that these have necessarily increased the costs of the total operation. Nevertheless, the Mittersill mine and ore dressing plant is still ranked among the most economic tungsten ore concentrate producers in the world, although the ore content ranges only around 0.4–0.7% WO_3 and the concentration is only possible by flotation.

The high economic level of the mine will be proven by the following. In contrast to nearly all other western mines, Mittersill was only closed for a short period from 1994 to 1995 despite extremely low world market prices. It should also be noted that the outstanding economy of the Mittersill mine has a double reason: low operating costs, and full integration with the conversion plant of the same company which is capable of converting low-grade ore concentrates of approximately 30% WO_3.

Mining started in 1976 as a 100% open pit mine, changing gradually to 100% underground until 1980. The following important regulations had to be obeyed:

- After closing the open pit operation the whole area had to be completely recultivated so that it could be reused as pasture. This also included the destruction of the access road built exclusively for the mine (Figs. 12.2 and 12.3).

FIGURE 12.1. Entrance to the Mittersill underground mine. By courtesy of Wolfram Bergbau und Hüttenges. m.b.H., Austria.

- In developing the underground mine, all facilities for the miners, warehouses, and maintenance shops, as well as the first crushing steps, had to be incorporated inside the underground mine (Fig. 12.4).
- An underground connection (a tunnel equipped with conveyer belts) had to be constructed in order not to disturb the tourist traffic by heavy trucks (Fig. 12.5).
- Special attention had to be paid to the wastewater quality leaving the tailings pond. Table 12.1 informs about the limits.
- The first tailings pond was located near the ore dressing plant. After 6 years of operation its capacity was exhausted and the only available area was 10 km away. This required the construction of a pipeline and pumping stations. Meanwhile, several sections of the new tailings pond have been completed and recultivated.

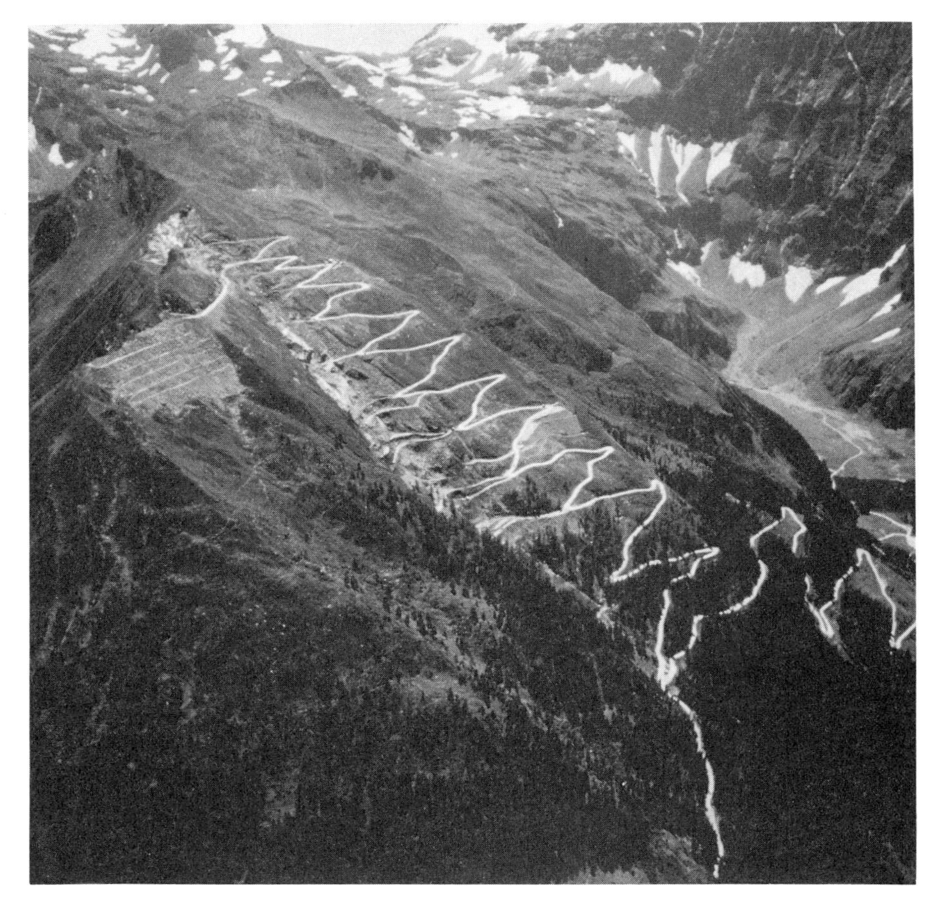

FIGURE 12.2. Specially constructed road to give access to the open pit mining area. By courtesy of Wolfram Bergbau und Hüttenges. m.b.H., Austria.

TABLE 12.1. Limits Pertaining to Wastewater Leaving the Tailings Pond (Mittersill Mine and Ore Dressing Plant)

Temperature	≤30°C
pH	6.5–8.5
Undissolved substances	≤100 ml/l
Settling substances	≤0.3 ml/l
Extractable substances	≤10 mg/l
Dissolved Fe	≤2 mg/l
Dissolved Al	≤2 mg/l

[a] By courtesy of Wolfram Bergbau und Hüttenges. m.b.H., Austria.

FIGURE 12.3. Recultivation of the open pit mine area. By courtesy of Wolfram Bergbau und Hüttenges. m.b.H., Austria.

FIGURE 12.4. Underground installed crushers. By courtesy of Wolfram Bergbau und Hüttenges. m.b.H., Austria.

An as yet undissolved problem is that, in the case of whole flotation, a huge amount of energy is necessary to finely disintegrate the ore, of which only about 1% is valuable. If no other valuable constituents are present, the separated gangue material is worthless and cannot be used at present for any other purpose. The only possible reuse is a partial back-filling in the caverns of the underground mine.

12.3. CHEMICAL CONVERSION [12.1, 12.2]

A chemical plant is always confronted with environmental problems and APT plants are no exception. We present as an example a plant operated by the same company as that mentioned above—the conversion plant of Wolfram Bergbau and Hüttenges. m.b.H. located in Bergla/Styria (Austria). The facility was constructed in the mid-1970s and has been in operation since 1977. Although it is a relatively modern plant, a series of new regulations have been introduced recently and are summarized in Table 12.2.

It can be seen that the discard of sodium sulfate is the biggest pollution problem caused by that plant. As described earlier in Chapter 5, this is still a worldwide problem especially in the case of scheelite digestion, requiring a large stoichiometric excess of sodium carbonate. Improvements for the future are necessary. Some can be found in the current literature and are being used on an industrial or pilot plant scale. Examples are

FIGURE 12.5. Underground conveyer belt connection to the ore dressing plant. By courtesy of Wolfram Bergbau und Hüttenges. m.b.H., Austria.

TABLE 12.2. Emission and Wastewater limits (Wolfram-hütte Bergla)

SO_2 emission (ore concentrate roasting)	$\leq 2.0\ g/Nm^3$
NH_3 emission (APT calcination)	$\leq 5.0\ mg/Nm^3$
NO_x emission (steam bloc)	$\leq 200\ mg/Nm^3$
Wastewater	
Temperature	$\leq 30\ °C$
pH	$\leq 6.5–8.5$
Al^{3+}	$\leq 3\ mg/l$
Co^{2+}	$\leq 1\ mg/l$
Cu^{2+}	$\leq 0.5\ mg/l$
Ni^{2+}	$\leq 0.5\ mg/l$
NH_4^+	$\leq 30\ mg/l$
F^-	$\leq 20\ mg/l$
SO_4^{2-}	$\leq 250\ mg/l$
Organic hydrocarbons	$\leq 10\ mg/l$

[a] By courtesy of Wolfram Bergbau und Hüttenges. m.b.H., Austria.

leach milling (performed in China) [12.3], which decreases the amount of sodium hydroxide for dissolution (see Section 5.2.6), and electrodialysis [12.4] of tungstate solutions to recycle sodium hydroxide and avoid chemical waste (see Section 5.2.4.7). These examples are mainly driven by the economy, but at the same time serve to improve the environment. (To some extent, both milling and electrodialysis increase electric power consumption, which itself creates environmental pollution elsewhere.)

12.4. POWDER METALLURGY

Powder metallurgical production methods for high-tech materials are very clean operations and do not generate any waste or emissions. Internal and external scrap recycling is excellently organized worldwide.

For information about hazardous substances in the production of tungsten alloys, such as thoriated tungsten, cobalt, or nickel-containing cemented carbides, see Chapter 14.

12.5. ENVIRONMENTAL CONSIDERATIONS ABOUT THE SUBSTITUTION OF HIGH SPEED STEEL BY HARDMETALS [12.5]

Not only will technical aspects direct the selection of tool materials in the future, but also ecological considerations. In regard to cutting tools, one may say that the tungsten content of high speed steels is an impurity in regard to steel scrap and is finally lost by dilution. Any loss in tungsten is connected with increased mining activities and chemical conversion creating environmental destruction, decrease in natural reserves, and pollution. In contrast, hardmetal tools and their valuable constituents with tungsten in the first place can be recycled to a high degree. A special advantage is the possibility of direct recycling, which has special environmental suitability.

REFERENCES FOR CHAPTER 12

12.1. M. Spross, *Erzmetall* **44** (1991), 438–442.
12.2. M. Spross, *Proc. Leobner Bergmannstag*, pp. 606–612 (1987).
12.3. K. Vadasdi, in: *The Chemistry of Non-Sag Tungsten* (L. Bartha, E. Lassner, W. D. Schubert, and B. Lux, eds.), pp. 45–60, Elsevier Science, Oxford (1995).
12.4. H. Li, M., Liu, and Z. Si, Chinese Patent 85100350.8 (1985).
12.5. E. Egami, in: *Proc. 7th Tungsten Symposium, Goslar*, pp. 13–23, ITIA, London (1996).

13

Economy

13.1. INTRODUCTION [13.1–13.10]

Tungsten's economy was generally ruled by the supply/demand situation, which itself was a consequence of the world's general economic situation. Booming industry was always associated with enhanced tungsten consumption and mine production as well as with the search for new deposits. Times of general recession always showed minimum of the consumption combined with mine closures. Wars reflected situations of highest industrial activity and always maximum of consumption and prices.

Supply and demand have not always been evident, especially during the cold war. Moreover, in tungsten trading, not only have economic considerations determined the rule but also political influences played an important role, due to the strategic importance of the metal. The accumulation of government stockpiles or their release have sometimes acted severely as a disturbance or help. For example, in the 1950s, when China's supply was disrupted during the Korean war, U.S. mine production was expanded to feed the government stockpile and also concentrates were bought from abroad. This program ended in 1959, and sales of excessive concentrates from stock started in 1965. Sometimes, the release of government stock has had a stabilizing effect on the market, by avoiding shortages.

Mine or market stocks whose size is partially unknown (China, CIS, and Hong Kong) sometimes create a distorted picture of the current situation. For example, huge market stocks and stocks of downstream products have been one reason for the abrupt decrease in consumption between 1992 and 1994, besides a general worldwide decrease in production. The extremely low tungsten price forced the rigid reduction of all inventories. Furthermore, these stocks are the reason why quite divergent figures very often occur in published statistics.

In addition to stocks, another uncertainty is tungsten supply by scrap. Trading of scrap is only partly published, so the amounts can only be estimated. Consequently, the consumption figures given subsequently reflect only the amount supplied by primary tungsten and not quantities originating from any scrap. An estimate discussed in Chapter 11 is that, worldwide, about 1/3 tungsten consumption originates from reclaimed tungsten (including all types of recycling). This means that in the MEC (market economy countries) the percentage is much higher than in the rest of the world.

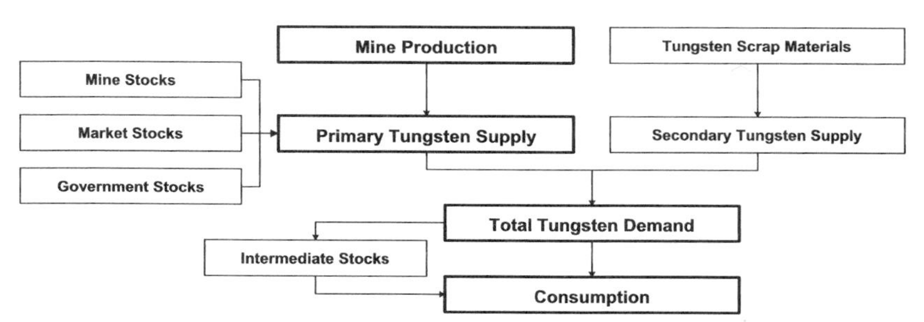

FIGURE 13.1. Parameters of tungsten economy.

Other parameters influencing tungsten's economy are shown schematically in Fig. 13.1. The different types of stock may evidently create a confusing picture of the situation at any given time. No direct relationship exists between mine production, supply, demand, and consumption. The supply may originate from mine production, from scrap, or from stocks. The demand must not be necessarily balanced with consumption, because economic reasons may dictate a build up or decrease of intermediate stocks.

13.2. WORLD MINE PRODUCTION

In Fig. 13.2, the share of different countries and continents or political areas in available tungsten ore deposits is shown. In former times, the percentage of mine production showed a similar picture, but since the late 1970s the situation has changed.

Due to the drastic fall in the price of tungsten, almost all MEC mines were closed. Figure 13.3 presents the price of tungsten for the period from 1960 to 1996 while the dates of important mine closures are indicated. This diagram is also of special interest because in some cases it allows a rough calculation of the break-even point of the mines. Mittersill mine is excluded from these examples, because its excellent economy is based on full integration from the mine to highly pure W and WC powders, while Panasqueira mine's special economy is due to the presence of Sn in the ore.

Within the CIS (community of independent states) in 1990, 10 mines were still working but today only one remains. Also in China, which was always the main supplier of

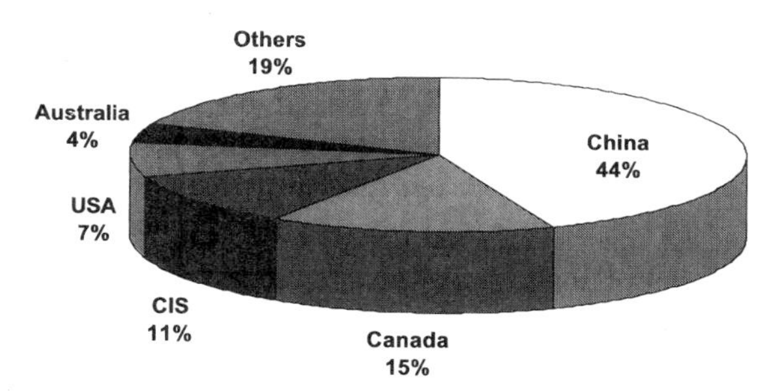

FIGURE 13.2. Share of different countries, continents, or political area in known minable tungsten deposits [13.13].

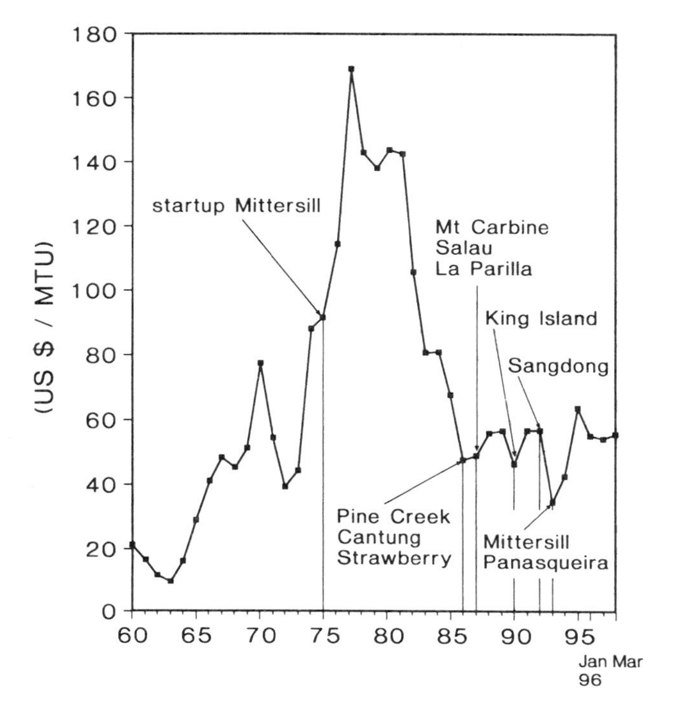

FIGURE 13.3. Tungsten price (LBM quotation) versus time (1960–1996), including closure dates of MEC mines [13.7].

tungsten ore concentrates, there was a decrease in mine production due to the loss of governmental subsidization since 1991 and increased economic consciousness. The annual output was 32,200 t in 1990 but decreased to 26.600 t in 1996. The decrease is based on a considerable reduction in the output of state-owned mines while locally operated mine production remained more or less constant. Nevertheless, China has tremendously increased its share in the worldwide supply of tungsten. In 1989 it was around 69%, and in 1995 it climbed to 80%.

It is of interest that in the 1960s and 1970s about one-half of the tungsten supply (49%) came from politically unstable countries (China included). In order to change this unsatisfactory situation, enhanced exploration in the MEC lead to mine openings and the share of the MEC shifted to 58% in 1980, while a further increase to 69% was expected in 1990. As explained above, the prediction went completely wrong and, in recent years the tungsten world has depended almost totally on China's supply, although today China cannot be regarded as a politically stable country. Recently, the CIS (also politically unstable) has became an important supplier, too.

Figure 13.4 shows the total world mine production from 1973 on and, in addition, a split is made in the supply by China, CIS, and others in more recent years.

From 1975–1980, a yearly increase in mine production of 8% could be observed. From then on, consumption remained between 40,000 and 50,000 t W/a until 1989, when a sharp decrease began.

It should be emphasized that the tremendous decrease in total production is not necessarily connected only with an equal reduction in W consumption. Supplies from different stocks of varying size have contributed toward demand.

This decrease in mine production since 1991 is based on several reasons. In the MEC, the low price necessitated the closure or mothballing of nearly all mines. In China the mines have lost their governmental subsidization since 1991 and, together with a growing awareness of the necessity to think economically, many mines were closed. Russia's consumption fell dramatically since 1991, but mine production proceeded at full capacity until 1994 with governmental support, which stopped in 1994. Huge stocks were accumulated during that period.

The end of the cold war had a strong worldwide impact on tungsten consumption, and consequently on mine production (see Section 13.3 for more details).

Any figures must be understood partly as estimates, because reports from China and the former Soviet Union have not always been precise or available. Nevertheless, the order of magnitude is correct and, since 1990, these countries have been more open and cooperative.

In the future the following picture seems likely. Chinese output will not exceed approximately 25,000 t W/a even at higher price levels. Further closures of state-owned mines are already planned. Estimates are about 4000 t W/a for the CIS and 6000 t W/a for the MEC. This corresponds to a sum of 35,000 t W/a. Even under much more favorable market conditions than today, the maximum world mine production would be around 40,000 t and this must be compared with a yearly consumption of 40,000 t, which will certainly increase.

Nonetheless one should keep in mind that predictions of future developments in any direction has always been made but have proved wrong in most cases, as shown by the following examples. In the early 1980s an annual 4% growth in W demand was postulated. However, between 1980 and 1990 there was no actual growth and later there was a decrease. Average worldwide demand forecasts were 70,000 t W/a for 1990 and 80,000 t

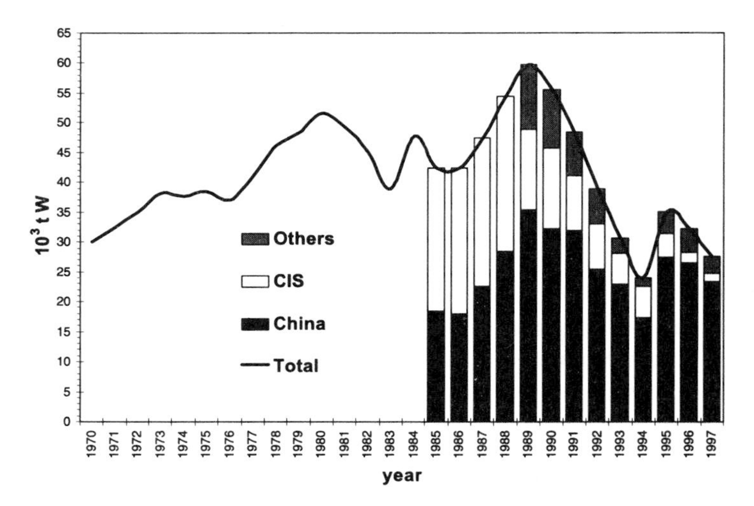

FIGURE 13.4. Total world mine production (1970–1997) [13.3, 13.7, 13.8].

W/a for the year 2000, but demand has remained at around 40,000 t W. A bigger share of MEC mines in total world production was anticipated, but actually China now supplies 80% of the world's demand.

These mistakes are mainly based on the fact that the world's political and economic development is no longer predictable.

13.3. PRICE OF TUNGSTEN

Historically, it might be interesting to note that the price for 1 kg of pure tungsten metal in 1860 was 1400 Deutsche Mark, between 1861 and 1884 it decreased to about 200 Deutsche Mark, and from 1884 only a technical grade existed on the market and ranged between 4 and 12 Deutsche Mark per kg up to 1902 [13.11, 3.12]. Meanwhile, its technical importance has grown enormously and the quality of the metal has improved extensively, but the current price of DM 30 per kg of tungsten powder is not so different from that of 1900; if we take into account inflation since 1900, it is quite cheaper!

In contrast to many other metals, no terminal market quotations exist for tungsten ore. Sales are negotiated between either producers and consumers, or merchants and users.

The London Metal Bulletin (LMB) publishes twice a week a price for wolframite ore concentrate. This published price reflects concluded transactions of the previous half week reported by producers, consumers, and traders. The price quotation is presented as a range between high and low, and is related to relatively clean standard-grade wolframite ore concentrate (>65% WO_3). The content of "dangerous" impurities (P, As, S, Ca), as well as lower WO_3 concentrations, decrease the price (for example, if the concentration of Ca exceeds 1%). Most likely, Ca is contained as scheelite and will depress the digestion yield.

The LMB quotation is based on a relatively small percentage of world production (about 10%). Internationally, tungsten trading prices are indicated in US $/MTU WO_3 for wolframite ore concentrates. MTU means "Metric Ton Unit" and corresponds to 10 kg WO_3. In the USA, prices indicated in US $/STU are quite common. STU stands for "Short Ton Unit", equaling 20 lb WO_3.

Price quotations for ferrotungsten are published by the London Metal Bulletin and by Metalsweek N.Y., in US $/kg W. Prices for scheelite ore concentrates always orientate around the wolframite price. Besides these price quotations, since 1989 quotations for APT have also been published (in MTU for Europe and STU for USA). Different price quotations for USA and Europe exist, because the U.S. market is somewhat stronger and not as competitive.

The necessity for APT quotations arose due to a change in raw material trading. In former times, the main traded tungsten raw materials were ore concentrates of wolframite and scheelite. In recent times, a change to APT took place because China, the biggest supplier of tungsten ore concentrates in the past, renewed its trading strategy and gave preference to the sale of more advanced (further upgraded) tungsten products, like APT, blue oxide, yellow oxide, and ferrotungsten.

In order to encourage the world in following the new policy, prices for the intermediates have been about the same as for ore concentrates in regard to W content, which means that it was cheaper to buy APT instead of buying ore concentrate and converting it themselves. The same happened with ferrotungsten. The "answer" of the

MEC has been to impose an import duty of 151% on W content in the USA and an antidumping duty of 37% in the EU.

The new policy converted China from the most important ore concentrate supplier to almost the main supplier of highly pure tungsten intermediates. This is demonstrated clearly in Fig. 13.5.

The diagram in Fig. 13.6 presents the yearly average LMB quotations since 1960 for wolframite, ferrotungsten, and APT (USA and Europe). The fluctuations are fairly large. With few exceptions, the APT price follows the fluctuations of the concentrate. In principle, it must be said that wars or warlike confrontations of a bigger extent have always been connected with price maxima. The strategic importance of tungsten caused a higher demand, and as a consequence the price went up. Tungsten as cemented carbide is necessary to manufacture weapons and military equipment like trucks, tanks, guns, etc. Consequently, a lot of ammunition like kinetic penetrators, scatter grenades, etc. are made of tungsten heavy metal alloys or hard metals.

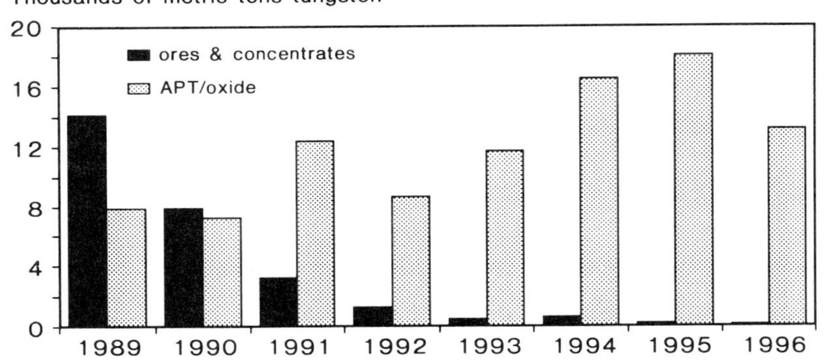

FIGURE 13.5. Change in Chinese tungsten supply (1989–1996) [13.7].

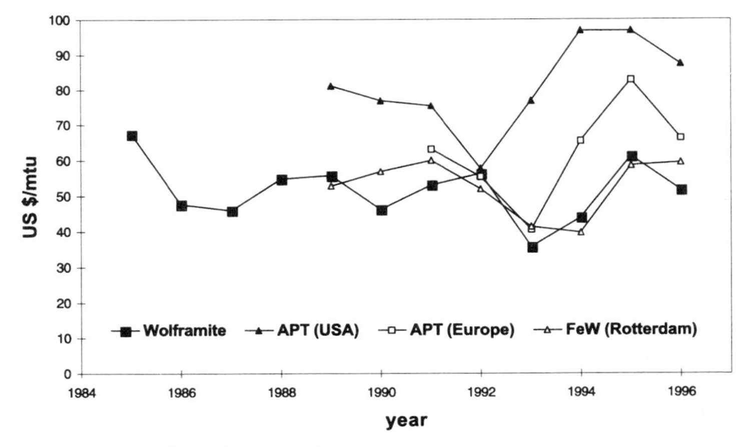

FIGURE 13.6. Metal Bulletin Quotations for wolframite concentrates, APT (USA and Europe), and ferrotungsten (1985–1996) [13.8].

Also, increased tungsten demand originating from peaceful industrial booms leads to price maxima and, correspondingly, recessions lead to minima. The historical maximum reached in 1977 was the consequence of increased consumption.

Until 1977, one can state that supply and demand dictated the price of tungsten. However, from then on the price control mechanism was no longer effective. Although annual tungsten consumption within the period from 1977 to 1989 increased further from <40,000 t W to 52,000 t W, the price fell continuously. Several circumstances were responsible for the new situation.

Up to 1977, China was the biggest and most important tungsten ore concentrate supplier. In its trading policy, China was always price-conscious and all relevant activities were carried out by state officials. The situation changed drastically in 1977 with the new liberalization policy. Producers were allowed to sell at least a certain percentage of their production on their own. This was the starting point of a historical price decline. In order to conduct business, the offers became cheaper and cheaper while the traders were pressuring in the same direction. The MEC was flooded by cheap Chinese ore concentrates. The price went further down, and with the closure of the MEC mines China's share in world tungsten supply increased.

The aforementioned change of raw material supply, from ore concentrates to APT, etc., starting in 1987, did not improve the price situation, but had a dangerous impact on MEC converters, as well as APT and ferrotungsten plants. Some of them were forced to close.

The collapse of the communist world together with the sudden fall in worldwide tungsten consumption lead to the historical minimum of the price of tungsten in 1993. Since then, there has been a slight recovery to levels like in the period between 1986–1992.

Here, too, any future prediction seems as worthless as the general opinion expressed at the 2nd International Tungsten Symposium, held in 1982 at San Francisco, that the LMB quotation will not fall below 100 US $/MTU.

13.4. SUPPLY AND DEMAND

As it can be seen in Table 13.1, that supply/demand statistics are worldwide more or less balanced. Primary tungsten supply to the MEC is illustrated in Fig. 13.7 (by suppliers). The share of China increased while the supply from the MEC decreased. The 1994 imports from CIS were additional. However, what is not at all obvious is the origin of the supply, either from stocks or from mine production. Therefore, another statistic (comparison of mine production and demand) shows a much less balanced situation (Fig. 13.8). The following short historical consideration should explain the situation between 1981 and 1996.

From 1982–1991, a large build-up of market inventories took place, originated by a steady oversupply from China. Between 1992 and 1994, these inventories were a major source of supply. In 1994, consumption overtook supply temporarily and increased the flow of stock materials from China and Russia. In 1995, supply overtook demand and a build-up of inventories occurred again. In 1995, the inventories were reduced.

Figure 13.9 gives a picture of the worldwide primary tungsten production/consumption situation from 1989 to 1994 for the three blocks MEC, China, and CIS. Figure 13.10

TABLE 13.1. Analysis of Supply and Demand (1990–1996) (W content in metric tons) [13.8]

	1990	1991	1992	1993	1994	1995	1996
Supply							
Imports of concentrates[a]	9950	7300	5800	2550	3150	5350	5350
MEC imports of concentrates from China	5600	4850	1200	400	200	150	50
MEC imports of products from China	12000	12850	9450	14950	19850	20300	16200
Subtotal	27550	25000	16450	17900	23200	25800	21600
Imports from CIS			200	400	4300	7000	4650
China's export to Eastern Europe and North Korea							
Concentrates	4500						
Products			200	500	900	800	50
China's domestic consumption	9900	10000	10500	11000	11000	11000	11000
Total	41950	35000	27350	29800	39400	44800	37300
Demand							
Europe	11000	7700	8650	8300	11750	13000	8200
USA	8250	11400	4250	5000	7450	8850	8500
Japan	4650	4450	3750	3350	5150	6500	5150
Other Western countries	3250	2850	1700	1700	2750	3000	3000
Subtotal	27150	26400	16350	18350	27100	31350	24850
China's concentrate import					400	1450	1400
China's domestic consumption	9900	10000	10500	11000	11000	11000	11000
China's export to Eastern Europe and North Korea	4500		200	500	900	800	50
Total	41550	36400	27050	29850	39400	44600	37300

[a] Other than from China and CIS, including supplies from stocks and from MEC production.

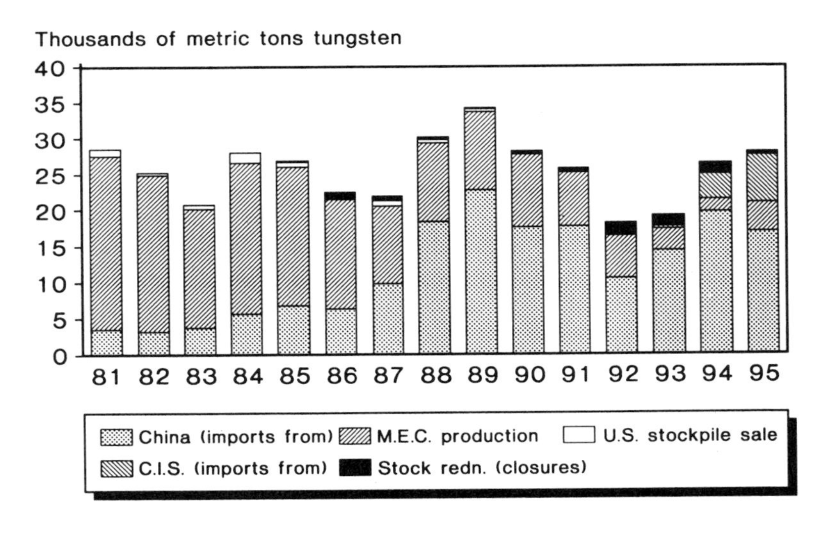

FIGURE 13.7. Primary tungsten supply [13.6].

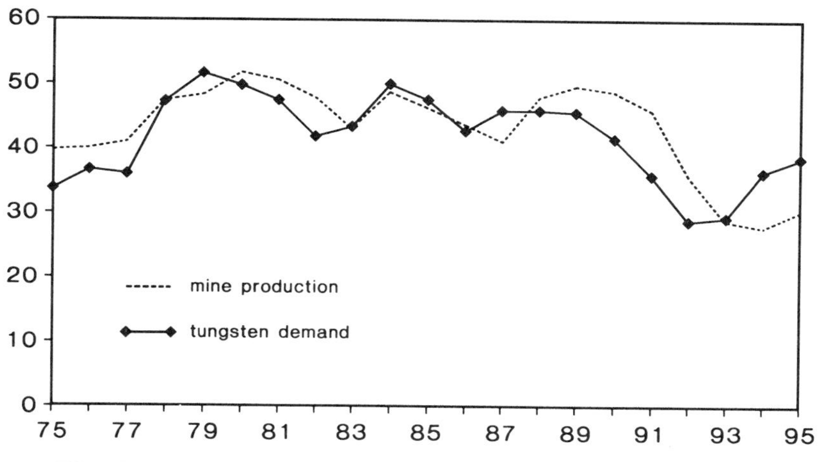

FIGURE 13.8. World tungsten mine production and demand (1975–1995) [13.7].

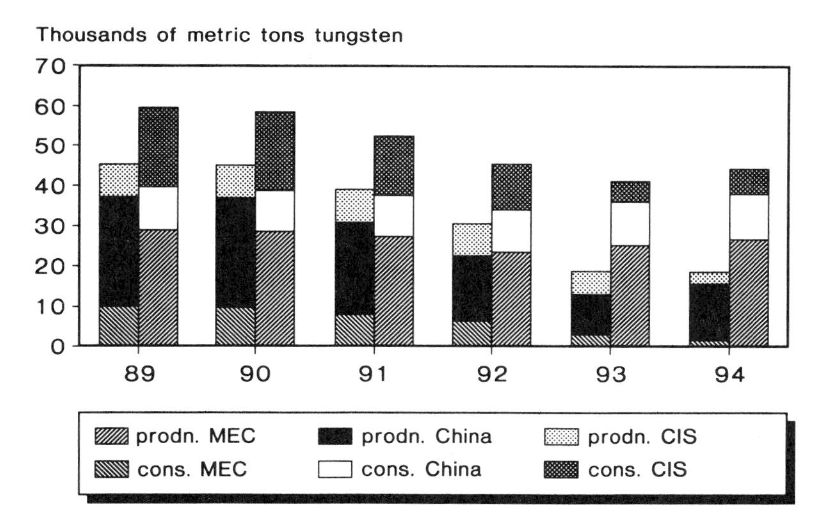

FIGURE 13.9. Primary tungsten production/consumption (worldwide) [13.6].

presents the primary tungsten supply/demand situation in the MEC and the cumulative inventory change between 1981 and 1994.

13.5. CONSUMPTION AND USE

The main proprotion of tungsten consumption is closely related to industrial processes. Sixty percent of W is consumed by the cemented carbides. These materials serve as tools for construction work (metals, ceramics, wood, and plastic), mining tools

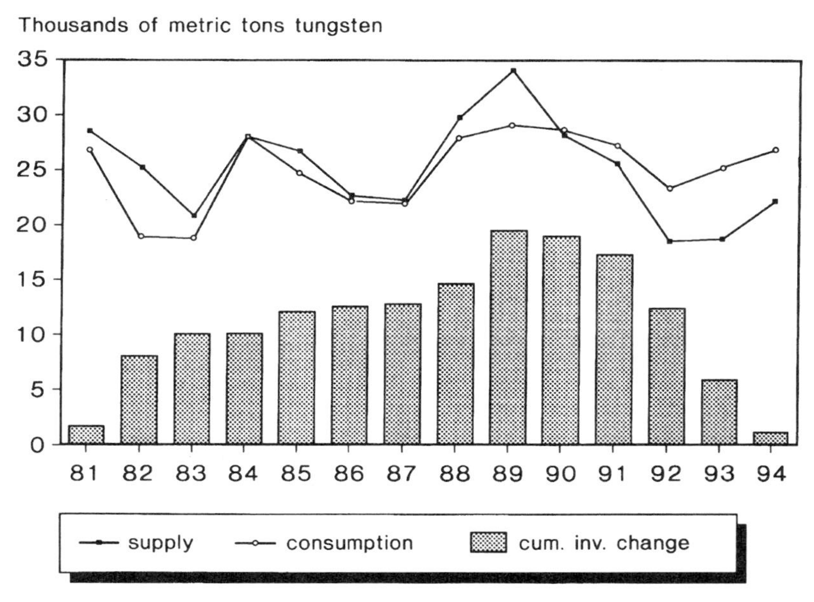

FIGURE 13.10. Primary tungsten supply/demand (MEC) [13.6].

(rock, oil, and water drilling), and as wear parts. All these applications are linked to industrial activities. The same is more or less also valid for the rest of tungsten consumers like alloys, pure tungsten, etc.

Historical tungsten consumption development is marked by the steady increase in industrialization and a broadening of the application of tungsten and its products. Fluctuations in tungsten consumption are due to booms or recessions. In average, a growth of 4% annually was predicted in the early 1980s. As already pointed out, predictions no longer fit in our world. Consequently, the situation is quite variable at different places of the world.

In the MEC over the last 15 years, the average growth of W consumption was only 1–2%. On the other hand, in the PR China the net growth was 5% and in countries of the former East Block consumption went down by 80%. The figures of these countries may reflect a wrong picture, because it is unknown how much of the consumption was fed by inventories.

Within a period of about 20 years, especially in the MEC, a reduction in W consumption was due to:

- Improvements in quality (longer lifetime of tools, coatings of hard metals, etc.).
- Substitution of tungsten or tungsten-containing products by other materials (cermets or ceramics instead of cemented carbides, depleted uranium instead of heavy metal).

A contrary development took place in China: the replacement of high speed steel by cemented carbides (the latter contains a higher percentage of W).

TABLE 13.2. Tungsten Consumption [13.3, 13.6, 13.8]

Year	10^3 t W		
	Source 1 [13.3]	Source 2 [13.6]	Source 3 [13.8]
1973	38.0		
1974	37.6		
1975	34.4		
1976	36.8		
1977	36.0		
1978	46.0		
1979	48.0		
1980	51.2		
1981	48.0		
1982	40.8		
1983	39.9		
1984	39.9		
1985		46.7	
1986		41.9	
1987		39.7	
1988		47.3	
1989	60.0	51.3	
1990	58.0	41.6	45.1
1991	57.0	36.4	39.3
1992	46.0	27.1	28.7
1993	41.5	29.9	
1994	44.5	39.5	
1995	44.6	43.5	
1996	39.5		

Consumptions in 1996 and 1997 range around 40,000 t W/a. Experts count on an annual growth of 4.4%, which can be split into 1.8% in MEC, 5.3% in China, and 19.9% in the CIS.

Table 13.2 shows primary tungsten consumption since 1973 and is also partly based on estimates. This is clearly reflected by the diverging figures.

The 1996 consumption by end use, for different areas, is presented in Table 13.3.

TABLE 13.3. Tungsten Consumption by End use—1996 (share in %) [13.8, 13.10]

	Western Europe	Japan	USA	China
WC	62	49	62	56
Special steel, superalloys	25	34	7	36
Mill products	9	11	25	5
Others	4	6	6	9

13.6. THE "INTERNATIONAL TUNGSTEN INDUSTRY ASSOCIATION" (ITIA)

The ITIA was formerly inaugurated in Brussels in February 1988, and is registered as an association with scientific aims under Belgian law.

The Association is a very useful instrument for the distribution and exchange of valuable tungsten-related information between the different groups and members. Some 60 companies from 19 countries are members of the Association including mining companies, processors, consumers, trading companies, and assayers. Any company involved in tungsten may join, subject to fulfilling the requirements laid down by the ITIA charter.

One can get a good impression of how small the "tungsten family" really is by taking into account that all major mines and convertors are included in the 60 members of the Association, besides traders, etc.

The principal aims of the ITIA are defined as follows:

- To promote the use of tungsten and tungsten products.
- To provide a regular meeting forum.
- To collate from different sources comprehensive statistics covering tungsten production, processing, consumption, and end uses. To ensure confidentiality, these statistics are submitted directly by members to an independent firm of accountants, who prepare a summary. Safeguards are taken to make certain that no individual company's data can be identified.
- To prepare market reports for each ITIA meeting.
- To collate and circulate on a regular basis information relating to the ITIA's activities and the tungsten industry, and to arrange for publication of a periodic bulletin.
- To organize symposia and seminars.
- To liaise and exchange information on a regular basis with other metal trade associations.
- To maintain contact with governments and professional organizations. For example, the ITIA is accredited to the UNCTAD (United Nations Committee on Trade and Development) Committee on Tungsten.

The International Tungsten Symposia, organized by the Association every third year, enable not only close personal contact of the members, but furthermore, by the presentation of technical and market-related papers the current situation of the tungsten industry and market is highlighted. The 7th Symposium was held in Goslar, Germany in 1996. With the exception of the 6th (Guangzhou, China), all papers as well as discussions have been printed and are available as proceedings.

ACKNOWLEDGMENT

Without the kind and generous supply of information by the ITIA and its secretary, Mr. Michael Maby, this chapter could not have been written in its present form.

REFERENCES FOR CHAPTER 13

13.1. P. T. Stafford, in: *Minerals, Facts and Problems* **2** (1983), 881–894.

13.2. A. P. Newey, in: *Proc. 3rd Int. Tungsten Symposium, Madrid*, pp. 19–33, MPR Publishing Services Ltd., Shrewsbury, England (1985).

13.3. T. F. Anstett, D. I. Bleiwas, and R. J. Hurdelbrink, *Bureau of Mines Information Circular*, IC 9025 (1985).

13.4. R. Kohli, in: *6th Int. Tungsten Symposium, Guangzhou*; transcript of papers: ITIA, London, UK (1993).

13.5. E. Doronkin, in: *ITIA Annual General Meeting, Huntsville, Al, USA*; transcript of papers: ITIA, London, UK (1994).

13.6. R. M. Bunting, "Tungsten Market Studies," in: *ITIA Annual General Meetings, Huntsville, Al, USA* (1994) and *Stockholm* (1995); transcript of papers: ITIA, London, UK.

13.7. M. Spross, *Berg und Hüttenmänn:sche Monatshefte* **141** (1996), 359–362.

13.8. M. Maby, "Market Reports," in: *ITIA Annual General Meetings, Huntsville, Al, USA* (1994), *Stockholm* (1995), *Goslar* (1996), and *Corning*, USA (1997); transcript of papers: ITIA, London, UK.

13.9. *ITIA Newsletters* 1993–1997.

13.10. Peng Pugang, in: *10th ITIA Annual General Meeting, Corning, USA* (1997).

13.11. A. Rössing, *Geschichte der Metalle*, Verlag Leonhard Simion, Berlin, 1901.

13.12. B. Neumann, *Die Metalle*, Verlag Wilhelm Knapp, Halle a.S. (1904).

13.13. "Tungsten Market, Update Analysis and Outlook," Roskill Information Services Ltd., p. 3, London (1995).

14

Tungsten and Living Organisms

14.1. INTRODUCTION [14.1, 14.2, 14.3, 14.4]

Metals play a very important role in the metabolism of living organisms. Ninety-six elements form the earth's crust, of which 22 are essential for life. Within the essential elements for humans we distinguish three overlapping groups (Table 14.1), the most numerous consisting of metals which are found in only low concentrations in living beings. By "low" we mean, for example, that the total amount of an element in the body of an adult human ranges, depending on the element, between one gram and some milligrams. The low concentration is the reason why some of the essential elements have been discovered to be essential by deficiency reactions of patients receiving long-term parenteral nutrition, and today we cannot be sure of knowing all the essential elements.

Until now, tungsten was not included in the group of essential elements like its near relatives chromium and molybdenum for higher developed beings. However, we know due to recent findings that it is essential for some simple living organisms (anaerobic bacteria), which contain tungsten enzymes.

Like many other metals, tungsten and its compounds exhibit various influences on living beings ranging from toxic, via indifferent, to beneficial. Generally speaking, between poisonous and healthy or medicinal the threshold is often very close. Sometimes, it is only the concentration which is decisive, or in other cases it is the valence state of the element, such as chromium, which is among the essential elements in the trivalent state, but is very toxic as hexavalent ion. In other cases, like barium, it is the solubility of a compound which differentiates between toxic and indifferent; the insoluble barium sulfate can be ingested in large quantities without any reaction while soluble salts are highly toxic.

TABLE 14.1. Essential Elements of the Human Body

Group I	Group II	Group III
Components of major structural elements (amino acids, fatty acids, nucleotides, $Ca_5(OH)(PO_4)_3$)	Responsible for maintaining ionic equilibria, electrochemical functions, and membrane potentials	Enzyme components (metalloenzymes and metal ion activated enzymes)
H, O, C, N, P, S, Ca	Na, K, Mg, Ca, Cl	Fe, Cu, Zn, Mn, Mo, Co, Cr, Se, V, Sn, Ni

The influence of tungsten on living organisms depends on three items:

The Kind of Living Organism. As noted earlier, for some it is essential and for others indifferent. Poisonous effects may occur in one species (rats) and not in others (mice). Allergic reactions of humans occur, but are rare.

The Type of Tungsten Compound. Metallic tungsten in powder form ingested by humans did not cause adverse effects as a substitute for barium sulfate in radiological examinations. Compared to most other heavy metals, tungsten and most of its compounds exhibit low toxicity, although there exist some acute toxic compounds like WCl_6 and WF_6. Effectively, it is not the tungsten but the immediate reaction with water or water vapor which generates the gaseous compounds H_2F_2 or HCl, respectively. They are highly toxic and very rapidly damage eyes and the respiratory tract.

Diluted tungstate solutions also may have beneficial influence on humans.

The combination of tungsten carbide and cobalt is responsible for the so-called hardmetal disease.

The Concentration of Tungsten. As for most substances, there exist concentration thresholds between toxicity, indifference, and positive influences.

The distribution of tungsten in the body was checked by experiments with animals (mostly mice and rats). Roughly, one-third of tungsten inhaled and deposited as WO_3 aerosol and one-half of ingested water-soluble tungstate are absorbed. Most of it soon leaves the body in the urine. The remaining amount is distributed among blood cells, spleen, kidneys, and bones. Three months after digestion or 6 months after inhalation, it is transferred almost completely to the bones (very low concentration). The biological half-life of tungsten in bone is 1100 d.

14.2. TUNGSTEN IN BACTERIA (TUNGSTEN ENZYMES) [14.3]

In the early 1970s, it was found that the growth of certain microorganisms are stimulated by the addition of tungstate. Ten years later the first naturally occurring tungsten-containing protein was isolated, and now six tungstoenzymes are known. The number is likely to increase in the future.

Tungsten-containing enzymes until now, can be found exclusively in microorganisms of the following types: methanogens, acetogens, and hyperthermophiles. All of them are obligate anaerobes and are more or less thermophilic. Their enzymes have similar catalytic properties. They catalyze redox reactions involving carboxylic $\rightarrow$ aldehyde conversions.

The essential role of molybdenum in a variety of fundamental biological processes has been known for a long time. Although tungsten and molybdenum have very similar atomic and ionic radii and comparable coordination chemistry, a substitution of Mo by W always causes an inactivation of the enzymes. This elucidates the enormous selectivity of biological, chemical reactions. In contrast, the substitution of W by Mo was possible in some cases without any loss in activity, but not in all cases.

The acetogens *Chlostridium thermoaceticum* and *Chlostridium formiaceticum* grow by sugar fermentation to produce acetate, but can also grow on CO_2 and H_2. In this metabolism, CO_2 is converted to acetate. In the first step, CO_2 reacts to the acetate methyl group and this reaction is catalyzed by the tungstoenzyme formate dehydrogenase (FDH), which contains W, Fe, Se, and S. Two kinds of FDH could be isolated: one containing Mo and the other W. The W species is named FDH2.

Another tungstoenzyme contained in acetogens is carboxylic acid reductase (CAR), which catalyzes the reduction of carboxylic acids to the corresponding aldehyde:

$$RCO_2H + 2H^+ + 2e^- \rightarrow RCHO + H_2O$$

In the methanogen *Methanococcus vanelli* growth, stimulation by tungstate was observed when formate was used as energy source, but not when it was growing on CO_2 and H_2. In contrast, the methanogen *Methanobacterium wolfei* grows only on CO_2 and H_2. The tungstoenzyme formyl methanofuran dehydrogenase (FMDH) catalyzes the first step in the conversion of CO_2 to CH_4 (reduction of CO_2 to formaldehyde), which will be bound to the methanofurane cofactor:

$$CO_2 + H_2 + 2e^- + MPR \rightarrow MPR\text{-}CHO + 2H^+$$

FMDH is a molybdoenzyme in most methanogens with the exception of the above, which contains a FMDHI (Mo) and a FMDHII (W).

Hyperthermophiles generally grow at >90 °C. Pyrococcus furiosus uses sugars as a carbon and energy source. The tungstoenzyme aldehyde ferredoxin oxidoreductase (AOR) catalyzes the oxidation of aldehydes:

$$RCOH + H_2O + Fdox \rightarrow RCO_2H + Fdred + 2H^+$$

Thermococcus litoralis gains energy from the fermentation of peptides. The tungstoenzyme formaldehyde ferredoxin oxidoreductase (FOR) catalyzes the formaldehyde oxidation to formate:

$$HCOH + H_2O + 2Fdox \rightarrow HCO_2H + 2Fdred + 2H^+$$

Some properties of the six tungsten enzymes known so far are summarized in Table 14.2. It is seen that some are very complex, having molecular weights of 300,000 and more, and contain high concentrations of iron (CAR and FDH2 from *Chlostridium*

TABLE 14.2. Properties of Tungsten Enzymes

			g Atom · mol^{-1}		
Microorganisms	Enzyme	Mol. weight	W	Fe	Opt. Temp. (°C)
Thermocuccus litoralis	FOR	280,000	3.3	16	>90
Pyrococcus furiosus	AOR	35,000	1.0	7	>90
Chlostridum formiaceticum	CAR	134,000	1.4	11	40
Chlostridum thermoaceticum	CAR	300,000	3.4	82	?
	FDH2	340,000	2.0	36	?
Methanobacterium wolfei	FMDHII	130,000	0.4	2–5	70

thermoaceticum). Of interest is that CAR from *Chlostridium thermoaceticum* and from *Chlostridium formiaceticum* are not identical.

All tungstoenzymes are located in the cytoplasmic actions of the cells and are very sensitive to inactivation by oxygen.

Moderately thermophylic organisms are not obligatorily dependent on tungsten, because if tungsten is omitted and molybdenum is offered instead they grow further. It seems not to be the case for hyperthermophiles, which require tungsten.

Tungstoenzymes play an important role in the metabolism of carbohydrates and peptides in the hyperthermophiles. Hyperthermophily is regarded as an ancient phenotype. It is assumed that the earliest life forms were hyperthermophilic and W-dependent, which later evolved into mesophilic molybdenum-dependent species and now are the overwhelming majority. As to the reason why W was required and not Mo, there exist two assumptions: W is present in volcanic areas where hyperthermophiles can be found (e.g., deep-sea hydrothermal vents); or alternatively, the better temperature stability of the valence state in question of the W compound.

14.3. TUNGSTEN AND ANIMALS [14.4, 14.5]

Investigations have been performed with rats, mice, and guinea pigs. About investigations related to the Hardmetal Disease, see Section 14.4.2.

14.3.1. Rats

Chronic Toxicity. 5 $\mu g \cdot g^{-1}$ W as Na_2SO_4 to drinking water caused slightly increased growth development and significantly decreased life span.

150 $\mu g \cdot g^{-1}$ W to drinking water increased the incidence of N-nitroso-N-methylurea induced mammary carcinoma.

Reproductive Toxicity. At a dose range without toxic effects to the maternal organism, the following influences could be observed: Increase in embryo lethality; decrease in ossification of bones; tungsten accumulation in the fetus; no significant retention in the placenta.

Toxico kinetics. 100 ppm W to adict containing 30 ppm Mo resulted in a decrease in the activity of sulfite oxidase and xanthine oxidase. Tungstate antagonizes the metabolic action of molybdate in the xanthine hydrogenase, sulfite oxidase, and aldehyde oxidase.

Toxicological Data. Tungsten LD_{50} (i.p.), 5000 $mg \cdot kg^{-1}$; sodium tungstate (s.c.), 223–255 $mg \cdot kg^{-1}$; phosphotungstic acid LD_{50} (oral), 3300 $mg \cdot kg^{-1}$; WC–8%Co LDLo (intratracheal), 75 $mg \cdot kg^{-1}$; WC–15%Co LDLo (intratracheal), 50 $mg \cdot kg^{-1}$.

In oral uptake, the toxicity was highest for Na_2WO_4, intermediate for WO_3, and lowest for $(NH_4)_2WO_4$.

14.3.2. Mice

Chronic Toxicity. 5 $\mu g \cdot g^{-1}$ W as Na_2WO_4 to drinking water caused no abnormal growth development and no decrease in life span.

Genotoxicity. 5 $\mu g \cdot g^{-1}$ W, as above, showed no tumoric effect.

Reproductive Toxicity. Ingestion of 0.1 ml of 25 $mmol \cdot l^{-1}$ Na_2WO_4 on day 3 and 8 of pregnancy had no effect on plantation and gave rise to increased frequency of resorption.

14.3.3. Guinea Pigs

Orally or injected, Na_2WO_4 solution caused anorexia, colic, in-coordination of movement, trembling, and dyspnea.

Intratracheal administration of WC–Co gave rise to no tumors.

14.3.4. Rabbits

0.005 to 5 $mg \cdot kg^{-1}$ sodium tungstate decreases the concentration of blood alkaline phosphatase.

14.4. TUNGSTEN AND HUMANS [14.2]

In comparison to the majority of other heavy metals, tungsten and most of its compounds possess very low toxicity, if at all. Intoxications occur rarely, almost exclusively by occupational exposure. As treatment in acute poisoning, Dimercaprol (British Antilewisite) may be useful.

14.4.1. Important Data

Average Tungsten Level in Blood is $5.8 \pm 3.5\ \mu g\ W \cdot l^{-1}$ ($1\ \mu g\ W \cdot l^{-1}$).

Average Tungsten Level in Bones is 0.25 ppb W.

Demand per Day is 1–15 μg.

Occupational Health. MAK/NIOSH (TLV-TWA): 1 $mg \cdot m^{-3}$ W or soluble W compounds, 5 $mg \cdot m^{-3}$ WC or other insoluble compounds.

Patch Test. In 853 tested workers, 2% showed irritant pustular reactions when exposed to Na_2WO_4.

14.4.2. Hardmetal Disease [14.6–14.9]

The association between parenchymal lung disease and hardmetal exposure was first observed in Germany and several later reports exist from other countries. An extensive summary of the related literature in regard to clinical surveys, epidemiological studies, clinical presentation, pathology, and hypotheses on pathogenesis was presented recently by D. Lison and co-workers [14.6, 14.7].

This parenchymal disease (fibrosing alveolitis) was only observed to occur among hardmetal (cemented carbide) and diamond workers, the clinical presentation of which may vary from subacute alveolitis to progressive fibrosis. Although at first WC was thought to be responsible, and later Co, now it is evident that only the presence of both hardmetal components induces hardmetal disease, characterized in the early stages by the presence of bizarre, multinucleated giant cells in the lung interstitium, alveolar lumen, and the broncho alveolar lavage. In contrast, parenchymal toxicity could never be observed when exposure was to Co alone or WC alone. Moreover, an increased risk of lung cancer was found in hardmetal workers.

Extensive investigations comparing exposure to WC and to Co separately, and to WC–Co mixtures or hardmetal powder, have been conducted recently [14.6–14.9]. The following methods have been applied:

- in vivo animal studies by intratracheal administration,
- in vitro cytotoxicity investigations using macrophages
- in vitro genotoxic activity determinations counting the DNA single strand breaks in human peripheral lymphocytes.

All these findings very clearly demonstrate that only the combination of cobalt with tungsten carbide is a necessary condition to induce severe alveolitis leading to fibrosis. The pulmonary response produced by hardmetal dust is much more pronounced than that caused by pure cobalt or cobalt compounds, while WC alone shows almost no effect. Hardmetal disease is not a consequence of one of the hardmetal components but is a result of interaction between Co and WC particles, producing toxic activated oxygen species, presumably hydroxyl radicals.

This physicochemical interaction was investigated by arachidonic acid peroxidation, ESR spectroscopy, and electrochemical measurements, and the following model is proposed: [14.7].

Although Co is thermodynamically able to reduce oxygen, due to its surface characteristics the reduction rate is extremely slow (resistance to oxidation). WC itself is also unable to react with oxygen, but has high electrical conductivity and an active surface. When both are in contact with each other, a surface electron transfer from Co to WC takes place leading to the reduction of oxygen by the migrating electrons and dissolution of an equivalent amount of oxidized Co as Co^{2+}. A schematic drawing is presented in Fig. 14.1.

A possible but not yet fully proven carcinogenity of hardmetal dust could result from a synergism between the activated oxygen species (genotoxicity—damage to DNA) and the already proven repair inhibition by Co^{2+}.

There do not exist any special values for permissible concentrations of hardmetal dust on the work place. Therefore, the MAK values for insoluble W and Co are valid ($5\,mg\,WC/m^3$ and $0.5\,mg\,Co/m^3$). The ITIA (see Section 13.6) is encouraging inducing enhanced activity in regard to acute toxicity tests to be conducted on tungsten compounds including hardmetals (ITIA Newsletter, June 1997).

14.4.3. Beneficial Influences [14.2, 14.10, 14.11]

In contrast to certain toxic or hazardous effects of tungsten metal or its compounds and alloys, very beneficial influences on human beings have also been described. In this case, an aqueous sodium tungstate solution is applied to the body. The roots of this knowledge—tungsten prolongs life or heals diseases—can be traced to the Middle and Far East. The first publication was written by Mukherjje in 1936 [14.6] and today there exists much relevant literature all written in Japanese. A compilation is given in the book *Tungsten sui Soda* by K. Kase also in Japanese [14.7].

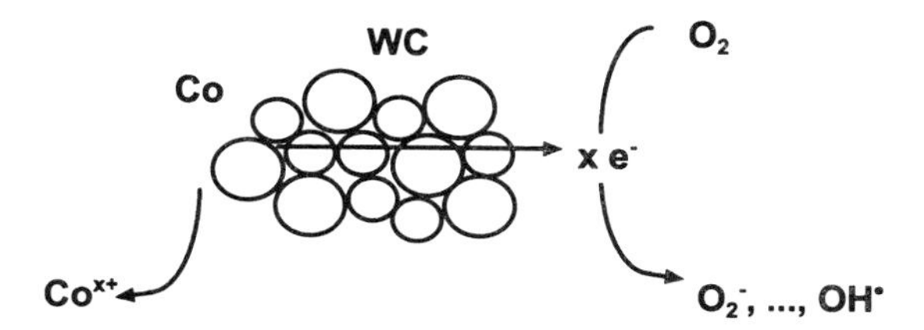

FIGURE 14.1. Mechanism of Co–WC interactive toxicity. See: D. Lison, "Health and Safety in the Hardmetal Industry," paper presented at the EPMA Meeting "Hard Materials—Production and Application," Grenoble, France (1995).

The therapy comprises application as an oral 2% aqueous sodium tungstate solution, or as a gargle, as eye droplets, as tincture to the skin, as inhalation vapor, or as subcutaneous injection. Experiments on 500 people in Japan showed remarkably successful results for a wide range of illnesses. According to the corresponding literature, Table 14.3 lists the types of diseases which have been healed, or at least in which positive influences could be noted. Moreover, it was found, that typical symptoms of aging, such as partial loss of memory, etc. can be considerably retarded. In oral therapy, a daily dose of 25 ml of the above solution corresponding to 500 mg Na_2WO_4, or 331 mg W, is recommended.

Unfortunately, no further systematic clinical tests have yet been performed. Personal experience over the years of one of the authors of the present book and of some friends have been thoroughly positive and satisfying.

TABLE 14.3. Experience with Sodium Tungstate Therapy (500 persons) [14.7]

Very successful	Cararact	Hangover
	Duodenum ulcer	Receding gum
	Halitosis	Stomach ulcer
Successful	Adenomatous hyperplasia	Heat rash
	Angina pectoris	Herpes
	Arrythmias	High blood pressure
	Asthma	Lumbago
	Beginning near-sightedness	Mercury poisoning
	Bruises	Migraine headache
	Burning injuries	Obesity
	Cancer	Parkinson's disease
	Chronic nose bleeding	Polyps
	Diminishing eyesight	Rheumatism
	Diabetes	Selenium poisoning
	Fungus tinea pedis	Sinusitis
	Gout	Sunburn
	Haemorrhoids	UV allergy
	Hair loss	
No influence	Circulatory system problems	
	Glaucoma	
	Retina detachment	

REFERENCES FOR CHAPTER 14

14.1. T. C. P. Stamp, "Mineral Metabolism," in: *Nutrition in the Clinical Management of Disease* (J. W. T. Dickerson and H. A. Lee, eds.), Edward Arnold, London (1988).

14.2. E. Lassner, in: *6th International Tungsten Symposium, Guangzhou*; transcript of papers: ITIA, London, UK (1993).

14.3. M. W. W. Adams, "Tungsten Proteins," in: *Encylopedia of Inorganic Chemistry* (R. Bruce King, ed.), Vol. 8, pp. 4284–4291, Wiley, Chichester, New York (1994).

14.4. E. Lassner, W. D. Schubert, E. Lüderitz, and H. U. Wolf, "Tungsten, Tungsten Alloys, and Tungsten Compounds," in *Ullmann's Encyclopedia of Industrial Chemistry*, Vol. A27, pp. 229–266, VCH Verlagsgesselschaft, Weinheim (1966).

14.5. G. D. Clayton, ed., *Patty's Industrial Hygiene and Toxicology*, CD-ROM, Wiley Interscience, New York (1996).

14.6. D. Lison, *Crit. Rev. Toxicol.* **26** (1996), 585–616.

14.7. D. Lison, R. Lauwerys, M. Demedts, and B. Nemery, *Eur. Respir. J.* **9** (1996), 1024–1028.

14.8. D. Anard, M. Kirsch-volders, A. Elhajouji, K. Belpaeme, and D. Lison, *Carcinogenesis* **18** (1997), 177–184.

14.9. D. Lison, P. Carbonelle, L. Mollo, R. Lauwerys, and B. Fubini, *Chem. Res. Toxicol.* **8** (1995), 600–606.

14.10. H. N. Mukherjee, *Calcutta Med. J.* **30** (1936), 452.

14.11. K. Kase, *Tungsten sui Soda*, Tokyo, ISBN-4-257-03314-2 (1991).

Index